The Smallest Anthropoids

AF890634

Developments in Primatology: Progress and Prospects

Series Editor:
Russell Tuttle
Department of Anthropology
The University of Chicago, IL, USA

For other titles published in this series, go to
www.springer.com/series/5852

Susan M. Ford · Leila M. Porter · Lesa C. Davis
Editors

The Smallest Anthropoids

The Marmoset/Callimico Radiation

Springer

Editors
Susan M. Ford
Southern Illinois University
Carbondale, IL
USA
sford@siu.edu

Leila M. Porter
Northern Illinois University
DeKalb, IL
USA
lmporter@niu.edu

Lesa C. Davis
Northeastern Illinois University
Chicago, IL
USA
lcdavis@neiu.edu

ISBN 978-1-4614-2446-8 e-ISBN 978-1-4419-0293-1
DOI 10.1007/978-1-4419-0293-1
Springer New York Dordrecht Heidelberg London

© Springer Science+Business Media, LLC 2009
Softcover reprint of the hardcover 1st edition 2009
All rights reserved. This work may not be translated or copied in whole or in part without the written permission of the publisher (Springer Science+Business Media, LLC, 233 Spring Street, New York, NY 10013, USA), except for brief excerpts in connection with reviews or scholarly analysis. Use in connection with any form of information storage and retrieval, electronic adaptation, computer software, or by similar or dissimilar methodology now known or hereafter developed is forbidden.
The use in this publication of trade names, trademarks, service marks, and similar terms, even if they are not identified as such, is not to be taken as an expression of opinion as to whether or not they are subject to proprietary rights.

Cover Illustration: Top: by Wallisen Tadashi Hattori; Bottom: by Leila M. Porter

Printed on acid-free paper

Springer is part of Springer Science+Business Media (www.springer.com)

Preface

The marmosets and callimicos are diminutive monkeys from the Amazon basin and Atlantic Coastal Forest of South America. The marmosets are the smallest anthropoid primates in the world, ranging in size from approximately 100 to 350 g (Hershkovitz 1977; Soini 1988; Ford and Davis 1992; Araújo et al. 2000); callimicos are not much bigger, at around 350–540 g (Ford and Davis 1992; Encarnación and Heymann 1998; Garber and Leigh 2001). Overwhelming genetic evidence, from both nuclear and mitochondrial DNA, now indicates that these taxa represent a unified clade within the callitrichid radiation of New World monkeys, a finding that was unthinkable to all but a few geneticists a decade ago (see review in Cortés-Ortiz, this volume Chap. 2). With increasing evidence that the earliest anthropoids were themselves small bodied (under the 0.8–1 kg threshold that marks all other living anthropoids; see Ross and Kay 2004), the ecology, behavior, reproductive stresses, and anatomical adaptations of the marmosets and callimicos provide the best living models with which to assess the types of adaptations that may have characterized early anthropoids.

When Anthony Rylands' *Marmosets and Tamarins: Systematics, Behaviour and Ecology* was published in 1993, contributions focused almost entirely on tamarins due to the scarcity of data on marmoset behavior and the almost total lack of knowledge about the enigmatic callimicos. Fortunately, this has changed (see Fig. 1). In the last 15 years, there has been an explosion of new information on the South American marmosets and callimicos. In 1977, in his seminal work, Philip Hershkovitz recognized only four species of marmosets: *Cebuella pygmaea* (the pygmy marmoset), *Callithrix jacchus* (the tufted-ear or common marmoset), *Callithrix argentata* (the bare-ear marmoset), and *Callithrix humeralifer* (the tassel-ear marmoset). Since then, many other species have been accepted, and the Amazonian non-pygmy marmosets have been raised to generic level, *Mico* (Rylands et al. 2000). An entirely new monotypic genus, *Callibella humilis,* was first discovered in 1998 (van Roosmalen et al. 1998) and raised to genus level in 2003 (van Roosmalen and van Roosmalen 2003). This is the first new genus of a New World monkey discovered since *Callimico* 100 years ago (Thomas 1904), and it remains enigmatic and poorly known (see, in particular, Aguiar and Lacher Jr Chap. 18, and Ford and Davis Chap. 21). Six other new species have been identified: *Mico nigriceps* (Ferrari and Lopes 1992), *M. mauesi* (Mittermeier et al. 1992), *M. marcai* (Alperin 1993), *M. saterei* (Silva Jr and

Fig. 1 Illustrations of the diminutive anthropoids of the marmoset-callimico radiation. Drawings by Stephen Nash, Conservation International

Fig. 1 (continued)

Fig. 1 (continued)

Fig. 1 (continued)

Noronha 1998), *M. acariensis* (van Roosmalen et al. 2000), and *M. manicorensis* (van Roosmalen et al. 2000). Ten others have been raised to full species status, including *Callithrix aurita, C. flaviceps, C. geoffroyi, C. kuhlii, C. penicillata, Mico chrysoleucus, M. emiliae, M. intermedius, M. leucippe, M. melanurus,* with at least one other pending: *Mico* cf. *emiliae* [Rondônia]. Therefore, this radiation of diminutive monkeys is currently recognized as comprising 4 genera and 22 species of marmosets plus one of callimicos. The taxonomic status, distribution, and descriptions for each of these is reviewed by Rylands et al. (Chap. 2), and the differences between them are discussed in a number of chapters, particularly in the Phylogeny and Anatomy sections of the volume.

We have also dramatically improved our understanding of their anatomy, locomotion, diet, ecology, and social behavior, much of which has been quite surprising. We now know that members of this radiation may form monogamous, polyandrous, and occasionally polygynandrous groups; *Cebuella* vertically clings but does not often leap; *Callibella* may not twin all of time; *Callimico* eats fungi, and may have twinned in the distant past; and the group displays measurable craniodental and postcranial differences related to the unique ways they exploit their respective habitats. In addition, many of these species are at risk or are highly endangered (see, in particular, de la Torre et al. Chap. 22 and Ferrari Chap. 23).

This volume is an outgrowth of a lunch conversation between the three co-editors and Paul Garber during the 2003 Annual Meeting of the American Association of Physical Anthropologists in Tempe, Arizona, when we realized that there was considerable new information available with which to reassess the complexities of callimico and marmoset behavior and anatomy. In 2005, the co-editors chaired a symposium at the Annual Meeting of the American Association of Physical Anthropologists in Milwaukee, Wisconsin (*Advances in Marmoset and Goeldi's Monkey (Callimico) Research: Anatomy, Behavioral Ecology, Phylogeny, and Conservation*). Expanded versions of the presentations from that symposium form the core of this volume, with the addition of a number of other important contributions.

This volume presents the work of primary researchers from three continents, and it explores these diminutive primates from several different perspectives. In Section I, phylogeny of the group is examined through genetics, morphology, distribution, and vocalization. Behavior and ecology are examined both in the wild and in captivity, in their reproductive, social, and cognitive dimensions (Section II) as well as ranging and locomotion (Section III). These include studies on four of the five genera (*Callimico, Callithrix, Mico,* and *Cebuella*; there are still no in depth studies on *Callibella*, although new information on ecology is included in Aguiar and Lacher Jr, Chap. 18). The functional and phylogenetic anatomy of all five genera is discussed in Section IV, including chapters on cranial anatomy, postcranial anatomy, and jaw mechanics and musculature associated with the marmosets' distinctive diet of exudativory. The volume closes with consideration of conservation issues and concerns (Section V).

For consistency throughout the volume, we have adopted certain taxonomic conventions (largely following Rylands et al. 2000; Rylands et al. Chap. 2). Although not all authors fully agree with the classificatory choices made here, all have agreed to use the following names in their chapters: all marmosets, tamarins, and callimicos are in the family Callitrichidae; no subfamily names are used; and any reference to the twinning callitrichids (excluding callimicos) is to the "marmosets and tamarins." The common name used for *Callimico* is callimicos, although they are often referred to as Goeldi's monkeys elsewhere. The first time any platyrrhine is mentioned in a chapter, both the scientific and common names are given; after that, individual author(s) may adopt either scientific or common terminology.

Much of the work on these smallest anthropoids has been published in languages other than English or in sources that can be difficult to acquire. It is our hope that by bringing the work of these authors together in one place and one language, with abstracts in both Spanish and Portuguese, this volume will provide readily accessible information on the evolution, behavior, adaptations, and conservation needs of the marmosets and callimicos both for people interested in these wonderful monkeys for themselves, and for those who discover that knowledge of the diminutive marmosets and callimicos aids their understanding of other organisms and the diverse ecosystems in which marmosets and callimicos live.

Acknowledgments The editors are particularly indebted to several individuals for advice, critique, and assistance, without which this volume could not have been completed. These include Dr. Anthony Rylands for leading a spirited and inspiring discussion after the original symposium; Drs. Paul Garber and Anthony Rylands (plus two anonymous reviewers) for suggesting some additional contributors and generally much-welcomed advice; Dr. Dan Gebo for critique of some chapters; the authors themselves, some of whom read and provided input on other chapters; Dr. Liliana Cortés-Ortiz and Mr. José Romulo Sanchez for translating most of the abstracts into Spanish; and Drs. Maria B.C. de Sousa and Gabriel Marroig, and Ms. Ana Carolina Castro for translating most into Portuguese. Several graduate students provided editorial assistance, including Rose Hores, Chihiro Shibata, Karen Wolf, and for the index: Elliott Forsythe and Aimee Hosemann. Conservation International and Dr. Russell Mittermeier underwrote the marvelous color plates by Stephen Nash, to whom we are particularly indebted. Drs. Jennifer Rehg and Tab Rasmussen provided the hospitality of their cottage for one memorable weekend of work. LCD is grateful for the organizational talents of Avril Murray and Felicia Elliott in the early stages of the book, and for the enduring gift of JM's support. SMF especially thanks her husband, Al Allen, for assistance with many illustrations and Al, David and Emily for their patience, love, and good humor (mostly) throughout the production of the book.

Carbondale, IL — Susan M. Ford
DeKalb, IL — Leila M. Porter
Chicago, IL — Lesa C. Davis

References

Aguiar JM, Lacher TE Jr (this volume) Cranial morphology of the dwarf marmoset *Callibella* in the context of callitrichid variability. In: Ford SM, Porter LM, Davis LC (eds) The smallest anthropoids: The marmoset/callimico radiation. Springer Press, New York, pp 355–380

Alperin R (1993) *Callithrix argentata* (Linnaeus, 1771): Considerações taxonômicas e descrição de subespécie nova. Bol Mus Para Emílio Goeldi, Sér Zool. 9(2):317–328

Araújo A, Arruda MF, Alencar AI, Albuquerque F, Nascimento MC, Yamamoto ME (2000) Body weight of wild and captive common marmosets (*Callithrix jacchus*). Int J Primatol 21:317–324

Cortés-Ortiz L (this volume) Molecular phylogenetics of the Callitrichidae with an emphasis on the marmosets and *Callimico*. In: Ford SM, Porter LM, Davis LC (eds) The smallest anthropoids: The marmoset/callimico radiation. Springer Press, New York, pp 3–24

de la Torre S, Yepez P, Snowdon C (this volume) Conservation status of pygmy marmosets in Ecuadorian Amazon. In: Ford SM, Porter LM, Davis LC (eds) The smallest anthropoids: The marmoset/callimico radiation. Springer Press, New York, pp 451–464

Ferrari SF (this volume) Marmosets and callimicos: conservation status and concerns. In: Ford SM, Porter LM, Davis LC (eds) The smallest anthropoids: The marmoset/callimico radiation. Springer Press, New York, pp 465–476

Ferrari SF, Lopes MA (1992) A new species of marmoset, genus *Callithrix* Erxleben 1777 (Callitrichidae, Primates) from western Brazilian Amazonia. Goeldi Zool (12):1–3

Ford SM, Davis LC (1992) Systematics and body size: Implications for feeding adaptations in New World monkeys. Am J Phys Anthropol 88:415–468

Garber PA, Leigh SR (2001) Patterns of positional behavior in mixed-species troops of *Callimico goeldii*, *Saguinus labiatus*, and *Saguinus fuscicollis* in northwestern Brazil. Am J Primatol 54:17–31

Hershkovitz P (1977) Living New World Monkeys (Platyrrhini) with an Introduction to Primates, vol 1. Chicago University Press, Chicago

Mittermeier RA, Schwarz M, Ayres JM (1992) A new species of marmoset, genus *Callithrix* Erxleben 1777 (Callitrichidae, Primates), from the Rio Maués region, state of Amazonas, Central Brazilian Amazonia. Goeldi Zool (14):1–17

Ross CF, Kay RF (eds) (2004) Anthropoid origins: New visions. Kluwer Academic/Plenum Publishing, New York

Rylands AB, (1993) Marmosets and Tamarins: Systematics, Behaviour and Ecology. Rylands AB (ed) Oxford University Press, Oxford

Rylands AB, Mittermeier RA, Coimbra-Filho AF (this volume) The systematics and distribution of the marmosets (*Callithrix, Callibella, Cebuella,* and *Mico*) and Callimico (*Callimico*) (Callitrichidae, Primates). In: Ford SM, Porter LM, Davis LC (eds) The smallest anthropoids: The marmoset/callimico radiation. Springer Press, New York, pp 25–61

Rylands AB, Schneider H, Langguth A, Mittermeier RA, Groves CP, Rodríguez-Luna E (2000) An assessment of the diversity of New World primates. Neotrop Primates 8(2):61–93

Silva Jr JS, Noronha M de A (1998) On a new species of bare-eared marmoset, genus *Callithrix* Erxleben, 1777, from central Amazonia, Brazil (Primates: Callitrichidae). Goeldi Zool (21):1–28

Soini P (1988) The pygmy marmosets, genus *Cebuella*. In: Mittermeier RA, Rylands AB, Coimbra-Filho AF, da Fonseca, GA (eds) Ecology and behavior of neotropical primates, vol 2. World Wildlife Fund, Washington DC, pp 79–129

Thomas O (1904) New *Callithrix*, *Midas*, *Felis*, *Rhipidomys*, and *Proechimys* from Brazil and Ecuador. Ann Mag Nat Hist 14(7):188–196

Van Roosmalen MGM, Van Roosmalen T (2003) The description of a new marmoset genus, *Callibella* (Callitrichinae, Primates), including its molecular phylogenetic status. Neotrop Primates 11(1):1–10

Van Roosmalen MGM, Van Roosmalen T, Mittermeier RA, Fonseca GAB da (1998) A new and distinctive species of marmoset (Callitrichidae, Primates) from the lower Rio Aripuanã, state of Amazonas, central Brazilian Amazonia. Goeldi Zool (22):1–27

Van Roosmalen MGM, Van Roosmalen T, Mittermeier RA, Rylands AB (2000) Two new species of marmoset, genus *Callithrix* Erxleben, 1777 (Callitrichidae, Primates), from the Tapajós/Madeira interfluvium, south central Amazonia, Brazil. Neotrop Primates 8(1):2–18

Contents

Contributors

John M. Aguiar
Texas A&M University, Dept. of Wildlife and Fisheries Sciences, 210 Nagle Hall, 2258 TAMU, College Station, Texas 77843-2258, USA
steelshard@new.tamu.edu

Ana Claudia Sales da Rocha Albuquerque
Departamento de Ciências Biológicas, Universidade Estadual do Rio Grande do Norte, Mossoró, Brazil
acsra@bol.com.br

Anuska Irene Alencar
Secretaria Municipal de Saúde, Natal, RN, Brazil
anuskaalencar@yahoo.com.br

Fred C. Anapol
Department of Anthropology, University of Wisconsin – Milwaukee, Milwaukee, WI 53201, USA
fred@uwm.edu

Gustl Anzenberger
Anthropologisches Institut der Universitaet Zuerich, Zuerich, Switzerland

Arrilton Araújo
Departamento de Fisiologia, Caixa Postal 1511, Universidade Federal do Rio Grande do Norte, Natal, RN, 59072-970, Brazil
arrilton@cb.ufrn.br

Maria de Fátima Arruda
Departamento de Fisiologia, Caixa Postal 1511, Universidade Federal do Rio Grande do Norte, Natal, RN, 59072-970, Brazil
arruda@cb.ufrn.br

Gregory E. Blomquist
Department of Anthropology, University of Illinois, Urbana, IL, USA

James M. Cheverud
Department of Anatomy & Neurobiology, Washington University School of Medicine, 660 S. Euclid Ave, Campus Box 8108,
Saint Louis, MO 63110, USA

Adelmar F. Coimbra-Filho
Rua Artur Araripe 60/901 Gávea, Rio de Janeiro 22451-020,
Rio de Janeiro, Brazil

Liliana Cortés Ortiz
Museum of Zoology, University of Michigan, 1109 Geddes Ave, Ann Arbor, MI 48109, USA
lcortes@umich.edu

Paulo E.G. Coutinho
Department of Genetics, Universidade Federal do Pará, Caixa Postal 8607, 66.075-970 Belém, PA, Brazil

Maria Adélia Oliveira Monteiro da Cruz
Laboratório de Ecofisiologia e Comportamento Animal, Departamento de Morfologia e Fisiologia Animal, Universidade Federal Rural de Pernambuco, Pernambuco, Brazil
adelia@ufrpe.br

Lesa C. Davis
Department of Anthropology, Northeastern Illinois University, Chicago, IL 60625, USA
lcdavis@neiu.edu

Stella de la Torre
Universidad San Francisco de Quito, Quito, Ecuador
stella@usfq.edu.ec

Paulo De Marco Jr.
Laboratório de Ecologia Quantitaviva, Universidade Federal de Viçosa, Minas Gerais, Brazil

Leslie Digby
Department of Biological Anthropology and Anatomy, Duke University, Durham, NC, USA
ldigby@duke.edu

Carolyn M. Eng
Muscle Physiology Laboratory, University of California, San Diego, Medical Teaching Facility, San Diego, CA 92093, USA
carolyn@muscle.ucsd.edu

Christopher G. Faulkes
School of Biological and Chemical Sciences, Queen Mary, University of London, London, UK
c.g.faulkes@qmul.ac.uk

Stephen F. Ferrari
Department of Biology, Universidade Federal de Sergipe, São Cristóvão, SE, Brazil
godbufu@yahoo.co.uk

Susan M. Ford
Department of Anthropology, Southern Illinois University, Carbondale, IL 62901, USA
sford@siu.edu

Paul A. Garber
Department of Anthropology, University of Illinois, 109 Davenport Hall, 607 South Mathews Ave, Urbana, IL 61801, USA
p-garber@uiuc.edu

Brooke A. Garner
Department of Anatomy – NEOUCOM, 4209 St Rt 44, Box 95, Rootstown, OH 44272, USA
bgarner@neoucom.edu

Evonnildo Gonçalves
Department of Genetics, Universidade Federal do Pará, Caixa Postal 8607, 66.075-970 Belém, PA, Brazil
ecostag@ufpa.br

Ludwig Huber
Department for Behavior, Neurobiology and Cognition, Faculty of Life Sciences, University of Vienna, Althanstrasse 14, 1090 Vienna, Austria
ludwig.huber@univie.ac.*at*

William L. Hylander
Department of Biological Anthropology and Anatomy, Duke University, Durham, NC 27710, USA
hylander@baa.duhs.duke.edu

Thomas E. Lacher Jr.
Center for Applied Biodiversity Science, Conservation International, 2011 Crystal Drive, Suite 500, Arlington, VA 22202, USA

Peter W. Lucas
Department of Anthropology, George Washington University, Washington, DC, USA
pwlucas@gwu.edu

Yumma Bernardo Maranhao Valle
Programa de Mestrado de Gestão e Políticas Ambientais, Departamento de Geografia, Universidade Federal de Pernambuco, UFPE, Brazil
yummavalle@gmail.com

Gabriel Marroig
Departamento de Biologia, Instituto de Biociências, Universidade de São Paulo, Caixa Postal 11.461, 05422-970 São Paulo, SP, Brasil
Gmarroig@ib.usp.br

Leonardo César de Oliveira Melo
Programa de Doutorado em Zoologia do Departamento de Sistemática e Ecologia, DSE da Universidade Federal de Paraíba, UFPB, Brazil
jacchus05@yahoo.com.br

Sérgio Lucena Mendes
Departamento de Ciências Biológicas, Universidade Federal do Espírito Santo, Av. Mal. Campos 1468, Maruipe, Vitória, ES, Brazil
slmendes1@gmail.com

Elytânia Menezes
Department of Genetics, Universidade Federal do Pará, Caixa Postal 8607, 66.075-970 Belém, PA, Brazil

Russell A. Mittermeier
Center for Applied Biodiversity Science, Conservation International, 1919 M Street NW, Washington DC 20036, USA

M.A.O. Monteiro da Cruz
Federal Rural University of Pernambuco, Pernambuco, Brazil

Amy L. Mork
Department of Anatomy – NEOUCOM, 4209 St Rt 44, Box 95, Rootstown, OH 44272
almork@neoucom.edu

Leila M. Porter
Department of Anthropology, Northern Illinois University, DeKalb, IL 60115, USA
lmporter@niu.edu

Jennifer A. Rehg
Department of Anthropology, Southern Illinois University Edwardsville, Box 1451, Edwardsville, IL, 62026, USA
jrehg@siue.edu

Julienne Rutherford
Department of Oral Biology, University of Illinois at Chicago, 801 S. Paulina St., Chicago, IL 60612, USA
ruther4d@uic.edu

Anthony B. Rylands
Center for Applied Biodiversity Science, Conservation International, 1919 M Street NW, Washington, DC 20036, USA
a.rylands@conservation.org

Ana Sallenave
Department of Anthropology, University of Illinois, Urbana, IL, USA

Wendy Saltzman
Department of Biology, University of California, Riverside, CA, USA

Daniel Schmitt
Department of Biological Anthropology and Anatomy, Duke University, Durham, NC 27710, USA
Daniel_schmitt@baa.duhs.duke.edu

Maria Paula C. Schneider
Department of Genetics, Universidade Federal do Pará, Caixa Postal 8607, 66.075-970 Belém, PA, Brazil

Artur Silva
Department of Genetics, Universidade Federal do Pará, Caixa Postal 8607, 66.075-970 Belém, PA, Brazil

Maria Bernardete Cordeiro de Sousa
Departamento de Fisiologia, Caixa Postal 1511, Universidade Federal do Rio Grande do Norte, Natal, RN, 59072-970, Brazil
mdesousa@cb.ufrn.br

Charles T. Snowdon
Department of Psychology, University of Wisconsin, Madison, WI, USA
snowdon@wisc.edu

Suzette Tardif
Southwest National Primate Research Center, San Antonio, TX, USA

Andrea B. Taylor
Departments of Community and Family Medicine and Evolutionary Anthropology, Duke University School of Medicine, Box 104002, Durham, NC, 27708, USA
andrea.taylor@duke.edu

Mônica M. Valença-Montenegro
Centro de Proteção de Primatas Brasileiros, IBAMA, Brazil
monica.montenegro@ibama.gov.br

Cecilia Veracini
Laboratori di Antropologia, Dipartimento di Biologia Evoluzionistica "L. Pardi", Via del Proconsolo, 12. CAP, Firenze, 50122, Italy
cpfveracini@yahoo.com

Jacques Marie Edme Vielliard
Departamento de Zoologia, Universidade Estadual de Campinas, São Paulo, Brazil

Chris Vinyard
Department of Anatomy – NEOUCOM, 4209 St Rt 44, Box 95, Rootstown, OH 44272, USA
cvinyard@neoucom.edu

Bernhard Voelkl
Department for Behavior, Neurobiology and Cognition, Faculty of Life Sciences, University of Vienna, Althanstrasse 14, 1090 Vienna, Austria

Christine E. Wall
Department of Biological Anthropology and Anatomy, Duke University, Durham, NC 27710, USA
Christine_wall@baa.duhs.duke.edu

Susan H. Williams
Department of Biomedical Sciences, Ohio University, Athens, OH 45701 USA
Williams@exchange.oucom.ohiou.edu

Maria Emília Yamamoto
Departamento de Fisiologia, Caixa Postal 1511, Universidade Federal do Rio Grande do Norte, Natal, RN, 59072-970, Brazil
emiliayamamoto@gmail.com

Pablo Yépez
Fundación VIHOMA, Quito, Ecuador
huati@yahoo.com

Dionisios Youlatos
Aristotle University of Thessaloniki, School of Biology, Department of Zoology, GR-54124 Thessaloniki
Greece, dyoul@bio.auth.gr

Section I
Phylogeny

Susan M. Ford, Leila M. Porter, Lesa C. Davis

From the initial discovery of *Callimico* (Thomas 1904), the relationship of callimicos to other platyrrhines, particularly to the various small-bodied callitrichids, has been an enigma. Throughout the 1970s, 1980s, and early 1990s, substantial morphological and behavioral data strongly supported the view that callimicos were the sister-group to the combined group of marmosets and tamarins. However, the few early molecular studies from this period argued for close ties between the callimicos and marmosets alone. This minority and generally disregarded view garnered substantial support through the 1990s and the 21st century from a growing number of molecular studies on a variety of aspects of the genome (reviewed in Cortés-Ortiz, Chap. 1). DNA has presented a very uniform and highly supported argument for the affinities of the marmosets with callimicos (with the tamarins more distant). Although there is growing consensus as to the callimico – marmoset clade, there is now increasing need to reassess relationships among different marmosets. The rapidity with which new species of marmosets have been recognized and/or discovered for the first time (see Ford et al. Preface and especially Rylands et al. Chap. 2) and our almost total lack of knowledge or even good specimens for many of these species, especially those from Amazonia, results in continued confusion over the relationships between the various marmosets, or even how many species and genera should be recognized.

At this stage, it is beneficial to have at least a road map to the taxa involved and the questions still before us. Cortés-Ortiz (Chap. 1) provides a complete overview and analysis of the genetic studies to date. Her consideration of the few but important differences between their results suggests how the evolution of different genes at different rates may result in the seeming conflicts present in the literature, yet in fact reflect a single highly resolved tree. In her view, the generic relationships are quite clear: (*Saguinus* (*Leontopithecus* (*Callimico* (*Callithrix* (*Callibella* (*Cebuella*/*Mico*)))))), although the positions of the tamarins (*Saguinus*) and golden lion tamarins (*Leontopithecus*) may be reversed. Insufficient data (genetic or morphological) exist for most species, however, to allow a species-level consideration

Susan M. Ford
Department of Anthropology, Southern Illinois University, Carbondale, IL 62901, USA
e-mail: sford@siu.edu

of relationships. The marmoset generic relationships and possible evolutionary scenarios are further considered by Marroig and Cheverud (Chap. 17), Aguiar and Lacher (Chap. 18), and Ford and Davis (Chap. 21).

Rylands et al. (Chap. 2) present a thorough review of the taxonomy of each of the 23 currently recognized species of marmosets and callimicos, as well as maps and details of their known distributions. They argue for recognizing four genera of marmosets (*Callithrix, Mico, Callibella,* and *Cebuella*), the view adopted throughout this book. They note that the highly constrained geographic ranges and high degree of differentiation of the Amazonian marmosets (of genus *Mico*) suggest that more species may need to be recognized or remain to be discovered. In addition, the contrast between the great diversity of *Mico* and the single recognized species for each of the widespread genera *Cebuella* and *Callimico*, which are often found patchily in widely separated areas, hints that far more real diversity and differentiation may exist in these genera than has been interpreted due to their rather uniform cryptic pelage coloration (agouti brownish-gray and black, respectively).

In the final contribution in this section, Mendes et al. (Chap. 3) use a variety of features of the call structure, particularly the long call, to examine the distinctness and possible relationships between the six identified species of the Atlantic Coastal forest marmosets, genus *Callithrix*. They find that the call structure of each of the six species is unique. Further, vocalizations support morphological and genetic separations of *Callithrix* into two species groups: the *C. aurita*/*C. flaviceps* groups and the *C. jacchus*, *C. geoffroyi*, *C. penicillata*, *C. kuhlii* group. However, similarities between the distinctive vocalizations of those in the latter group do not conform to affinities suggested by morphology; they discuss several historical factors related to species recognition that may have influenced the evolution of call structure.

While studies of *Callithrix* like that of Mendes et al. have examined the complex evolutionary history and relations of the species within this genus, there have been no comparable studies of species and populations of their closest relatives in the Amazon (*Callimico, Mico, Callibella, Cebuella*). The endangered status of many of these taxa make this work both urgent and problematic – specimens and DNA samples are rarely available across their geographic range.

Reference

Thomas O (1904) New *Callithrix*, *Midas*, *Felis*, *Rhipidomys*, and *Proechimys* from Brazil and Ecuador. Ann Mag Nat Hist 14(7):188–196

Chapter 1
Molecular Phylogenetics of the Callitrichidae with an Emphasis on the Marmosets and *Callimico*

Liliana Cortés-Ortiz

Abstract In the last two decades analyses of DNA sequence data have significantly increased our understanding of the phylogenetic relationships among callitrichid primates. The current taxonomic synthesis includes seven genera within the family Callitrichidae: *Saguinus*, *Leontopithecus*, *Callimico*, *Mico*, *Cebuella*, *Callibella* and *Callithrix*. The relationships among genera are relatively well defined by molecular data and can be described as follows: (*Saguinus* (*Leontopithecus* (*Callimico* (*Callithrix* (*Callibella* (*Cebuella*/*Mico*)))))). Nonetheless, relationships among *Leontopithecus*, *Callimico*, and *Saguinus* are less clear and show different patterns depending on the data utilized to infer these relationships; this could reflect the relatively rapid divergence of these lineages. The sister relationship of *Callimico* to the marmosets (*Callithrix, Callibella, Cebuella,* and *Mico*) is highly supported by molecular and cytogenetic studies. Within this clade (*Callimico*/marmosets), only a few studies have focused on determining relationships at the intrageneric level, with only two genera analyzed to date. These studies have produced uncertain results concerning the taxonomic distinctiveness of the putative species and their phylogenetic relationships. In this chapter I review, synthesize, and analyze the molecular evidence that has been used to elucidate the taxonomy within the Callitrichidae.

Resumen Los análisis de datos de secuencias de ADN han aumentado significativamente nuestra comprensión de las relaciones filogenéticas entre los primates calitrícidos en las últimas dos décadas. La síntesis taxonómica actual incluye siete géneros dentro de la familia Callitrichidae: *Saguinus*, *Leontopithecus*, *Callimico*, *Mico*, *Cebuella*, *Callibella* y *Callithrix*. Las relaciones entre los distintos géneros se han definido relativamente bien a través del uso de marcadores moleculares y pueden ser descritas de la siguiente manera: (*Saguinus* (*Leontopithecus* (*Callimico* (*Callithrix* (*Callibella* (*Cebuella*/*Mico*)))))). Sin embargo, las relaciones entre

L. Cortés-Ortiz (✉)
Museum of Zoology, University of Michigan, 1109 Geddes Ave, Ann Arbor, MI, 48109, USA
e-mail: lcortes@umich.edu

S.M. Ford et al. (eds.), *The Smallest Anthropoids*, Developments in Primatology: Progress and Prospects, DOI 10.1007/978-1-4419-0293-1_1,
© Springer Science+Business Media, LLC 2009

Leontopithecus, *Callimico* y *Saguinus* son menos claras y muestran distintos patrones dependiendo de los datos ultilizados para inferir dichas relaciones; esto puede ser el reflejo de una divergencia relativamente rápida entre estos linajes. La relación de taxa hermanos entre los callimicos y las marmosetas (*Callithrix, Callibella, Cebuella,* y *Mico*) está fuertemente apoyada por estudios moleculares y citogenéticos. Dentro de este clade (*Callimico*/marmosetas), sólo unos pocos estudios se han enfocado en determinar las relaciones a nivel intragenérico, con solamente dos géneros analizados hasta la fecha. Estos estudios han proporcionado resultados inciertos en relación a la distinción taxonómica de las supuestas especies y sus relaciones filogenéticas. En el presente trabajo se revisa, sintetiza y analiza la evidencia molecular que ha sido utilizada para elucidar la taxonomía de los calitrícidos.

Resumo Análises de dados de sequências de ADN aumentaram significativamente nosso conhecimento das relações filogenéticas entre os primatas calitriquídeos nas últimas duas décadas. A síntese taxonômica atual inclui sete gêneros dentro da família Callitrichidae: *Saguinus, Leontopithecus, Callimico, Mico, Cebuella, Callibella* e *Callithrix*. As relações entre os gêneros são relativamente bem definidas por dados moleculares e podem ser descritas como: (*Saguinus (Leontopithecus (Callimico (Callithrix (Callibella (Cebuella/Mico*)))))). No entanto, as relações entre *Leontopithecus, Callimico* e *Saguinus* são menos claras e mostram padrões diferentes dependendo dos dados utilizados para inferir o parentesco; isto pode refletir a divergência relativamente rápida destas linhagens. O relacionamento de *Callimico* como grupo-irmão dos sagüis (*Callithrix, Callibella, Cebuella*, e *Mico*) tem um suporte alto em estudos citogenéticos e moleculares. Dentro deste clado (*Callimico*/sagüis), poucos estudos têm focado na determinação das relações de parentesco no nível intra-genérico, com somente dois gêneros analisados até o momento. Estes estudos têm produzido resultados relevantes sobre a diferenciação taxonômica de supostas espécies e seu relacionamento filogenético. Neste capítulo eu faço uma revisão, síntese, e análise da evidência molecular que têm sido usada para elucidar a taxonômia dentro de Callitrichidae.

1.1 Introduction

The Callitrichidae are diminutive-sized primates from South and Central America. This group includes the marmosets (*Callithrix, Mico, Cebuella,* and *Callibella*), tamarins (*Saguinus*), lion tamarins (*Leontopithecus*), and callimicos (*Callimico*). Although this group was traditionally considered to constitute the family Callitrichidae, Rosenberger (1981) proposed to include this group as the subfamily Callitrichinae within the family Cebidae, as sister of the Cebinae (*Cebus* and *Saimiri*), separating them from the larger New World monkeys, belonging to the family Atelidae. This classification was followed by many researchers, but there is still no complete agreement on whether family or subfamily status should be used (see Rylands et al. Chapt. 2 this volume). Given that there is no consensus, and to be consistent with the rest of the book, in this chapter the group composed of marmosets, tamarins, lion tamarins, and callimicos is designated as the family Callitrichidae (see Preface). In one of the more recent treatises on primate taxonomy, Groves (2001) argued against the validity of the name Callitrichidae Thomas 1903, and referred to it as a synonym of Hapalidae Gray, 1821. Nonetheless, he later argued for continued use of Callitrichidae based on the "prevailing usage" of the name (Brandon-Jones and Groves 2002; see also Groves 2005; Rylands et al. Chapt. 2 this volume).

Callimico is a monotypic genus with *C. goeldi* as its only species. The position of this genus in relation to the other Callitrichidae has been controversial. Morphologically, *Callimico* possesses numerous character states that resemble those of marmosets and tamarins (e.g., presence of claws in all digits except in the hallux and a relatively small body weight). However, other morphological and physiological character states differ from those of marmosets and tamarins and resemble character states of the other ceboids (e.g., presence of third molar and reproduction by single births). Given this "intermediate" appearance, Hershkovitz (1977) placed *Callimico* in its own family, the Callimiconidae, separating it from his Cebidae (*Saimiri*, *Aotus*, *Callicebus*, *Alouatta*, *Pithecia*, *Chiropotes*, *Cacajao*, *Cebus*, *Ateles*, *Lagothrix* and *Brachyteles*) and the Callitrichidae (*Cebuella*, *Callithrix*, *Leontopithecus* and *Saguinus* in his terminology). However, most subsequent researchers placed *Callimico* as the more basal member of the Callithrichidae based on external characteristics (Pocock 1920), dental and skeletal morphology (Ford 1986, Hill 1959, Kay 1990), and vocalizations (Snowdon 1993) (see Pastorini et al. 1998 for a complete review of the history of *Callimico* taxonomy).

In the last two decades our understanding of the evolutionary history of primates improved with the development of techniques to amplify and sequence fragments of DNA from different sources (e.g., pieces of tissue, blood, hair follicles, feces, etc.). The ability to obtain and analyze DNA sequences has had a strong impact on systematics, and phylogenetic trees reconstructed from molecular sequence data can be compared to the established relationships based on other types of data (i.e., morphological, protein, cytogenetic data). Furthermore, improvements in methods for tree construction have helped to better depict the phylogenetic relationships inferred from the DNA sequence data.

In particular, the understanding of the phylogenetic relationships among New World primates has drastically changed since the early 1990s when the first set of DNA sequence data (Schneider et al. 1993) was produced for platyrrhine taxa. Earlier, Neotropical primate relationships were inferred primarily from morphological data (e.g., Coimbra-Filho and Mittermeier 1972; Ford 1980, 1986; Hershkovitz 1963, 1977, 1984; Hill 1957, 1960, 1962; Napier and Napier 1967; Rosenberger 1981; Rosenberger and Coimbra-Filho 1984; Rosenberger and Strier 1989). In many cases, previous classifications have been fully supported by molecular evidence (e.g., the monophyletic status of the Platyrrhini), but in others the relationships inferred from molecular data are in clear discordance with those derived from morphological data (e.g., the placement of *Callimico* in relation to all other callitrichids, see Sect. 1.3.2). The main aim of this chapter is to synthesize and discuss the rapidly growing molecular data that have shed light on the relationships of the extant callitrichids, emphasizing the position of *Callimico* within the family and the generic relationships of the marmosets. In a subsequent chapter, Rylands et al. review the taxonomy and geographical distribution of *Callimico* and marmoset species.

1.2 The Molecular Data

1.2.1 Types of Data

Two types of molecular data have been used to infer New World primate phylogenetic relationships: electrophoretic protein polymorphism (e.g., Cronin and Sarich 1975; Meireles et al. 1998; Sampaio et al. 1996) and DNA sequence data. Although sequence data can be more precise and offer more resolution than allozyme data, the high cost and effort involved in obtaining these types of data usually limit analyses to very few loci. On the other hand, allozyme studies generate allele frequency data for multiple loci, but the variation that can be detected at this level is limited and inferences of phylogenetic relationships from these data can be weak (see, e.g., Avise 1994, 2000).

Most molecular phylogenetic analyses of Neotropical primates are based on DNA sequence data. Two kinds of DNA are used for phylogenetic reconstruction of metazoans: nuclear (nucDNA) and mitochondrial DNA (mtDNA). Mitochondrial and nuclear genomes evolve at highly different rates; on average metazoan mitochondrial genes evolve about tenfold faster than nuclear genes (Brown et al. 1979, 1982). The high rate of mtDNA evolution and the fact that mtDNA is maternally inherited and does not recombine, make this type of marker very useful for analyzing the relationships among recently diverged taxa. Nonetheless, mtDNA becomes saturated (multiple nucleotide changes in the same position) earlier than nucDNA, which generally results in the emergence of homoplasious characters and the loss of phylogenetic signal. Therefore, although mtDNA is commonly used to infer phylogenetic relationships of conspecific and congeneric primates, it may be uninformative to infer relationships at higher taxonomic levels.

Different nuclear gene regions show differences in patterns and rates of evolution. For example, exons (coding regions of genes that are ultimately translated into a protein) generally display lower rates of nucleotide substitutions than introns (noncoding regions that are not translated into proteins and occur between exons), which are usually not constrained by selective pressures. Thus, noncoding regions such as introns can be useful to depict the phylogenetic relationships at lower taxonomic levels than exon sequence data, but still may not have enough resolution to reliably reconstruct the phylogenetic relationships of recently derived taxa (e.g., species). It is therefore not surprising that studies based on nucDNA and mtDNA sequence data have provided different phylogenetic hypotheses of New World primates, particularly regarding the relationships within the Callitrichidae.

1.2.2 Gene Trees Versus Species Trees

The assumption that the phylogenetic relationships represented in a tree derived from DNA sequence data (i.e., a gene tree) match the true pattern of branching events that occurred during the diversification of a group of organisms (i.e., the species tree) is not always true. Apart from the possible lack of phylogenetic signal of mitochondrial and nuclear DNA sequence data at certain taxonomic levels (see Sect. 1.2.1) and the appropriateness of the methods and models used to reconstruct gene trees, there are a number of reasons why gene trees may not accurately represent the true species tree.

First, many genes are members of gene families (sets of genes in one genome that have arisen from one or more duplications of an ancestral gene) and so if nonhomologous (i.e., nonorthologous) gene copies are used to reconstruct a gene tree, the phylogeny will be misleading because it will reflect the relationships of the gene family members and not of the species. Therefore, it is important to ensure that the gene sequences that are analyzed are orthologous or that phylogenetic analyses are limited to single copy genes.

Another reason for incongruence among gene trees and species trees relates to lineage sorting and coalescence. Lineage sorting refers to the random sorting of alleles at a polymorphic locus within descendant species during speciation (looking forward in time), whereas coalescence refers to the point (looking backward in time) at which alleles of different taxa "coalesce" to a single ancestral allele. The stochastic sorting in daughter species of ancestral polymorphisms present in their common ancestor can result in coalescences of allelic lineages of these daughter species that do not coincide with the timing of the splitting of these taxa. This will give rise to gene trees that are discordant with the species tree (Fig. 1.1). The probability of obtaining an erroneous tree is quite high if the interval between the first and second species splittings is short. Lineage sorting and coalescence may explain the incongruent topology of the IRBP gene tree representing *Callimico* as closer to *Leontopithecus* than to the marmosets (Barroso et al. 1997; Mundy and Kelly 2001), compared to the evidence from most other nuclear (ε-globin gene – Harada et al. 1995; Schneider et al. 1993; intron 11 of the vWF – Chaves et al. 1999; β-2M genes – Canavez et al. 1999a, b; G6PD introns D and E – von Dornum and Ruvolo 1999; intronic transferring gene – Tagliaro et al. 2000; SRY gene – Moreira 2002; MC1R gene – Mundy and Kelly 2003) and mitochondrial (Horovitz and Meyer 1995, Horovitz et al. 1998, Pastorini et al. 1998, Tagliaro et al. 2000) gene trees, in which the *Callimico*/marmosets clade is highly supported (see Sect. 1.3.2).

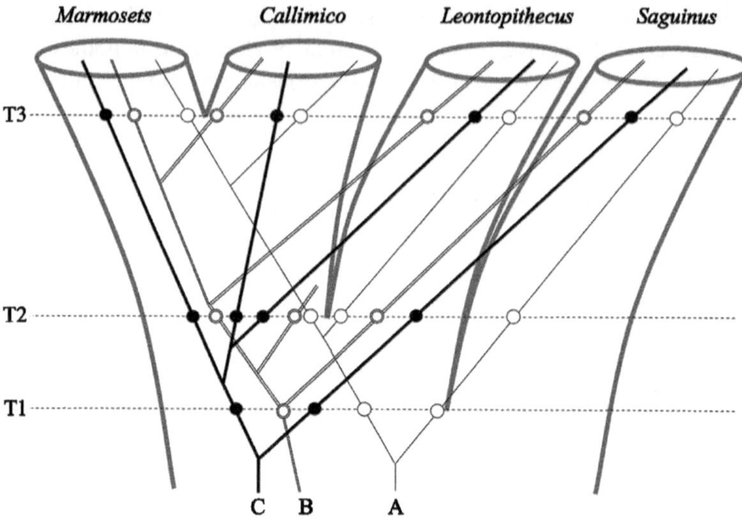

Fig. 1.1 Schematic representation of one possible explanation of how the incongruence of the IRBP gene tree with other nuclear gene trees might have arisen. Gray branches represent the "true" species tree. A, B, and C represent different nuclear genes, and circles represent alleles for each gene at different hypothetical points in time (T1, T2, and T3) when speciation events took place during the evolution of the lineage. For gene A the coalescent time for alleles predates each speciation event, reflecting the true phylogeny. For gene B coalescence time for alleles either predates or coincides with speciation events. One allele is lost after T2 and one led to the origin of the marmoset and *Callimico* alleles. In gene C (hypothetically what may have occurred with IRBP), the coalescence time for the allele that originated the marmoset's allele predates the coalescence time of the alleles from the *Callimico* and *Leontopithecus* lineages, yielding a misleading phylogenetic signal

1.2.3 Loci Used for Neotropical Primate Phylogenies

Since the beginning of the use of molecular data to address New World primate taxonomy in the early 1990s, the relationships among the Callitrichidae have been a focus of study. Several data sets of nuclear and mitochondrial loci have been created and analyzed for this purpose. Some of the studies have provided evidence on the relationships among the major callitrichid lineages, whereas others have targeted particular clades, but very few have focused on the intrageneric and intraspecific levels. A total of 11 nuclear gene regions have been analyzed to date: (1) different regions of the epsilon (ε) globin gene (Harada et al. 1995; Porter et al. 1997a, b; Schneider et al. 1993); (2) intron 1 of the interphotoreceptor retinoid-binding protein or IRBP (Barroso et al. 1997; Mundy and Kelly 2001; Schneider et al. 1996); (3) β-2-microglobulin or β-2M (Canavez et al. 1999a); (4) glucose-6-phosphate dehydrogenase or G6PD (von Dornum and Ruvolo 1999); (5) intron 11 of the von Willebrand factor or vWF (Chaves et al. 1999); (6) intronic regions of the transferrin gene (from exon 4 to exon 6) (Tagliaro et al. 2000); (7) sex-determining gene SRY (Moreira 2002); (8) Melanocortin-1 receptor or MC1R (Mundy and Kelly 2003); and (9–11) intronic sequences (including Alu elements) of the lysozyme gene (LYS) between exons 3/4, of stem cell tyrosine kinase (STK1) between exons 9/10, and of heparin-binding EGF-like growth factor precursor (HBGF) between exons 3/4 (Singer et al. 2003). Ten regions of mitochondrial DNA sequence data have also been used to explore the relationships within the Callitrichidae, including: (1) the 16S ribosomal gene (Horovitz and Meyer 1995); (2) cytochrome *b* (Moreira et al. 1996); (3) mitochondrial control region or D-loop (Tagliaro et al. 1997, van Roosmalen and van Roosmalen 2003); (4) 12S ribosomal gene (Horovitz et al. 1998); (5–8) NADH dehydrogenase subunit 4 (ND4), $tRNA_{His}$, $tRNA_{Ser}$, and $tRNA_{Leu}$ genes (Pastorini et al. 1998); (9) ND1 gene (Tagliaro et al. 2000); and (10) COII gene (Sena et al. 2002). Table 1.1 summarizes the DNA sequence evidence for the relationships within callitrichids.

1.3 Callitrichid Relationships

1.3.1 The Callitrichid Genera

Currently there are seven recognized genera within the Callitrichidae: *Saguinus* (the tamarins), *Leontopithecus* (lion tamarins), *Callimico* (callimicos), *Callithrix* (the Atlantic Forest marmosets), *Cebuella* (pygmy marmosets), *Callibella* (dwarf marmosets), and *Mico* (Amazonian marmosets other than *Cebuella* and *Callibella*). *Cebuella* was originally recognized as a different genus from *Callithrix*, which was considered to include two main groups: the Jacchus-group represented by marmosets from the Atlantic forest, the *cerrado,* and the *caatinga*, and the Argentata-group including the marmosets of the Amazonian forest and part of the *cerrado* (Hershkovitz 1977). However, numerous studies based on morphological (Natori 1994; Rosenberger 1981; Rosenberger and Coimbra-Filho 1984), karyological (Canavez et al. 1996), and molecular (Barroso et al. 1997; Canavez et al. 1999a, b; Chaves et al. 1999; Porter et al. 1997b; Schneider et al. 1996; Tagliaro et al. 1997, 2000) evidence argued that the pygmy marmosets should be included in the genus *Callithrix*. The reason for this change is based on the conclusion that *Cebuella* is more closely related to the Argentata-group than the latter is to its congeneric Jacchus-group, leaving *Callithrix* (in the classic sense) as a paraphyletic genus. Nonetheless, Rylands et al. (2000) believed that *Cebuella pygmaea* deserves generic status due to its distinctive morphological and ecological features. Therefore in order to recognize the distinctiveness of pygmy marmosets and to avoid paraphyletic genera, Rylands et al. kept *Cebuella* as a valid genus and separated *Callithrix* into two different genera: *Callithrix* for the Atlantic forest marmosets (formerly Jacchus-group) and *Mico* for the Amazonian marmosets (formerly Argentata-group) (see also van Roosmalen and van Roosmalen 2003).

Table 1.1 Summary of molecular studies that have provided evidence for the relationships among the Callitrichidae. Studies were arranged by publication date

Gene(s)	Number of nucleotides analyzed	Authors	*Callimico* as sister taxon of the marmosets[a,b]	*Leontopithecus*/*Saguinus* as a sister clade of the marmosets[a,b]	*Leontopithecus*/*Callimico* as a sister clade of the marmosets[a,b]	*Leontopithecus* as basal genus[a,b]	*Saguinus* as basal genus[a,b]	*Cebuella* within *Callithrix* (sensu Hershkovitz)[a,b]
Electrophoretic analyses (albumin and transferrin)	(2 loci)	Cronin and Sarich 1975	Yes	No	No	Unresolved	Unresolved	NA (only support as sister to *Callithrix*)
ε-globin gene	1,800	Schneider et al. 1993	Yes (69.6%)	Yes (77%)	No	No	No	NA (99% support as sister to *C. jacchus*)
ε-globin gene	1,800	Harada et al. 1995	Yes (68.9%)	Yes (75.8%)	No	No	No	NA (100% support as sister to *C. jacchus*)
Mitochondrial 16S ribosomal gene	542	Horovitz and Meyer 1995	Yes (no support provided)	No	No	Yes (no support provided)	No	NA (only as sister to *C. jacchus*)
Mitochondrial cytochrome *b*	301	Moreira et al. 1996	NA	No	NA	No	Yes (low support)	NA
IRBP intron 1	1,800	Schneider et al. 1996	No	No	Yes (92%)	No	Yes (highly supported)	NA (100% support as sister to *C. jacchus*)
ε-globin gene + IRBP intron 1	1,487 ε-globin + 1,800–1,900 IRBP	Schneider et al. 1996	Yes (64%)	No	No	No	Yes (low support)	NA (100% support as sister to *C. jacchus*)

Mitochondrial control region (D-loop)	920	Tagliaro et al. 1997	NA	NA	NA	NA	NA	Yes (75–91% support for grouping with species of the Argentata-group = *Mico*)
ε-globin gene (5′ region, exons 1, 2 and 3, and introns 1 and 2)	~ 3,100–3,600	Porter et al. 1997a	Yes (72%)	Yes (58%)	No	No	No	NA (100% support as sister to *C. jacchus*)
IRBP intron 1 + ε-globin gene	~1,800	Barroso et al. 1997	No	No	Yes (96%)	No	Yes (highly supported)	Yes (100%)
ε-globin gene [5′ region ("upstream"), exons 1, 2 and 3, and introns 1 and 2 ("downstream")]	~ 3,100–3,600	Porter et al. 1997b	Yes (70%) (only down-stream data)	Yes (59%) (only downstream data)	No	No	No	Yes (100%) (94% support for grouping with *C.* (=*M.*) *argentata* and *C.* (=*M.*) *mauesi*)
Mitochondrial 12S and 16S ribosomal gene	1,493 (aligned)	Horovitz et al. 1998	Yes (12)[c]	No	No	No	Yes	NA (support only as sister to *C. jacchus*)

(continued)

Table 1.1 (continued)

Gene(s)	Number of nucleotides analyzed	Authors	*Callimico* as sister taxon of the marmosets[a,b]	*Leontopithecus*/*Saguinus* as a sister clade of the marmosets[a,b]	*Leontopithecus*/*Callimico* as a sister clade of the marmosets[a,b]	*Leontopithecus* as basal genus[a,b]	*Saguinus* as basal genus[a,b]	*Cebuella* within *Callithrix* (sensu Hershkovitz)[a,b]
Mitochondrial + nuclear genes + morphological data	1,493 mtDNA (16S and 12S) Schneider et al. 1996 nuclear data and 76 morphological characters	Horovitz et al. 1998	Yes (18)[c]	No	No	Yes	No	NA (support only as sister to *C. jacchus*)
Electrophoretic analyses	(20 loci)	Meireles et al. 1998	NA	Yes (70%)	No	No	No	No (65% support as sister taxon to all other marmosets)
Mitochondrial ND4 gene (partial) and tRNAHis, tRNASer, and tRNALeu genes	887 (710 of ND4 and 177 of tRNAs)	Pastorini et al. 1998	Yes (BP 80%–JK 99%)	Yes (BP 62%–JK 79%)	No	No	No	NA (98–99% support as sister to *C. jacchus*)
Tandemly combined ε-globin gene (Schneider et al. 1993), mitochondrial 16S ribosomal gene (Horovitz and Meyer 1995), and ND4 and tRNA genes	~3,229	Pastorini et al. 1998	Yes (96%–98%)	Yes (76%–81%)	No	No	No	NA (100% support as sister to *C. jacchus*)

β2-microglobulin gene, exons 1, 2, and 3	885	Canavez et al. 1999a	Yes (85% MP–90% NJ)	No	No	No	Yes (highly supported)	Yes (99% MP–96% NJ) (77% MP–88% NJ support for grouping with *C.* (=*M.*) *argentata*, *C.* (=*M.*) *emiliae* and *C.* (=*M.*) *humeralifer*
β2-microglobulin gene. 5′ flanking region, exons 1, 2, and 3 and portions of introns 1, 2, and 3	1,706 [5′ flanking region (nt 1–91), exon 1 (nt 92–195), and a 5′ portion of intron 1 (nt 196–275). 3′ portion of intron 1 (nt 276–324), exon 2 (nt 325–593), intron 2 (nt 594–1,429), exon 3 (1,430–1,457), and a 58 portion of intron 3 (nt 1,458–1,706).	Canavez et al. 1999b	Yes (92% MP & NJ)	No	No	No	Yes (highly supported)	Yes (100% MP & NJ) (76% MP–97% NJ support for grouping with *C.* (=*M.*) *emiliae* and *C.* (=*M.*) *humeralifer*)

(continued)

Table 1.1 (continued)

Gene(s)	Number of nucleotides analyzed	Authors	*Callimico* as sister taxon of the marmosets[a,b]	*Leontopithecus*/*Saguinus* as a sister clade of the marmosets[a,b]	*Leontopithecus*/*Callimico* as a sister clade of the marmosets[a,b]	*Leontopithecus* as basal genus[a,b]	*Saguinus* as basal genus[a,b]	*Cebuella* within *Callithrix* (sensu Hershkovitz)[a,b]
Glucose-6-phosphate dehydrogenase (G6PD) introns D and E	1,286 (aligned) 571 Intron D and 715 Intron E.	von Dornum and Ruvolo 1999	Yes (80%)	No	No	No	Yes (highly supported)	NA
von Willebrand Factor gene, intron 11	922	Chaves et al. 1999	Yes (70–71%)	No	No	No	Yes (99–100%)	Yes (99–98%) (76–87 % support as sister taxa to the Argentata-group = *Mico*)
Mitochondrial ND1 gene and transferrin gene, two introns	1,472 bp (transferrin gene introns) and 951 ND1 gene	Tagliaro et al. 2000	Yes (76%, nucDNA) No support with mtDNA data	Yes (mtDNA data but low support)	No	Yes (nucDNA data but low support)	No	Yes (100% nucDNA data and 99% mtDNA data)
Interphotoreceptor retinoid-binding protein (IRBP)	1,816 (aligned)	Mundy and Kelly 2001	No	No	Yes (81%)	No	Yes (highly supported)	Yes (100%) (62% support grouping with species of the Jacchus-group)

Interphotoreceptor retinoid-binding protein (IRBP), ε-globin, glucose-6-phosphate dehydrogenase (G6PD) and β-2-microglobulin (β-2M)	6,763 (in tandem)	Schneider et al. 2001	Yes (98%)	No	No	No	Yes (highly supported)	NA (100% support as sister to *C. jacchus*)
Mitochondrial COII	549	Sena et al. 2002	NA	NA	NA	NA	NA	No (100% support grouping it with all the other marmosets)
SRY gene	838 (aligned)	Moreira 2002	Yes (100%)	No	No	Unresolved	Unresolved	No (100% support grouping it with all the other marmosets)
Melanocortin-1 receptor (MC1R)	951	Mundy and Kelly 2003	Yes (60%)	No	No	No	Yes (Moderately supported)	Yes (99%) (sister taxa to species of the Argentata-group = *Mico*, but not support provided for this topology)

(continued)

Table 1.1 (continued)

Gene(s)	Number of nucleotides analyzed	Authors	*Callimico* as sister taxon of the marmosets[a,b]	*Leontopithecus*/*Saguinus* as a sister clade of the marmosets[a,b]	*Leontopithecus*/*Callimico* as a sister clade of the marmosets[a,b]	*Leontopithecus* as basal genus[a,b]	*Saguinus* as basal genus[a,b]	*Cebuella* within *Callithrix* (sensu Hershkovitz)[a,b]
Alu elements and intron sequences	2,071 (concatenated intron sequences)	Singer et al. 2003	Yes (92%)	No	No	Yes (Moderately supported)	No	Not applicable (just as sister of *C. jacchus*, *C. geoffroyi* and *C. penicillata*)
Mitochondrial control region (D-loop)	902	van Roosmalen and van Roosmalen 2003	NA	NA	NA	NA	NA	NA (100% support as sister to *C. jacchus*)

[a]*NA* not analyzed
[b]Boostrap supports for particular topologies are given in between brackets when available
[c]Numbers of unambiguous characters supporting the node (clade)

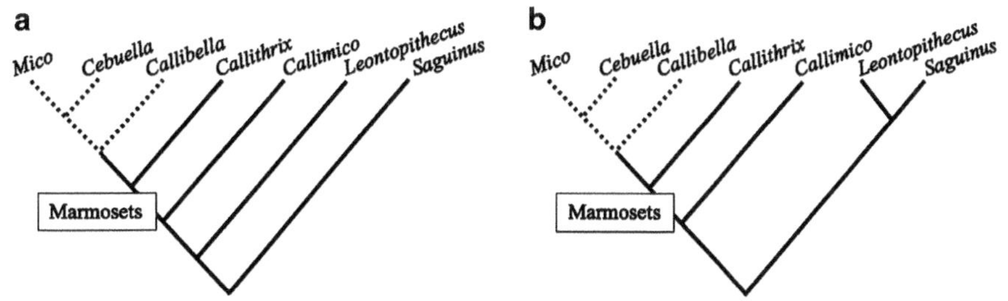

Fig. 1.2 Alternative cladograms of callitrichid phylogeny based on molecular data. (**a**) *Saguinus* as basal to all Callitrichidae, (**b**) *Leontopithecus* and *Saguinus* form a sister clade to the marmoset/*Callimico* clade. *Dotted lines* indicate low support for the relationships among *Mico, Cebuella*, and *Callibella*

Callibella is a very recently named genus (van Roosmalen and van Roosmalen 2003) that was initially recognized as a new species of marmoset, *Callithrix humilis* (van Roosmalen et al. 1998), then identified as a distinct genus in 2003. van Roosmalen and van Roosmalen (2003:4) pointed out that morphologically, *Callibella* is a "very small marmoset... slightly larger in size than *Cebuella*... but sharing more physical and behavioral characteristics with *Mico*" (see Aguiar and Lacher 2003, Chapt. 18 this volume; Ford and Davis Chapt. 21 this volume; van Roosmalen et al. 1998; and van Roosmalen and van Roosmalen 2003 for a more detailed description). Mitochondrial control region sequence data suggest that *Callibella* is basal to the rest of the Amazonian marmosets (including *Mico* and *Cebuella*, Fig. 1.2); nonetheless the branching support for the tree topology only moderately supports this arrangement (van Roosmalen and van Roosmalen 2003). More molecular studies on different mtDNA and nucDNA sequence data may be necessary to fully support the placement (and validation) of this new marmoset genus.

1.3.2 The Position of Callimicos (The Goeldi's Monkey), Callimico

As discussed above, morphological data has consistently been interpreted to indicate that *Callimico* is the basal member of the callitrichid clade. Although sharing the smaller size and claws exhibited by all callitrichids, callimicos retain primitive platyrrhine singleton births, third molars, and the hypocone on upper third molars (although reduced) (Ford 1986; Kay 1990; Rosenberger 1981). In the 1970s Cronin and Sarich (1975) used immunological data to show that the marmosets and *Callimico* formed a clade, but the classification of *Callimico* as basal to all other callitrichids prevailed well into the 1990s. Around that time, cytogenetic and molecular analyses refuted the basal position of *Callimico* and supported a sister relationship to the marmosets (*Callithrix/Cebuella/Mico/Callibella*) (Schneider et al. 1993; Seuánez et al. 1989). Since then, substantial evidence from karyotypic studies and DNA sequence data have supported the sister relationship of *Callimico* and the marmosets (Table 1.1); virtually none of the numerous studies support a basal position. Only analyses of the IRBP intron 1 sequence data (Barroso et al. 1997; Mundy and Kelly 2001; Schneider et al. 1996) have contradicted the placement of *Callimico* as a sister relative of the marmosets, placing it as a sister of the lion tamarins (*Leontopithecus*). However, as has been stated by Mundy and Kelly (2001), it is likely that this discrepancy is the result of stochastic lineage sorting from a polymorphic ancestral locus (see Sect. 1.2.2). Figure 1.1 shows the possible scenario of lineage sorting in which the coalescence of most of the other genes currently analyzed has occurred around the same time as the lineage divergence, and thus these genes give the same phylogenetic signal as the species tree, whereas the coalescence time of the IRBP gene alleles predated the speciation event that led to the separation of the *Leontopithecus* and *Callimico*/marmosets lineages.

Acceptance of the strongly supported relationship between *Callimico* and the marmosets requires a reconsideration of the evolutionary history of the loss of the third molars, loss of the hypocone, and the development of the unique callitrichid reproductive strategy including twinning and associated specializations in placentation (see Rutherford and Tardif, Chapt. 16 this volume). As Pastorini et al. (1998) discuss, either these features have evolved twice (particularly unlikely for the highly unique reproductive features of the marmosets and tamarins) or else they have been secondarily lost in callimicos. Resolving these issues will require careful anatomical and physiological studies (such as Oerke et al. 2000; Martin et al. 2005).

1.3.3 Relations of **Saguinus** *and* **Leontopithecus**

None of the morphological, cytogenetic, or molecular data have been able to unequivocally place either *Leontopithecus* or *Saguinus* as a closer relative of the *Callimico*/marmosets clade. Cladistic analyses of morphological traits supported the placing of *Callimico* as the most basal taxon within Callitrichidae (Ford 1986; Kay 1990; Rosenberger 1981; see Sect. 2.3.2). Depending on the type of morphological data analyzed, *Leontopithecus* and *Saguinus* were either placed together in a sister clade to the marmosets (Byrd 1981; Ford 1986; Hill 1959; Rosenberger and Coimbra-Filho 1984), positioned *Leontopithecus* (Snowdon 1993) or *Saguinus* (Kay 1990) as the closer relative to the marmosets, or just left these two lineages in an unresolved polytomy with marmosets (Kay 1994). Cytogenetic data either place *Saguinus* as basal to all Callitrichidae (Nagamachi et al. 1999, Neusser et al. 2001) or leave *Saguinus* in an undefined polytomy with *Leontopithecus* and the marmosets (Gerbault-Serreau et al. 2004).

Molecular data, in addition to always placing *Callimico* closest to the marmosets, have been more consistent in placing *Saguinus* as basal to all Callitrichidae. This is true using both nuclear (Barroso et al. 1997; Canavez et al. 1999a; Mundy and Kelly 2001, 2003; Schneider et al. 2001; von Dornum and Ruvolo 1999) and mitochondrial (Horovitz et al. 1998) sequence data, although some of the gene trees have more support than others for this placement. Nonetheless, some genes, including the nuclear epsilon-globin gene (Harada et al. 1995; Porter et al. 1997a; Schneider et al. 1993) and mitochondrial genes ND4 and tRNA (Pastorini et al. 1998) and ND1 (Tagliaro et al. 2000), place *Saguinus* and *Leontopithecus* together as a sister clade of the *Callimico*/marmosets clade, as do some morphological data.

The difficulties in finding congruent trees showing the phylogenetic relationships at the higher level within the Callitrichidae may be due to what may have been a short time of divergence, such that there are very short internodes during the divergence of the three lineages that led to *Saguinus*, *Leontopithecus* and *Callimico*/marmosets (see Barroso et al. 1997; Mundy and Kelly 2001, 2003; Porter et al. 1997a; and Tagliaro et al. 2000 for the internode genetic distances separating these lineages with mitochondrial and nuclear data). With short internodes like the ones observed, the mtDNA data have a greater chance of being congruent with the history of speciation, but their resolution is likely to be obscured by homoplasy at this higher taxonomic level (especially if the divergence time for all three major lineages is ancient). In contrast, gene trees of more slowly evolving nuclear loci may not resolve the relationships due to a lack of substitutions during the short internode or they may be incongruent among different loci due to the stochastic effects of lineage sorting (see McCracken and Sorenson 2005 for the same type of problem in duck phylogeny). Complete mtDNA sequences combined with data from a very large number of nuclear loci may be the only way to resolve such types of trichotomies (McCracken and Sorenson 2005), including the *Saguinus* – *Leontopithecus* – *Callimico*/marmosets trichotomy.

1.4 Discussion and Final Remarks

The use of molecular techniques to sequence and analyze DNA data has supported previous views and established new hypotheses of phylogenetic relationships among the different callitrichid taxa. Relationships at the generic level have been sustained by nucDNA and mtDNA sequence data, and the best supported topology shows the following relationships among them: (*Saguinus* (*Leontopithecus* (*Callimico* (*Callithrix* (*Callibella* (*Cebuella*/*Mico*)))))) (Fig. 1.2a). Nonetheless, some analyses do not hold up the basal position of *Saguinus*, but instead imply that *Saguinus* and *Leontopithecus* are sister taxa (see Table 1.1, Fig. 1.2b).

Although relationships at the generic level among marmosets (*Callithrix*, *Callibella*, *Cebuella*, and *Mico*) and *Callimico* are quite well defined, few studies have focused on relationships at the intrageneric and intraspecific levels. Those studies that have used mtDNA to address these relationships included data from some (but not all) of the currently recognized species (or subspecific populations), and very often taxa are represented by data of single individuals, of individuals from the same locality, or of captive individuals of unknown origin. Within the *Callimico*/marmoset clade, the only taxa that have been analyzed at intrageneric levels are *Callithrix* and *Mico* (Sena et al. 2002; Tagliaro et al. 1997, 2000; van Roosmalen and van Roosmalen 2003); the other three genera are assumed to be monotypic (but see Rylands et al. Chapt. 2 this volume). The fragments of mitochondrial DNA that were utilized in these studies did not provide enough resolution to support or refute the taxonomic distinctiveness of many of the putative species and their phylogenetic relationships within each genus, although they support a monophyletic origin for both genera. Amazon marmosets (*Callibella*, *Cebuella*, and *Mico*) group together and seem to be reciprocally monophyletic clades. Nonetheless, the relationships among the species of *Mico* (the only one of these three genera with multiple species currently recognized) show paraphyletic origins of some species (Sena et al. 2002; van Roosmalen and van Roosmalen 2003). The same is true for many of the putative species of the Atlantic Forest marmosets (genus *Callithrix*) that show very low sequence variation among them (Tagliaro et al. 1997, 2000) and rather seem to follow the patterns of variation found among populations (or possibly subspecies) of other Neotropical primates (see, e.g., Cortés-Ortiz et al. 2003 for a comparison with the variation among individuals within *Alouatta palliata* and *A. belzebul* clades). This lack of phylogenetic signal can also be a consequence of hybridization, which has been reported to occur frequently among the Atlantic Forest species of marmosets (Coimbra-Filho et al. 1993, Mendes 1997). A more complete mtDNA data set and a better representation of all putative species within the two genera could provide better resolution on the phylogenetic relationships at the intrageneric level and could validate or refute the taxonomic distinctiveness of the currently recognized species. Furthermore, the use of more variable molecular markers, such as microsatellites, could help to analyze genetic structure and levels of gene flow among populations of the same and different species.

Molecular sequence data can be costly to obtain and many times they fully support previous views of the relationships based on morphological data. Therefore, an intelligent way to use these types of data would be to apply the molecular approach for phylogenetic inference in situations where morphological variation is limited or the homology of morphological features is unclear (Avise 1994). The high cost of molecular techniques is a limitation especially in Latin America, where all the Neotropical primates actually occur. Furthermore, transporting biological samples or highly valuable museum specimens for research in/out of the limits of the countries of origin is usually a very complicated issue. The best and most efficient alternative to advance our knowledge of the evolutionary history of Neotropical primates, and particularly of the Callitrichidae, would be to develop collaborative projects involving different groups performing molecular, cytogenetic, morphological, and behavioral studies to target controversial areas still present in the relationships within and among taxa. Furthermore, the placement of DNA sequence data in public databases (e.g., GenBank) also helps to use these types of data wisely, avoids duplication, and allows different analyses of the data and comparisons with different taxonomic groups. Unfortunately, many researchers do not publicly release their sequences, despite using publicly available data in their studies. These types of situations have been minimized because of the requirement of many journals that sequences discussed in manuscripts be made available in public databases prior to publication; however, there are still many journals that have not considered the importance of this practice. These steps will be particularly important to resolve the remaining issues (see Fig. 1.2) of whether *Saguinus* or *Saguinus/Leontopithecus* (as a clade) is closest to the *Callimico/*marmosets clade, the relations among *Mico/Cebuella/Callibella*, and the intrageneric species relationships within both *Callithrix* and *Mico.*

References

Aguiar JM, Lacher TE Jr (2003) On the morphological distinctiveness of *Callithrix humilis* van Roosmalen et al., 1998. Neotrop Prim 11:11–18

Aguiar JM, Lacher Jr TE (this volume) Cranial morphology of the dwarf marmoset *Callibella* in the context of callitrichid variability. In: Ford SM, Porter LM, Davis LC (eds) The smallest anthropoids: The marmoset/callimico radiation. Springer Press, New York, pp 355–380

Avise JC (1994) Molecular markers, natural history and evolution. Chapman and Hall, New York

Avise JC (2000) Phylogeography. The history and formation of species. Harvard University Press, Cambridge, Massachusetts

Barroso CML, Schneider H, Schneider MPC, Sampaio I, Harada ML, Czelusniak J, Goodman M (1997) Update on the phylogenetic systematics of New World monkeys: further DNA evidence for placing the pygmy marmoset (*Cebuella*) within the genus *Callithrix*. Int J Primatol 18:651–674

Brandon-Jones D, Groves CP (2002) Neotropical primate family-group names replaced by Groves (2001) in contravention of Article 40 of the International Code of Zoological Nomenclature. Neotrop Prim 10:113–115

Brown WM, George M Jr, Wilson AC (1979) Rapid evolution of animal mitochondrial DNA. Proc Natl Acad Sci USA 76:1967–1971

Brown WM, Prager EM, Wang A, Wilson AC (1982) Mitochondrial DNA sequences of primates: tempo and mode of evolution. J Mol Evol 18:225–239

Byrd KE (1981) Sequences of dental ontogeny and callitrichid taxonomy. Primates 22:103–118

Canavez FC, Alves G, Fanning TG, Seuánez HN (1996) Comparative karyology and evolution of the Amazonian *Callithrix* (Platyrrhini, Primates). Chromosoma 104:348–357

Canavez FC, Moreira MAM, Simon F, Parham P, Seuánez HN (1999a) Phylogenetic relationships of the Callitrichinae (Platyrrhini, Primates) based on beta2-microglobulin DNA sequences. Am J Primatol 48:225–236

Canavez FC, Moreira MAM, Ladasky JJ, Pissinatti A, Parham P, Seuánez HN (1999b) Molecular phylogeny of New World primates (Platyrrhini) based on beta2-microglobulin DNA sequences. Mol Phylogenet Evol 12:74–82

Chaves R, Sampaio I, Schneider MPC, Schneider H, Page SL, Goodman M (1999) The place of *Callimico goeldii* in the callitrichine phylogenetic tree: evidence from von Willebrand Factor Gene Intron II sequences. Mol Phylogenet Evol 13:392–404

Coimbra-Filho AF, Mittermeier RA (1972) Taxonomy of the genus *Leontopithecus* Lesson, 1840. In: Bridgwater DD (ed) Saving the lion marmoset. Wild Animal Propagation Trust, Wheeling, West Virginia, pp 7–22

Coimbra-Filho AF, Pissinatti AF, Rylands AB (1993) Experimental multiple hybridism among *Callithrix* species from eastern Brazil. In: Rylands AB (ed) Marmosets and tamarins: systematics, ecology, and behaviour. Oxford University Press, Oxford, pp 95–120

Cortés-Ortiz L, Bermingham E, Rico C, Rodríguez-Luna E, Sampaio I, Ruiz-García M (2003) Molecular systematics and biogeography of the Neotropical monkey genus, *Alouatta*. Mol Phylogenet Evol 26:64–81

Cronin JE, Sarich VM (1975) Molecular systematics of the New World monkeys. J Hum Evol 4:357–375

Ford SM (1980) Callitrichids as phyletic dwarfs and the place of Callitrichidae in the Platyrrhini. Primates 21:31–43

Ford SM (1986) Systematics of the New World monkeys. In: Swindler DR, Erwin J (eds) Systematics, evolution, and anatomy. Alan R. Liss, New York, pp 739–135

Ford SM, Davis LC (this volume) Marmoset postcrania and the skeleton of the dwarf marmoset, *Callibella humilis*. In: Ford SM, Porter LM, Davis LC (eds) The smallest anthropoids: The marmoset/callimico radiation. Springer Press, New York, pp 411–448

Gerbault-Serreau M, Bonnet-Garnier A, Richard F, Dutrillaux B (2004) Chromosome painting comparison of *Leontopithecus chrysomelas* (Callitrichine, Platyrrhini) with man and its phylogenetic position. Chromosome Res 12:691–701

Groves CP (2001) Primate taxonomy. Smithsonian Institution Press, Washington, DC

Groves CP (2005) Order Primates. In: Wilson DE, Reeder DM (eds) Mammal species of the world: a taxonomic and geographic reference, vol 1, 3rd edn. Johns Hopkins University Press, Baltimore, pp 111–184

Harada ML, Schneider H, Schneider MPC, Sampaio I, Czelusniak J, Goodman M (1995) DNA evidence on the phylogenetic systematics of New World monkeys: support for the sister-grouping of *Cebus* and *Saimiri* from two unlinked nuclear genes. Mol Phylogenet Evol 4:331–349

Hershkovitz P (1963) Primates. Comparative anatomy and taxonomy, [volume] V, Cebidae, part B. A Monograph, by W. C. Osman Hill. Edinburgh University Press. 1962, xxix 537 pp., 31 pls., 94 figs., 3 maps. $32.00. A critical review with a summary of the volumes on New World Primates. Am J Phys Anthropol 21:391–398

Hershkovitz P (1977) Living New World monkeys (Platyrrhini). University of Chicago Press, Chicago

Hershkovitz P (1984) Taxonomy of squirrel monkeys, genus *Saimiri,* (Cebidae, Platyrrhini): a preliminary report with description of a hitherto unnamed form. Am J Primatol 4:209–243

Hill WCO (1957) Primates comparative anatomy and taxonomy III. Hapalidae. Edinburgh University Press, Edinburgh

Hill WCO (1959) The anatomy of *Callimico goeldii* (Thomas). A primitive American primate. Trans Amer Phil Soc 49:1–116

Hill WCO (1960) Primates comparative anatomy and taxonomy IV. Cebidae part A. Edinburgh University Press, Edinburgh

Hill WCO (1962) Primates comparative anatomy and taxonomy V. Cebidae part B. Edinburgh University Press, Edinburgh

Horovitz I, Meyer A (1995) Systematics of New World monkeys (Platyrrhini, Primates) based on 16S mitochondrial DNA sequences: a comparative analysis of different weighting methods in cladistic analysis. Mol Phylogenet Evol 4:448–456

Horovitz I, Zardoya R, Meyer A (1998) Platyrrhine systematics: a simultaneous analysis of molecular and morphometrical data. Am J Phys Anthropol 106:261–281

Kay RF (1990) The phyletic relationships of extant and fossil Pitheciinae (Platyrrhini, Anthropoidea). J Hum Evol 19:175–208

Kay RF (1994) “Giant” tamarin from the Miocene of Colombia. Amer J Phys Anthropol 95:333–353

Martin RD, Oerke AK, Hodges JK (2005) New light on relationships of *Callimico* from reproductive biology [Abstract]. Am J Phys Anthropol Suppl 40:146

McCracken KG, Sorenson MD (2005) Is homoplasy or lineage sorting the source of incongruent mtDNA and nuclear gene trees in the stiff-tailed ducks (*Nomonyx-Oxyura*)? Syst Biol 54:35–55

Meireles CM, Czelusniak J, Sampaio I, Schneider H, Ferrari SF, Coimbra-Filho AF, Pissinatti A, Muniz JAPC, Ferreira HS, Schneider MPC (1998) Electrophoretic polymorphisms and their taxonomic implications in Callitrichini (Primates, Platyrrhini). Biochem Genet 36:229–244

Mendes SL (1997) Hybridization in free-ranging *Callithrix flaviceps* and the taxonomy of the Atlantic forest marmosets. Neotropical Primates 5:6–8

Moreira MAM (2002) SRY Evolution in Cebidae (Platyrrhini: Primates). J Mol Evol 55:92–103

Moreira MAM, Almeida CAS, Canavez F, Olicio R, Seuánez HN (1996) Heteroduplex mobility assays (HMAs) and analogous sequence analysis of a cytochrome *b* region indicate phylogenetic relationships of selected callitrichids. J Hered 87:456–460

Mundy N, Kelly J (2001) Phylogeny of lion tamarins (*Leontopithecus* spp) based on interphotoreceptor retinol binding protein sequences. Am J Primatol 54:33–40

Mundy N, Kelly J (2003) Evolution of a pigmentation gene, the Melanocortin-1 receptor, in primates. Am J Phys Anthropol 121:67–80

Nagamachi CY, Pieczarka JC, Muniz JA, Barros RM, Mattevi MS (1999) Proposed chromosomal phylogeny for the South American primates of the Callitrichidae family (Platyrrhini). Am J Primatol 49:133–152

Napier JS, Napier PH (1967) A handbook of living primates. Academic Press, London

Natori M (1994) Craniometral variation among Eastern Brazilian marmosets and their systematic relationships. Primates 35:167–176

Neusser M, Stanyon R, Bigoni F, Wienberg J, Müller S (2001) Molecular cytotaxonomy of New World monkeys (Platyrrhini) – comparative analysis of five species by multi-color chromosome painting gives evidence for a classification of *Callimico goeldii* within the family of Callitrichidae. Cytogenet Cell Genet 94:206–215

Oerke AK, Martin RD, Hodges JK (2000) Ultrasonography in Goeldi's monkey (*Callimico goeldii*): reproductive data with evolutionary significance [Abstract]. Folia Primatol 71:204

Pastorini J, Forstner MRJ, Martin RD, Melnick DJ (1998) A reexamination of the phylogenetic position of *Callimico* (Primates) incorporating new mitochondrial DNA sequence data. J Mol Evol 47:32–41

Pocock RI (1920) On the external characters of the South American monkeys. Proc Zool Soc Lond 1920:91–113

Porter CA, Page SL, Czelusniak J, Schneider H, Schneider MPC, Sampaio I, Goodman M (1997a) Phylogeny and evolution of selected primates as determined by sequences of the e-globin locus and 5' flanking regions. Int J Primatol 18:261–295

Porter CA, Czelusniak J, Schneider H, Schneider MP, Sampaio I, Goodman M (1997b) Sequences of the primate epsilon-globin gene: implications for systematics of the marmosets and other New World primates. Gene 205:59–71

Rosenberger AL (1981) Systematics: the higher taxa. In: Coimbra-Filho AF, Mittermeier RA (eds) Ecology and behavior of Neotropical primates, vol 1. Academia Brasileira de Ciências, Rio de Janeiro, pp 9–27

Rosenberger AL, Coimbra-Filho AF (1984) Morphology, taxonomic status and affinities of the lion tamarins, *Leontopithecus* (Callitrichinae, Cebidae). Folia Primatol 42:149–179

Rosenberger AB, Strier KB (1989) Adaptive radiation of the ateline primates. J Hum Evol 18:717–750

Rutherford J, Tardif S (this volume) Mother's little helper? The placenta and its role in intrauterine maternal investment in the common marmoset (*Callithrix jacchus*). In: Ford SM, Porter LM, Davis LC (eds) The smallest anthropoids: The marmoset/callimico radiation. Springer Press, New York, pp 301–329

Rylands AB, Mittermeier RA, Coimbra-Filho AF (this volume) The systematics and distribution of the marmosets (*Callithrix, Callibella, Cebuella,* and *Mico*) and callimico (*Callimico*)

(Callitrichidae, Primates). In: Ford SM, Porter LM, Davis LC (eds) The smallest anthropoids: The marmoset/callimico radiation. Springer Press, New York, pp 25–61

Rylands AB, Schneider H, Langguth A, Mittermeier RA, Groves CP, Rodríguez-Luna E (2000) An assessment of the diversity of New World primates. Neotrop Prim 8:61–93

Sampaio I, Schneider MP, Schneider H (1996) Taxonomy of the *Alouatta seniculus* group: biochemical and chromosome data. Primates 37:65–73

Schneider H, Schneider MPC, Sampaio I, Harada ML, Stanhope M, Czelusniak J, Goodman M (1993) Molecular phylogeny of the New World monkeys (Platyrrhini, Primates). Mol Phylogenet Evol 2:225–242

Schneider H, Sampaio MIC, Harada ML, Barroso CML, Schneider MPC, Czelusniak J, Goodman M (1996) Molecular Phylogeny of the New World monkeys (Platyrrhini, Primates) based on two unlinked nuclear genes: IRBP intron 1 and Â-globin sequences. Amer J Phys Anthropol 100:153–179

Schneider H, Canavez FC, Sampaio I, Moreira MA, Tagliaro CH, Seuánez HN (2001) Can molecular data place each neotropical monkey in its own branch? Chromosoma 109:515–523

Sena L, Vallinoto M, Sampaio I, Schneider H, Ferrari SF, Cruz Schneider MP (2002) Mitochondrial COII gene sequences provide new insights into the phylogeny of marmoset species groups (Callitrichidae, Primates). Folia Primatol 73:240–251

Seuánez HN, Forman L, Matayoshi T, Fanning TG (1989) The *Callimico goeldii* (Primates, Platyrrhini) genome: karyology and middle repetitive (LINE-1) DNA sequences. Chromosoma 98:389–395

Singer SS, Schmitz J, Schwiegk C, Zischler H (2003) Molecular cladistic markers in New World monkey phylogeny (Platyrrhini, Primates). Mol Phylogenet Evol 26:490–501

Snowdon CT (1993) A vocal taxonomy of the callitrichids. In: Rylands AB (ed) Marmosets and tamarins. Systematics, behaviour, and ecology. Oxford University Press, Oxford, pp 78–94

Tagliaro CH, Schneider MPC, Schneider H, Sampaio IC, Stanhope MJ (1997) Marmoset phylogenetics, conservation perspectives, and evolution of the mtDNA control region. Mol Biol Evol 14:674–684

Tagliaro CA, Schneider MPC, Schneider H, Sampaio IC, Stanhope MJ (2000) Molecular studies of *Callithrix pygmaea* (Primates, Platyrrhini) based on transferrin intronic and ND1 regions: implications for taxonomy and conservation. Genet Mol Biol 23:729–737

van Roosmalen MGM, van Roosmalen T (2003) The description of a new marmoset genus, *Callibella* (Callitrichinae, Primates), including its molecular phylogenetic status. Neotrop Prim 11:1–10

van Roosmalen MGM, van Roosmalen T, Mittermeier RA et al (1998) A new and distinctive species of marmoset (Callitrichidae, Primates) from the lower Rio Aripuanã, state of Amazonas, central Brazilian Amazonia. Goeldiana Zoologia 22:1–27

von Dornum M, Ruvolo M (1999) Phylogenetic relationships of the New World monkeys (Primates, Platyrrhini) based on nuclear G6PD DNA sequences. Mol Phylogenet Evol 11:459–476

Chapter 2
The Systematics and Distributions of the Marmosets (*Callithrix*, *Callibella*, *Cebuella*, and *Mico*) and Callimico (*Callimico*) (Callitrichidae, Primates)

Anthony B. Rylands, Adelmar F. Coimbra-Filho, and Russell A. Mittermeier

Abstract The New World primate family Callitrichidae includes seven genera of marmosets, tamarins, lion tamarins and callimico. They are small, arboreal, diurnal, insectivore/frugivores of the forests, chaco, and scrub of tropical Central and South America. Some 60 species and subspecies of the family Callitrichidae are now recognized, 22 of which are considered to be marmosets, the "short-tusked" genera with dental and behavioral adaptations for tree-gouging and exudate-feeding. The marmosets are divided into four taxonomic groups, which we recognize as genera: *Callithrix* (eastern Brazilian marmosets of the Jacchus-group), *Cebuella* (the Amazonian pygmy marmosets), *Callibella* (the Amazonian dwarf marmoset), and *Mico* (the Amazonian marmosets of the Argentata-group). Studies over the last decade have demonstrated that Goeldi's monkey or callimico is a sister species to these marmosets. Here we review the most recent information concerning the taxonomy of these 23 species and what is known of their geographic distributions in the wild.

Resumen La familia de primates del Nuevo Mundo Callitrichidae incluye siete géneros de marmosetas, tamarinos, tamarinos leones y calimicos. Son primates pequeños, arbóreos, diurnos, insectívoros/frugívoros del bosque, chaco y monte de Centro y Sudamérica tropical. De las 60 especies y subespecies de la familia Callitrichidae se reconocen actualmente, 22 de las cuales son consideradas como marmosetas, el género "colmillo corto" con adaptaciones dentales y de comportamiento para excavación en árboles y alimentación de exudados. Las marmosetas son divididas en cuatro grupos taxonómicos, los cuales reconocemos como géneros: *Callithrix* (marmosetas del este de Brasil del grupo *Jacchus*), *Cebuella* (marmosetas pigmeas Amazónicas), *Callibella* (marmosetas Amazónicas enanas) y

A.B. Rylands (✉)
Center for Applied Biodiversity Science, Conservation International,
1919 M Street NW, Washington, DC, 20036, USA
e-mail: a.rylands@conservation.org

S.M. Ford et al. (eds.), *The Smallest Anthropoids*, Developments in Primatology: Progress and Prospects, DOI 10.1007/978-1-4419-0293-1_2,
© Springer Science+Business Media, LLC 2009

Mico (marmosetas del grupo *Argentata*). Los estudios en la última década han demostrado que los calimicos (o monos Goeldi) son una especie hermana de estas marmosetas. En el presente estudio revisamos la información más reciente concerniente a la taxonomía de estas 23 especies y lo que se conoce de sus distribuciones geográficas en estado silvestre.

Resumo A família Callitrichidae de primatas do Novo Mundo inclui sete gêneros de sagüis e micos. Eles são pequenos, arbóreos, de hábitos diurnos, frugívoros/insetivoros ocorrendo em florestas, chaco e cerrado da América Central e Sul tropical. Cerca de 60 espécies e subespécies da família Callitrichidae são reconhecidas, 22 das quais são consideradas sagüis e micos dos gêneros de "presas-baixas" com adaptações dentais e comportamentais para raspar árvores e comer exudatos. Este grupo é dividido em quatro grupos taxonômicos que nós reconhecemos como gêneros: *Callithrix* (sagüis do leste brasileiro do grupo jacchus), *Cebuella* (sagüi-leãozinho amazônico), *Callibella* (o sägui amazônico anão) e *Mico* (o sagüis amazônicos do grupo argentata). Estudos desta última década têm demonstrado que o mico de Goeldi ou callimico é uma espécie irmã dos outros sagüis. Aqui nós revemos a informação mais recente relacionada a taxonomia destas 23 espécies e o que nós sabemos das suas distribuições geográficas na natureza.

2.1 Introduction

The marmosets, tamarins, lion tamarins, and callimico of the family Callitrichidae comprise a remarkable radiation of small, arboreal, diurnal, variously gum-feeding, insectivore (faunivore)/frugivores of the forests, chaco, and scrub of tropical Central and South America. In 1977, Hershkovitz published his groundbreaking opus reviewing and synthesizing the taxonomy and distributions and all that was known of the morphology and biology of the family. He recognized 46 species and subspecies, 12 of which were marmosets (*Cebuella* and *Callithrix*). Some 60 species and subspecies of the family Callitrichidae are now recognized, 22 of which are marmosets (Rylands et al. 2000; Groves 2001, 2005). The differences in these numbers arise from the description of new species and subspecies, the recognition of the validity of some forms discounted by Hershkovitz (1977), the elimination of a subspecies of saddleback tamarin (*Saguinus fuscicollis*) resulting from evidence that it was a hybrid (Peres et al. 1996), and the inclusion of *Callimico goeldii* (Hershkovitz listed it as a separate family, the Callimiconidae) (see Table 2.1). These 60 callitrichids represent some 30% of the extant New World primates (Rylands et al. 2000).

In the case of the marmosets, seven new forms have been described since 1977, and we recognize three forms which Hershkovitz (1977) did not – (tentatively) a subspecies of the pygmy marmoset (*Cebuella pygmaea niveiventris* Lönnberg, 1940); Wied's black-tufted-ear marmoset (*Callithrix kuhlii* Coimbra-Filho, 1985) considered by Hershkovitz (1977) to be a natural hybrid; and Snethlage's marmoset

Table 2.1 The New World marmosets and callimico

Species	Common name
Family Callitrichidae	
Subfamily Callitrichinae	
Callithrix Erxleben, 1777	Atlantic forest marmosets
Callithrix jacchus (Linnaeus, 1758)	Common marmoset
Callithrix penicillata (Geoffroy Saint-Hilaire, 1812)	Black-tufted-ear marmoset
Callithrix kuhlii Coimbra-Filho, 1985	Wied's black-tufted-ear marmoset
Callithrix geoffroyi (Humboldt, 1812)	Geoffroy's tufted-ear marmoset
Callithrix aurita (Geoffroy Saint-Hilaire, 1812)	Buffy-tufted-ear marmoset
Callithrix flaviceps (Thomas, 1903)	Buffy-headed marmoset
Callibella Van Roosmalen and Van Roosmalen, 2003	Dwarf marmoset
Callibella humilis (Van Roosmalen et al., 1998)	Black-crowned dwarf marmoset
Cebuella Gray, 1866	Pygmy marmosets
Cebuella pygmaea pygmaea (Spix, 1823)	Western pygmy marmoset
Cebuella pygmaea niveiventris Lönnberg, 1940	Eastern pygmy marmoset
Mico Lesson, 1840	Amazonian marmosets
Mico argentatus (Linnaeus, 1766)	Silvery marmoset
Mico leucippe (Thomas, 1922)	Golden-white bare-ear marmoset
Mico melanurus (Geoffroy Saint-Hilaire, 1812)	Black-tailed marmoset
Mico marcai (Alperin, 1993)	Marca's marmoset
Mico intermedius (Hershkovitz, 1977)	Aripuanã marmoset
Mico emiliae (Thomas, 1920)	Snethlage's marmoset
Mico nigriceps (Ferrari and Lopes, 1992)	Black-headed marmoset
Mico cf. *emiliae* [Rondônia][a]	Rondônia marmoset
Mico humeralifer (Geoffroy Saint-Hilaire, 1812)	Black-and-white tassel-ear marmoset
Mico chrysoleucus (Wagner, 1842)	Golden-white tassel-ear marmoset
Mico mauesi (Mittermeier et al., 1992)	Maués marmoset
Mico saterei (Silva and Noronha, 1998)	Sateré marmoset
Mico manicorensis (Van Roosmalen et al., 2000)	Manicoré marmoset
Mico acariensis (Van Roosmalen et al., 2000)	Rio Acarí marmoset
Callimico Miranda Ribeiro, 1912	Goeldi's monkey
Callimico goeldii (Thomas, 1904)	Goeldi's monkey

[a]Not yet formally described

(*Mico emiliae* [Thomas, 1920]), considered by Hershkovitz (1977) to be merely a dark form of the silvery marmoset, *Callithrix argentata argentata* (Linnaeus, 1766) (now in *Mico*).

The predominant classification for the platyrrhines in the twentieth century was based on their separation into two families, the Callitrichidae (*Cebuella*, *Callithrix*, *Saguinus* and *Leontopithecus*) and Cebidae (the remaining genera), with *Callimico* being placed in either of the two, or in its own family (Dollman 1933; Hill 1957; Hershkovitz 1977). This system was maintained in all of the major syntheses published till the 1980s (e.g., Simpson 1945; Hill 1957, 1960, 1962; Cabrera 1957; Napier and Napier 1967; Hershkovitz 1977). It was the morphological studies of Rosenberger (1981, 1984; see also Rosenberger et al. 1990) that initiated a major change in thinking regarding the higher taxonomy of this group. His thesis involved placing the marmosets, tamarins and callimico in a subfamily (Callitrichinae), in a redefined Cebidae, which otherwise included squirrel monkeys (*Saimiri*) and capuchin monkeys (*Cebus*), the two comprising the Cebinae. This arrangement and the slight variations of it were subsequently amply reinforced and justified by a spate of genetic studies (e.g., Schneider et al. 1993, 1996; Harada et al. 1995; Nagamachi et al. 1996, 1999; Schneider and Rosenberger 1996). All established platyrrhine classifications today accept the affinity of *Cebus*, *Saimiri* and the marmosets, tamarins and callimico. Some place them in separate families (Rylands et al. 2000) and others, as subfamilies of the Cebidae (Groves 1993, 2001, 2005).

Hershkovitz (1977) recognized just two genera of marmosets – one species of pygmy marmoset (*Cebuella pygmaea*) and three species of what he called "true marmosets" – *Callithrix argentata,* with three subspecies; *C. humeralifer,* also with three subspecies forming the "Argentata-group" in the Amazon; and *C. jacchus,* with five subspecies forming the "Jacchus-group" in central and eastern Brazil, and the Brazilian Atlantic forest. However, the pygmy marmoset's diminutive size led him to believe that "the ancestral form of *Cebuella* must have stemmed from near the very base of the ancestral callitrichid stock" (p 450), indicating that it was no more closely related to the "true marmosets" than it was to the tamarins (*Saguinus*) and lion tamarins (*Leontopithecus*). He believed, as such, that it was a "false marmoset." Considerable discussion in the 1970s and 1980s concluded, however, that small size in callitrichids is derived rather than primitive (Leutenegger 1973, 1980; Ford 1980; Rosenberger 1981, 1984), and morphological affinities (especially in the dental adaptation for tree-gouging and gum-feeding) and numerous genetic studies have placed *Cebuella* as a close sister species to the Amazonian marmosets, even to the extent of questioning its status as a separate genus (Rosenberger 1984; Rosenberger and Coimbra-Filho 1984; Natori 1986, 1994; Natori and Shigehara 1992; Barroso 1995; Moreira et al. 1996; Schneider and Rosenberger 1996; Schneider et al. 1996; Barroso et al. 1997; Tagliaro et al. 1997, 2001; Porter et al. 1997; Canavez et al. 1999a). As such, Rosenberger (1984) proposed a single genus (*Callithrix*) with subgenera for *Cebuella*, the Amazonian *Callithrix* (Argentata-group), and the eastern Brazilian *Callithrix* (Jacchus-group). This was not upheld, however, in the classification presented in Rosenberger et al. (1990). Groves (2001) followed the recommendation of Rosenberger (1984) in grouping all the marmosets in the genus *Callithrix*, with the following subgenera: *Cebuella* for the pygmy marmosets, *Mico* for the Amazonian marmosets (Argentata-group), and *Callithrix* for the eastern Brazilian marmosets (Jacchus-group) (Table 2.2).

Groves' (2005) most recent listing separated out the dwarf marmoset as the subgenus *Calibella* [*sic*]. Intermediate in size between the pygmy marmoset and the "true marmosets," it was first described by van Roosmalen et al. (1998) in the genus *Callithrix*. Further research provided evidence, however, for its classification in a distinct genus, *Callibella* van Roosmalen and van Roosmalen, 2003.

Table 2.2 A comparison of recent taxonomies of the marmosets and callimico according to Hershkovitz (1977), Groves (1993, 2001, 2005), Mittermeier et al. (1988), Rylands et al. (1993), Rylands et al. (2001) and Rylands (this paper)

Hershkovitz (1977)	Mittermeier et al. (1988), Rylands et al. (1993)	Groves (1993, 2001, 2005)	Rylands et al. (2000), this paper
Family Callitrichidae	Family Callitrichidae	Family Cebidae	Family Callitrichidae
		Subfamily Callitrichinae[a]	Subfamily Callitrichinae
Atlantic forest marmosets – Jacchus-group	Atlantic forest marmosets – Jacchus-group	Subgenus Callithrix – Atlantic marmosets	Atlantic forest marmosets – Jacchus-group
Callithrix jacchus jacchus	*Callithrix jacchus*	*Callithrix jacchus*	*Callithrix jacchus*
Callithrix jacchus penicillata	*Callithrix penicillata*	*Callithrix penicillata*	*Callithrix penicillata*
	Callithrix kuhli[b]	*Callithrix kuhlii*[b]	*Callithrix kuhlii*[b]
Callithrix jacchus geoffroyi	*Callithrix geoffroyi*	*Callithrix geoffroyi*	*Callithrix geoffroyi*
Callithrix jacchus aurita	*Callithrix aurita*	*Callithrix aurita*	*Callithrix aurita*
Callithrix jacchus flaviceps	*Callithrix flaviceps*	*Callithrix flaviceps*	*Callithrix flaviceps*
		Subgenus *Calibella* – Dwarf marmoset	Dwarf marmoset
		Callithrix humilis[c]	*Callibella humilis*[c]
Pygmy marmoset	Pygmy marmoset	Subgenus *Cebuella* – pygmy marmosets	Pygmy marmosets
Cebuella pygmaea	*Cebuella pygmaea*	*Callithrix pygmaea pygmaea*	*Cebuella pygmaea pygmaea*
		Callithrix pygmaea niveiventris	*Cebuella pygmaea niveiventris*
Amazonian marmosets – Argentata-group	Amazonian marmosets – Argentata-group	Subgenus *Mico* – Amazonian marmosets	Amazonian marmosets – Argentata-group
Callithrix argentata argentata	*Callithrix argentata argentata*	*Callithrix argentata*	*Mico argentatus*
Callithrix argentata leucippe	*Callithrix argentata leucippe*	*Callithrix leucippe*	*Mico leucippe*
Callithrix argentata melanura	*Callithrix argentata melanura*	*Callithrix melanura*	*Mico melanurus*
Callithrix humeralifer intermedius	*Callithrix humeralifer intermedius*	*Callithrix intermedia*	*Mico intermedius*
	Callithrix emiliae	*Callithrix emiliae*	*Mico emiliae*
	Callithrix nigriceps	*Callithrix nigriceps*	*Mico nigriceps*
		Callithrix marcai	*Mico marcai*

(continued)

Table 2.2 (continued)

Hershkovitz (1977)	Mittermeier et al. (1988), Rylands et al. (1993)	Groves (1993, 2001, 2005)	Rylands et al. (2000), this paper
Callithrix humeralifer humeralifer	*Callithrix humeralifer humeralifer*	*Callithrix humeralifera*	*Mico humeralifer*
Callithrix humeralifer chrysoleuca	*Callithrix humeralifer chrysoleuca*	*Callithrix chrysoleuca*	*Mico chrysoleucus*
		Callithrix mauesi	*Mico mauesi*
		Callithrix saterei	*Mico saterei*
		Callithrix manicorensis	*Mico manicorensis*
		Callithrix acariensis	*Mico acariensis*
			Mico cf. *emiliae*[d]
Family Callimiconidae			
Callimico goeldii	*Callimico goeldii*	*Callimico goeldii*	*Callimico goeldii*

[a] Groves (2001) used the name Hapalinae Gray, 1821. Groves (2005) reverted to Callitrichinae (see Brandon-Jones and Groves 2002)

[b] Argued by Hershkovitz (1977) to be a hybrid of the forms *penicillata* and *geoffroyi*

[c] Listed in the genus *Callithrix* by Rylands et al. (2000) and Groves (2001). Groves (2001) listed it in the subgenus *Mico*, whereas Groves (2005) placed it in the subgenus *Calibella* [*sic*]. The genus *Callibella* was described by Van Roosmalen and Van Roosmalen (2003)

[d] This form was described as *C. emiliae* by de Vivo (1985). See text on *Mico melanurus* and Rylands et al. (1993)

Aguiar and Lacher (2003, see also Chapt. 18 this volume) found it to be distinct from all other marmoset and tamarin taxa in its cranial and mandibular morphology. In their genetic analysis, van Roosmalen and van Roosmalen (2003) found that the marmosets were clearly separated into Amazonian (*Callibella*, *Cebuella* and *Mico*) and eastern Brazilian (*Callithrix*) clades. *Callibella* was basal in the Amazonian marmoset clade, with *Cebuella* branching off subsequently, followed by the radiation of "true marmosets" of the Argentata-group.

Both morphological and genetic studies have suggested that the pygmy marmoset is more closely related to the Amazonian marmosets than the Amazonian marmosets are to the Atlantic forest marmosets (Tagliaro et al. 1997, 2001; Chaves et al. 1999; Ford and Davis Chapt. 21 this volume; but see Marroig and Cheverud Chapt. 17 this volume). The same, it seems, is true for the dwarf marmoset (van Roosmalen and van Roosmalen 2003; Aguiar and Lacher 2003, Chapt. 18 this volume; Ford and Davis Chapt. 21 this volume). For this reason, to avoid paraphyly, neither *Cebuella* nor *Callibella* should be recognized as distinct genera, unless *Mico* (for the Amazonian Argentata-group marmosets) is as well (Groves 2004). Rylands et al. (2000) were of the opinion that *Cebuella pygmaea* should be distinguished as a distinct genus and for this reason raised all the subgenera of Groves (2001, 2005) to full genera (Table 2.1).

Napier and Napier (1967) placed *Callimico* firmly in the Callitrichidae and, although Hershkovitz (1977) gave it its own family (Callimiconidae), there is now ample evidence that *Callimico* and the marmosets, tamarins and lion tamarins form a monophyletic group. Rosenberger (1981) placed *Callimico* in its own tribe, the Callimiconini, in the Callitrichinae. It was thought to be basal to the callitrichids (for reviews see Ford and Davis 1992; Martin 1992; Pastorini et al. 1998), the evidence coming from morphological and physiological data, even though immunological (Cronin and Sarich 1978) and genetic evidence (Seuánez et al. 1989) had hinted that it was, in fact, a sister group to *Callithrix*. Rosenberger (1984) and Martin (1992) considered this unlikely, largely due to the unique system of twinning. Rosenberger (1984) wrote that a *Callimico-Callithrix* sister-grouping "requires the diphyly of the callitrichins, the parallel evolution of dizygotic twinning, third molar loss, and hypocone reduction in *Callithrix*, or a less likely series of reversals in the genus *Callimico*," and he concluded that "neither scenario seems possible" (p 173). Subsequent genetic evidence, however, has repeatedly confirmed the findings of Cronin and Sarich (1978) and Seuánez et al. (1989). Pastorini et al. (1998), Chaves et al. (1999), Canavez et al. (1999a, b) and Neusser et al. (2001) all demonstrated that *Callithrix* (sensu Groves 2005) and *Callimico* were more closely related to each other than *Saguinus* or *Leontopithecus* are to *Callithrix* (for review see Pastorini et al. 1998, and Cortés-Ortiz Chapt. 1 this volume). Schneider and Rosenberger (1996) placed *Callimico* in the Tribe Callitrichini with the other callitrichin genera. Note that placing *Callimico* in a separate subfamily is not valid due to the sister grouping of *Callimico* and *Callithrix* (unless *Saguinus* and *Leontopithecus* are also separated out at the subfamily level; see Groves 2004).

2.2 Distributions and Some Notes on the Taxonomy of the Species

Here, we provide brief accounts of the distributions and some taxonomic notes for callimico and each of the marmosets listed in Table 2.1. The distributions are mapped in Figs. 2.1–2.5.

2.2.1 *Genus* Callithrix *Erxleben, 1777*

2.2.1.1 *Callithrix jacchus* (Linnaeus, 1758) Common Marmoset

Type locality: America, restricted to Pernambuco, Brazil, by Thomas (1911) (Hershkovitz 1977).

The common marmoset occurs in the scrub forest (forest patches in dry caatinga thorn scrub) and the Atlantic forest to the north-east of Brazil, in the states of Alagoas, Pernambuco, Paraíba, Rio Grande do Norte, Ceará, Piauí, Maranhão,

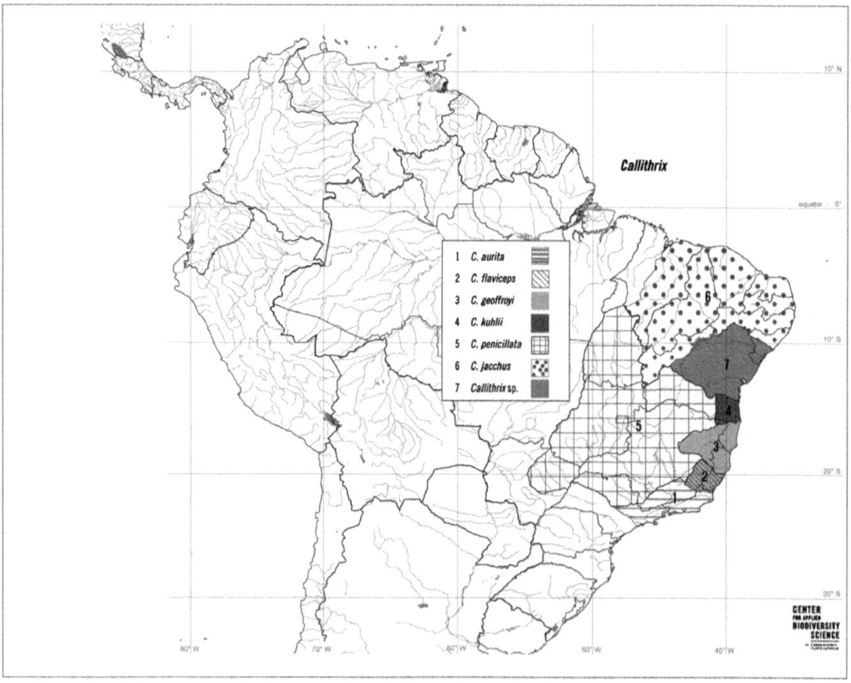

Fig. 2.1 The distributions of the Jacchus-group marmosets: *Callithrix jacchus, C. penicillata, C. kuhlii, C. geoffroyi, C. aurita*, and *C. flaviceps*

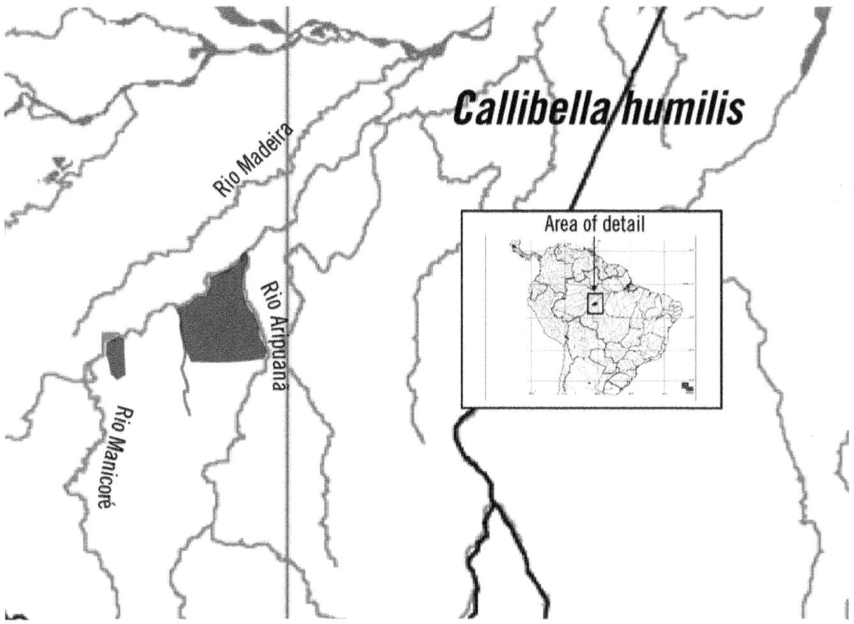

Fig. 2.2 The distribution of the dwarf marmoset, *Callibella humilis*

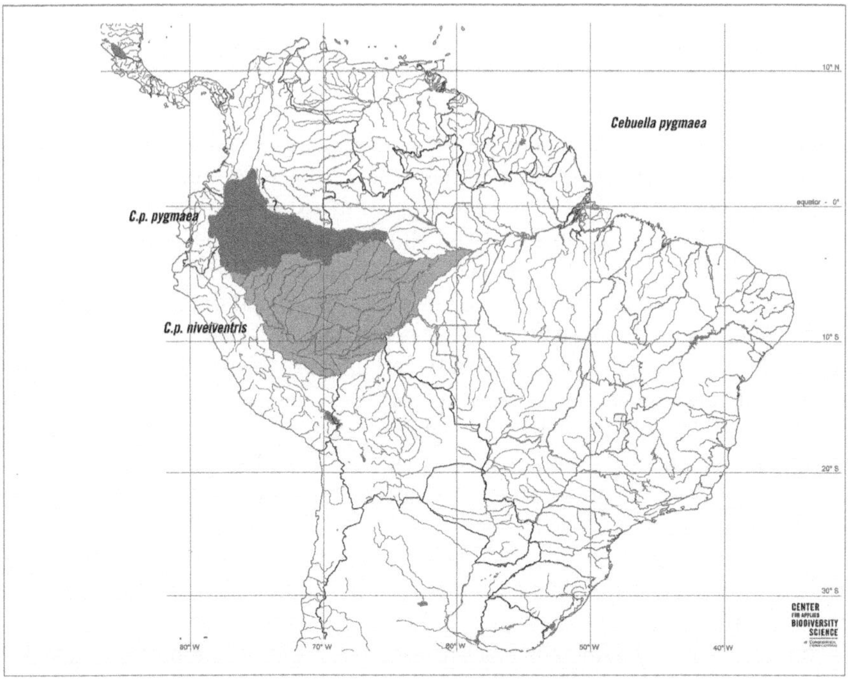

Fig. 2.3 The distributions of the pygmy marmosets: *Cebuella pygmaea pygmaea* and *C. p. niveiventris*

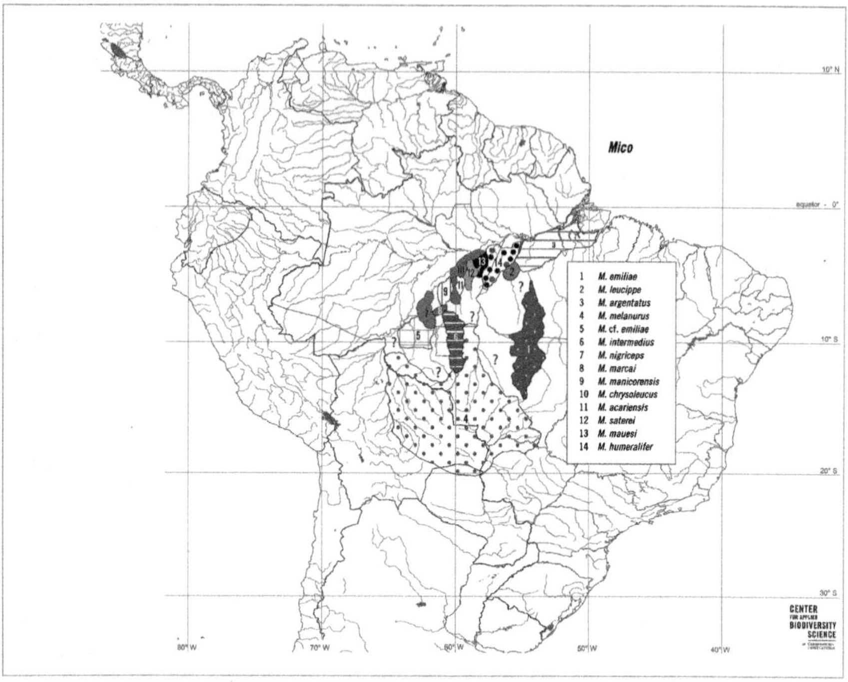

Fig. 2.4 The distributions of the Argentata-group marmosets: *Mico argentatus*, *M. leucippe*, *M. melanurus*, *M. intermedius*, *M. emiliae*, *M. nigriceps*, *M. marcai*, *M. humeralifer*, *M. chrysoleucus*, *M. mauesi*, *M. saterei*, *M. manicorensis*, *M. acariensis* and *M.* cf. *emiliae*

Bahia, and possibly northeastern Tocantins, originally extending south as far as the Rio São Francisco and its west (left) bank tributary, the Rio Grande (about 11°30′S). Hershkovitz (1977) indicated that it also probably extends north-west into the state of Maranhão, to the left bank of the Rio Parnaíba and the Serra do Valentim (Hershkovitz, 1977). Hershkovitz (1977) extended the distribution no further west than the middle reaches of the Rio Grande (left bank) and the upper Rio Parnaíba (right bank), with a lacuna between these points and the Rio Tocantins. Silva (1999) reported on localities in Maranhão and Piauí marking the northwestern limit to its range, and determined that, as Hershkovitz (1977) had indicated, it extends to the left bank of the Rio Parnaíba, probably as far as the interfluvium of the Rios Itapecurú and Mearim, south of the city of São Luis. The black-handed tamarin, *Saguinus niger*, occurs to the west. Flesher (2001) recorded *C. jacchus* in the Serra das Mangabeiras at the headwaters of the Rio Parnaíba in Piauí, approximately 10°S, 46°W. South of the Serra das Mangabeiras it is possible that the Serra Geral de Goiás marks the divide with *Callithrix penicillata* (the black-tufted-ear marmoset) to the west.

It has spread into numerous other regions as a result of introductions outside of its original range, south of the Rio São Francisco, accompanying the destruction and degradation of the Atlantic coastal forest and its associated ecosystems (Coimbra-Filho and Câmara 1996). Introduced and recent populations include

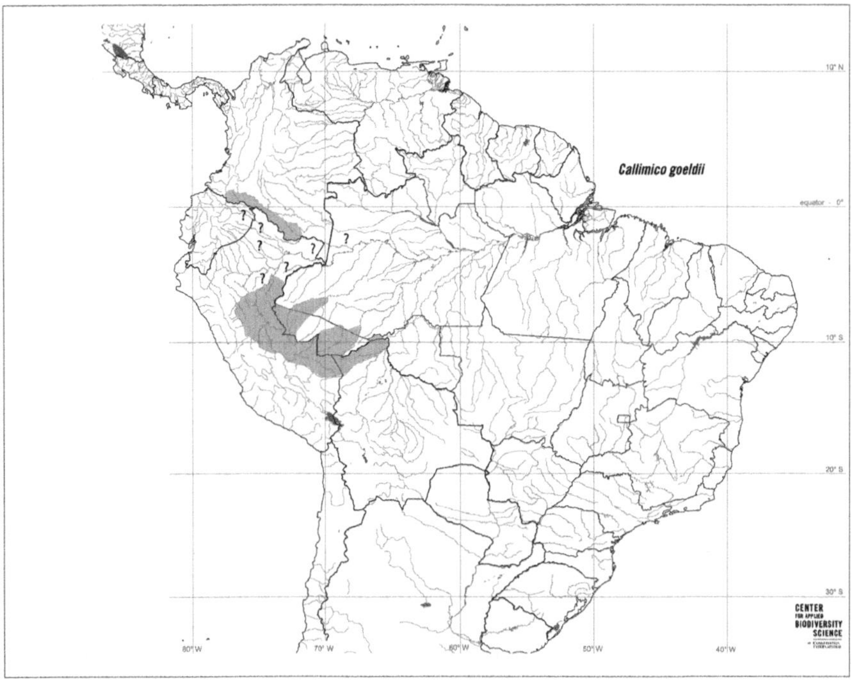

Fig. 2.5 The distribution of callimico, *Callimico goeldii*

those in the state of Sergipe and the north and north-east of Bahia, including the "Recóncavo da Bahia" (Alonso et al. 1987), the state of Rio de Janeiro in south-east Brazil (Coimbra-Filho 1984; Ruiz-Miranda et al. 2000), and the Ilha de Santa Catarina in southern Brazil (Santos et al. 2005). They have also established themselves in Buenos Aires. Alonso et al. (1987) indicated that the Recóncavo da Bahia shows a relatively narrow zone of mixing between *C. penicillata* and *C. jacchus*. Coimbra-Filho et al. (1991/1992) and Coimbra-Filho and Câmara (1996), however, have shown that this region was originally forested, and argue that the destruction of the natural vegetation over vast areas since the European discovery of Brazil in 1500, along with frequent and repeated introductions, certainly of *C. jacchus* but probably also of *C. penicillata*, has resulted in a confused picture of hybrids between these species and between *C. penicillata* and Wied's black-tufted-ear marmoset, *C. kuhlii* (see Coimbra-Filho et al. 1993). They argue that pure *C. kuhlii* was the original form occurring there.

2.2.1.2 *Callithrix penicillata* (Geoffroy Saint-Hilaire, 1812) Black-Tufted-Ear Marmoset

Type locality: Brazil, restricted to Lamarão, Bahia. Restriction of type locality attributed to Thomas (1904) by de Ávila-Pires (1969) and to Thomas (1911) by Groves (2001). The exact locality of "Lamarão, near Bahia" is a little confused (the city of Salvador was called Bahia in the past). In the distribution map of Hershkovitz (1968, p 567), Lamarão is placed in the north-central region of the state of Bahia on the uppermost reaches of the Rio Itapicurú (locality 292 in Hershkovitz [1968, p 567], and was listed as locality 292d, "Lamarão, upper Rio Itapicurú, 10°46′S, 40°21′W, 490 meters, *Callithrix penicillata penicillata*, A. Robert, May–June, 1903, at 300 meters" by Hershkovitz [1975, p 168, 1977, p 937]). Napier (1976, p 8) gave the coordinates for the type locality as "10°45′S, 40°20′W, 300 meters," probably read from the map of Hershkovitz (1968). Kinzey (1982) gives the same coordinates as those of Hershkovitz (1977), which place this locality about 320 km north-west as the crow flies from what was Bahia, today the city of Salvador, capital of the state of Bahia. De Vivo (1991) made no reference to the location of Lamarão. We have been unable to identify, however, any reference to a "Lamarão" on the upper Rio Itapicurú (e.g., Brazil IBGE 1972). A town called Lamarão, however, does exist on the railway-line midway between the towns of Água Fria (south) and Serrinha (north), 11°45′S, 38°53′W, north-west of Salvador, about 140 km as the crow flies (Vanzolini and Papavero 1968; Brazil IBGE 1972). Paynter and Traylor (1991) also give this as the locality that Alphonse Robert visited in 1903: "Lamarão, Bahia, 291 m, on railroad 140 km NW of Salvador, eastern Bahia." An atlas in the British Museum (*Stieler's Hand-Atlas*, Gotha: Justus Perthes 1905) was used by Oldfield Thomas, and it contains numerous annotations in his own hand. He underlined the town of Lamarão, indicating the probability that this is the correct locality where Alphonse Robert collected the series of *C. penicillata* that he studied and that comprise the type series for the species.

C. penicillata has a very wide distribution, occurring in the cerrado region of east central Brazil, in the states of Bahia, Minas Gerais, Goiás, the south-west tip of Piauí, Maranhão and the north of São Paulo, north of the Rios Tieté and Piracicaba (Hershkovitz 1977). In the north, it would seem that it is restricted to the south of the Rio Grande and Rio São Francisco (*C. jacchus* occurring to the north of the Rio Grande), although de Vivo (1991) identified two skins in the Museu Nacional, Rio de Janeiro, from the north-east coast of Maranhão, at Miritiba (now called Humberto dos Campos), which, he indicated, extends its range right through eastern Maranhão, along the left bank of the Rio Parnaíba. The large gap (some 850 km) between the next northernmost locality to the south (Canabrava, Rio Tocantins, Goiás, locality 275a of Hershkovitz 1977, p 490) and this northern Maranhão locality indicates that they were probably introduced animals. They were not located by Hershkovitz (1977) and were presumed by him to be *C. jacchus*, following de Ávila-Pires (1969). Silva (1999) carried out surveys in Maranhão and Piauí and did not report the occurrence of *C. penicillata*, only *C. jacchus*. The western limits of its range would seem to be marked by the Rio Araguaia, south from around 8°S in the region of the Serra das Cordilheiras, extending into the northeast of the state of Mato Grosso Sul, east of the Serra de Maracaju to the level of the Rios Pardo or Taquaraçú, the west (right) bank tributaries of the Rio Paraná.

Surveys in the north of the state of Minas Gerais have shown that *C. penicillata* extends its range through the region between the upper Rio São Francisco and the

Rio Jequitinhonha, along the western slopes of the Serra do Espinhaço. *C. penicillata* occurs on both sides of the Rio Jequitinhonha as far east as the Rio Araçuaí, a south (right) bank tributary of the upper Jequitinhonha, beyond which it is restricted to the north of the river, with *Callithrix geoffroyi* (Geoffroy's marmoset) occurring to the south (Rylands et al. 1988), the result of a recent introduction (around 1975) in the vicinity of Belmonte (Coimbra-Filho, unpubl. data). *C. penicillata* is typically of the cerrado region of Minas Gerais (in the central, southwest, west, and north of the state). Those parts originally covered by the Atlantic coastal forest in the east and south-east (the Zona da Mata) are the domain of *C. geoffroyi*, *C. flaviceps* (the buffy-headed marmoset), and *C. aurita* (the buffy-tufted-ear marmoset) that occurs in part of the Rio Doce valley. However, with the destruction of the forest, and also resulting from introductions (misguided release of confiscated animals), *C. penicillata* is taking a hold, and probably replacing, other species in numerous localities east and south of its original range (see, e.g., Olmos and Martuscelli 1995). This is happening in the Rio Doce State Park, and is possibly also the case of *C. penicillata* reported by de Vivo (1991; see also Coimbra-Filho, 1984) from the Itatiaia National Park straddling the border of the states of Rio de Janeiro and Minas Gerais. In both cases *C. aurita* is the species naturally occurring in the area.

2.2.1.3 *Callithrix kuhlii* Coimbra-Filho, 1985 Wied's Black-Tufted-Ear Marmoset

Type locality. Near the mouth of Rio Belmonte (= Rio Jequitinhonha) Bahia, Brazil (Hershkovitz 1977).

De Vivo (1991) argued that individual and clinal variation in pelage color of *C. penicillata* invalidated any separation of these coastal forest marmosets of southern Bahia. Hershkovitz (1977) argued that it is a hybrid of *C. penicillata* and *C. geoffroyi*. However, these marmosets are distinct from *C. penicillata* in their pelage color in both infants and adults (Coimbra-Filho 1985), and experimental hybrids reared at the Rio de Janeiro Primate Centre (CPRJ) failed to produce any forms similar to *C. kuhlii* (Coimbra-Filho 1984; Coimbra-Filho et al. 1993). Hybrid *geoffroyi* × *penicillata* in the wild, notably along the Rio Piracicaba, southeastern Minas Gerais, vary in their mix of characters and do not approximate to *C. kuhlii* (Passamani et al. 1997). The *kuhlii* phenotype does not show the intermediate characters of a *C. geoffroyi* × *C. penicillata* hybrid (it is closer in appearance to the former species than the latter) and is constant throughout its known range, where it has been observed. Natori (1990) studied the dental morphology (postcanine) of *C. kuhlii*, *C. penicillata*, and *C. geoffroyi* and, on these grounds, also argued that *C. kuhlii* is not a hybrid but a distinct species. *Callithrix kuhlii* is also quite distinct from other Jacchus-group marmosets in cranial morphology (Marroig et al. 2004) and vocalizations (Mendes 1997b, Mendes et al. Chapt. 3 this volume).

Callithrix kuhlii occurs between the Rio de Contas and Rio Jequitinhonha in southern Bahia, just entering the northeasternmost tip of the state of Minas Gerais

(Santos et al. 1987; Rylands et al. 1988). The western boundary is not well known, but is undoubtedly defined by the inland limits of the Atlantic coastal forest. I. B. Santos (in Rylands et al. 1988) observed hybrids of *C. penicillata* and *C. kuhlii* in the region of Almenara, Minas Gerais, the left bank of the Rio Jequitinhonha (16°41′S, 40°51′W).

Surveys in 1986/1987 by Oliver and Santos (1991) demonstrated the presence of forms intermediate in appearance between *C. kuhlii* and *C. penicillata* north from the Rio de Contas, along the coast up to the regions of Valença and Nazaré, just south of the city of Salvador (Mittermeier et al. 1988). Individuals observed by Rylands near Nazaré, just south of the city of Salvador, lacked the white frontal blaze, and, although retaining the pale cheek patches typical of *kuhlii,* were paler grey. A photograph of the marmoset from Valença, Bahia, north of the Rio de Contas, is provided in Mittermeier et al. (1988, p 19). The variation in pelage coloration of the marmosets in this region is considerable, but Coimbra-Filho et al. (1991/1992) showed that true *C. kuhlii* extended north through coastal Bahia into the state of Sergipe as far as the Rio São Francisco, in the recent past. The present-day confusion has arisen from the widespread forest destruction, most marked and nearly total in Sergipe, and the introductions and invasions of *C. jacchus* and *C. penicillata.*

2.2.1.4 *Callithrix geoffroyi* (Humboldt, 1812) Geoffroy's Marmoset

Type locality: Brazil, restricted to near Victoria, "between the Rios Espírito Santo and Jucú" by Cabrera (1957), who attributes the restriction of the type locality to Wied-Neuwied (1826). Hershkovitz (1977) notes that the names of these rivers are synonyms.

Geoffroy's marmoset occurs in the state of Espírito Santo and the forested eastern and northeastern part of Minas Gerais, in the north as far as the Rios Jequitinhonha and Araçuaí and in the south to near the state border of Espírito Santo and Rio de Janeiro (de Ávila-Pires 1969; Hershkovitz 1977; Coimbra-Filho 1984; Rylands et al. 1988). The populations just south of the Rio Jequitinhonha resulted from animals released near its mouth, at Belmonte, around 1975 (Coimbra-Filho 1986c). From there it spread eastward, and today it occurs in the gallery forests throughout the region of dry thorn scrub (*caatinga*) of the middle reaches of the river (Rylands et al. 1988). De Vivo (1991) limits it to the east of the Serra do Espinhaço in Minas Gerais. It has been recorded from the eastern slopes of Serra do Cipó, a southerly section of the Serra do Espinhaço range, at an altitude of 1274 m (Oliveira et al. 2003). Hybrid populations of *C. penicillata* and *C. geoffroyi* have been observed in some parts of the Serra da Piedade along the Rio Piracicaba, affluent of the upper Rio Doce, where the Atlantic coastal forest gives way to the cerrado (Coimbra-Filho et al. 1993; Passamani et al. 1997). The range of *C. geoffroyi* overlaps with *C. flaviceps* (see below) in southern Espírito Santo (south of the Rio Doce) and south-east Minas Gerais. *C. geoffroyi*, however, is generally restricted to lowland areas, below 500–700 m, and *C. flaviceps* to altitudes above

400–500 m (Coimbra-Filho 1971; Coimbra-Filho et al. 1981). Hershkovitz (1977) asserted that the highest recorded locality for *C. geoffroyi* is Santa Teresa, 659 m above sea level, but Mendes (1993, 1997a) has observed mixed bands of *C. geoffroyi* and *C. flaviceps* at altitudes of 800 m. Hybrid populations have been recorded for intermediate elevations (Mendes 1993, 1997a).

2.2.1.5 *Callithrix aurita* (Geoffroy Saint-Hilaire, 1812) Buffy-Tufted-Ear Marmoset

Type locality: Brazil, restricted to the vicinity of Rio de Janeiro, Guanabara, by Vieira (1944) (Hershkovitz 1977).

Coimbra-Filho (1986a, 1986b, 1990, 1991; Coimbra-Filho et al. 1993) has argued that *C. aurita* and *C. flaviceps* are so closely related that they should be considered subspecies. Close similarities exist in their dental morphology (Natori 1986), behavior, pelage (infants of the two forms are practically identical in appearance), and vocalizations (Mendes 1997b, c, Mendes et al. Chapt. 3 this volume). Evidently, hybrid *C. aurita* and *C. flaviceps* can be found in Carangola, Minas Gerais (Ferrari and Mendes 1991; Coimbra-Filho et al. 1993). Ferrari et al. (1996b) studied and reviewed the ecology and behavior of *C. aurita* and *C. flaviceps* groups and concluded that, although they are undoubtedly very similar, the "comparison appears to have done more to re-emphasize the enormous flexibility underlying the behavioral ecology of the marmosets as a whole than clarify the relationships between these two taxa in particular" (p 167).

Callithrix aurita occurs in the montane rain forests of south-east Brazil, in the southern part of the state of Minas Gerais, the state of Rio de Janeiro, and the east and north-east of the state of São Paulo (see Coimbra-Filho 1986b; Olmos and Martuscelli 1995; Brandão and Develey 1998; Ferrari et al. 1996b). Hershkovitz (1977) marks the northern limit in Minas Gerais as the Rio Muriaé, but it also occurs to the north in the Rio Doce State Park in Minas Gerais (Mittermeier et al. 1982), and hybrids (with *C. flaviceps*) have been recorded at Carangola in the Serra do Brigadeiro, Minas Gerais (Ferrari and Mendes 1991; Cosenza and de Melo 1998). Hershkovitz (1977) indicated the southeasternmost locality to be the Rio Ribeira de Iguapé in São Paulo. Olmos and Martuscelli (1995), however, failed to find evidence for this. They reported that extensive fieldwork (1982–1995) in such localities as the Fazenda Intervales State Park, Alto Ribeira State Park, Ilha do Cardoso and Carlos Botelho, and the Jureia Ecological Station and the municipalities of Juquitiba and Miracatú in the Serra da Paranapicaba, consistently failed to find *C. aurita*. They proposed the southern limit to be near the city of São Paulo, north of the junction of the Rios Pinheiros and Tietê. The Rio Tieté forms the southernmost boundary, and the most southerly record is close to Ipanema (23°26′S, 47°36′W), today Araçoiaba da Serra (the type locality for *Leontopithecus chrysopygus*). From there it extends west between the upper reaches of the Rios Tieté/Piracicaba. Again the exact limits are unclear, but believed by Olmos and Martuscelli (1995) to be the junction of these two rivers.

Brandão and Develey (1998) carried out surveys to understand better the range of *C. aurita*. Although generally believed to be largely montane in its range (600–1200 m according to Olmos and Martuscelli [1995] and 500–800 m according to Rylands [1994]), museum specimens have been collected in the the foothills of the Serra do Mar, south of Rio de Janeiro: Pedra Blanca, municipality of Paratí at 80 m, and Mambucaba, municipality of Angra dos Reis at 100 m (Brandão and Develey 1998). Coimbra-Filho (1991) and Mendes (1993) also indicated that it occurred elsewhere in lowland Rio de Janeiro, including the north-east, but is, probably, extinct there today. All recent records are montane. Brandão and Develey (1998) carried out extensive surveys of the lowland coastal forests of São Paulo and Rio de Janeiro and were unable to obtain evidence of the species' existence anywhere except at Mambucaba, where they found one in captivity and observed a group at 165 m.

This marmoset has been recorded north of the Rio Paraíba do Sul at the following sites: Mogi-Guaçú (Rio Mogi-Guaçú) by R. A. Mittermeier (unpubl.) and Muskin (1984); Alfenas, upper Rio Grande, in Minas Gerais (Hershkovitz 1977; Muskin 1984); Vargem Grande, São Paulo (Muskin 1984); Fazenda Monte Alegre, Monte Belo, Minas Gerais (Muskin 1984) and in the vicinity of Viçosa, Minas Gerais (Mendes 1993); Serra do Capanema, Rio de Janeiro (21°03′S, 42°03′W) (Mendes 1993); Fazenda João Abdo, Rio de Janeiro (21°27′S, 41°56′W) (Mendes 1993). The westernmost locality shown by Hershkovitz (1977, p 490) is Boracéia, north-east of Bauru, on the upper Rio Tieté (22°10′S, 48°45′W), but Olmos and Martuscelli (1995) found this to be an outlier and suggested that the locality really refers to the Boracéia Biological Station near the headwaters of the Rio Tietê.

2.2.1.6 *Callithrix flaviceps* (Thomas, 1903) Buffy-Headed Marmoset

Type locality: Engenheiro Reeve (now Rive), municipality of Alegre, southwestern Espírito Santo, eastern Brazil, altitude 500 m (Hershkovitz 1977).

As discussed in the case of *C. aurita*, Coimbra-Filho (1986a, 1986b, 1990) has argued that *C. flaviceps* could well be considered a subspecies of *C. aurita*. The distribution of *C. flaviceps* is described by Hershkovitz (1977), Coimbra-Filho et al. (1981), and Coimbra-Filho (1986a). It occurs in the Serra da Mantiqueira in southern Espírito Santo, south of the Rio Doce at least to the state boundary with Rio de Janeiro (and, in the past, possibly in the north of the state of Rio de Janeiro, in the municipalities of Natividade, Porciuncula and the north of Bom Jesus do Itabapoãna when they were forested). It extends west into eastern Minas Gerais in scattered localities in the highly fragmented forests of the Rio Manhuaçu basin as far as Manhuaçú (40°02′W), as noted by Coimbra-Filho (1986a) and Coimbra-Filho et al. (1981). Ferrari and Mendes (1991) and Mendes (1993) reviewed the distribution of *C. flaviceps*. Hirsch (2003; Hirsch et al. in prep.) obtained records in Minas Gerais which have extended its known range somewhat to the north and west, toward the east (right bank) of the Rio Doce (Fazenda Saet [19°43′S, 42°26′W] and the Fazenda do Eraldo A. Alves [19°45′S, 42°25′W], both at an altitiude of 270 m, and about 10 km from the east bank of the Rio Doce, in the municipality of Pingo d'Água). Hirsch (2003) also refers to two localities which would extend the range a little further south in Minas Gerais, but they have still to be confirmed and may be hybrids with *C. aurita* (the left bank of the Rio Matipó, municipality of Abre Campo, and the Córrego Jurumirim, the left bank of the Rio Casca, municipality of Rio Casca).

2.2.2 Genus **Callibella** *Van Roosmalen and Van Roosmalen, 2003*

2.2.2.1 *Callibella humilis* (Van Roosmalen et al., 1998) Dwarf marmoset

Type locality: West bank of the lower Rio Aripuanã, one kilometer south of the settlement of Nova Olinda, 41 km southwest of the town of Novo Aripuanã, Amazonas state, Brazil. The region is located in south-central Amazonia, Brazil, south of the Rio Amazonas, and east of the Rio Madeira. Coordinates 05°30′63″S, 60°24′61″W. Altitude 45 m (van Roosmalen et al. 1998).

According to van Roosmalen et al. (1998) and van Roosmalen and van Roosmalen (2003), the dwarf marmoset has a very restricted range along the west bank of the Rio Aripuanã, from its mouth, just southwest of the town of Novo Aripuanã, south at least to the village of Tucunaré, and west, along the right bank of the Rio Madeira to the mouth of the Rio Mataurá, and the right bank of the Rio Uruá. They speculated that the southern limit is probably marked by the headwaters of the Rios Mariepauá and Arauá. An isolated population was also found along the middle of Rio Atininga, about 50 km southwest of the presumed southern limit of the main population, about 10 km east of the Rio Manicoré. The range of *Callibella humilis* is entirely within that hypothesized for *Mico manicorensis* (the Manicoré marmoset).

2.2.3 Genus **Cebuella** *Gray, 1866*

2.2.3.1 *Cebuella pygmaea pygmaea* (Spix, 1823) Western Pygmy Marmoset

Type locality: Tabatinga, Rio Solimões, Amazonas, Brazil (Napier 1976; Hershkovitz 1977).

Although da Cruz Lima (1945) and Napier (1976) recognized the two subspecies of pygmy marmoset listed here, Hershkovitz (1977) did not, and they were generally ignored until van Roosmalen and van Roosmalen (1997) argued for their validity. The difference is mainly in the color of the underparts. Van Roosmalen and van Roosmalen (1997) described the ventral surface of the Spix's type specimen as ochraceous, whereas that of *niveiventris* Lönnberg, 1940, is whitish in the chest, belly and inner side of the arms and legs. Napier (1976) distinguished the two as follows: *C. p. pygmaea* – "back mottled greyish-brown; underparts and inside of limbs yellowish-brown"; *C. p. niveinventris* – "back paler and more greyish, especially posteriorly; underparts and inside of limbs white, sharply demarcated from upper parts" (p 12). The distribution was clearly delimited by van Roosmalen and van Roosmalen (1997), but Groves (2001, 2005), although listing the two subspecies, agreed with Hershkovitz (1977) that "the picture is more complicated." The range of *C. p. pygmaea* includes Amazonian Ecuador, and de la Torre (2000), who knows the species well, described the throat, chest and belly as yellowish white. Lönnberg (1940) based his description of *niveiventris* on a comparison with five specimens (of *C. p. pygmaea*) from João Pessoa, on the upper Rio Juruá, a locality which van Roosmalen and van Roosmalen (1997) include well within the range of *niveiventris*. Hershkovitz (1977) discussed the variation in the color of the fur of the underparts of pygmy marmosets from various localities, and concluded that it is individually and locally variable and cannot be used alone to support the subspecific status of *niveiventris*. We describe the distributions of the two subspecies separately based on van Roosmalen and van Roosmalen (1997), but with no strong conviction that the taxonomic arrangement is valid. This is a clear case where further enquiry is needed.

Following van Roosmalen and van Roosmalen (1997), *C. p. pygmaea* occurs in the upper Amazon basin, north of the Rio Solimões in Brazil and west from the Rio Japurá, and south of the Río Caquetá in Colombia, Amazonian Ecuador and Peru, north of the Solimões-Amazonas-Marañon, and east (left bank) of the Rio Pastaza in Peru. The easternmost localities in Brazil include the Paraná do Aranapu and Paraná do Jarauá on the lower Rio Japurá. Hernández-Camacho and Cooper (1976) and Defler (2004) recorded that it is well known in the south of the Río Caquetá in Colombia, but that reports of its occurrence further north in the upper Río Guaviare region remain to be confirmed. The only evidence to date is a captive specimen believed to have been obtained from Cano Morrocoy on the south bank of the Río Guaviare. Izawa (1975) reported that it is absent from the Río Peneya, north of the Río Caquetá; during further surveys he was unable to confirm its presence anywhere else north of the Río Caquetá, although in a later publication, Izawa (1979) indicated that it may occur on the Río Orteguaza, a northern tributary of the uppermost reaches. It would evidently be the subspecies throughout the Ecuadorian Amazon (de la Torre 2000). In Peru, Aquino and Encarnación (1994) extend the range to the west of the Río Pastaza, to the region of the Cerro Campanquiz and the basin of the Río Santiago, south to the Río Mayo in the Department of San Martín, a left bank tributary of the Río Huallaga, and to both sides of the Río Marañon. The range south of the Río Marañon to the west of the Rio Huallaga could belong to either *pygmaea* or *niveiventris*, according to the distribution proposed by van Roosmalen and van Roosmalen (1997).

2.2.3.2 *Cebuella pygmaea niveiventris* Lönnberg, 1940 Eastern Pygmy Marmoset

Type locality: Lago Ipixuna, Rio Solimões, east of the Rio Tefé, Brazil, 3°52′S, 63°52′W (Napier 1976).

The basis for recognizing *niveiventris* is discussed in the text on *C. p. pygmaea* above. Following van Roosmalen and van Roosmalen's (1997) hypothesis, the eastern pygmy marmoset would be the form south of the Rio Solimões-Amazonas-Marañon and east of the lower Río Huallaga and middle to upper Río Ucayali. Aquino and Encarnación (1994) indicated a larger range in Peru, occupying the entire Amazonian lowlands and Andean foothills east of the Río Mayo and the Río Huallaga above the Río Mayo, and including the Río Pachitea and the Río Ucayali basins, south to the upper Río Purus and the basins of the Rió Madre de Dios and Río de las Piedras and the Río Tambopata. From there it extends east into northern Bolivia to the region of Cobija (Freese et al. 1982; Buchanan-Smith et al. 2000). In Bolivia, Izawa (1979) and Izawa and Bejarano (1981) confined it to the north and west of the Ríos Orthon and Manuripi, northern tributaries of the Río Madre de Dios. Although Brown and Rumiz (1986) doubted that it occured as far south as the Río Manuripi and limited its distribution to the north of the Río Tahuamanu, Buchanan-Smith et al. (2000) confirmed its presence south of the river along the Rio Muyumanu. The easternmost record obtained by Buchanan-Smith et al. (2000) was at Santa Rosa on the Río Abunã. Its presence in northern Bolivia indicates that it should occur in parts of eastern Acre, including the Ríos Acre and uppermost Abunã, not indicated by Hershkovitz (1977). This was confirmed by Bicca-Marques and Calegaro-Marques (1995). Van Roosmalen and van Roosmalen (1997) observed pygmy marmosets between the lower Rios Purus and Madeira, and indicated a range extending south at least to the Rio Ipixuna (a right bank tributary of the Rio Purus). The Rio Abunã is a left bank tributary of the Rio Madeira, so it is reasonable to believe that pygmy marmosets occur throughout the interfluvium of the Rios Purus and Madeira south to the Rio Abunã. The southernmost locality reported so far is the Manu National Park, approximately 12°S (Soini 1988).

2.2.4 Genus Mico *Lesson, 1840*

As discussed above, the genus *Mico* recognized here includes all of the Amazonian marmosets along with the black-tailed marmoset (*M. melanurus*), which were formerly considered members of the genus *Callithrix* in the Argentata-group as defined by Hershkovitz (1977).

2.2.4.1 *Mico argentatus* (Linnaeus, 1771) Silvery Marmoset

Type locality: Pará, Brazil, restricted by de Carvalho (1965) to Cametá, left bank of lower Río Tocantins (Hershkovitz 1977). (Note: *argentata* used with the feminine genus "*Callithrix*" is here changed to *argentatus* to agree with the masculine *Mico*.)

Mico argentatus occurs south of the Rio Amazonas, in relatively flat, lowland forest, between the mouth of the Rio Tocantins in the east and the Rios Tapajós and Cuparí (an eastern tributary) in the west (Ferrari and Lopes Ferrari 1990; Ferrari and Lopes 1996; Pimenta and Silva 2005), extending south to the Rio Irirí as far as the lower Rio Curuá (Hershkovitz 1977). Ferrari and Lopes Ferrari (1990) (see also Ferrari 1993a; Ferrari and Lopes 1996) argued that its restricted range (lowland floodplain) east of the Rio Tocantins is due to habitat differences and sympatry with the black-handed tamarin *Saguinus niger* (a wider ranging species extending west to the Rio Xingú and east to the Rio Parnaíba). Ferrari (1993b) indicated that *S. niger* has the competitive edge in the forests on the relatively nutrient-poor soils of the Brazilian Shield, and that *M. argentatus* was a "newcomer" resulting from a Holocene range expansion of the genus. The southernmost record listed by Hershkovitz (1977) is the type locality, Maloca, upper Rio Curuá, of an individual with a "blackish crown and grayish brown back" described by Thomas (1920) as *Hapale emiliae*, illustrated by da Cruz Lima (1945) as *Callithrix emiliae* (Snethlage's marmoset), and recognized tentatively here as a separate species. Da Cruz Lima (1945) pointed out that *emiliae* "bears a closer resemblance to the form *melanura* of Mato Grosso than to the typical form of the marginal river zone of the Amazon." De Ávila-Pires (1986) also argued for the validity of *Callithrix argentata emiliae* on the basis of specimens obtained further south, on the Rio Peixoto de Azevedo (see *M. melanurus* and *M. emiliae* below). *Mico argentatus* does not occur south of Belo Monte on the Rio Xingú (Transamazon highway) and is restricted to the north of the Tucuruí dam reservoir on the Rio Tocantins (Ferrari and Lopes Ferrari 1990; Ferrari and Lopes 1996). This restricts the range of *M. argentatus* well to the north of the mouth of the Rio Irirí on the eastern bank of the Xingú, with the southern limits being somewhere between the Rios Cuparí and Irirí to the west of the Rio Xingú, as indicated by Hershkovitz (1977; see also Martins et al. 1988).

2.2.4.2 *Mico leucippe* (Thomas, 1922) Golden-White Bare-Ear Marmoset

Type locality: Pimental, right bank of the Rio Tapajós, below mouth of Rio Jamanxim, Pará, Brazil (Hershkovitz 1977).

Mico leucippe was considered by de Carvalho (1959) to be a subspecies of *Callithrix chrysoleuca*, but it was placed as a subspecies of *C. argentata* by Hershkovitz (1977). It is similar to *M. argentatus*, predominantly white, but with a tail and feet of pale gold. The face and ears are largely unpigmented (Hershkovitz 1977). It is known only from a small area in the state of Pará, between the Rios Cuparí and Tapajós (right bank of the Rio Tapajos), south to the Rio Jamanxim (Hershkovitz 1977; Pimenta and Silva 2005).

2.2.4.3 *Mico emiliae* (Thomas, 1920) Snethlage's Marmoset

Type locality: Maloca, upper Rio Curuá, upper Rio Irirí, Rio Xingú, Pará, Brazil (Thomas 1920).

This marmoset, named by Thomas (1920) as *Hapale emiliae*, was considered a dark form of *Callithrix argentata argentata* by Hershkovitz (1977). It was recognized by da Cruz Lima (1945), Cabrera (1957) and Hill (1957), however, as a distinct subspecies of *C. argentata*, and de Ávila-Pires (1986) also argued its validity on the basis of three skins obtained from the Rio Peixoto de Azevedo, well to the south of the type locality. Cabrera (1957) described its distribution as the south of the state of Pará, possibly entering contiguous parts of the state of Mato Grosso. De Ávila-Pires (1986) was more exact, indicating that it occurs south from the Rio Irirí (*C. a. argentata* occurring to the north – confirmed by Martins et al. 1988), at least as far south as the southern (left) margin of the Rio Peixoto de Azevedo, an eastern tributary of the Rio Teles Pires. Martins et al. (1988) recorded it on the left bank of the Rio Irirí, south from its mouth. Pimenta and Silva (2005) recorded it from the Serra do Cachimbo and the right bank of the upper Teles Pires, a significant extension of the range to the west. The southern limits would evidently not be beyond the headwaters and upper Rio Paraguai, approximately 14°30′S, where *M. melanurus* has been registered for a number of localities (Hershkovitz 1977; de Vivo 1985). De Ávila-Pires (1986) suggested that the Rio Teles Pires marks the western limit of its range. Martins et al. (1988) indicated that *C. emiliae* is limited to the west (left) bank of the lower Rio Irirí, with an undescribed *C. argentata* subspecies occurring between the Rios Irirí and Xingú. These authors also indicated that no marmoset occurs east of the Rio Xingú above the mouth of the Rio Irirí. The distribution of *Mico emiliae* has been confused somewhat by its alignment with a similar, if slightly darker, form in the state of Rondônia by de Vivo (1985), referred to here as *Mico* cf. *emiliae* (also discussed in the text on *M. melanurus*).

2.2.4.4 *Mico melanurus* (Geoffroy Saint-Hilaire, 1812) Black-Tailed Marmoset

Type locality: Brazil, restricted to Cuyabá (= Cuiabá) by Allen (1916) (Hershkovitz 1977). (Note: *melanura* used with the feminine genus "*Callithrix*" is here changed to *melanurus* to agree with the masculine *Mico*.)

The most widespread of the Argentata-group marmosets, *M. melanurus* is the only one to occur naturally outside of Brazil, extending south as it does through the Pantanal of Mato Grosso into Bolivia and Paraguay. Hershkovitz (1977) indicated the Rio Tacuarí in Brazil and the headwaters of the Río Mamoré in Bolivia as the southern limit of its distribution, but Stallings and Mittermeier (1983) and Stallings (1985) recorded it also from the northeastern Paraguayan chaco, extending the known range to approximately 20°S. In Bolivia, it occurs east of the Río Mamoré, in the Departments of Beni and Santa Cruz (Brown and Rumiz 1986).

According to Hershkovitz (1977), in Brazil it occurs to the east of the Rio Madeira, from the mouth of the Rio Aripuanã extending south to beyond the Rio Guaporé and west to the Rio Roosevelt (Hershkovitz 1977). However, field research and the discovery of a number of distinct new marmosets has modified the range he proposed. No evidence has been forthcoming for its occurrence between the Rios Aripuanã and Roosevelt (the range of *M. intermedius*, the Aripuanã marmoset). It does occur on the east bank of the Rio Aripuaná, north at least to 10°S, and probably west to the Rio Juruena, or the Rio Teles Pires, where de Ávila-Pires (1986) predicted that it would meet the range of the form *M. emiliae*. Hershkovitz's (1977) proposal for its occurrence west of the Rio Aripuanã-Roosevelt was based on three localities. The first was the Foz do Rio Castanho (near the junction of the Rios Roosevelt, Guariba, and Aripuanã in the state of Amazonas) (locality 197b, p 569, Hershkovitz 1977). This is the type locality of the distinct form *M. marcai* (Marca's marmoset) described by Alperin (1993). According to de Vivo (1985), the marmosets at the other two localities indicated by Hershkovitz (1977: 214b, mouth of the Rio Jiparaná, upper Rio Madeira; and 214c, Urupá, Rio Jiparaná) would not be the form *melanura*, but *Callithrix (=Mico) emiliae* based on their similarity (although they are darker) to the marmosets from the Rio Curuá, in Pará (the type locality of *M. emiliae* [Thomas 1920]). The "*Callithrix emiliae*" of de Vivo (1985), however, occurs to the west of the range of *M. melanurus* and, if aligned with the *Callithrix emiliae* of Thomas (1920), listed by da Cruz Lima (1945), Cabrera (1957), and de Ávila-Pires (1986), it would indicate a disjunct distribution, being separated by typical *M. melanurus* between the Rios Aripuanã and Juruena (or Teles Pires). The Rondônia marmoset of de Vivo (1985, 1991) is distinct from *melanurus* in its paler color (less brownish dorsum) and the lack of the distinct pale thigh stripe. It is listed below as *Mico* cf. *emiliae*.

2.2.4.5 *Mico marcai* (Alperin, 1993) Marca's Marmoset

Type locality: Mouth of the Rio Castanho (= Rio Roosevelt), a left bank tributary of the Rio Aripuanã, state of Amazonas, Brazil (Alperin 1993). Alperin (2002) discusses the type locality.

This form was first described as *Callithrix argentata marcai* Alperin 1993, from three skins in the Museu Nacional, Rio de Janeiro. It is distinct from *Mico leucippe* and *M. argentatus* in having a marked coloration of the mantle; from *M. melanurus*, in not having the white patches on the hips, and the white patch on the forehead; and from *M. emiliae*, in having pale hands and feet, and the dark brown forehead. It is known only from these specimens, collected by the Rondon Commission in April 1914. Its range is unknown but probably extends south along the left bank of the Rio Roosevelt and at least part of the way north to meet, somewhere, the southern limits of the range of *M. manicorensis* (the Manicoré marmoset). Ferrari (1993c, 1994) reported the collection of an adult female "*C. emiliae*" on the east bank of the Rio dos Marmelos opposite the Tenharin Indian settlement (on the west bank, 07°57′S, 62°03′W). (For the correct location of Tenharin, see Ferrari 1994.) Ferrari (1993c) said it was easily distinguished from *M. nigriceps*, the black-headed marmoset (collected on the west bank at the same location) by the lack of pigmentation on the facial skin. It would seem that Ferrari (1993c) presumed the identity of this animal to be *C. emiliae* based on de Vivo (1985,

1991) who stated that *C. emiliae* occurred on the left (west) bank of the Rio Aripuanã: a belief arising from his interpretation of the identity of the marmoset of the Rio Castanho, here listed as *Mico marcai* (Alperin, 1993). The true identity of the marmoset from the east bank of the Rio dos Marmelos at Tenharin, however, has yet to be determined in the light of this.

2.2.4.6 *Mico intermedius* (Hershkovitz, 1977) Aripuanã Marmoset

Type locality: Near the mouth of Rio Guariba, the left bank of Rio Aripuanã, south-eastern Amazonas, Brazil (Hershkovitz 1977).

M. intermedius is similar to *M. melanurus* in aspects such as the distinct pale thigh stripe, similarly colored hindquarters, dark crown (just a little paler than *melanurus*), and the lack of an ear-tuft (it has a rudimentary tuft from behind the pinna and not the well-developed tuft from within and around the pinna as in *Mico humeralifer,* the Santarém marmoset). The face is variably depigmented (some individuals have quite dark-greyish faces), the forequarters are paler, and varying parts of the tail are pale off-white rather than black. It occurs between the Rios Roosevelt and Aripuanã, including the entire basin of the Rio Guariba. *M. intermedius* and *M. melanurus* are not sympatric between the Rios Aripuanã and Roosevelt as was indicated by Hershkovitz (1977). The exact southern limits are not known, but they are probably around the headwaters of these two rivers.

2.2.4.7 *Mico nigriceps* (Ferrari and Lopes, 1992) Black-Headed Marmoset

Type locality: Lago dos Reis (7°31′S, 62°52′W, = Lago Paraiso), 17 km east of Humaitá, Amazonas, Brazil, on the Trans-Amazon highway BR-230 (right, or east, bank of the Rio Madeira) (Ferrari and Lopes 1992; Ferrari 1993c, 1994).

This marmoset is darker than the form *M.* cf. *emiliae* described by de Vivo (1985) from adjacent Rondônia, and differs in the pigmentation of the face and ears, pheomelanization of the forelimbs, mantle and ventrum, a brown rather than grey dorsum, an orange/russet coloration of the posterior limbs, and pale hips and upper thighs. It is known from two localities separated by little more than 50 km, the paratype locality being Calama (8°03′S, 62°53′W), Rondônia, Brazil (right or east bank of the Rio Madeira), east of the Rio Jiparaná. This marmoset is believed to occur between the Rio dos Marmelos in the north and east, the Rio Madeira in the west and the Rio Jiparaná in the south, in the state of Rondônia, Brazil. Ferrari (1993c) reported on the capture of two adult male *M. nigriceps* at the Tenharin Indian settlement, on the west bank of the Rio dos Marmelos (07°57′S, 62°03′W). (The location of Tenharin on the map, Fig. 2.1, in Ferrari [1993c] was incorrect. The correct location was shown in Ferrari [1994].) Ferrari and Lopes (1992) and Ferrari (1993c) argued that it is unlikely to extend further west than the Rios Madeira and Jiparaná, or east to the Rios

Aripuanã and Roosevelt, and the southeastern limits are defined by an area of savanna vegetation at the headwaters of the Rio dos Marmelos and along the middle of Rio Jiparaná.

2.2.4.8 *Mico* cf. *emiliae* (de Vivo, 1985) Rondônia Marmoset

Type locality: This marmoset was first described by de Vivo (1985) as *Callithrix emiliae* (Thomas, 1920), the type locality of which is Maloca, upper Rio Curuá, upper Rio Irirí, Rio Xingú, Pará, Brazil. As argued here (see also the text on *M. melanurus*), we consider it a distinct form yet to be given a type specimen (and, as such, a type locality). De Vivo (1985; map 1, p 104) refers specifically to four localities for his definition of *C. emiliae*: the type locality of *Hapale emiliae* Thomas, 1920 (locality 11); the Foz do Rio Castanho (locality 10; subsequently attributed to *M. marcai* by Alperin 1993); Ji-paraná, Rondônia (10°52′S, 61°57′W) (locality 13); and Nova Brasília, Rondônia (10°52′S, 61°57′W) (locality 14). The last two are candidates for the type locality.

De Vivo (1985) was the first to alert the scientific community to this marmoset of the state of Rondônia, Brazil. De Vivo (1985) provided a detailed description of a specimen from Nova Brasília. He believed it was *Callithrix emiliae* (Thomas, 1920), being similar in pelage color if just a little darker. The Foz do Rio Castanho specimens, later described as *Callithrix argentata marcai* by Alperin (1993), De Vivo said were similar, and indicated, as a result, that it occurred east of the Rio Madeira and north to the west (left) bank of the Rio Aripuanã to the mouth of the Rio Roosevelt. De Vivo (1985) made no mention of the two other localities indicated by Hershkovitz (1977: 214b, mouth of the Rio Jiparaná, upper Rio Madeira; and 214c, Urupá, Rio Jiparaná), but his map indicated that they would also be "*C. emiliae*." The discovery of *M. nigriceps* (Ferrari and Lopes, 1992) means that the range of de Vivo's (1985) "*C. emiliae*" is restricted to the left bank of the Rio Jiparaná.

It is almost certain that these Rondônia marmosets should be considered a new species, similar to, but distinct from, both *M. emiliae* and *M. melanura*. Nagamachi et al. (1996, 1997, 1999) analysed the karyotype, and Sena et al. (2002) carried out a phylogenetic analysis of mitochondrial cytochrome oxidase II gene sequences that substantiated the distinctiveness of de Vivo's Rondônia marmoset. They reported a range delimited to the north and west by the Rios Mamoré-Madeira, and Jiparaná and to the south by the Serra dos Pacáas Novos, where they indicated it may be parapatric with *M. melanura*, which is "typically found in the savannah-like, rather than rainforest ecosystems, that predominate in southern Rondônia" (p 81). Ferrari et al. (1995) failed to find any evidence of the occurrence of marmosets during surveys of the Guajará-Mirim State Park in west central Rondônia. Likewise no marmosets were recorded at Pimenta Bueno on the upper Rio Jiparaná (Ferrari et al. 1996a). The range of this marmoset is evidently much smaller than previously thought. Conversely, the saddleback tamarin, *Saguinus fuscicollis weddelli*, once believed to be restricted to the west of the Rio Madeira (Hershkovitz 1977), was found at both localities. Ferrari et al. (1996a) pointed out that this indicates a much larger range in this portion of the Amazon than that portrayed by Rylands et al. (1993), and discussed the contrasting habitat preferences and competitive abilities of the two species (see also Lopes and Ferrari 1994) similar to that recorded by Ferrari and Lopes (1996) for *M. argentatus* and *Saguinus niger*.

2.2.4.9 *Mico humeralifer* (Geoffroy Saint-Hilaire, 1812) Santarém Marmoset

Type locality: Brazil, restricted to Paricatuba, the left bank of the Rio Tapajós, near the mouth, Pará, Brazil (Hershkovitz 1966).

Field research since the publications of Hershkovitz (1977) and Rylands et al. (1993) has divided up and diminished the supposed distribution of the Santarém marmoset. Hershkovitz's (1977) range is now shared by four marmosets, being occupied by *M. mauesi* (the Maués marmoset), *M. saterei* (the Sateré marmoset), and *M. acariensis* (the Rio Acarí marmoset) besides *M. humeralifer*. According to our current understanding, *M. humeralifer* occurs south of the Rio Amazonas, between the Rio Maués (and possibly its tributary, the Rio Parauari) in the west and the Rio Tapajós in the east. The southern limit is not known, but it may be in the region of the Rio Paracari. The southernmost locality for the Santarém marmoset plotted by Hershkovitz (1977) was Vila Braga (4°25′S) on the Trans-amazon highway, just north of the Amazônia National Park. A pale orange-brown marmoset very similar to *M. humeralifer* obtained from the Rio Arapiuns in the northern part of the range was photographed by Mittermeier et al. (1988, p 20) in the collection of the Belém Primate Centre. This may well be a new and as yet unregistered species.

2.2.4.10 *Mico chrysoleucus* (Wagner, 1842) Golden-White Tassel-Ear Marmoset

Type locality: Borba, lower Rio Madeira, Amazonas, Brazil (Hershkovitz 1977). (Note: *chrysoleuca* used with the feminine genus "*Callithrix*" is here changed to *chrysoleucus* to agree with the masculine *Mico*.)

M. chrysoleucus is almost completely white, with an unpigmented face and long white ear-tufts. The body is pale gold to whitish and the tail, fore- and hindlimbs are golden to orange. Very little known, it occurs in a sliver south of the Rio Amazonas, between the Rios Madeira and lower Aripuanã in the west and the Rio Canumã (= Cunumã) in the east (Hershkovitz 1977; Silva and Noronha 1996). It occurs on the north (left) bank of the Paraná Urariá (*M. mauesi* occurs on the opposite side of the Urariá). Silva and Noronha (1996) observed *M. chrysoleucus* at Santa Bárbara, on the left bank of the Rio Canumã. The southernmost locality is Prainha, a short distance north of the mouth of the Rio Roosevelt, on the east (right) bank of the Rio Aripuanã. It is probable that Prainha is near the southern limit to its distribution, which may be marked by the headwaters of Rio Sucundurí, Serra do Sucundurí toward 8°S.

2.2.4.11 *Mico mauesi* (Mittermeier et al., 1992) Maués Marmoset

Type locality: West bank of the Rio Maués-Açú, directly across the river from the town of Maués, Amazonas state, Brazil. Located in central Brazilian Amazonia, south of the Rio Amazonas and between the Rio Madeira and the Rio Tapajós (3°23′S, 57°46′W) (Mittermeier et al. 1992).

In the original description, the Maués marmoset was known only from the immediate vicinity of the type locality, but local people informed that it occurred along the Rio Maués to the south of the type locality and to the west as far as the Paraná Urariá and Rio Abacaxis (Mittermeier et al. 1992). Silva and Noronha (1995) reported a further locality: Santa Maria, on the right of the lower Rio Abacaxis, municipality of Nova Olinda do Norte, state of Amazonas (3°54′S, 58°46′W). They also obtained reports of *M. mauesi* occurring in the vicinity of the town of Abacaxis, on the right bank of the Rio Abacaxis (3°55′S, 58°45′W), a few kilometers downriver from Santa Maria. It was reported not to occur at São João on the left bank of the Rio Marimari, near its confluence with the Rio Abacaxis (3°57′S, 58°48′W), or at two other localities on the west bank of the Rio Abacaxis. Silva and Noronha (1995) reported the occurrence of a bare-eared marmoset on the west bank of the Rio Abacaxis, which they later described (1998) as *Callithrix saterei*.

2.2.4.12 *Mico saterei* (Silva and Noronha, 1998) Sateré Marmoset

Type locality: Mouth of the Rio Canumã, the right bank of the lower Rio Canumã, in front of its confluence with the Paraná Urariá (3°59′50.8″S, 59°05′36.7″W).

M. saterei occurs in the interfluvium of the Rios Abacaxis (in the east) and Canumã-Sucunduri (in the west), right bank tributaries of the Rio Madeira. It is a most unusual-looking marmoset, which lacks ear-tufts despite being surrounded by marmosets displaying extravagantly hirsute pinnae (*humeralifer*, *chrysoleucus* and *mauesi*). *M. acariensis* occurs on the opposite (left) bank of the Canumã-Sucundurí, and *M. mauesi*, on the opposite (right) bank of the Rio Abacaxis. The lack of ear tufts is significant when considering Hershkovitz's (1977) original separation between the bare-ear marmosets (subspecies of *argentata*) and the tassel-ear marmosets (subspecies of *humeralifer*). The black-tailed marmoset and the marmosets east of the Rio Tapajós are certainly largely or entirely bare-eared, but so too are the forms recently discovered in the interfluvium of the Tapajós and Madeira (*saterei*, *nigriceps*, *manicorensis*, the Rondônia "*emiliae*," and *marcai*). *M. intermedius* and *M. acariensis* have small thin tufts growing from around, not within, the pinna. A notable aspect of *M. saterei* is the distinctive pale thigh stripe, shared most particularly with *melanurus*, *intermedius*, and *acariensis*, and hinted at by the hip patch in *mauesi* and *humeralifer* to the north of *saterei*. Only the entirely pale *M. chrysoleucus* lacks it (see Van Roosmalen et al. 2000, Fig. 2.3). These marmosets all occur between the Rio Roosevelt-Aripuanã-Madeira and the Rio Tapajós.

2.2.4.13 *Mico manicorensis* (Van Roosmalen et al., 2000) Manicoré Marmoset

Type locality: Seringal São Luis, east bank of the middle Rio Madeira, in the vicinity of the town of Manicoré, state of Amazonas, Brazil. This region is located in south central Amazonia, Brazil, south of the Rio Amazonas, east of the Rio Madeira, and west of the lower Rio Aripuanã. Coordinates for the type locality are 05°50′28″S, 61°18′19″W, altitude 45 m (van Roosmalen et al. 2000).

The Manicoré marmoset is known from the west bank of the lower Rio Aripuanã from the mouth, west as far as the Rio Manicoré, and south toward its headwaters. Despite the map provided by van Roosmalen et al. (2000), it evidently does not reach the confluence with the Rio Roosevelt – the type locality of *M. marcai*. The southern limits are probably marked by the headwaters of the Rio Mataurá or Rio Arauá, about 7°S.

2.2.4.14 *Mico acariensis* (Van Roosmalen et al., 2000) Rio Acarí Marmoset

Type locality: A small settlement on the right bank of the lower Rio Acarí, close to the confluence with the Rios Sucundurí and Canumã, state of Amazonas, south central Amazonia, Brazil. Coordinates for the type locality are 05°07′08″S, 60°01′14″W (Van Roosmalen et al. 2000).

Mico acariensis has not been observed in the wild, but according to local settlers it is known from the right bank of the lower Rio Acarí, and it presumably occurs through the interfluvium of the Rios Acarí (in the west) and Sucundurí (to the east), south perhaps to a contact zone with *M. melanurus* between the Rios Aripuanã and Juruena.

2.2.5 *Genus* Callimico *Miranda Ribeiro, 1912 Callimico or Goeldi's monkey*

2.2.5.1 *Callimico goeldii* (Thomas, 1904) Callimico or Goeldi's Monkey

Type locality: Rio Iaco, Acre, Brazil.

This is a monotypic genus, but speculation persists regarding the possibility of there being more than one species or subspecies. Vàsàrhelyi (2002) examined the genetic structure of the founder stock of captive callimicos and concluded that more than one cryptic subspecies or species may be represented. *Callimico goeldii* occurs in the upper Amazon from the Rio Caquetá in Colombia, south through the

Peruvian Amazon and the extreme western Amazon of Brazil into the Pando region of northern Bolivia (Hershkovitz 1977). Hershkovitz (1977) predicted that it should occur in the Ecuadorian Amazon, but none have been found there to date. Despite its wide range, callimico is notoriously patchy in its distribution and is evidently absent over a large part of the locality. In Colombia, it occurs from the base of the Cordillera Oriental of the Andes in the Department of Putumayo between the Ríos Putumayo and Caquetá east at least to the mouth of the Río Cahuinarí, a right bank affluent of the Caquetá. It is not known to occur in the Colombian trapezius (Hernández-Camacho and Cooper 1976; Defler 2004). In Peru, it is evidently limited largely to the eastern Amazon. Hershkovitz (1977) mapped numerous localities south of the Río Napo, along the lower and middle of the Río Ucayali and the Río Tapiche. The westernmost locality given by Hershkovitz (1977, map p 864) is on the Río Marañon, but it is listed in the gazetteer as "Apaga (Rio), enters Río Putumayo from south at approximately 4°42′S, 77°10′W, P Soini, April 1970, sight record." The coordinates would seem to be right, but the description of the locality wrong, and the Río Marañon is excluded from the distribution map of Aquino and Encarnación (1994). They have callimico definitely occurring only south of the lower Ucayali (from the mouth of the Rio Blanco), extending to both sides of the Ucayali at about 6°S, and south along the Andean foothills to the Ríos Pachitea and Madre de Dios. *Callimico* occurs in the Manu National Park (Aquino and Encarnación 1994). From there it extends east into extreme northern Bolivia, north of the Río Tahuamanu (Buchanan-Smith et al. 2000; Christen and Geissmann 1994). Christen and Geissmann (1994) reported seeing callimico south of the Río Nareuda, indicating it occurs in the south as far as the Rio Muyumanu. Buchanan-Smith et al. (2000) found no evidence of its occurrence south of the Tahuamanu-Nareuda. *Callimico* occurs in a small part in the south-west Brazilian Amazon in the state of Acre, through the Serra do Divisor south of the upper Rio Juruá to the Rio Gregório (state of Amazonas), to the Rio Iaco (above the Rio Acre) on the south (right) bank of the upper Purus, and into the Madeira basin along the Rio Abunã in the state of Rondônia (Hershkovitz 1977; Ferrari et al. 1999; Lopes and Rehg 2003).

2.3 Some Reflections and a Summary

The last three decades have seen some major changes in the way we perceive and catalogue the South American tamarins, marmosets, and callimico. Hershkovitz (1977) summarized our understanding of this group in the late 1970s with *Cebuella* as a primitive ancestral form (his interpretation being that the evolutionary tendency in this group of primates was to increase in size), *Callimico* as an outgroup family of its own, and three species of "true" marmosets – *Callithrix argentata* with three subspecies, *C. humeralifer* with three, and *C. jacchus* with five. Adjustments to this taxonomy came from the realization that the group as a whole was tending to decrease, not increase, in body size (Ford 1980; Leutenegger

1980; Martin 1992), from genetic studies, both karyological and molecular that provided us with revealing insights as to the affinities and differences of the various forms, from morphological studies validating the classification of the Atlantic forest marmosets as species (and the validity of the form *kuhlii*), and from the discovery of numerous new species. We have now come to recognize a diversity of marmosets that is more comparable to that of *Saguinus* than was previously documented by Hershkovitz (1977). Hershkovitz recognized the existence of 12 marmosets (including *Cebuella*) and 33 tamarins of the genus *Saguinus*. Groves (2005), on the other hand, was able to list 21 marmosets (placing *Mico*, *Cebuella* and *Callibella* as subgenera of *Callithrix*) and 32 tamarins. It was the exploration of the Tapajós-Madeira interfluvium that brought to light eight previously unknown marmosets since the description of *Callithrix* (now *Mico*) *intermedius* by Hershkovitz (1977). Further surveys will probably reveal yet more in the coming years. No "new" species of tamarin has been described since *Saguinus nigricollis hernandezi* Hershkovitz, 1982.

As pointed out to us by Susan Ford, it is curious that the pygmy marmoset, *Cebuella*, the dwarf marmoset, *Callibella*, and callimico have failed to produce the diversity of forms that characterize *Callithrix*, *Mico*, and *Saguinus*. This is in part dependant on our vision as to the generic separation of these forms and what comprises a "marmoset radiation" – for Groves (2001, 2005) the question would have no meaning because the pygmy marmosets and the dwarf marmosets are a component of the marmoset radiation, not a separate radiation in themselves. How we classify these animals inevitably affects the way we attempt to explain their evolution and adaptive radiations.

It would seem evident that while all have evolved to occupy the niche of a small arboreal faunivore-frugivore, their evolutionary diminution in size being associated with an increased breeding rate, one of the key gauges to our understanding of this group is their capacity to substitute fruits for plant exudates – something which undoubtedly evolved in an ancestral montane (as opposed to lowland) form in south-east Brazil, that suffered periodic (or even a generalized) scarcity in fruits, and which is today represented by the forms *Callithrix aurita* and *C. flaviceps*. Extreme specialization in tree-gouging to obtain gums evolved twice: *Callithrix jacchus* and *C. penicillata* in highly seasonal and dry forests of the north-east and central Brazil and *Callibella* and *Cebuella pygmaea* in the Amazon (Rylands 1984). The forces driving these two extremes of specialization were, of course, different. For *C. jacchus* and *C. penicillata*, it allowed them to occupy highly seasonal and dry forests in central and north-east Brazil, that lack fruits for long periods of the year. For the pygmy and dwarf marmosets, it was the occupation of a niche that could avoid competition with tamarins (saddleback and moustached), or other marmosets, in the case of *Callibella* – parapatric or sympatric speciation. Their diminutive size and minute home ranges exclude any possibility of their having access to their preferred fruits year round.

The case of *Callimico* (and what must have been the loss of the callithrichid twinning unique in mammals, if we are to believe the conclusions of the geneticists) remains a mystery, but the answer, if it is available to us, will be found in a profound understanding of the forest's resources – what its habitat has to offer in space and time, in dispersion and abundance, that *Callimico* has specifically evolved to exploit. Its enigmatically patchy distribution may well be a reflection of a habitat (sensu lato) abundant in the past and now disappearing (Izawa 1979).

Acknowledgments ABR is most grateful to Susan Ford, Lesa Davis and Leila Porter for their invitation to participate in the symposium "Advances in Marmoset and Goeldi's Monkey (*Callimico*) Research: Anatomy, Behavioral Ecology, Phylogeny, and Conservation," during the 74th Annual Meeting of the American Association of Physical Anthropologists, Milwaukee, Wisconsin, USA, 6–9 April 2005. Our thanks are also due to Kimberly Meek, Center for Applied Biodiversity Science at Conservation International, for drawing the maps.

References

Aguiar JM, Lacher TE Jr (2003) On the morphological distinctiveness of *Callithrix humilis* Van Roosmalen et al., 1998. Neotrop Primates 11(1):11–18

Aguiar JM, Lacher Jr TE (this volume) Cranial morphology of the dwarf marmoset *Callibella* in the context of callitrichid variability. In: Ford SM, Porter LM, Davis LC (eds) The smallest anthropoids: The marmoset/callimico radiation. Springer Press, New York pp 355–380

Allen JA (1916) Mammals collected on the Roosevelt Brazilian Expedition, with field notes by Leo E. Miller. Bull Am Mus Nat Hist 35(30):559–610

Alonso C, de Faria DS, Langguth A, Santee DF (1987) Variação da pelagem na área de intergradação entre *Callithrix jacchus* e *Callithrix penicillata*. Rev Brasil Biol 47(4):465–470

Alperin R (1993) *Callithrix argentata* (Linnaeus, 1771): considerações taxonômicas e descrição de subespécie nova. Bol Mus Para Emílio Goeldi Sér Zool 9(2):317–328

Alperin R (2002) Sobre a localidade tipo de *Mico marcai* (Alperin, 1993). Neotrop Primates 10(3):126–128

Aquino R, Encarnación F (1994) Primates of Peru/Los primates del Perú. Prim Rep (40):1–127

Barroso CML (1995) Filogenia molecular da subfamília Callitrichinae (sensu Rosenberger 1981). Universidade Federal do Pará, Belém, Doctoral Thesis

Barroso CML, Schneider H, Schneider MPC, Sampaio I, Harada ML, Czelusniak J, Goodman M (1997) Update on the phylogenetic systematics of New World monkeys: further DNA evidence for placing the pygmy marmoset (*Cebuella*) within the genus *Callithrix*. Int J Primatol 18(4):651–674

Bicca-Marques JC, Calegaro-Marques C (1995) Updating the known distribution of the pygmy marmoset (*Cebuella pygmaea*) in the state of Acre, Brazil. Neotrop Primates 3(2):48–49

Brandão LD, Develey PF (1998) Distribution and conservation of the buffy-tufted-ear marmoset, *Callithrix aurita*, in lowland coastal Atlantic forest, south-east Brazil. Neotrop Primates 6(3):86–88

Brandon-Jones D, Groves CP (2002) Neotropical primate family-group names replaced by Groves (2001) in contravention of Article 40 of the International Code of Zoological Nomenclature. Neotrop Primates 10(3):113–115

Brazil IBGE (1972) Brasil. Carta Internacional do Mundo ao Milionesimo. Departamento de Documentação e Divulgação Geográfica e Cartográfica, Fundação Instituto Brasileiro de Geografia e Estatística (IBGE), Ministério do Planejamento e Coordenação Geral, Rio de Janeiro. Scale 1:1,000,000

Brown AD, Rumiz DI (1986) Distribución de los primates en Bolivia. In: de Mello MT (ed) A primatologia no Brasil–2. Sociedade Brasileira de Primatologia, Brasília, pp 335–363

Buchanan-Smith HM, Hardie SM, Caceres C, Prescott MJ (2000) Distribution and forest utlization of *Saguinus* and other primates of the Pando Department, northern Bolivia. Int J Primatol 21(3):353–378

Cabrera A (1957) Catalogo de los mamíferos de América del Sur. Rev Mus Argent Cienc Nat Bernardino Rivadavia 4(1):1–307

Canavez FC, Moreira MAM, Simon F, Parham P, Seuánez HN (1999a) Phylogenetic relationships of the Callitrichinae (Platyrrhini, Primates) based on beta2-microglobulin DNA sequences. Am J Primatol 48(3):225–236

Canavez FC, Moreira MAM, Ladasky JJ, Pissinatti A, Parham P, Seuánez HN (1999b) Molecular phylogeny of New World primates (Platyrrhini) based on beta2-microglobulin DNA sequences. Molec Phylogen Evol 12(1):74–82

Chaves R, Sampaio I, Schneider MPC, Schneider H, Page SL, Goodman M (1999) The place of *Callimico goeldii* in the callitrichine phylogenetic tree: Evidence from von Willebrand Factor Gene Intron II sequences. Mol Phylogen Evol 13:392–404

Christen A, Geissmann T (1994) A primate survey in northern Bolivia, with special reference to Goeldi's monkey, *Callimico goeldii*. Int J Primatol 15(2):239–274

Coimbra-Filho AF (1971) Os sagüis do gênero *Callithrix* da região oriental brasileira e um caso de duplo-hibridismo entre três de suas formas (Callithricidae, Primates). Rev Brasil Biol 31:377–388

Coimbra-Filho AF (1984) Situação atual dos calitriquídeos que ocorrem no Brasil (Callitrichidae–Primates). In: de Mello MT (ed) A Primatologia no Brasil. Sociedade Brasileira de Primatologia, Brasília, pp 15–33

Coimbra-Filho AF (1985) Sagüi-de-Wied *Callithrix kuhli* (Wied, 1826). FBCN/Inf Rio de Janeiro 9(4):5

Coimbra-Filho AF (1986a) Sagüi-da-serra *Callithrix flaviceps* (Thomas, 1903). FBCN/Inf Rio de Janeiro 10(1):3

Coimbra-Filho AF (1986b) Sagüi-da-serra-escuro *Callithrix aurita* (É. Geoffroy, 1812). FBCN/Inf Rio de Janeiro 10(2):3

Coimbra-Filho AF (1986c) Sagüi-de-cara-branca *Callithrix geoffroyi* (Humboldt, 1812). FBCN/Inf Rio de Janeiro 10(3):3

Coimbra-Filho AF (1990) Sistemática, distribuição geográfica e situação atual dos símios brasileiros (Platyrrhini–Primates). Rev Brasil Biol 50:1063–1079

Coimbra-Filho AF (1991) Apontamentos sobre *Callithrix aurita* (É. Geoffroy, 1812) um sagüi pouco conhecido. In: Rylands AB, Bernardes AT (eds) A primatologia no Brasil–3. Sociedade Brasileira de Primatologia and Fundação Biodiversitas, Belo Horizonte, pp 145–158

Coimbra-Filho AF, Câmara I de G (1996) Os limites originais do bioma mata atlântica na região Nordeste do Brasil. Fundação Brasileira para a Conservação da Natureza, Rio de Janeiro

Coimbra-Filho AF, Mittermeier RA, Constable ID (1981) *Callithrix flaviceps* (Thomas, 1903) recorded from Minas Gerais, Brazil (Callitrichidae, Primates). Rev Brasil Biol 41(1):141–147

Coimbra-Filho AF, Rylands AB, Pissinatti A, Santos IB (1991/1992) The distribution and conservation of the buff-headed capuchin monkey, *Cebus xanthosternos*, in the Atlantic forest region of eastern Brazil. Primate Conserv (12–13):24–30

Coimbra-Filho AF, Pissinatti A, Rylands AB (1993) Experimental multiple hybridism among *Callithrix* species from eastern Brazil. In: Rylands AB (ed) Marmosets and tamarins: systematics, ecology, and behaviour. Oxford University Press, Oxford, pp 95–120

Cortés-Ortiz L (this volume) Molecular phylogenetics of the Callitrichidae with an emphasis on the marmosets and *Callimico*. In: Ford SM, Porter LM, Davis LC (eds) The smallest anthropoids: The marmoset/callimico radiation. Springer Press, New York pp 3–24

Cosenza BAP, de Melo FR (1998) Primates of the Serra do Brigadeiro State Park, Minas Gerais, Brazil. Neotrop Primates 6(1):18–20

Cronin JE, Sarich VM (1978) Marmoset evolution: the molecular evidence. In: Gengozian N, Deinhardt F (eds) Primates in Medicine, vol 10. S Karger, Basel, pp 12–19

da Cruz Lima E (1945) Mammals of Amazônia, vol. 1. General introduction and primates. Contribuições do Museu Paraense Emílio Goeldi de História Natural e Etnografia, Belém do Pará

de Ávila-Pires FD (1969) Taxonomia e zoogeografia do gênero 'Callithrix' Erxleben, 1777 (Primates, Callithricidae). Rev Brasil Biol 29(1):49–64

de Ávila-Pires FD (1986) On the validity of and geographical distribution of *Callithrix argentata emiliae* Thomas, 1920 (Primates, Callithricidae). In: de Mello MT (ed) A primatologia no Brasil–2. Sociedade Brasileira de Primatologia, Brasília, pp 319–322
de Carvalho CT (1959) Sobre a validez de *Callithrix leucippe* (Thos.) (Callithricidae, Primates). Pap Avuls Dept Zool Sec Agric S Paulo 13(27):317–320
de Carvalho CT (1965) Comentários sobre os mamíferos descritos e figurados por Alexandre Rodrigues Ferreira em 1790. Arq Zool 12:7–70
de la Torre S (2000) Primates de la Amazonía del Ecuador/Primates of Amazonian Ecuador. Sociedad para la Investigación y Monitoreo de la Biodiversidad (SIMBOE), Quito
de Vivo M (1985) On some monkeys from Rondônia, Brasil (Primates: Callitrichidae, Cebidae). Pap Avuls Zool S Paulo 4:1–31
de Vivo M (1991) Taxonomia de *Callithrix* Erxleben, 1777 (Callitrichidae, Primates). Fundação Biodiversitas, Belo Horizonte
Defler TR (2004) Primates of Colombia. Tropical Field Guide Series, Conservation International, Washington, DC
Dollman G (1933) Primates. Ser. 3. British Museum (Natural History), London
Ferrari SF (1993a) The adaptive radiation of Amazonan callitrichids (Primates, Platyrrhini). Evolución Biológica 7:81–103
Ferrari SF (1993b) Ecological differentiation in the Callitrichidae. In: Rylands AB (ed) Marmosets and tamarins: systematics, behaviour and ecology. Oxford University Press, Oxford, pp 314–328
Ferrari SF (1993c) An update on the black-headed marmoset, *Callithrix nigriceps* Ferrari and Lopes, 1992. Neotrop Primates 1(4):11–13
Ferrari SF (1994) The distribution of the black-headed marmoset, *Callithrix nigriceps*: A correction. Neotrop Primates 2(1):11–12
Ferrari SF, Lopes Ferrari MA (1990) A survey of primates in central Pará. Bol Mus Para Emílio Goeldi sér Zool 6(2):169–179
Ferrari SF, Lopes MA (1992) A new species of marmoset, genus *Callithrix* Erxleben 1777 (Callitrichidae, Primates) from western Brazilian Amazonia. Goeldiana Zoologia (12):1–3
Ferrari SF, Lopes MA (1996) Primate populations in eastern Amazonia. In: Norconk MA, Rosenberger AL, Garber PA (eds). Adaptive radiations of neotropical primates. Plenum Press, New York, pp 53–67
Ferrari SF, Mendes SL (1991) Buffy-headed marmosets 10 years on. Oryx 25(2):105–109
Ferrari SF Lopes MA, da Cruz Neto EH, Silveira MAES, Ramos EM, Ramos PCS, Tourinho DM, Magalhães NFA (1995) Primates and conservation in the Guajará-Mirim State Park, Rondônia, Brazil. Neotrop Primates 3(3):81–82
Ferrari SF, Iwanaga S, da Silva JL (1996a) Platyrrhines in Pimenta Bueno, Rondônia, Brazil. Neotrop Primates 4(4):151–153
Ferrari SF, Corrêa HKM, Coutinho PEG (1996b) Ecology of the southern marmosets (*Callithrix aurita* and *Callithrix flaviceps*) – how different, how similar? In: Norconk MA, Rosenberger AL, Garber PA (eds) Adaptive radiations of neotropical primates. Plenum Press, New York, pp 157–171
Ferrari SF, Iwanaga S, Ramos EM, Messias MR, Ramos PCS, da Cruz Neto EH (1999) Expansion of the known distribution of Goeldi's monkey (*Callimico goeldii*) in south-western Brazilian Amazonia. Folia Primatol 70:112–116
Flesher K (2001) Primates of the Chapada das Mangabeiras, Piauí, Brasil: a northern extension to the range of *Alouatta caraya*. Neotrop Primates 9(1):19–22
Ford SM (1980) Callitrichids as phyletic dwarfs and the place of Callitrichidae in the Platyrrhini. Primates 21:31–43
Ford SM, Davis LC (1992) Systematics and body size: implications for feeding adaptations in New World monkeys. Am J Phys Anthropol 88:415–468
Ford SM, Davis LC (this volume) Marmoset postcrania and the skeleton of the dwarf marmoset, *Callibella humilis*. In: Ford SM, Porter LM, Davis LC (eds) The smallest anthropoids: The marmoset/callimico radiation. Springer Press, New York pp 411–448

Freese CH, Heltne PG, Castro RN, Whitesides GH (1982) Patterns and determinants of monkey densities in Peru and Bolivia, with notes on distributions. Int J Primatol 3(1):53–90

Geoffroy Saint-Hilaire É (1812) Tableau des quadrumanes, ou des animaux composant le premier ordre de la classe des mammifères. Annls Mus Hist. Nat Paris 19:85–122

Groves CP (1993) Order Primates. In: Wilson DE, Reeder DM (eds) Mammal species of the world: a taxonomic and geographic reference, 2nd edn. Smithsonian Institution Press, Washington, DC, pp 243–277

Groves CP (2001) Primate taxonomy. Smithsonian Institution Press, Washington, DC

Groves CP (2004) The what, why and how of primate taxonomy. Int J Primatol 25(5):1105–1126

Groves CP (2005) Order Primates. In: Wilson DE, Reeder DM (eds) Mammal species of the world: a taxonomic and geographic reference, vol 1, 3rd edn. Johns Hopkins University Press, Baltimore, pp 111–184

Harada ML, Schneider H, Schneider MPC, Sampaio I, Czelusniak J, Goodman M (1995) DNA evidence on the phylogenetic systematics of New World monkeys: support for the sister-grouping of *Cebus* and *Saimiri* from two unlinked nuclear genes. Molec Phylogen Evol 4(3):331–349

Hernández-Camacho J, Cooper RW (1976) The non-human primates of Colombia. In: Thorington RW Jr, Heltne PG (eds) Neotropical primates: field studies and conservation. National Academy of Sciences, Washington, DC, pp 35–69

Hershkovitz P (1966) On the identification of some marmosets Family Callithricidae (Primates). Mammalia 30(2):327–332

Hershkovitz P (1968) Metachromism or the principle of evolutionary change in mammalian tegumentary colors. Evolution 22:556–575

Hershkovitz P (1975) Comments on the taxonomy of Brazilian marmosets (*Callithrix*, Callitrichidae). Folia Primatol 24:137–172

Hershkovitz P (1977) Living New World monkeys (Platyrrhini) with an introduction to primates, vol 1. Chicago University Press, Chicago

Hershkovitz P (1982) Subspecies and geographic distribution of black-mantle tamarins *Saguinus nigricollis* Spix (Primates: Callitrichidae). Proc Biol Soc Wash 95(4):647–656

Hill WCO (1957) Primates: comparative anatomy and taxonomy III. Edinburgh University Press, Edinburgh, Hapalidae

Hill WCO (1960) Primates: comparative anatomy and taxonomy IV. Cebidae part A. Edinburgh University Press, Edinburgh

Hill WCO (1962) Primates: comparative anatomy and taxonomy V. Cebidae part B. Edinburgh University Press, Edinburgh

Hirsch A (2003) Avaliação da fragmentação do habitat e seleção de áreas prioritárias para a conservação dos primatas na bacia do rio Doce, Minas Gerais, através da aplicação de um sistema de informações geográficas". Universidade Federal de Minas Gerais, Belo Horizonte, Doctoral thesis

Hirsch A, Toledo PP, Brito BFA de, Rylands AB (In preparation) New records enlarge the geographic distribution of *Callithrix flaviceps*

Humboldt [FH]A (1812) Tableau synoptique des singes de l'Amérique. In: Humboldt [F H]A, Bonpland A [JA]. Recueil d'observations de zoologie et d'anatomie comparée, faites dans l'océan Atlantique, dans l'intérieur du nouveau continent et dans la mer du sud pendant les années 1799[–]1803. Premier volume. Deuxième partie. Observations de zoologie et d'anatomie comparée. Schoell and Dufour & Co, Paris, 368pp

Izawa K (1975) Foods and feeding behaviour of monkeys in the upper Amazon basin. Primates 16(3):295–316

Izawa K (1979) Studies on the peculiar distribution pattern of *Callimico*. Kyoto Univ Overseas Res Rep New World Monkeys (1979):23–41

Izawa K, Bejarano G (1981) Distribution ranges and patterns of nonhuman primates in western Pando, Bolivia. Kyoto Univ Overseas Res Rep New World Monkeys (1981):1–12

Kinzey WG (1982) Distribution of primates and forest refuges. In: Prance GT (ed) Biological diversification in the tropics. Columbia University Press, New York, pp 455–482

Leutenegger W (1973) Maternal-fetal weight relationships in primates. Folia Primatol 20:280–294
Leutenegger W (1980) Monogamy in callitrichids: A consequence of phyletic dwarfism? Int J Primatol 1(1):95–98
Lönnberg E (1940) Notes on marmosets. Ark Zool 32A(10):1–22
Lopes MA, Ferrari SF (1994) Foraging behavior of a tamarin group (*Saguinus fuscicollis weddelli*) and interactions with marmosets (*Callithrix emiliae*). Int J Primatol 15(3):373–387
Lopes MA de OA, Rehg JA (2003) Observations of *Callimico goeldii* with *Saguinus imperator* in the Serra do Divisor National Park, Acre, Brazil. Neotrop Primates 11(3):181–183
Marroig G, Cropp S, Cheverud JM (2004) Systematics and evolution of the Jacchus group of marmosets (Platyrrhini). Am J Phys Anthropol 123:11–22
Marroig G, Cheverud JM (this volume) Size and shape in callimico and marmoset skulls: allometry and heterochrony in the morphological evolution of small anthropoids. In: Ford SM, Porter LM, Davis LC (eds) The smallest anthropoids: The marmoset/callimico radiation. Springer Press, New York pp 331–353
Martin RD (1992) Goeldi and the dwarfs: the evolutionary biology of the small New World monkeys. J Hum Evol 22:367–393
Martins ES, Ayres JM, Valle MBR do (1988) On the status of *Ateles belzebuth marginatus* with notes on other primates of the Irirí river basin. Primate Conserv (9):87–91
Mendes SL (1993) Distribuição geográfica e estado de conservação de *Callithrix flaviceps* (Primates: Callitrichidae). In: Yamamoto ME, de Sousa MBC (eds) A primatologia no Brasil – 4. Universidade Federal do Rio Grande do Norte (UFRN), Sociedade Brasileira de Primatologia, Natal, pp 139–154
Mendes SL (1997a) Hybridization in free-ranging *Callithrix flaviceps* and the taxonomy of the Atlantic forest marmosets. Neotrop Primates 5(1):6–8
Mendes SL (1997b) Padrões biogeográficas e vocais em *Callithrix* do Grupo *Jacchus* (Primates, Callitrichidae). Universidade Estadual de Campinas, Campinas, São Paulo, Doctoral thesis
Mendes SL (1997c) Vocalizations in Atlantic forest marmosets, *Callithrix*. Neotrop Primates 5(4):116–117
Mendes SL, Vielliard JME, De Marco Jr P (this volume) The vocal identity of the *Callithrix* species (Primates, Callitrichidae). In: Ford SM, Porter LM, Davis LC (eds) The smallest anthropoids: The marmoset/callimico radiation. Springer Press, New York pp 63–84
Mittermeier RA, Coimbra-Filho AF, Constable ID, Rylands AB, Valle CMC (1982) Conservation of primates in the Atlantic forest region of eastern Brazil. Int Zoo Yearb 22:2–17
Mittermeier RA, Rylands AB, Coimbra-Filho AF (1988) Systematics: species and subspecies – an update. In: Mittermeier RA, Rylands AB, Coimbra-Filho AF, da Fonseca GAB (eds) Ecology and behavior of neotropical primates, vol 2. World Wildlife Fund, Washington, DC, pp 13–75
Mittermeier RA, Schwarz M, Ayres JM (1992) A new species of marmoset, genus *Callithrix* Erxleben, 1777 (Callitrichidae, Primates), from the Rio Maués region, state of Amazonas, Central Brazilian Amazonia. Goeldiana Zoologia (14): 1–17
Moreira MAM, Almeida CAS, Canavez F, Olicio R, Seuánez HN (1996) Heteroduplex mobility assays (HMAs) and analogous sequence analysis of a cytochrome *b* region indicate phylogenetic relationships of selected callitrichids. J Hered 87:456–460
Muskin A (1984) Field notes and geographic distribution of *Callithrix aurita* in eastern Brazil. Am J Primatol 7:377–380
Nagamachi CY, Pieczarka JC, Barros RMS, Schwarz M, Muniz JAPC, Mattevi MS (1996) Chromosomal relationships and phylogenetic and clustering analyses on genus *Callithrix*, group *argentata* (Callitrichidae, Primates). Cytogenet Cell Genet 72:331–338
Nagamachi CY, Pieczarka JC, Schwarz M, Barros RMS, Mattevi MS (1997) Comparative chromosomal study of five taxa of genus *Callithrix*, group *Jacchus* (Playtrrhini, Primates). Am J Primatol 41(1):53–60
Nagamachi CY, Pieczarka JC, Muniz JAPC, Barros RMS, Mattevi MS (1999) Proposed chromosomal phylogeny for the South American primates of the Callitrichidae family (Platyrrhini). Am J Primatol 49:133–152

Napier PH (1976) Catalogue of the primates in the British Museum (Natural History). Part I. Families Callitrichidae and Cebidae, British Museum (Natural History), London

Napier JS, Napier PH (1967) A handbook of living primates. Academic Press, London

Natori M (1986) Interspecific relationships of *Callithrix* based on dental characters. Primates 27(3):321–336

Natori M (1990) Numerical analysis of the taxonomical status of *Callithrix kuhli* based on the measurements of the postcanine dentition. Primates 31(4):555–562

Natori M (1994) Craniometrical variations among eastern Brazilian marmosets and their systematic relationships. Primates 35(2):167–176

Natori M, Shigehara N (1992) Interspecific differences in lower dentition among eastern Brazilian marmosets. J Mammal 73(3):668–671

Neusser M, Stanyon R, Bigoni F, Wienberg J, Müller S (2001) Molecular cytotaxonomy of New World monkeys (Platyrrhini)–Comparative analysis of five species by multi-color chromosome painting gives evidence for a classification of *Callimico goeldii* within the family of Callitrichidae. Cytogenet Cell Genet 94(34):206–215

Oliveira L de C, Câmara EMVC, Hirsch A, Paschoal AMO, Alvarenga RM, Belarmino MG (2003) *Callithrix geoffroyi* (Primates Callitrichidae) and *Alouatta caraya* (Primates: Atelidae) in the Serra do Cipó National Park, Minas Gerais, Brazil. Neotrop Primates 11(2):86–89

Oliver WLR, Santos IB (1991) Threatened endemic mammals of the Atlantic forest region of south-east Brazil. Wildl Preserv Trust Special Scientific Report 4:1–125

Olmos F, Martuscelli P (1995) Habitat and distribution of buffy tufted-ear marmoset *Callithrix aurita* in São Paulo State, Brazil, with notes on its natural history. Neotrop Primates 3(3):75–79

Passamani M, Aguiar LMS, Machado R, Figueiredo E (1997) Hybridization between *Callithrix geoffroyi* and *C. penicillata* in southeastern Minas Gerais, Brazil. Neotrop Primates 5(1):9–10

Pastorini J, Forstner MRJ, Martin RD, Melnick DJ (1998) A reexamination of the phylogenetic position of *Callimico* (Primates) incorporating new mitochondrial DNA sequence data. J Molec Evol 47(1):32–41

Paynter RA Jr, Traylor MA Jr (1991) Orinithological gazetteer of Brazil, vol 1, A–M. Harvard University, Cambridge, MA, Musem of Comparative Zoology

Peres CA, Patton JL, da Silva MNF (1996) Riverine barriers and gene flow in Amazonian saddleback tamarins. Folia Primatol 67(3):113–124

Pimenta FE, Silva JS Jr (2005) An update of the distribution of primates of the Tapajós-Xingu interfluvium, Central Amazonia. Neotrop Primates 13(2):23–28

Porter CA, Czelusniak J, Schneider H, Schneider MPC, Sampaio I, Goodman M (1997) Sequences of the primate epsilon-globin gene: Implications for systematics of the marmosets and other New World primates. Gene 205(1–2):59–71

Rosenberger AL (1980) Gradistic views and adaptive radiation of platyrrhine primates. Z. Morph Anthrop 71(2):157–163

Rosenberger AL (1981) Systematics: the higher taxa. In: Coimbra-Filho AF, Mittermeier RA (eds) Ecology and behavior of neotropical primates, vol 1. Academia Brasileira de Ciências, Rio de Janeiro, pp 9–27

Rosenberger AL (1984) Aspects of the systematics and evolution of the marmosets. In: de Melo MT (ed) A primatologia no Brasil. Sociedade Brasileira de Primatologia, Brasília, pp 159–180

Rosenberger AL, Coimbra-Filho AF (1984) Morphology, taxonomic status and affinities of the lion tamarins, *Leontopithecus* (Callitrichinae, Cebidae). Folia Primatol 42:149–179

Rosenberger AL, Setoguchi T, Shigehara N (1990) The fossil record of callitrichine primates. J Hum Evol 19:209–236

Ruiz-Miranda CR, Affonso AG, Martins A, Beck B (2000) Distribuição do sagüi (*Callithrix jacchus*) nas áreas de ocorrência do mico-leão-dourado (*Leontopithecus rosalia*) no estado do Rio de Janeiro. Neotrop Primates 8(3):98–101

Rylands AB (1984) Exudate-eating and tree-gouging by marmosets (Callitrichidae, Primates). In: Chadwick AC, Sutton SL (eds) Tropical rain forest: The Leeds Symposium. Leeds Philosophical and Literary Society, Leeds, UK, pp 155–168

Rylands AB (1994) Sagüi-da-serra-escuro, *Callithrix aurita* (É. Geoffroy, 1812). In: da Fonseca GAB, Rylands AB, Costa CMR, Machado RB, Leite YLR (eds) Livro vermelho dos mamíferos brasileiros ameaçados de extinção. Fundação Biodiversitas, Belo Horizonte, pp 47–54

Rylands AB, Spironelo WR, Tornisielo VL, Lemos de Sá RM, Kierulff MCM, Santos IB (1988) Primates of the Rio Jequitinhonha valley, Minas Gerais, Brazil. Primate Conserv (9):100–109

Rylands AB, Coimbra-Filho AF, Mittermeier RA (1993) Systematics, distributions, and some notes on the conservation status of the Callitrichidae. In: Rylands AB (ed) Marmosets and tamarins: systematics, behaviour and ecology. Oxford University Press, Oxford, pp 11–77

Rylands AB, Schneider H, Langguth A, Mittermeier RA, Groves CP, Rodríguez-Luna E (2000) An assessment of the diversity of New World primates. Neotrop Primates 8(2):61–93

Santos CV, Luz KP, Sant'anna FS (2005) As três espécies de primates do gênero *Callithrix* (*C. jacchus*, *C. penicillata* e *C. geoffroyi*) introduzidos na Ilha de Santa Catarina – SC: a importância de pesquisa na implantação do manejo. In: Sociedade Brasileira de Primatologia (ed) Programa e livro de resumos: XI Congresso Brasileiro de Primatologia. Pontificia Universidade Católica do Rio Grande Sul, Porto Alegre, 13–18 February, 2005, p 59

Santos IB, Mittermeier RA, Rylands AB, Valle C (1987) The distribution and conservation status of primates in southern Bahia, Brazil. Primate Conserv (8):126–142

Schneider H, Rosenberger AL (1996) Molecules, morphology, and platyrrhine systematics. In: Norconk MA, Rosenberger AL, Garber PA (eds) Adaptive radiations of neotropical primates. Plenum Press, New York, pp 3–19

Schneider H, Schneider MPC, Sampaio I, Harada ML, Stanhope M, Czelusniak J, Goodman M (1993) Molecular phylogeny of the New World monkeys (Platyrrhini, Primates). Molec Phylogenet Evol 2(3):225–242

Schneider H, Sampaio I, Harada ML, Barroso CML, Schneider MPC, Czelusniak J, Goodman M (1996) Molecular phylogeny of the New World monkeys (Platyrrhini, Primates) based on two unlinked nuclear genes: IRBP Intron 1 and epsilon-globin sequences. Am J Phys Anthropol 100:153–179

Sena L, Vallinoto M, Sampaio I, Schneider H, Ferrari SF, Schneider MPC (2002) Mitochondrial COII gene sequences provide new insights into the phylogeny of marmoset species groups (Callitrichidae, Primates). Folia Primatol 73(5):240–251

Seuánez HN, Forman L, Matayoshi T, Fanning TG (1989) The *Callimico goeldii* (Primates, Platyrrhini) genome: karyology and middle repetitive (LINE-1) DNA sequences. Chromosoma 98:389–395

Silva Jr JS (1999) Novos dados sobre ocorrências e uso de habitat pelo sagüi-do-nordeste, *Callithrix jacchus* (Primates: Callitrichidae). In: Sociedade Brasileira de Primatologia (ed) Livro de resumos: IX Congresso Brasileiro de Primatologia. Museu de Biologia Mello Leitão, Santa Teresa, 25–30 July 1999, p 77

Silva Jr JS, Noronha M de A (1995) A new record for *Callithrix mauesi* Mittermeier, Schwarz and Ayres, 1992. Neotrop Primates 3(3):79–81

Silva Jr JS, Noronha M de A (1996) Discovery of a new species of marmoset in the Brazilian Amazon. Neotrop Primates 4(2):58–59

Silva Jr JS, Noronha M de A (1998) On a new species of bare-eared marmoset, genus *Callithrix* Erxleben, 1777, from central Amazonia, Brazil (Primates: Callitrichidae). Goeldiana Zoologia (21):1–28

Simpson GG (1945) The principles of classification and a classification of mammals. Bull Am Mus Nat Hist 85:1–350

Soini P (1988) The pygmy marmoset, genus *Cebuella*. In: Mittermeier RA, Rylands AB, Coimbra-Filho AF, da Fonseca GAB (eds) Ecology and behavior of neotropical primates, vol 2. World Wildlife Fund, Washington, DC, pp 79–129

Spix J (1823) Simiarum et Vespertilionum Brasiliensium species novae ou Histoire Naturelle des Espèces Nouvelles de Singes et de Chauves-souris observées et recueillies pendant le voyage dans l'intérieur du Brésil. Typis Francisci Seraphici Hubschmanni, Monachii

Stallings JR (1985) Distribution and status of primates in Paraguay. Primate Conserv (6):51–58.

Stallings JR, Mittermeier RA (1983) The black-tailed marmoset (*Callithrix argentata melanura*) recorded from Paraguay. Am J Primatol 4(2):159–163

Tagliaro CH, Schneider MPC, Schneider H, Sampaio IC, Stanhope MJ (1997) Marmoset phylogenetics, conservation perspectives, and evolution of the mtDNA control region. Mol Biol Evol 14(6):674–684

Tagliaro CH, Schneider MPC, Schneider H, Sampaio IC, Stanhope MJ (2001) Molecular studies of *Callithrix pygmaea* (Primates, Platyrrhini) based on Transferrin intronic and ND1 regions: implications for taxonomy and conservation. Genet Mol Biol 23(4):729–737

Thomas O (1904) New *Callithrix*, *Midas*, *Felis*, *Rhipidomys*, and *Proechimys* from Brazil and Ecuador. Ann Mag Nat Hist 14(7):188–196

Thomas O (1911) The mammals of the tenth edition of Linnaeus: An attempt to fix the types of the genera and the exact bases and localities of the species. Proc Zool Soc Lond 1911:20–158

Thomas O (1920) On mammals from the lower Rio Amazonas in the Goeldi Museum, Pará. Ann Mag Nat Hist 9(6):266–283

Thomas O (1922) On the systematic arrangement of the marmosets. Ann Mag Nat Hist 9(9):196–199

Van Roosmalen MGM, Van Roosmalen T (1997) An eastern extension of the geographical range of the pygmy marmoset, *Cebuella pygmaea*. Neotrop Primates 5(1):3–6

Van Roosmalen MGM, Van Roosmalen T (2003) The description of a new marmoset genus, *Callibella* (Callitrichinae, Primates), including its molecular phylogenetic status. Neotrop Primates 11(1):1–10

Van Roosmalen MGM, Van Roosmalen T, Mittermeier RA, Fonseca GAB (1998) A new and distinctive species of marmoset (Callitrichidae, Primates) from the lower Rio Aripuanã, state of Amazonas, central Brazilian Amazonia. Goeldiana Zoologia 22:1–27

Van Roosmalen MGM, Van Roosmalen T, Mittermeier RA, Rylands AB (2000) Two new species of marmoset, genus *Callithrix* Erxleben, 1777 (Callitrichidae, Primates), from the Tapajós/Madeira interfluvium, south central Amazonia, Brazil. Neotrop Primates 8(1):2–18

Vanzolini PE, Papavero N (1968) Índice dos topônimos contidos na carta do Brasil 1: 1 000 000 do IBGE [Fundação Instituto Brasileiro de Geografia e Estatística]. Fundação de Ampáro à Pesquisa do Estado de São Paulo (FAPESP), São Paulo [Preface dated December 1967]

Vàsàrhelyi K (2002) The nature of relationships among founders in the captive population of Goeldi's monkey (*Callimico goeldii*). Evol Anthropol 11(suppl 1):155–158

Vieira C da C (1944) Os símios do estado de São Paulo. Pap Avuls. Dept Zool Sec Agric S Paulo 4:1–31

Wied-Neuwied M (1826) Beiträge zur Naturgeschichte von Brasilien, vol. 2. Landes-Industrie-Comptoirs, Weimar, p 620

Chapter 3
The Vocal Identity of the *Callithrix* Species (Primates, Callitrichidae)

Sérgio Lucena Mendes, Jacques Marie Edme Vielliard, and Paulo De Marco Jr.

Abstract We have recorded long calls of free ranging groups of six *Callithrix* species, looking for species-specific parameters in the physical structure of the vocalizations. This comparative study is the first time a behavioral character (vocalization) has been incorporated in a study on *Callithrix* taxonomy. The long calls were analyzed in a digital sonograph and statistical comparisons performed, using analysis of variance. The data show that each of the six analyzed taxa can be distinguished on the basis of its call structure, mainly using the first long call note or syllable. In agreement with analyses based on morphological and genetic data, the vocalizations support the division of *Callithrix* into two groups: a jacchus-group and an aurita-group. The first group encompasses *C. jacchus*, *C. geoffroyi*, *C. penicillata*, and *C. kuhlii*, and the second, *C. aurita*, and *C. flaviceps*. Vocal parameters are able to distinguish all the taxa within each group, even *C. flaviceps* and *C. kuhlii*, whose specific status has been questioned. Differences between the four taxa of the jacchus-group suggested by vocalization data are not identical to those inferred on the basis of some morphological data. It is possible that vocalization differences are also influenced by historical secondary contact between populations and could have evolved as mechanisms of specific recognition and reproductive isolation. The results indicate that vocalization is a behavioral character that can be used to supplement other taxonomic data to study the phylogeny of the Callitrichidae and to improve our understanding of the evolution of the reproductive isolation of natural populations.

Resumen Se grabaron vocalizaciones de canto de largo alcance de seis especies de *Callithrix* con la finalidad de comparar los patrones vocales de seis especies, para buscar caracteres capaces de diferenciarlos. Las vocalizaciones fueron analizadas en fonógrafo digital y tratadas estadísticamiente por medio de análisis de discrepancia. Los análisis demostraron que cada una de las seis especies pueden ser diferenciadas con base a la estructura de los cantos, principalmiente si se utiliza la primera nota

S.L. Mendes (✉)
Departamento de Ciências Biológicas, Universidade Federal do Espírito Santo, Vitória, ES, Brazil
e-mail: slmendes1@gmail.com

S.M. Ford et al. (eds.), *The Smallest Anthropoids*, Developments in Primatology: Progress and Prospects, DOI 10.1007/978-1-4419-0293-1_3,
© Springer Science+Business Media, LLC 2009

o sílaba. De acuerdo con los análisis basados en datos morfológicos y genéticos, las vocalizaciones apoyan la división de *Callithrix* en dos grupos: grupo *jacchus* y el grupo *aurita*. El primero comprende *C. jacchus*, *C. geoffroyi*, *C. penicillata* y *C. kuhlii*, y el segundo incluye *C. aurita* y *C. flaviceps*. Los parámetros vocales pueden distinguir todas las especies dentro de cada grupo, incluyendo a *C. flaviceps* y a *C. kuhlii*, cuyos estados específicos han sido cuestionados. Las diferencias entre las cuatro especies del grupo *Jacchus* sugeridas por los datos de vocalización no son idénticas con las inferidas en base a los datos morfológicos. Es posible que dichas diferencias también están influenciadas por la historia de contacto secundario entre dichas poblaciones, y pueden haber evolucionado como mecanismos de reconocimiento poblacional y aislamiento reproductivo. Los resultados indican que la vocalización es un carácter del comportamiento que puede ser utilizado para complementar otros datos taxonómicos para estudiar la filogenia de los Callithrichidae y para mejorar nuestro entendimiento sobre la evolución del aislamiento reproductivo en poblaciones naturales.

Resumo Foram gravadas vocalizações de longo alcance (canto) de *Callithrix* de vida livre, com a finalidade de comparar os padrões vocais de seis *taxa*, em busca de caracteres capazes de diferenciá-los. As vocalizações foram analisadas em sonógrafo digital e tratadas estatisticamente através de análises de variância. As análises sonoras demonstraram que cada um dos seis *taxa* pode ser diferenciado com base na estrutura dos seus cantos, principalmente utilizando-se a primeira nota, ou sílaba. Coerente com os dados morfológicos e genéticos, as vocalizações apoiam a divisão de Callithrix em dois grupos: *jacchus* e *aurita*. O primeiro compreende *C. jacchus*, *C. geoffroyi*, *C. pencillata* e *C. kuhlii*, e o segundo *C. aurita* e *C. flaviceps*. Dentro de cada grupo há parâmetros relacionados à duração e freqüência das notas capazes de diferenciar todos os *taxa*, inclusive *C. flaviceps* e *C. kuhlii*, cujos *status* taxonômicos tem sido questionados. As diferenças nos parâmetros acústicos entre os quatro *taxa* de *Callithrix* do grupo *jacchus* não estão claramente relacionadas às supostas distâncias filogenéticas inferidas com base em dados morfológicos. É possível que essas diferenças também tenham relação com a história das zonas de contato secundário entre esses *taxa*, sendo reforçadas como mecanismos de reconhecimento populacional e isolamento reprodutivo. Os resultados deste estudo indicam que a vocalização é um caráter comportamental que pode ser usado para suplementar outras técnicas de taxonomia de Callithrichidae e no entendimento da evolução do isolamento reprodutivo em populações naturais.

3.1 Introduction

Hershkovitz (1977) and Vivo (1991) divided the marmosets of the genus *Callithrix* Erxleben, 1777, into two species groups: the argentata-group and the jacchus-group. The first occupy part of the Amazon basin and the river Paraguay basin, while the second occupy the center and east of Brazil. On the basis of morphological data, Rosenberger (1981) and Natori (1994) argued that *Cebuella*, the pygmy marmoset, should be included in the genus *Callithrix*, an argument supported by genetic researchers (Barroso et al. 1997; Schneider et al. 1996; Schneider and Rosenberger 1997; Tagliaro et al. 1997; Canavez et al. 1999; Moreira and Seuanez 1999; Singer et al. 2003). Overall, these data suggested that *Cebuella* is more closely related to the Amazonian marmosets (argentata-group) than the latter are to the eastern Brazilian marmosets (jacchus-group). However, Rylands et al. (2000) believed that *Cebuella* deserves generic status, and placed argentata-group and jacchus-group marmosets in different genera: Amazonian marmosets were placed in the genus *Mico* Lesson, 1840, restricting the genus *Callithrix* Erxleben, 1777, to the central and eastern Brazilian marmosets. This taxonomy is adopted in this work, which examines relationships within central and eastern Brazilian *Callithrix*.

The monophyly of eastern Brazilian *Callithrix* has been supported by morphological, biogeographical, and genetic evidence (Hershkovitz 1977; Mittermeier and Coimbra-Filho 1981; Vivo 1991; Meireles et al. 1992; Natori 1994; Sampaio et al. 1996; Tagliaro et al. 1997). The division of *Callithrix* into two groups, a jacchus-group and an aurita-group, is implicit in Hershkovitz (1977, who considered *C. flaviceps* as a pale form of *C. aurita*), and is corroborated by morphological (Natori 1986, 1994; Marroig et al. 2004) and genetic data (Tagliaro et al. 1997; Sena et al. 2002). All of the *Callithrix* of the center and east of Brazil were treated as a unique polytypic species by Hershkovitz (1977), who recognized five forms, based on their geographic distribution and supposed intergradations in appearance: *jacchus* (the common marmoset), *aurita* (the buffy tufted-ear marmoset), *geoffroyi* (Geoffroy's tufted-ear marmoset), *penicillata* (the black tufted-ear marmoset), and *flaviceps* (the buffy-headed marmoset). Mittermeier and Coimbra-Filho (1981), Vivo (1991), Rylands et al. (1993, 2000), and Marroig et al. (2004) considered these as good species. However, there remains uncertainty about the number of valid taxa, specifically the number of morphologically discreet and geographically defined populations, as well as the specific validity of each taxon (see Rylands et al. this volume, Chap. 2).

Hershkovitz (1977) stated that the south Bahian specimens, identified as *Callithrix kuhlii* (Wied's black tufted-ear marmoset) by Wied (1826), were hybrids between *C. penicillata* and *C. geoffroyi*. However, Mittermeier and Coimbra-Filho (1981) and Coimbra-Filho (1990) stated that *C. kuhlii* is a valid taxon and, possibly, a good species, a view corroborated by Natori's (1990) analysis of dental characters. Vivo (1991), following Hershkovitz, did not recognize this species, suggesting that the series collected in southern Bahia represented a small dark *C. penicillata*, but craniometrical data of Natori (1994) further supports the specific status of *C. kuhlii*. Moreover, Marroig et al. (2004), analyzing cranial characters, concluded that *C. kuhlii* is morphologically distinct from all other marmoset taxa, including *C. geoffroyi* and *C. penicillata*, which were previously hypothesized to be the parental populations that formed *C. kuhlii* by hybridization. Another taxonomic polemic encompasses *Callithrix aurita* and *C. flaviceps*, which are treated as good species in some papers (Mittermeier and Coimbra-Filho 1981; Vivo 1991; Rylands et al. 2000, this volume, Chap. 2), while in others, it has been argued that *C. flaviceps* should be considered subspecific to *C. aurita* (Coimbra-Filho 1986a, b, 1993).

Vocalizations have been used as a taxonomic tool in several studies of Old World primates, including prosimians of the genus *Galago* (Zimmermann et al. 1988; Courtenay and Bearder 1989; Masters 1991; Ambrose 2003), cercopithecids such as *Colobus* (Oates and Trocco 1983), *Cercopithecus* (Struhsaker 1970; Gautier 1978), *Cercocebus* (Quris 1980; Waser 1982), and *Presbytis* (Hohmann 1989), and the Hylobatidae (Marshall and Marshall 1976; Haimoff et al. 1982; Brockelman and Schilling 1984; Chen et al. 2004). Vocalization is an important form of communication among primates, especially among arboreal species, because dense forests generally make visual communication over long distances more difficult (Altmann 1967). Vocal signals are considered resistant to changes in the immediate environment and therefore can serve as good specific and genetic markers (Struhsaker 1970). This is especially true of long distance communication signals, which are often emitted by adult males (Masters 1991).

The long calls of callitrichids seem to be used in various contexts, e.g., territory defense, intra-group cohesion, contact between separated individuals, and attraction of sexual mates (Snowdon 1993). Snowdon demonstrated that callitrichids have high pitched calls (between 5 and 10 kHz), and that taxonomic arrangements based on morphology are reflected in their vocalization structure. From a comparison of the long calls of the golden lion tamarins *Leontopithecus rosalia*, *L. chrysopygus* and *L. chrysomelas*, Snowdon et al. (1986) concluded that vocalization structure can show interesting parallels with morphologic characters, and they reinforced the use of bioacoustic studies to supplement more traditional techniques in taxonomy.

Epple (1968) described in detail the vocalizations of *Callithrix jacchus* and stated that the main group of vocalizations is comprised of their contact calls, including the "long calls." Besides other functions, the long calls are used for inter-group communication, during interactions that can involve adults and older juveniles. Epple suggested that, under natural conditions, these interactions and calls function to show that the territory is occupied.

Given the probable function of the long calls of marmosets as a mechanism of long distance inter-group communication, they may play an important role in identifying con-specifics. This study examines whether there are species-specific characteristics in the physical structure of the long calls of marmosets in the genus *Callithrix*. The objective is to present a comparative study of the vocal patterns of each taxon, adding for the first time a systematic analysis of a behavioral character to the discussion of *Callithrix* taxonomy.

3.2 Methodology

3.2.1 The Study Area

Due to the wide geographic distribution of *Callithrix*, we recorded marmosets at every locality we had access to, trying to reach all six species, and we made an additional effort to record the same species in different localities. Table 3.1 lists 22

Table 3.1 Number of long calls analyzed in this study for each recording locality

Species	Locality	Coordinates	Recorded by	Long calls analyzed
C. aurita	Anhembi, SP	22°40′S, 48°10′W	S.L. Mendes	34
C. aurita	Pádua Sales, SP	22°18′S, 47°00′W	S.L. Mendes	11
C. aurita	Mogi Guaçu, SP	22°20′S, 47°00′W	J. Vielliard	14
C. aurita	Serra Brigadeiro, MG	20°40′S, 42°28′W	F.R. Melo	04
C. flaviceps	Est. Biol. Caratinga, MG	19°44′S, 41°49′W	S.L. Mendes	46
C. flaviceps	Res. Biol. A. Ruschi, ES	19°50′S, 40°32′W	S.L. Mendes	23
C. flaviceps	Pque Estadual Pedra Azul, ES	20°25′S, 41°00′W	S.L. Mendes	05
C. geoffroyi	Res. Biol Duas Bocas, ES	20°15′S, 40°30′W	S.L. Mendes	15
C. geoffroyi	Res. Nat. CVRD, ES	19°10′S, 40°00′W	S.L. Mendes	11
C. geoffroyi	Pque Museu Biol. M. Leitão, ES	19°55′S, 40°36′W	S.L. Mendes	40
C. geoffroyi	Alto Rio Novo, ES	19°03′S, 41°02′W	S.L. Mendes	05
C. jacchus	Est. Ecológica de Tapacurá, PE	07°58′S, 35°10′W	J. Vielliard	21
C. jacchus	Mata de Capatuba, Sapé, PB	07°03′S, 35°10′W	J. Vielliard	08
C. jacchus	Campus UFPB, João Pessoa, PB	07°10′S, 34°50′W	J. Vielliard	01
C. jacchus	Quebrangulo, AL	09°20′S, 36°29′W	J. Vielliard	05
C. jacchus	Nísia Floresta, RN	06°05′S, 35°12′W	D. Santee	39
C. kuhlii	Entre Camacã e Mascote, BA	15°32′S, 39°20′W	SLM/B. Brito	07
C. kuhlii	Est. Exper. Lemos Maia, BA	15°15′S, 39°05′W	SLM/B. Brito	16
C. penicillata	Parque Ecológico Americana, SP	22°45′S, 47°20′W	S.L. Mendes	10
C. penicillata	Pque Zoológico B. Horizonte, MG	19°49′S, 43 57′W	B.F.A. Brito	06
C. penicillata	Pque Mangabeiras, B. Horizonte, MG	19°49′S, 43 57′W	B.F.A. Brito	20
C. penicillata	Horto Municipal, Campinas, SP	22°54′S, 47°05′W	S.L. Mendes	03

localities where good recordings of free-ranging marmoset long calls were obtained, including recordings provided by other researchers. The recordings were carried out mainly during the first hours of daylight, when the marmosets are more active vocally. The marmosets were attracted using playback of previously recorded *Callithrix* vocalizations. We used a cassette player and a tape recording of vocalizations of *C. geoffroyi* and *C. flaviceps*, which were reproduced with the aid of a 80 W power tweeter.

To attract the marmosets, we used repeat sequences of five to ten long calls, intercalated by silent periods of about 2 min, waiting for a vocal answer or an approach.

The *Callithrix* vocal answers were recorded with a Uher 4200 recorder or a Marantz PMD-201 recorder, and a directional microphone Sennheiser ME 88 with a module K3-U.

3.2.2 Methods

The vocalizations were analyzed with the digital sonograph UNISCAN II, of the Unicamp Bioacoustic Laboratory, in "real time." The long call, also called phrase, is a compound of one or more notes. We consider long calls discrete sequences, separated from one another by periods longer than the duration of each call. A single note, which is also called a syllable, is a continual sonorous unit easily identified on the sonogram. In the few cases where there was a doubt if a certain note sequence included one or more long calls, we did not analyze the vocalization.

We analyzed only good quality long calls, taking the following measures directly from the sonograms: total number of notes, duration of each note and of entire call, intervals between notes, and for each of the first three notes: minimal frequency, maximal frequency, initial frequency, and final frequency. The frequency measurements were taken with 8 Hz of precision, and temporal measurements with 12 ms of precision. From these measures of frequency, three new parameters were calculated: mean frequency for each note, frequency modulation of each note, and frequency amplitude of the long call. Mean frequency was computed as the arithmetical average between the initial frequency and the final frequency of each note. The frequency modulation was computed as the difference between the final frequency and the initial frequency of each note, divided by the note duration. The long call frequency amplitude is the difference between maximal frequency and the minimal frequency, using all the notes. From these data, we looked at the general features of the long calls, the patterns of change from note to note within a long call, and the characteristics of the first note of each call.

The general features of long calls analyzed for each species are the mean number of notes, the total duration of the long call, the overall minimal frequency, the overall maximal frequency, and the frequency amplitude. In order to verify differences between the species, analysis of variance (ANOVA) was performed on the data transformed to square roots (Snedecor and Cochran 1989). The multiple comparison test of Tukey was used to compare the differences between species pairs.

The first three notes were used to compare the patterns of change across notes in the long calls of each species, as the majority of long calls have, at most, three notes. ANOVA tests were used to compare changes in the following parameters: duration of each note, intervals between notes, initial frequency of each note, mean frequency of each note, and frequency modulation of each note.

Previous analysis (Mendes 1997a) has shown that the first note is less variable and more typical for each species, so we used only the first note of each long call to draw further comparison in the vocal structure of species, thereby avoiding the use of redundant information, and facilitating statistical treatment. Species differences were compared based on the following parameters of the first note: note duration, interval between the first and second note, initial frequency, mean frequency, and frequency modulation.

In many cases, it was not possible to identify whether the recorded vocalizations were from one or more individuals of a given group; therefore we grouped the vocalizations by group recorded in the field. To avoid a sampling bias due to the different number of recordings of each group in the field, we used a nested ANOVA, considering the groups as the inferior level in the hierarchy and the species as the superior level. The test considers the species as samples and the groups as sub-samples.

3.3 Results

3.3.1 Long Call General Structure

Sonograms showing representative long calls of each of the six species are presented in Fig. 3.1, and basic data are reported in Table 3.2. The visual analysis of these sonograms demonstrates some differences between species, which were analyzed statistically. The peculiarities of the vocalizations of *C. aurita* and *C. flaviceps* are particularly interesting; their uniqueness when compared to the vocalisations of other *Callithrix* is evident enough to be easily recognized by human ears. In these two species, the recorded long calls fall into two variant patterns, which were emitted in the same context of playback response. In one pattern, the long calls are comprised of about three to four long notes of similar format, always with descendent modulation. In the other pattern, those long notes are followed by short notes, with an inverted "U" or "V" format, progressively shorter, lower pitched, and weaker. This pattern was not observed in any other species, clearly distinguishing these two species from the others.

3.3.2 Number of Notes per Long Call

The number of notes per long call is variable in all the species, but there are significant differences between *C. jacchus* and the other species in the mean number of notes, with *C. jacchus* having the fewest. *Callithrix aurita* is similar only to

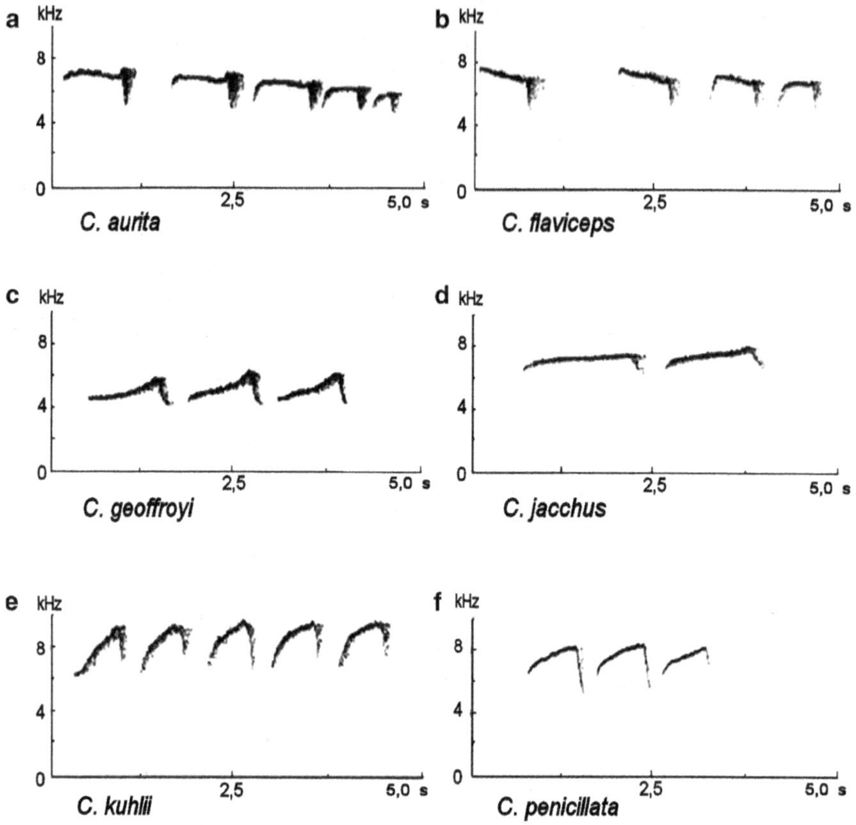

Fig. 3.1 Sonograms of representative long calls of *Callithrix*

C. kuhlii, in having the highest mean number of notes (Table 3.3). There are no differences in the mean number of notes per long call between the other four species, at the 5% significance level.

Despite the intraspecific variation, each species tends to present a predominant pattern in the number of notes. *Callithrix aurita* and *C. flaviceps* have the largest variation in the number of notes, as some long calls present long notes and others present long and short notes. More than 50% of the *C. aurita* long calls included two to six notes, while more than 50% of the *C. flaviceps* long calls included one to four notes. In *C. geoffroyi*, 85% of the long calls included two to four notes. In *C. jacchus*, 80% of the long calls included only one or two notes, and this species rarely emitted long calls with more than four notes. About 70% of the *C. kuhlii* long calls included three or four notes, differing from *C. geoffroyi* whose long calls presented a tendency to a smaller number of notes. The character of the long calls of *C. penicillata* with regard to the number of notes is similar to that of *C. geoffroyi*, differing mainly by a larger number of long calls with only one note.

Table 3.2 Parameters of vocalizations, measured from sonagrams (mean values)

Long call trait[a]	*C. aurita*	*C. flaviceps*	*C. geoffroyi*	*C. jacchus*	*C. kuhlii*	*C. penicillata*
Mean no. notes	3.9	2.8	2.8	1.8	3.0	2.9
Duration Note 1	798	823	1,154	1,371	653	1,213
Duration Note 2	686	892	844	1,083	499	820
Duration Note 3	580	814	800	862	497	693
Interval between Notes 1–2	482	841	323	369	253	320
Interval between Notes 2–3	284	580	303	312	259	285
Interval between Notes 3–4	199	418	298	404	258	279
Note 1 initial frequency	7.37	7.95	6.01	7.38	7.19	6.81
Note 2 initial frequency	7.33	7.55	5.86	7.30	7.06	6.79
Note 3 initial frequency	7.02	7.32	5.96	7.70	7.06	6.71
Note 1 frequency modulation	−0.17	−0.73	0.56	0.66	1.53	0.44
Note 2 frequency modulation	−0.24	−0.55	1.43	1.62	1.94	0.78
Note 3 frequency modulation	−0.19	−0.44	2.11	1.58	2.02	0.98

[a]Durations in milliseconds (ms), frequencies in kilohertz (kHz)

Table 3.3 Mean number of notes per long call in *Callithrix* species and similarities between them according to the Tukey test

	Number of notes	Standard error	*N*	Similarity[a]
C. aurita	3.9	0.16	63	a
C. flaviceps	2.8	0.15	74	f
C. geoffroyi	2.8	0.15	74	f
C. jacchus	1.8	0.15	74	j
C. kuhlii	3.0	0.26	24	a, f
C. penicillata	2.9	0.21	39	f

[a]Same letters in similarity column indicate that the number of notes is statistically similar at the 5% significance level

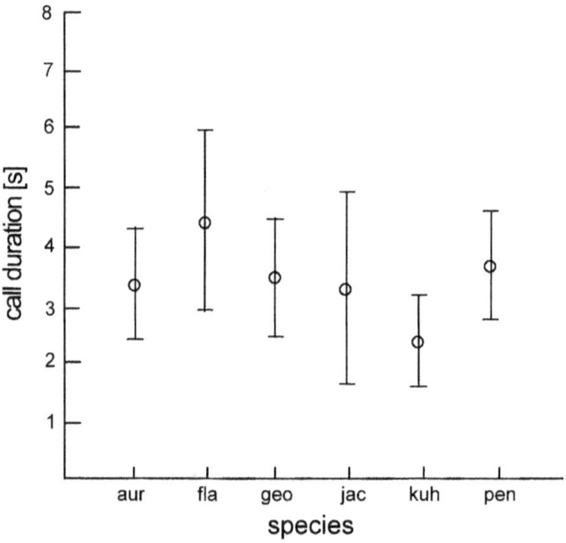

Fig. 3.2 Mean duration of long calls of *C. aurita* (aur), *C. flaviceps* (fla), *C. geoffroyi* (geo), *C. jacchus* (jac), *C. kuhlii* (kuh), and *C. penicillata* (pen)

3.3.3 Duration of Long Calls

The long call mean duration was about 3 s. *Callithrix kuhlii* had the shortest mean and *C. flaviceps* had the longest; each was significantly different from all the other species. The remaining four species did not exhibit significant differences in their long call duration (see Fig. 3.2).

3.3.4 Frequency Parameters of Long Calls

The mean of minimal frequencies varied from about 5 to about 7 kHz. The mean of maximal frequencies varied from about 7 to about 9 kHz. Although there was overlap in the frequency bands used by each species, some groups were clearly distinguished (see Figs. 3.3 and 3.4).

Mean minimal frequency groups *C. aurita*, *C. flaviceps* and *C. geoffroyi*, with a frequency band around 5 kHz, and *C. jacchus*, *C. kuhlii* and *C. penicillata, with a minimal frequency* of around 6 kHz or more. Mean maximal frequency distinguishes *C. geoffroyi*, presenting the lowest values, as well as *C. jacchus*, presenting the highest ones, from the remaining four species.

Frequency amplitudes demonstrate large overlaps between the species and large variations around the averages. Only *C. flaviceps* is clearly distinguishable, with a mean frequency amplitude much higher than in the other species.

3.3.5 Comparison of the First Three Notes of the Long Call

Figures 3.5–3.9 demonstrate that the first three notes of the long calls of *Callithrix* species are similar, but some parameters are distinct. The first note tends to be more characteristic and less variable within species.

The long call first note is the longest one, and the other notes get progressively shorter in most marmosets (Fig. 3.5, Table 3.4). The exceptions are *Callithrix kuhlii*,

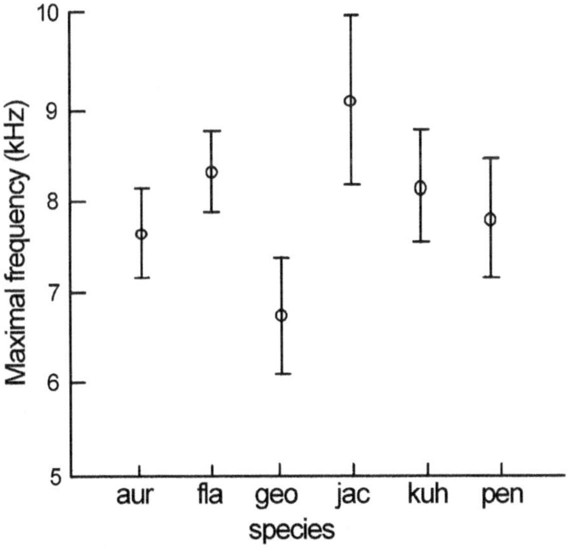

Fig. 3.3 Mean of minimal frequencies of long calls of *C. aurita* (aur), *C. flaviceps* (fla), *C. geoffroyi* (geo), *C. jacchus* (jac), *C. kuhlii* (kuh), and *C. penicillata* (pen)

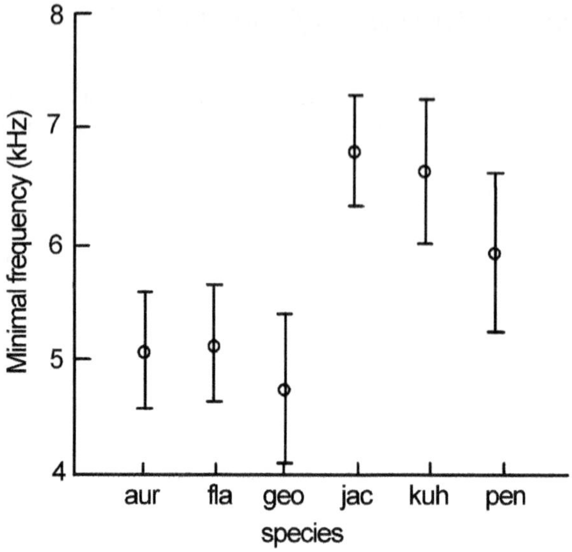

Fig. 3.4 Mean of maximal frequencies of long calls of *C. aurita* (aur), *C. flaviceps* (fla), *C. geoffroyi* (geo), *C. jacchus* (jac), *C. kuhlii* (kuh), and *C. penicillata* (pen)

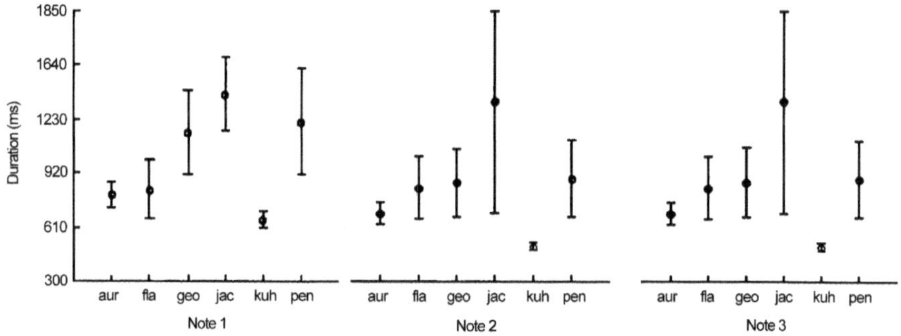

Fig. 3.5 Mean of duration of first three notes of long calls of *C. aurita* (aur), *C. flaviceps* (fla), *C. geoffroyi* (geo), *C. jacchus* (jac), *C. kuhlii* (kuh), and *C. penicillata* (pen)

whose notes did not differ significantly in duration ($P=0.175$), and *C. jacchus*, whose Notes 2 and 3 vary significantly enough to preclude the definition of a clear pattern.

The interval between notes tends to get smaller in the *C. aurita* and *C. flaviceps* vocalizations, but this does not occur in the other species, where no significant difference was found at the 5% level of significance (see Fig. 3.6). The apparent differences in the interval between Notes 3 and 4 in *C. jacchus* is due to a small sample, as only a few long calls of this species have more than three notes.

Another difference between the *C. aurita* plus *C. flaviceps* pair and the other four species is the average of initial frequencies. In the first two species, subsequent notes tend to present lower initial frequencies, while in the others there are no differences between the initial frequencies of notes at the 5% significance level (see Fig. 3.7).

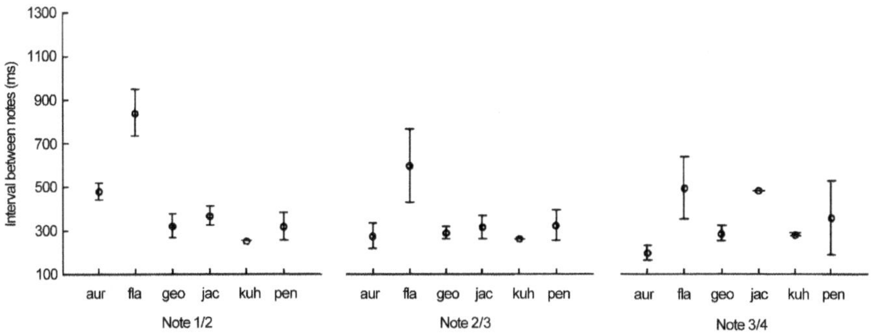

Fig. 3.6 Mean of intervals between the first four notes of long calls of *C. aurita* (aur), *C. flaviceps* (fla), *C. geoffroyi* (geo), *C. jacchus* (jac), *C. kuhlii* (kuh), and *C. penicillata* (pen)

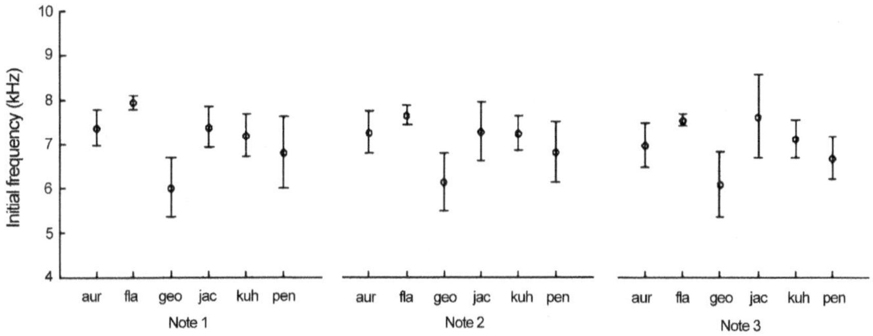

Fig. 3.7 Mean of initial frequencies of the first three notes of long calls of *C. aurita* (aur), *C. flaviceps* (fla), *C. geoffroyi* (geo), *C. jacchus* (jac), *C. kuhlii* (kuh), and *C. penicillata* (pen)

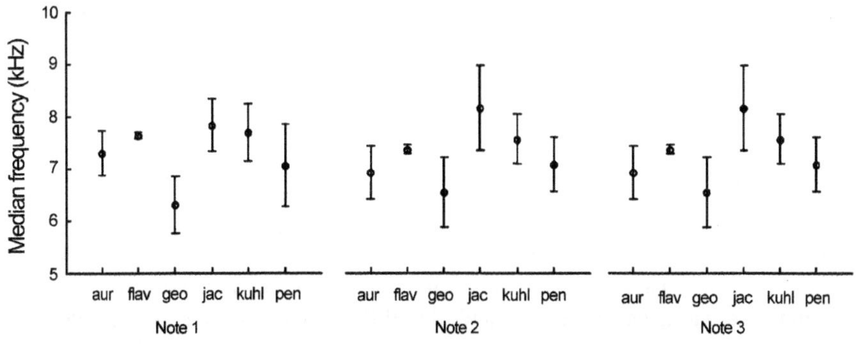

Fig. 3.8 Mean of median frequencies of the first three notes of long calls of *C. aurita* (aur), *C. flaviceps* (fla), *C. geoffroyi* (geo), *C. jacchus* (jac), *C. kuhlii* (kuh), and *C. penicillata* (pen)

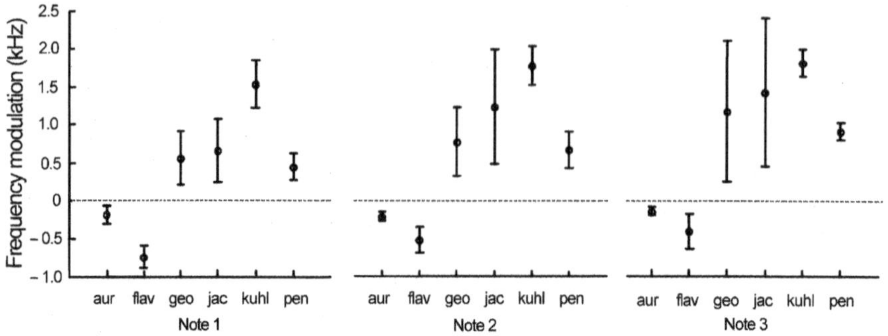

Fig. 3.9 Mean of frequency modulations of the first three notes of long calls of *C. aurita* (aur), *C. flaviceps* (fla), *C. geoffroyi* (geo), *C. jacchus* (jac), *C. kuhlii* (kuh), and *C. penicillata* (pen)

Table 3.4 Changes between notes within the long calls of *Callithrix* species

Changes from Notes 1–3	*C. aurita*	*C. flaviceps*	*C. geoffroyi*	*C. jacchus*	*C. kuhlii*	*C. penicillata*
In note duration	Get shorter	Get shorter	Get shorter	Variable	Same	Get shorter
In interval between	Decrease	Decrease	Same	Same	Same	Same
In initial frequency	Decrease	Decrease	Same	Same	Same	Same
In mean frequency	Decrease	Decrease	Increase (not sig.)	Increase (not sig.)	Increase (not sig.)	Increase (not sig.)
In frequency modulation	Same	Reduce	Increase	Increase	Same	Increase

In *C. aurita* and *C. flaviceps*, subsequent notes tend to present lower mean frequencies, while in the other species there is a tendency for an increase in the mean frequency along the call, but the differences are variable and not significant (see Fig. 3.8).

In *C. aurita* and *C. kuhlii,* there are no significant differences in the modulation between the first three notes at the 5% significance level. However, in *C. flaviceps,* the subsequent notes tend to be less modulated, while *C. geoffroyi*, *C. jacchus* and *C. penicillata* show the opposite pattern (see Fig. 3.9).

3.3.6 Attributes of Note 1

Although comparisons of the first three notes of the long call demonstrate differences between species that appear suitable for inter-specific comparison, in Notes 2, 3, and beyond, the variance increases, and the differences between species decrease. Therefore, it is practical and informative to compare the species' vocalizations focusing only on the attributes of the first note, the least variable (see Table 3.5).

Table 3.5 Parameters of Note 1 for long calls of *Callithrix*[a]: note duration (*T*), interval between notes (DT), initial frequency (IF), mean frequency (MF), frequency modulation (FM)

Species	*T*	SE	*N*	DT	SE	*N*	FI	SE	*N*	MF	SE	*N*	FM	SE	*N*
C. aurita	798	22	63	482	22	62	7.37	0.04	62	7.30	0.05	62	–0.17	0.06	62
C. flaviceps	823	27	74	841	26	54	7.95	0.05	74	7.64	0.06	74	–0.73	0.08	74
C. geoffroyi	1,154	31	74	323	30	69	6.01	0.06	74	6.30	0.07	74	0.56	0.09	74
C. jacchus	1,371	37	74	369	36	49	7.38	0.07	74	7.83	0.08	74	0.66	0.11	74
C. kuhli	653	35	23	253	34	22	7.19	0.07	23	7.69	0.08	23	1.53	0.11	23
C. penicillata	1,213	28	39	320	30	32	6.81	0.05	39	7.05	0.06	39	0.44	0.08	39

[a]Durations in milliseconds (ms), frequencies in kilohertz (kHz)

3.3.6.1 Note 1 Duration

Note 1 mean duration is shortest in *C. kuhlii* (653 ms) and longest in *C. jacchus* (1,371 ms), both differing significantly from the other species (see Table 3.5, Fig. 3.5). According to the Tukey test of multiple comparison, Note 1 mean duration is significantly different between all the species pairs at the 5% level of significance, except between *C. aurita* and *C. flaviceps* and between *C. geoffroyi* and *C. penicillata*. Note 1 tends to be shorter as the number of notes per long call increases (see Table 3.6).

3.3.6.2 Interval Between Notes 1 and 2

The interval between Notes 1 and 2 is longer in *Callithrix flaviceps* than in all the other species (see Table 3.5, Fig. 3.6), and the differences are highly significant. *Callithrix aurita* also exhibits significantly larger intervals than the others except for *C. jacchus*, in which while still longer, the difference is not significant. The remaining three species do not differ significantly from each other.

3.3.6.3 Note 1 Initial Frequency

The lowest mean initial frequency is found in *C. geoffroyi* (6.01 kHz), followed by *C. penicillata* (6.81 kHz). These differ significantly from each other and from the other species (see Fig. 3.7). *C. flaviceps* has the highest initial frequency (7.95 kHz), significantly differing from the others. *Callithrix kuhlii* (7.19 kHz), *C. aurita* (7.37 kHz) and *C. jacchus* (7.38 kHz) are not significantly different at the 5% significance level.

3.3.6.4 Note 1 Mean Frequency

The lowest mean frequency is found in *C. geoffroyi* (6.30 kHz), which differs significantly from all the other species (see Table 3.5, Fig. 3.8). *Callithrix penicillata* (7.05 kHz) and *C. aurita* (7.30 kHz) also differ from the others, while *C. flaviceps* and *C. kuhlii*, which vocalize at a higher frequency, do not differ significantly.

Table 3.6 Regression analyses between the number of notes and duration of Note 1 in *Callithrix* (*F* is the Anova *F*-statistic and *N* is the sample size)

Species	Slope	Slope standard error	*F*	*P*-value	*N*
C. aurita	−31.97	12.32	6.72	0.012	60
C. flaviceps	−80.61	9.86	66.75	<0.001	82
C. geoffroyi	−56.80	30.38	3.49	0.065	81
C. jacchus	−145.81	29.17	24.98	<0.001	29
C. kuhlii	−66.00	17.39	14.39	0.001	22
C. penicillata	−236.30	44.38	28.34	<0.001	23

3.3.6.5 Note 1 Frequency Modulation

Callithrix aurita and *C. flaviceps* present a descending frequency modulation during Note 1, which is clearly distinct from the other species (Table 3.5, Fig. 3.9). Though descending, the long call modulation of *C. aurita* (–0.07 kHz/ms) is small and significantly different from that of *C. flaviceps* (–0.73 kHz/ms). *Callithrix kuhlii*, with the highest modulation (1.53 kHz/ms), significantly differs from all the other species. *Callithrix geoffroyi*, *C. jacchus* and *C. penicillata* did not differ from one another.

3.3.7 Synthesis of the Differences Between Species

C. aurita and *C. flaviceps* differ from the other species mainly in three aspects of the general structure of their vocalizations: (1) long notes are usually followed by short notes, progressively lower and weaker, with a peculiar modulation; (2) the long notes have a descending modulation, opposite to those of the other species; and (3) the interval between notes gets progressively shorter. *Callithrix flaviceps* differs from *C. aurita* in the longer interval between Notes 1 and 2, in the higher frequency modulation, and in its higher pitched long call. The interval between the first two notes of *C. flaviceps* is almost double that of *C. aurita*.

Callithrix geoffroyi differs from the other species mainly in its lower frequency vocalization in all parameters. *Callithrix jacchus* has the highest pitched vocalization (measured by mean frequency), a significantly lower number of notes, and a longer duration of the first note compared to the other species. *Callithrix kuhlii* is the most distinctive species, as it emits significantly shorter, more modulated notes, and it does not present differences in the duration or the modulation between subsequent notes along the vocalization. Furthermore, its vocal parameters present the smallest variation. *Callithrix penicillata* most closely approaches *C. geoffroyi* in terms of the general structure of its vocalization, not differing in the number of notes or in the notes' temporal patterns. However, *C. penicillata* differs from *C. geoffroyi* and the other species in the initial frequency and mean frequency, which are low but not as low as in *C. geoffroyi*. In synthesis, all of the six studied species present statistically distinct vocal parameters that allow for species identification.

3.4 Discussion

The analyzed vocalizations show that the long call structure of *Callithrix* is generally similar to those of the majority of other previously studied callitrichids (see Snowdon 1993). There is a great similarity between the long calls of different

Callithrix species, mainly in terms of frequency and total duration. It is interesting that the duration of notes is inversely proportional to the number of notes in a call, resulting in little variation in the total duration of long calls. Possibly, there are historical processes and selective forces confining long call structure to a relatively restricted range of variation. For example, longer calls can carry more precise information, but they can also increase the probability of attracting predators (Marler 1955).

In all of the analyzed species, Note 1 is the most characteristic component of the long call, as it is the most differentiated and least variable note in structure. This suggests that the specific recognition message is already encoded in the long call's first note; therefore, later notes are less significant, and there is tolerance for greater variation in the number of notes and aspects of later notes in the long calls of members of the same species. It is possible that the function of the other notes is signal reinforcement or carrying other information, such as emotional state, sex, and identity of the vocalizer. Jones et al. (1993) also pointed out the similarity between notes when comparing the first two syllables (or notes) of *C. jacchus* long calls.

It is relevant that long call structure is coherent with the division of *Callithrix* into two groups (the jacchus-group and the aurita-group), originally based on morphologic and genetic data (Natori 1986, 1994; Tagliaro et al. 1997; Sena et al. 2002; Marroig et al. 2004). One distinct general pattern of long calls, including descendent modulation, the presence of short notes, and the presence of a peculiar modulation at the end of the long call, is restricted to *C. aurita* and *C. flaviceps* (the aurita-group species). The uniqueness of this pattern, easily perceived even by human ears, reinforces that there is a phylogenetic component to the long calls of *Callithrix*.

Although *C. aurita* and *C. flaviceps* have a clear vocal similarity, the long call structures of these species are different from each other. The long calls of *C. flaviceps* were recorded at three different regions: at Santa Teresa region, Espirito Santo state, in a contact zone with *C. geoffroyi*; at the Parque Estadual de Pedra Azul, Espirito Santo state, in the middle of its geographic distribution area; and at the Estação Biológica de Caratinga, close to a contact zone with *C. aurita*. Even so, the duration of the intervals between notes and the frequency modulation presented consistent differences from that of *C. aurita*. The long calls of *C. aurita* were mainly recorded in São Paulo State, but the samples obtained at Serra do Brigadeiro, in Minas Gerais State, close to the contact zone with *C. flaviceps*, presented a pattern similar to that of the samples from São Paulo State. This suggests that the vocal parameters that differentiate the species of the aurita-group are consistent, even considering the wide geographic distribution of *C. aurita*.

Coimbra-Filho (1986a, b, 1993) suggested that *C. flaviceps* should be treated as subspecies of *C. aurita* on the basis of similarities in the pelage patterns, ecological adaptations, ontogeny, and vocalizations. Despite these broad similarities, the differences in the pelage coloration of these species suggest that *C. flaviceps* is a geographically and morphologically distinct taxon (Vivo 1991). Although it hybridizes with *C. aurita* in contact zones, it also hybridizes with *C. geoffroyi*, and some degree of hybridization occurs at all studied contact zones of *Callithrix* species

(Mendes 1997b). While these two share some ecological adaptations, other *Callithrix* species pairs also share similarities, such as *C. geoffroyi* and *C. kuhlii* inhabiting dense lowland tropical forest at southern Bahia, and *C. jacchus* and *C. penicillata* inhabiting more open ecosystems, including *caatinga* and *cerrado* (Rylands and Faria 1993; Mendes 1997a). The present study reinforces differences between *C. aurita* and *C. flaviceps* on the basis of long call structures, which clearly distinguish the two. They are as distinct from each other as the members of the jacchus-group, which are widely treated as good species. Thus, there is no sufficient justification to consider *C. aurita* and *C. flaviceps* as the same species.

The jacchus-group includes a larger number of species, and there is a greater overlap of acoustic parameters. Even with less divergence in the vocal structure between some species of this group, there are consistent peculiarities in the long call of each species. The greatest vocal proximity was found between *C. geoffroyi* and *C. penicillata*. There was no significant difference in the temporal parameters of their long calls, but they differ in the frequency parameters.

It is important to point out the uniqueness of the vocalization of *C. kuhlii*, from southern Bahia, especially considering that Hershkovitz (1977) suggested that specimens from southern Bahia could be hybrids between *C. penicillata* and *C. geoffroyi*, and Vivo (1991) treated *kuhlii* as conspecific with *C. penicillata*. The vocal parameters indicate that not only does *C. kuhlii* have its own identity, but its vocal structure is also not intermediate between those of *C. penicillata* and *C. geoffroyi*. Thus, vocally, *C. kuhlii* is a very distinguishable taxon, corroborating previous morphologic studies that indicated the uniqueness of this species (Natori 1990, 1994; Marroig et al. 2004).

Some vocal differences between the species are subtle to human ears but can certainly be heard by the marmosets, because of their capacity to discriminate patterns much less distinct, such as sexual differences in the long calls (Norcross and Newman 1993). Considering the long calls as signals of species recognition, the vocal parameters appear to corroborate the treatment of all the six taxa as valid species. However, it is important to note that in the Callitrichidae, there are reports of sub-specific and population differences in the structure of vocalizations. In addition to species differences in the long call structure, Hodun et al. (1981) found subspecific differences in the vocalization of *Saguinus fuscicollis*; significant differences in the acoustic structures of long calls between populations of *Saguinus labiatus labiatus* were also reported by Maeda and Masataka (1987). It is possible that some *Saguinus* taxa which have been treated as subspecies may be valid species, but it is premature to use vocalization alone to define a callitrichid species.

Similarities found among the jacchus-group species with regard to vocalisation do not support the phylogenetic arrangement proposed by Natori (1994). In contrast, they are coherent with the jacchus-group relationships proposed by Marroig et al. (2004), which indicated a close proximity between *C. geoffroyi* and *C. penicillata*. The vocal structure of *C. penicillata* is much closer to that of *C. geoffroyi* than to any other *Callithrix*.

Other factors, in addition to phylogenetic history, may be driving the evolution of the long call structure in *Callithrix*. For instance, vocalization may serve as reinforcement of reproductive isolation. We posit a model of allopatric speciation in *Callithrix* (Vivo 1991; Marroig et al. 2004), with a later establishment of secondary contact zones where some hybridization may occur (Mendes 1997b). Where the hybrids have lower fitness, selective pressure will favor the fixation of specific recognition signals. The long call is a signal that attracts reproductive mates (Snowdon et al. 1986), and we demonstrate it has acquired species-specific characteristics that could reinforce reproductive isolation.

Therefore, we expect a higher degree of differentiation between the pairs of species that have shared more historical contact between their populations. This may explain, for instance, the great difference between the long calls of neighboring *C. geoffroyi* and *C. kuhlii*. These species inhabit the same kind of ecosystem with adjacent geographic distribution, without important ecological and geographic barriers, and they have similar general patterns of coloration. They could be subject to strong selective pressures favoring the fixation of very distinct vocal signals for specific mate recognition. Moreover, *Callithrix kuhlii* has a restricted geographic distribution of probable contact zones with *C. geoffroyi*, *C. penicillata* and *C. jacchus* (Mendes 1997a), which could reinforce the development of a clear vocal identity, a process that has led its long call to become highly distinctive from those of the other, related jacchus-group species.

Studies on the geographic variation of vocal parameters and playback experiments could help to test the hypothesis of vocalization as a mechanism of population recognition and reproductive isolation in callitrichids. This kind of study would enhance the use of vocalization as a behavioral taxonomic character, as well as provide important information on the mechanisms involved in the differentiation and evolution of primate natural populations.

Acknowledgements We acknowledge the collaboration of Dwain Santee, Braz Cosenza, Fabiano R. Melo, Vânia Diego, Bernardo F. A. Brito, Karen B. Strier, and Anthony B. Rylands, and the support of Unicamp, Museu de Biologia Prof. Mello Leitão, Reserva Natural da CVRD, Parque Estadual de Pedra Azul, Estação Biológica de Caratinga, Parque Ecológico de Americana, Reserva Biológica Augusto Ruschi, Reserva Biológica de Duas Bocas, Estação Experimental Lemos Maia, and Estação Ecológica de Tapacurá. We are also indebted to Susan Ford, who made very helpful suggestions on the manuscript.

References

Altmann SA (1967) The structure of primate communication. In: Altmann SA (ed) Social communication among primates. University of Chicago Press, Chicago, pp 325–362

Ambrose L (2003) Three acoustic forms of Allen's galagos (Primates; Galagonidae) in the Central African region. Primates 44:25–39

Barroso CML, Schneider H, Schneider MP, Sampaio I, Harada ML, Czelusniak J, Goodman M (1997) Update on the phylogenetic of New World monkeys: further DNA evidence for placing the pygmy marmoset (*Cebuella*) within the genus *Callithrix*. Int J Primatol 18(4):651–674

Brockelman WY, Schilling D (1984) Inheritance of stereotyped gibbon calls. Nature 312:634–636

Canavez FC, Moreira MAM, Simon F, Parham P, Seuanez HN (1999) Phylogenetic relationships of the Callitrichinae (Platyrrhini, Primates) based on β2-microglobulin DNA sequences. Am J Primatol 48:225–236

Chen HC, Geissmann T, Chen PC (2004) A survey of the taxonomic status of captive gibbons in Taiwan. Raffles Bull Zool 52(1):265–269

Coimbra-Filho AF (1986a) Sagui-da-serra, *Callithrix aurita* (Thomas, 1903). Informativo FBCN 10(1):3

Coimbra-Filho AF (1986b) Sagui-da-serra-escuro, *Callithrix aurita* (E. Geoffroy, 1812). Informativo FBCN 10(2):3

Coimbra-Filho AF (1990) Sistemática, distribuição geográfica e situação atual dos símios brasileiros (Platyrrhini-Primates). Rev Bras Biol 50(4):1063–1079

Coimbra-Filho AF, Pissinati A, Silva RR, Rylands AB (1993) Experimental multiple hybridism and natural hybrids among *Callithrix* species from eastern Brazil. In: Rylands AB (ed) Marmosets and tamarins: systematics, ecology and behaviour. Oxford University Press, Oxford, pp 95–120

Courtenay DO, Bearder SK (1989) The taxonomic status and distribution of bushbabies in Malawi with emphasis on the significance of vocalizations. Int J Primatol 10(1):17–34

Epple G (1968) Comparative studies on vocalization in marmoset monkeys (Hapalidae). Folia Primatol 8:1–40

Gautier JP (1978) Répertoire sonore de *Cercopithecus cephus*. Z Tierpsychol 46:113–169

Haimoff EH, Chivers DJ, Gittins SP, Whitten T (1982) A phylogeny of gibbons (*Hylobates* spp.) based on morphological and behavioural characters. Folia Primatol 39:213–237

Hershkovitz P (1977) Living New World monkeys (Platyrrhini), vol 1. University of Chicago Press, Chicago

Hodun A, Snowdon CT, Soini P (1981) Subspecific variation in the long calls of the tamarin, *Saguinus fuscicollis*. Z Tierpsychol 57:97–110

Hohmann G (1989) Comparative study of vocal communication in two Asian leaf monkeys, *Presbytis johnii* and *Presbytis entellus*. Folia Primatol 52:27–57

Jones BS, Harris DHR, Catchpole CK (1993) The stability of vocal signature in phee call of the common marmoset, *Callithrix jacchus*. Am J Primatol 31:67–75

Maeda T, Masataka N (1987) Locale specific vocal behaviour of the tamarin (*Saguinus l. labiatus*). Ethology 75:25–30

Marler P (1955) Characteristics of some animal calls. Nature 176:6–8

Marroig G, Cropp S, Cheverud JM (2004) Systematics and evolution of the *jacchus* group of marmosets (Platyrrhini). Am J Phys Anthropol 123:11–22

Marshall JT, Marshall ER (1976) Gibbons and their territorial songs. Science 193:235–237

Masters JC (1991) Loud calls of *Galago crassicaudatus* and *G. garnettii* and their relation to habitat structure. Primates 32:153–167

Meireles CMM, Sampaio MIC, Schneider MPC (1992) Protein variation, taxonomy and differentiation in five species of marmosets (genus *Callithrix* Erxleben, 1777). Primates 33:227–238

Mendes SL (1997a) Hybridization in free ranging *Callithrix flaviceps* and the taxonomy of the Atlantic forest marmosets. Neotrop Primates 5(1):6–8

Mendes SL (1997a) Padrões Biogeográficos e Vocais em *Callithrix* do grupo *jacchus* (Primates, Callitrichidae). Ph.D Thesis, Universidade Estadual de Campinas, Campinas, Brazil

Mittermeier RA, Coimbra-Filho AF (1981) Systematics: species and subspecies. In: Coimbra-Filho AF, Mittermeier RA (eds) Ecology and behavior of neotropical primates, vol 1. Academia Brasileira de Ciencias, Belo Horizonte, pp 29–111

Moreira MAM, Seuanez HN (1999) Mitochondrial pseudogenes and phyletic relationships of *Cebuella* and *Callithrix* (Platyrrhini, Primates). Primates 40(2):353–364

Natori M (1986) Interspecific relationships of *Callithrix* based on the dental characters. Primates 27(3):321–336

Natori M (1990) Numerical analysis of systematic status of *Callithrix kuhli* based on measurements of the postcanine dentition. Primates 31(4):555–562

Natori M (1994) Craniometrical variation among eastern Brazilian marmosets and their systematic relationships. Primates 35(2):167–176

Norcross JL, Newman JD (1993) Context and gender-specific differences in the acoustic structure of common marmoset (*Callithrix jacchus*) phee calls. Am J Primatol 30:37–54

Quris R (1980) Emission vocale de forte intensité chez *Cercocebus galeritus agilis*: structure, caractéristiques spécifiques e individuelles, modes d'émission. Mammalia 44(1):35–50

Rosenberger AL (1981) Systematics: the higher taxa. In: Coimbra-Filho AF, Mittermeier RA (eds) Ecology and behavior of neotropical primates, vol 1. Academia Brasileira de Ciencias, Belo Horizonte, pp 9–27

Rylands AB, Faria DS (1993) Habitats, feeding ecology, and home range size in the genus *Callithrix*. In: Rylands AB (ed) Marmosets and tamarins: systematics, ecology and behaviour. Oxford University Press, Oxford, pp 262–272

Rylands AB, Schneider H, Langguth A, Mittermeier RA, Groves CP, Rodriguez-Luna E (2000) An assessment of the diversity of New World Primates. Neotrop Primates 8(2):61–93

Rylands AB, Mittermeier RA, Coimbra-Filho AF (this volume) The systematics and distribution of the marmosets (*Callithrix*, *Callibella*, *Cebuella*, and *Mico*) and Callimico (*Callimico*) (Callitrichidae, Primates). In: Ford SM, Porter LM, Davis LC (eds) The smallest anthropoids: the marmoset/callimico radiation. Springer, New York pp 25–61

Sampaio IS, Schneider MP, Barroso CML, Schneider H (1996) Polymorphism of phosphogluconate dehydrogenase (PGD) in New World monkeys: taxonomic significance. Rev Bras Genét 19(1):87–92

Schneider H, Sampaio I, Harada ML, Barroso CML, Schneider MPC, Czelusniak J, Goodman M (1996) Molecular phylogeny of the New World monkeys (Platyrhini, Primates) based on two unlinked nuclear genes: IRBP Intron 1 and epsilon-globin sequences. Am J Phys Anthropol 100:153–179

Sena L, Vallinoto M, Sampaio I, Schneider H, Ferrari SF, Scneider MPC (2002) Mitochondrial COII gene sequences provide new insights into the phylogeny of marmoset species groups (Callitrichidae, Primates). Folia Primatol 73(5):240–251

Singer SS, Schmitz J, Schwiegk C, Zischler H (2003) Molecular cladistic markers in New World monkey phylogeny (Platyrrhini, Primates). Mol Phylogenet Evol 26:490–501

Snedecor GW, Cochran WG (1989) Statistical methods, 8ath edn. Iowa State University Press, Iowa

Snowdon CT (1993) A vocal taxonomy of the callitrichids. In: Rylands AB (ed) Marmosets and tamarins: systematics, behaviour, and ecology. Oxford University Press, Oxford, pp 78–94

Snowdon CT, Hodun A, Rosenberger AL, Coimbra-Filho AF (1986) Long-call structure and its relation to taxonomy in lion tamarins. Am J Primatol 11:253–261

Struhsaker TT (1970) Phylogenetic implications of some vocalizations of *Cercopithecus* monkeys. In: Napier JR, Napier PH (eds) Old World monkeys: evolution, systematics, and behavior. Academic, New York, pp 365–444

Tagliaro CH, Schneider MPC, Schneider H, Sampaio IC, Stanhope MJ (1997) Marmoset phylogenetics, conservation perspectives, and evolution of the mtDNA control region. Mol Biol Evol 14(6):674–684

Vivo M (1991) Taxonomia de *Callithrix* Erxleben, 1777 (Callitrichidae, Primates). Fundação Biodiversitas, Belo Horizonte

Waser PM (1982) The evolution of male loud calls among mangabeys and baboons. In: Snowdon CT, Brown CH, Petersen MR (eds) Primate communication. Cambridge University Press, Cambridge, pp 117–143

Wied M zu (1826) Beiträge zur Naturgeschichte von Brasilien. II Band. Landes-Industrie-Comptoirs, Weimar, Germany

Zimmermann E, Bearder SK, Doyle GA, Andersson AB (1988) Variations in vocal patterns in Senegal and South African lesser bushbabies and their implications for taxonomic relationships. Folia Primatol 51:87–105

Section II
Reproductive, Social, and Cognitive Behavior

Leila M. Porter, Susan M. Ford, Lesa C. Davis

Callitrichid social behavior represents an unusual balance between highly cooperative behavior and intense female reproductive competition. Callitrichids are unique among primates in their tendency to form groups in which a single dominant female monopolizes reproduction, and in the frequent help that group members provide for transporting and sharing food with the dominant female's offspring. Work in the last decade, both in the wild and in captivity, has dramatically increased both our realization of the variation in social groups and reproductive behaviors in callitrichids as well as our understanding of some of the conditions under which different strategies are pursued and may have evolved. While characterized by relatively small and simple brains, their cognitive abilities may belie the link between "intelligence" and brain complexity. Many of these issues are explored in this section for *Callimico* and *Callithrix* and indicate the importance of collecting similar types of data for the other marmosets.

Although communal care is generally assumed to have evolved in response to the high energy costs of raising twin offspring, Porter and Garber (Chap. 4) demonstrate that callimico mothers with single infants can maintain polyandrous groups in which multiple adult males assist with infant care. Thus, although the energy constraints of raising twins may have led initially to the rise of cooperative care among callitrichids, it cannot explain the maintenance of this behavior among callimicos. Faulkes et al. (Chap. 5), using a combination of behavioral and genetic data, explore possible explanations for cooperative care among marmosets. They demonstrate that members of marmoset groups cooperate in infant care even when the infants are unrelated, leading them to conclude that additional mechanisms besides kin selection are needed to explain this behavior among marmosets.

Leila M. Porter
Department of Anthropology, Northern Illinois University, DeKalb, IL, 60115, USA
e-mail: lmporter@niu.edu

Susan M. Ford
Department of Anthropology, Southern Illinois University, Carbondale, IL, 62901, USA
e-mail: sford@siu.edu

Additionally in this section, several authors demonstrate that demand for communal caregiving influences the reproductive fitness of marmoset females. Yamomoto et al. (Chap. 6) found that dominant females may not always be able to monopolize reproduction in their groups. In these cases, females must share the help of males in caring for multiple sets of twins, a situation that leads to reduced fitness for the dominant female. Digby and Saltzman (Chap. 7) further demonstrate that in these situations, dominant females may use infanticide to reduce competition for caregivers, reducing the subordinate females' fitness to zero. As a likely result of this intense competition among females for breeding positions, males are more cooperative than females. For example, Ferrari (Chap. 8) demonstrates that although agonism is rare among marmoset groups, a multi-tiered dominance hierarchy is present in marmoset groups, with the top tier occupied by a dominant breeding female, the middle tier by adult males, and the lowest tier by subordinate females. Sousa et al. (Chap. 9) further demonstrate that male marmosets are more willing to remain in groups and work cooperatively than are subordinate females. As a result, their data indicate that females emigrate from groups more frequently than males, and that females are better able to cope with the stress of isolation that occurs during emigration than males.

In addition to having highly complex social interactions, callitrichids must also learn to forage in a complex environment. Huber and Voelkl (Chap. 10) demonstrate that callitrichids have a detailed spatial memory of the location of food resources and are able to track resources that move over time. In addition, they demonstrate that callitrichid infants learn what foods to eat through social facilitation, and that these animals are capable of imitating one another to learn complex foraging skills. In sum, these chapters demonstrate that callimicos and marmosets have adapted numerous mechanisms for maximizing fitness that include a complicated mixture of both highly competitive and highly cooperative strategies.

Chapter 4
Social Behavior of Callimicos: Mating Strategies and Infant Care

Leila M. Porter and Paul A. Garber

Abstract Callimico (*Callimico goeldii*) is the only species of tamarin or marmoset to regularly produce one infant, possibly representing a derived trait. Studies of adult–infant interactions among callimicos, therefore, offer important insights into the evolution of cooperative infant care in callitrichids. We studied a group of callimicos consisting of one adult female (FH), two adult males (MR and MJ), and one young female (FS), at a Bolivian field site. Opportunistic data on infant care and mating behavior were taken for 3 months following FS's birth. Then, for 8 months, behavioral data and nearest neighbor distances were collected using focal animal sampling for 1,199 observation hours. Nearest neighbor distances among group members varied significantly ($p<0.001$): on average the breeding female was found 2.7 m from her infant, 2.8 m from adult male MJ, and 3.5 m from adult male MR. The frequency of grooming behaviors varied significantly among adults ($p<0.001$): FH was groomed most (8.4% of observations) and spent the least time grooming others (3.4%), while MR was groomed least (5.9%) and groomed others most (10.7%). Although FH was observed mating with both males within 2 months after the birth of FS, she did not give birth to additional offspring. All adults shared food with and carried the infant. Adults also shared food with each other. Aggressive behavior among adults was extremely rare ($n=8$) and occurred six times in a feeding context and twice in a non-feeding context. These data demonstrate that despite producing singletons, callimicos like many other callitrichids are characterized by a polyandrous mating pattern, spatially cohesive groups, and high levels of social cooperation.

Resumen *Callimico* (*Callimico goeldii*) es la única especie de tamarinos o marmosetas que regularmente produce sólo un infante, posiblemente representando un rasgo derivado. Por tanto, estudios de las interacciones adulto-infante entre los calimicos ofrecen importantes ideas sobre la evolución cooperativa del cuidado del

L.M. Porter (✉)
Department of Anthropology, Northern Illinois University, DeKalb, IL, 60115, USA
e-mail: lmporter@niu.edu

S.M. Ford et al. (eds.), *The Smallest Anthropoids*, Developments in Primatology: Progress and Prospects, DOI 10.1007/978-1-4419-0293-1_4,
© Springer Science+Business Media, LLC 2009

infante en los callitrícidos. Estudiamos un grupo de calimicos, que consistía en una hembra adulta (FH), dos machos adultos (MR y MJ) y una hembra joven (FS), en una localidad en Bolivia. Datos sobre cuidado de infantes y conducta de apareamiento fueron tomados de manera oportunista durante tres meses seguido al nacimiento de FS. Posteriormente, información sobre el comportamiento y distancias del vecino más cercano fueron recolectadas durante ocho meses utilizando un muestreo animal-focal, y completando 1,199 horas de observación. Las distancias más cercanas entre los miembros del grupo variaron significativamente ($p<0.001$): en promedio la hembra reproductora fue encontrada 2.7 m de su infante, 2.8 m del macho adulto MJ y 3.5 m del macho adulto MR. La frecuencia de la conducta de acicalamiento varió significativamente entre adultos ($p<0.001$): FH fue el más acicalado (8.4% de las observaciones) y paso el menor tiempo acicalando a otros (3.4%), mientras que MR fue el menos acicalado (5.9%) y acicaló más a otros (10.7%). A pesar de que se observó apareamiento entre FH y ambos machos dentro de los dos meses siguientes al nacimiento de FS, no se produjeron crías adicionales. Todos los adultos compartieron comida y cargaron al infante. Los adultos también intercambiaron comida entre ellos. Comportamiento agresivo entre los adultos fue extremadamente raro ($N=8$) y ocurriendo en seis ocasiones en contexto de alimentación y en dos ocasiones en contextos no alimenticios. Esta información demuestra que, a pesar que producir hijos únicos, los calimicos como muchas de las otras callitricidos son caracterizados por patrones de comportamiento poliándrico, grupos espacialmente cohesivos y niveles altos de cooperación social.

Resumo Sagüi-de-goeldi (*Callimico goeldii*) é a única espécie dentre os micos e sagüis que produz, regularmente, um infante, representando possivelmente uma característica derivada. Portanto, estudos sobre as interações adultos-infantes entre sagüis-de-goeldi oferecem importantes informações sobre a evolução do cuidado cooperativo em calitriquídeos. Nós estudamos um grupo de sagüis-de-goeldi, formado por uma fêmea adulta (FH) dois machos adultos (MR e MJ) e uma fêmea jovem (FS) em uma estação de campo na Bolívia. Foram coletados de modo oportunístico dados sobre o cuidado à prole e sobre o comportamento de acasalamento 3 meses após o nascimento de FS. A seguir, foram coletados, durante 8 meses, os dados comportamentais e a distância entre os animais mais próximos usando o método focal durante 1,199 horas de observação. As menores distâncias entre os membros do grupo variaram significativamente ($p<0.001$): em média a fêmea reprodutora ficou a 2.7 m do seu infante, 2.8 do macho adulto MJ e 3.5 m do outro macho adulto MR. A freqüência de catação social variou significativamente entre adultos ($p<0.001$): FH foi a que recebeu mais catação (8.4% das observações) e gastou o menor tempo catando os outros animais (3.4%) enquanto MR foi catado menos (5.9%) e catou mais os outros animais (10.7%). Embora FH tenha acasalado com ambos os machos nos dos dois meses após o nascimento de FS, ela não deu cria adicionais. Todos os adultos partilharam alimento com os infantes e os carregaram. Os adultos também partilharam alimento entre si. Os comportamentos agressivos entre adultos foram muito raros ($n=8$), sendo 6 deles no contexto alimentar.

Os dados demonstram que a despeito de gerar filhotes únicos, sagüis-de-goeldi, da mesma forma que muitos outros calitriquídeos, são caracterizados por um padrão poliândrico de acasalamento, sendo os grupos espacialmente coesos e com níveis elevados de cooperação social.

4.1 Introduction

Callitrichids (tamarins, lion tamarins, callimicos, and marmosets) are remarkable among New World primates in the flexibility and variability of their mating systems, the potential for biannual births, lactational anestrus, and the extensive care that group members provide for the growth of infants and juveniles (called communal or cooperative care). Studies of all species in the wild have demonstrated that within a single population there can be monogamous, polyandrous, and multi-male/multi-female matings (Dietz and Baker 1993; Garber 1994; Goldizen 1990; Nievergelt et al. 2000). Here we distinguish between mating which is defined in terms of the set of individuals that copulate, and breeding which is defined in terms of the individuals that contribute genes to the next generation (parents). Highly variable mating also appears to occur among callimico (*Callimico goeldii*) populations, despite the fact that it is the only callitrichid species that regularly gives birth to single offspring rather than twins. *Callimico* groups with one and two breeding females have been reported in the wild (reviewed in Porter 2003), although the details of social relations within these groups have not been well documented.

In *Callithrix*, *Mico*, *Cebuella* and *Saguinus*, a single dominant female has the potential to prevent breeding (but not mating) among subordinate female group members through hormonal suppression leading to ovulatory failure (Abbott et al. 1993). Hormonal suppression is not always successful however, and if subordinate females manage to reproduce, dominant females may resort to infanticide to eliminate the subordinate females' infants (see Digby and Saltzman this volume, Chap. 7; also Yamamoto et al. this volume, Chap. 6). In contrast, lion tamarins (*Leontopithecus*), like callimicos, lack the ability for hormonal suppression and dominant females may instead use aggression to disrupt breeding by other females (French and Inglett 1989). Data based on captive studies reveal that callimico groups with two breeding females are unstable, due to high levels of aggression between the resident adult females (Carroll 1988; French and Inglett 1989; Hardie 1995).

Captive studies of pair formation in callimicos indicate that males initiate affiliative behaviors and display behaviors more frequently than females, suggesting that males are more concerned with maintaining sexual-social bonds than females (Carroll 1985). In addition, Pryce et al. (2002) found that in newly formed captive groups containing two adults of each sex, the male that chose to mate exclusively with the dominant female became the group's resident male, whereas males that mated with both the dominant and subordinate females were later rejected from the group. In contrast, dominant and subordinate females presented to both males, and in two out of three groups the dominant females copulated with both males. Given evidence of sex-based differences in social behavior in captive callimicos, it is important to determine whether these behavioral patterns also exist in wild groups and their impact on patterns of infant care, infant survivorship, and mating strategies.

Regardless of whether mating is monogamous, polyandrous, or polygynous, in all callitrichids studied, adult males and non-breeding adult females help to transport and provide for the groups' infants. Cooperative caregiving is considered to be essential for the infants' survival (Tardif 1995). Non-maternal transport of infants can begin immediately after their birth, and food provisioning can begin at 1 month and continue for several months after weaning (Garber 1997; Yamamoto 1993). Several studies have shown that infant survival increases positively with the number of adult male helpers present in the group (Garber 1997; Garber et al. 1984; Heymann and Soini 1999; Koenig 1995; Price 1992). In addition, helpers reduce the costs of infant care to the mother, giving callitrichid females the potential to give birth twice per year. Biannual births are possible due to post-partum ovulation, which allows a female to nurse an infant and become pregnant at the same time (Jurke 2002; Ziegler et al. 1989). Although in captivity, females in several callitrichid species frequently give birth to twin infants twice per year, in the wild such high reproductive output is less common, particularly among *Saguinus*, *Leontopithecus*, and possibly *Mico* (Porter and Garber 2004).

4.1.1 Cooperative Infant Care

Cooperative infant caregiving appears to benefit infants and mothers in several ways, including reduced vulnerability to predators, reduced travel costs, and active food provisioning. These benefits enhance the ability of mothers to invest more of their nutritional resources into infant growth and/or returning to reproductive condition (Tardif 1994). However, the benefits that nonreproductive group members gain from helping in infant care are less clear. Steudel (2000, p. 21) reports that for mammals, "the increase in locomotor costs incurred when carrying loads of between 7 and 27% of body mass was directly proportional to the increase in mass." In the case of marmosets, for example, by 1 month of age a *single* infant weighs approximately 15% of adult body mass, and by 2 months of age it weighs approximately 25% of adult body mass. Field observations indicate that a tamarin or marmoset carrying twin offspring is often the last group member to arrive at a feeding site (Goldizen 1987; Garber pers. comm.). Moreover, when transporting infants, a tamarin or marmoset carrier is characterized by reduced leaping abilities (Schradin and Anzenberger 2001a), increased energetic costs (Tardif 1994, 1995), and a reduction in time spent foraging and feeding (Dietz et al. 1994; Sanchez et al. 1999). Given the potentially high costs to caregivers, under what conditions are they likely to obtain benefits associated with allocare?

The reproductive success of fathers and adult siblings may increase from allocare through inclusive fitness benefits (Hamilton 1964). Recent studies of tamarins (Huck et al. 2005) indicate that group members are often close relatives of infants, with a coefficient of relatedness of 0.31. Alternatively, group members may provide allocare as part of a system of reciprocity (Trivers 1971). Recently Hauser et al. (2003) have shown that callitrichids have the ability to remember individuals and past social interactions, making conditions for individuals to receive reciprocal benefits possible. In addition, infants may be attractive to all group members who provide allocare for a dominant female as a form of social currency that permits helpers to remain well integrated members of the group until either they can usurp the breeding position or find a vacant breeding position in a neighboring group (Dunbar 1995; Goldizen et al. 1996). Finally, infant transport and food sharing have been argued to function as a courtship strategy by which males improve current mating access to females (Price 1992). Unlike all other primate taxa, mating and infant care are temporally correlated in callitrichids due to the fact that females resume ovulation soon after the birth of offspring. At present there is little empirical evidence in support of a courtship hypothesis in any tamarin and marmoset species studied in captivity (Tardif and Bales 1997). However, wild female callitrichids may copulate with several adult males, and it is possible that any male forming even a temporary socio-sexual bond with a breeding female will assist in infant care.

Callimico is an anomaly among the callitrichids in that it is the only species that regularly gives birth to single rather than twin offspring (Porter and Garber 2004). Production of single rather than twin infants greatly reduces the cost of infant care, as a single callimico infant weighs approximately the same as a single callitrichid infant in a litter of two (Hartwig 1996). In callimico groups, unlike all other callitrichid groups except for lion tamarins, cooperative care does not begin immediately upon an infant's birth. In callimicos it is delayed for 10–15 days in the wild (Masataka 1981; Porter 2003) and up to 27 days in captivity (Jurke and Pryce 1994; Schradin and Anzenberger 2001b). Once allocare begins, however, group members help with infant transport and share food with the infant as in other callitrichids (Masataka 1981; Porter 2003). This help likely enables breeding females to reduce postnatal investment in infants as callimicos, like other callitrichids, are capable of post-partum ovulation and biannual births (Porter 2003).

In captivity, callimico infants grow and reach sexual maturity faster than other callitrichids (Garber and Leigh 1997). High prenatal investment by mothers and high postnatal investment by other group members in single rather than twin offspring may help to make this accelerated development possible (Promislow and Harvey 1990). Accelerated development may reduce predation risk which tends to be higher for infants than adults (Janson and van Schaik 1993), and may allow callimicos to begin reproductive careers earlier than other callitrichids.

The goal of our research was to document social relations among individuals of a wild callimico group. In particular we examined which individuals provide infant care and what type of care was provided. Based on current evidence it appears that ancestral callimicos were characterized by reproductive twinning and that the evolution of single births represents a derived trait in modern callimico populations (see review in Porter and Garber 2004). Thus, data on adult–infant interactions among callimicos offer important insights into the evolution of cooperative infant care in the only callitrichid species that appears to have lost the ability to twin. In addition, we examined adult male–female interactions to assess potential breeding strategies used by callimicos to increase their reproductive success.

4.2 Methods

Research was conducted in northern Bolivia, in the Department of the Pando, at field camp "Callimico" (UTM Zone 19, 0498129 W, 8737913 S, ca. 280 m a.s.l.), 1 km from the Tahuamanu Biological Field Station (formerly known as San Sebastian (Porter, 2001)), 2-km north of the Río Tahuamanu, 42-km east of the border of Peru, and 52-km southwest of Cobija, the capitol of the Pando (for details on the study site see Porter, 2001).

A group of callimicos was habituated during a 6-month period from March to August, 2002. In September 2002, it was possible to follow the group, but consistent data collection was not possible as the monkeys remained shy. From October through mid-November 2002, only one data collector could consistently identify group members; therefore, only data from his observations are presented for this period. From mid-November 2002 through August 2003, individuals were consistently identifiable by all data collectors; therefore, a different group member was followed each day to provide an equal number of observation days per group member per month. The group consisted of one adult female (FH) and two adult males (MR and MJ) when the habituation process began. A female infant (FS) was born into the group in August 2002, raising the group size to four.

We collected data on focal individuals using point samples (Altmann 1974) at 5-min intervals from the time the group left its sleeping site in the morning until it retired to its sleeping site at night. On some days observation times were shorter, however, due to loss of the group, torrential rain, or other factors. No data were collected during the month of January, 2003. In total 1,375 observation hours were collected across the study period.

The following data were recorded on the focal animals during observations: their general activity (e.g., resting, traveling, eating, foraging), details of their general activity (e.g., grooming while resting), and foods that were eaten. Food share was defined as one animal offering food or passively allowing food to be taken by another individual. Food rob was defined as one animal taking food from another who resisted (including loud vocalizations and rapid movement). Food types were noted using the following categories: fungus, nectar, fruit, plant exudate, arthropod, and vertebrate.

Mean monthly nearest neighbor distances were compared using paired sample *t*-tests between individuals. Chi-squared tests were conducted to compare grooming behavior among adults and diet among all group members. Wilcoxon rank tests were used to compare the distances between the infant and adults for 100 randomly selected samples throughout the study.

4.3 Results

4.3.1 Adult Relations

Callimicos form cohesive social groups. On average FH was found 2.8 m from MJ and 3.8 m from MR. Monthly mean distances of the spatial proximity of the two males to FH were not significantly different ($t = 1.693$, $n = 9$, $p = 0.129$). However, on average, in November MJ was 1.9 m closer to FH than MR, and in December MJ was 4.8 m closer to the adult female than MR (see Fig. 4.1). Monthly mean distances of the two males from FH were not significantly different when examined in paired t-tests by activity (resting $t = 1.679$, $n = 9$, $p = 0.132$; traveling $t = 1.816$, $n = 9$, $p = 0.107$; eating $t = 1.307$, $n = 9$, $p = 0.228$).

The frequency of grooming behaviors varied significantly among adults ($X^2 = 18.5$, df = 4, $p < 0.001$): FH was groomed most (8.7% of observations) and spent the least time grooming others (3.8%), whereas MR was groomed least (6.2%) and groomed others most (10.9%; see Fig. 4.2).

During September FH was observed copulating on 3 days (9/12, 9/17, 9/19) and on 1 day in October (10/15). On one occasion (9/17) the male was not identified, but on the other three occasions the male was MR. In November (11/25) FH copulated with MJ. MR was not observed to copulate during this month. Grooming data from November through January indicate that MJ groomed FH more frequently than MR, although data from other months are needed to test whether these differences are statistically significant (see Fig. 4.3). Despite the fact that copulations were

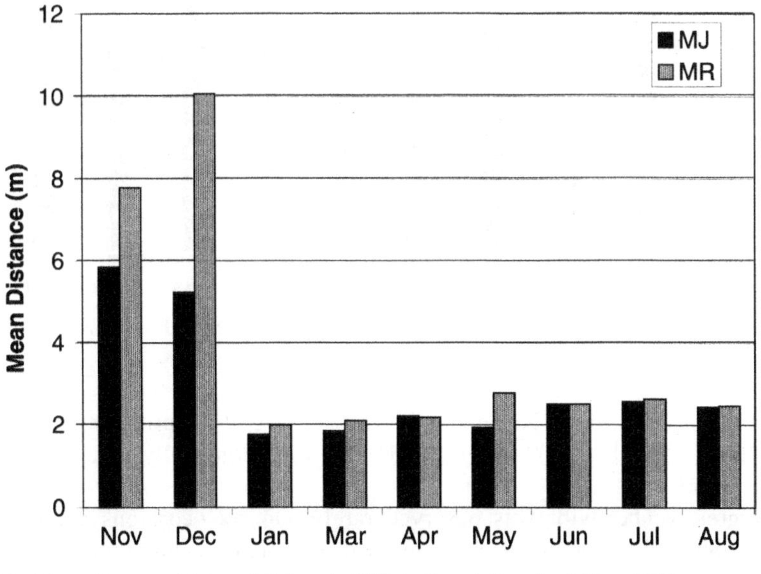

Fig. 4.1 Mean distances of the adult female (FH) to the adult males MR and MJ

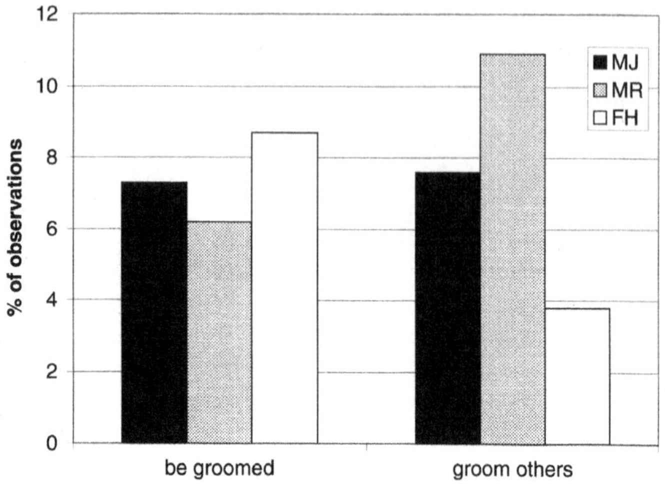

Fig. 4.2 Percentage of observations adults were engaged in grooming behavior

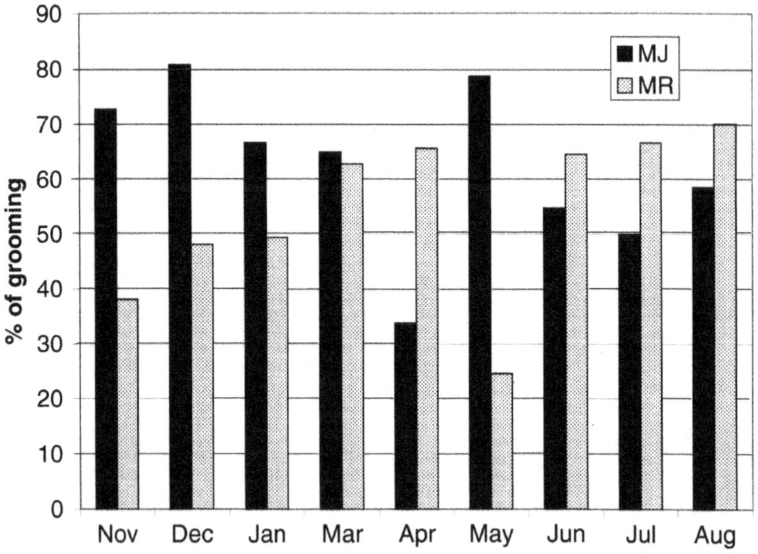

Fig. 4.3 Percentage of grooming of the adult female (FH) done by males MJ and MR

observed in October and November, FH did not give birth during the remainder of the study (December 2002–August 2003). MJ was additionally observed to mount FS in March.

Aggressive behavior was observed rarely. On six occasions adults robbed food from each other; however, none of these episodes ended with any sign of physical injury. Two incidences of aggression were observed in a non-food context.

The first case occurred in September (9/19) when FH attacked MJ. The second observation of aggressive behavior was observed on 12/24 when MJ attacked MR. No physical injuries resulted from either attack. In contrast to aggressive behaviors, cooperative behaviors were more frequent; for example, all adults shared food with one another ($n=6$), groomed one another and coordinated travel, feeding, and resting activities.

4.3.2 Adult–Infant Relations

FS was born at the end of the habituation period, but as observations were not conducted between August 10 and September 9, 2002, her exact birth date is unknown. Based on the size of FS on September 9 however, we estimate that she was born some time around August 22. Opportunistic observations in September indicate that FS was already walking independently and eating solid food offered to her by adults. Focal animal observations in November and December demonstrate that both males played with FS, and that all adults shared food with FS (including fruits, insects, and fungi) from November through January. FS continued to nurse and/or have nipple contact through December, but not in subsequent months. Thus FS was completely weaned approximately 4 months after her birth.

In terms of adult-offspring spatial proximity, mean monthly distances of each adult to the young female were very similar (MR vs. MJ: $t=0.238$, df=7, $p=0.820$; MR vs. FH: $t=-0.058$, df=6, $p=0.956$; MJ vs. FH: $t=0.836$, df=6, $p=0.435$; see Fig. 4.4), suggesting an important role for each group member in caring for the group's lone young. Mean monthly distances between adults and FS appear to vary

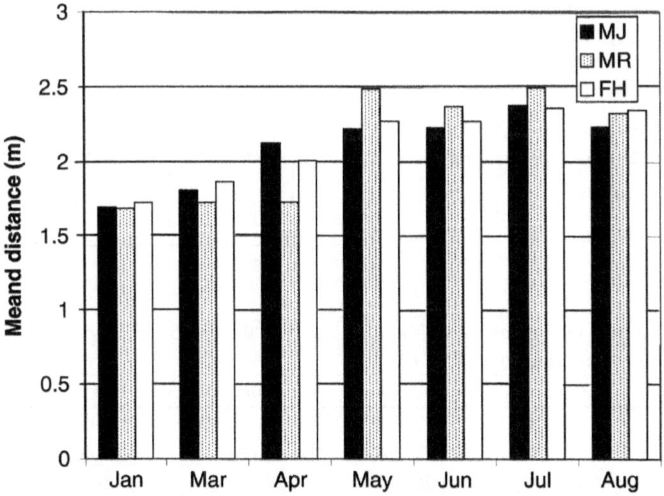

Fig. 4.4 Distance of the juvenile (FS) to adult group members (she was not transported during these months)

as she matured: adults were closer to FS in January through April (mean 1.82 m) when she was 5–8 months old, than in May through August (mean 2.34 m) when she was 9–12 months old.

In a test of 100 randomly selected points we found no differences among adults in proximity to the infant ($p>0.05$ for all comparisons: MJ-MR $Z=-0.048$; FH-MR $Z=-1.56$, FH-MJ $Z=-1.55$) when all behaviors are grouped together. Although slight, there were differences in the proximity of the adults to the offspring when examined by activity (see Table 4.1): FH traveled in closer proximity to FS than either male (FH vs. MJ: $t=3.95$, $p<0.01$, df=6; FH vs. MR $t=5.596$, $p<0.001$, df=6), and the males rested in closer proximity to FS than did FH (MJ-FH: $t=3.439$, $p<0.05$, df=6; MJ-MR; MJ-MR: $t=0.279$, $p=0.79$, df=6). No significant differences in the distances between adults and FS were found during feeding activities.

4.3.3 *Female and Male Diets*

Overall the diet of callimicos from mid-November to August was 39% fungi, 31% fruit, 14% arthropods, 14% exudates, and 2% other. Feeding records for each individual were summed by food category (see Table 4.2). The adult female FH ate the most fruits and exudates and the least amount of fungi, whereas MR ate the least amount of fruits and arthropods and the most fungi ($X^2=25.54$, df=12, $p=0.012$).

Mean nearest neighbor distances during feeding were less than 3.5 m for each group member (based on averages for individual group members) and all food types (see Fig. 4.5). When group averages were calculated by food type, the data indicate that animals are furthest apart during fruit feeding (group average distance 2.7 m), closer during arthropod and fungus eating (2.4 m), and closest during exudate feeding (2.2 m).

Table 4.1 Mean distance (m) of offspring FS to each adult during different activities

	Feed	Rest	Travel
FH	2.01	1.69	2.25
MJ	2.1	1.6	2.49
MR	2.15	1.61	2.54

Table 4.2 Feeding records for each individual listed as the number of feeding observations and in parentheses as a percentage of the total diet

	Fruit	Fungi	Arthropods	Exudates	Other
FH	58 (33)	52 (30)	24 (14)	40 (23)	2 (1)
FS	52 (24)	97 (46)	31 (15)	30 (14)	3 (1)
MJ	42 (22)	96 (50)	21 (11)	30 (17)	3 (1)
MR	63 (29)	92 (42)	34 (16)	25 (12)	2 (1)
Group total	215 (31)	337 (42)	110 (14)	125 (16)	10 (1)

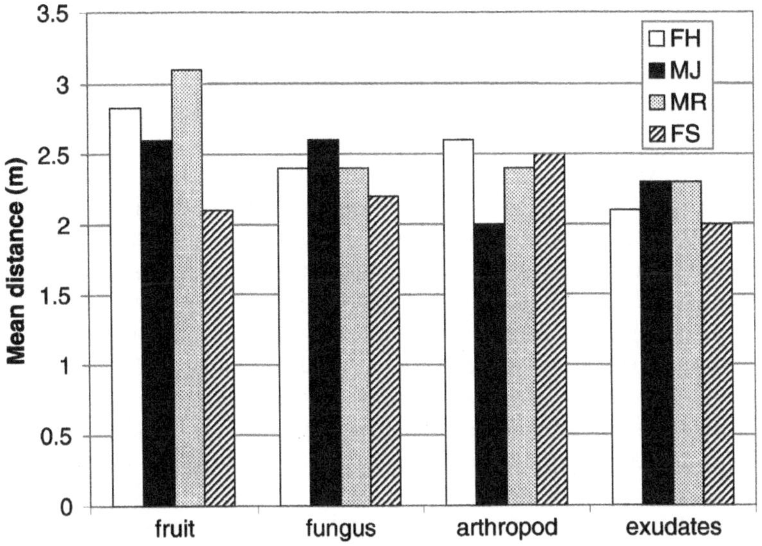

Fig. 4.5 Mean distance of individuals to the other group members while feeding on different food types

On average the offspring FS was more distant to other group members when feeding on arthropods than when feeding on plant resources (see Fig. 4.5). Although the adult female FH spent more time consuming fruits and exudates than other group members, these dietary differences did not correspond with inter-individual distances during feeding: mean nearest neighbor distances for FH during feeding were similar to those of other group members.

4.4 Discussion

These data demonstrate that despite producing singletons, callimicos like all callitrichids studied to date are characterized by spatially cohesive social groups and high levels of co-operative and co-coordinated behavior. All adults shared food with the infant and each other and remained in close proximity with one another throughout the year. As males groomed the female more than she groomed them, it appears that males invest more time and effort in developing and maintaining strong social relations than do females, a pattern that also has been documented in captive callimico groups (Carroll 1985).

One adult male (MJ) mated with the adult female and later was observed mounting her daughter. Given that studies on captive callitrichids indicate mating avoidance between fathers and daughters (Abbott 1993), it is likely that MJ either was a recent migrant into the group and/or had not previously mated with FH.

During opportunistic observations in September 2002, MR was identified as the male copulating with FH. The presence of two copulating males indicates that the breeding pattern of this group is likely to be polyandrous.

In November and December MJ was found to be in closer proximity to FH than was MR, and to groom FH more than MR. In addition, in December MJ attacked MR, although we note that this was the only observation of male-male aggression other than food robbing. These observations suggest the possibility that in polyandrous groups, one male may attempt to mate guard the female when she is in estrus. Although female callimicos show no visible (at least to the observer) signs of estrus, they mark their tails and branches with scent marks from their urine and anogenital glands, and these chemical clues may provide the males with information about the females' reproductive status (Heymann 2003). Mate guarding may represent a behavioral tactic used by some callitrichid males to maintain priority access to females during their most fertile period. Evidence of male mate guarding has been reported previously in some tamarin and marmoset species (reviewed in Huck et al. 2004). In addition, it may be possible that an older or dominant adult male can reduce fertility in subordinate males through hormonal suppression of sperm production (Abbott 1993; Garber 1997; Soini 1982). We plan to collect genetic samples from callimico groups in the future to determine whether males who mate guard are also characterized by increased reproductive success.

All adults in the group helped transport, share food with, and groom the infant. All three group members maintained close proximity to the infant, with the adult female traveling closest, and the adult males resting closest. Spatial proximity to the infant decreased over the study, as is expected for callimicos who reach sexual maturity by 57 weeks in captivity (Beck et al. 1982; Carroll 1982; Dettling and Pryce 1999).

Differences were found in the diets of group members. The breeding female devoted more time consuming fruits and exudates and less time consuming fungi than any other group member. Nutritional analyses indicate that fungi consumed by callimicos are a low quality resource, as they are high in fiber and low in protein and sugars (Hanson et al. 2006), and therefore the adult female appeared to be consuming a higher quality diet than other group members. This strategy may be necessary to overcome the high energetic costs of gestation and lactation (Lee and Bowman, 1995).

Overall, these data demonstrate that callimicos are found in highly cooperative groups, with extensive allocare, and polyandrous matings. Although the presence of these behaviors in other callitrichids has generally been attributed to the high costs to females of raising twins, this explanation is insufficient to account for their appearance in callimico groups, and suggests that they additionally serve to maximize infant and juvenile survivorship in non-twinning species. Further studies of callimicos' social behavior are needed to test the effects of communal care on their growth and survival rates to help determine how the return to a pattern of single births influences reproductive and life history strategies in this species.

Acknowledgments This research was supported by the Chicago Zoological Society, the Primate Action Fund of Conservation International, and the University of Illinois Urbana-Champaign.

Thanks to Edilio Nacimento, Julio Flores, and Fransisco Brilhante, for their help with habituation, Edilio Nacimento, Ruddy Cuadiay, Ruht Cuadiay, Carlos Amotari, and Erica Berloz, for their help with data collection, and Sarah Garner for her help with data entry. PAG thanks Sara and Jenni for their insights on more and less effective infant care strategies.

References

Abbott D (1993) Social conflict and reproductive suppression in marmoset and tamarin monkeys. In: Mason W, Mendoza S (eds) Primate social conflict. State University of New York Press, Albany, pp 331–372

Abbott D, Barrett J, George L (1993) Comparative aspects of the social suppression of reproduction in female marmosets and tamarins. In: Rylands AB (ed) Marmosets and tamarins: systematics, behaviour and ecology. Oxford University Press, Oxford, pp 152–163

Altmann J (1974) Observational study of behavior: sampling methods. Behavior 49:227–265

Beck B, Anderson D, Ogden J, Rettberg B, Brejla C, Scola R, Warneke M (1982) Breeding the Goeldi's monkey at Brookfield Zoo, Chicago. Intl Zoo Ybk 22:106–114

Carroll J (1982) Maintenance of the Goeldi's monkey at Jersey Wildlife Preservation Trust. Intl Zoo Ybk 22:101–105

Carroll J (1985) Pair bonding in the Goeldi's Monkey. Dodo J Wildl Preserv Trust 22:57–71

Carroll J (1988) The stability of multifemale groups of Goeldi's monkey, *Callimico goeldii* in captivity. Dodo 25:37–43

Dettling A, Pryce C (1999) Hormonal monitoring of age at sexual maturation in female Goeldi's Monkeys (*Callimico goeldii*) in their family groups. Am J Primatol 48:77–83

Dietz J, Baker A (1993) Polygyny and female reproductive success in golden lion tamarins, *Leontopithecus rosalia*. Anim Behav 46:1067–1078

Dietz J, Baker A, Miglioretti D (1994) Seasonal variation in reproduction, juvenile growth, and adult body mass in golden lion tamarins (*Leontopithecus rosalia*). Am J Primatol 34:115–132

Digby L, Saltzman W (this volume) Balancing cooperation and competition in callitrichid primates: examining the relative risk of infanticide across species. In: Ford SM, Porter LM, Davis LC (eds) The smallest anthropoids: the marmoset/callimico radiation. Springer, New York, pp 135–153

Dunbar R (1995) The mating system of callitrichid primates: II. The impact of helpers. Anim Behav 50:1071–1089

French J, Inglett B (1989) Female-female aggression and male indifferences in response to unfamiliar intruders in lion tamarins. Anim Behav 37:487–497

Garber P (1994) Phylogenetic approach to the study of tamarin and marmoset social systems. Am J Primatol 34:199–219

Garber P (1997) One for all and breeding for one: cooperation and competition as a tamarin reproductive strategy. Evol Anthropol 5:187–199

Garber P, Leigh S (1997) Ontogenetic variation in small-bodied New World primates Implications for patterns of reproduction and infant care. Folia Primatol 68:1–22

Garber P, Moya L, Malaga C (1984) A preliminary study of the moustached tamarin monkey (*S. mystax*) in northeastern Peru: questions concerned with the evolution of a communal breeding system. Folia Primatol 42:17–32

Goldizen A (1987) Facultative polyandry and the role of infant-carrying in wild saddle-backed tamarins (*Saguinus fuscicollis*). Behav Ecol Sociobiol 20:99–109

Goldizen A (1990) A comparative perspective on the evolution of tamarin and marmoset social systems. Intl J Primatol 11:63–83

Goldizen A, Mendelson J, van Vlaardingen M, Terborgh J (1996) Saddle-back tamarin (*Saguinus fuscicollis*) reproductive strategies: evidence from a 13-year study of a marked population. Am J Primatol 38:57–83

Hamilton W (1964) The genetical evolution of social behavior II. J Theoret Biol 7:17–32

Hanson A, Hall M, Porter L, Lintzenich B (2006) Composition and nutritional characteristics of fungi consumed by *Callimico goeldii* in Pando, Bolivia. Intl J Primatol 27:323–346
Hardie S (1995) Do subordinate female *Callimico* disperse from their social groups? Folia Primatol 64:192–195
Hartwig W (1996) Perinatal life history traits in New World monkeys. Am J Primatol 40:99–130
Hauser M, Chen M, Chen F, Chuang E (2003) Give unto others: genetically unrelated cotton-top tamarin monkeys preferentially give food to those who altruistically give food back. Proc R Soc Lond B 270:2363–2370
Heymann E (2003) Scent marking, paternal care, and sexual selection in callitrichines. In: Jones C (ed) Sexual selection and reproductive competition in primates: new perspectives and direction. Am J Primatol, Oklahoma, pp 305–325
Heymann E, Soini P (1999) Offspring number in pygmy marmosets, *Cebuella pygmaea*, in relation to group size and the number of adult males. Behav Ecol Sociobiol 46:400–404
Huck M, Löttker P, Heymann E (2004) Proximate mechanisms of reproductive monopolization in male moustached tamarins (*Saguinus mystax*). Am J Primatol 64:39–56
Huck M, Löttker P, Böhle U, Heymann E (2005) Paternity and kinship patterns in polyandrous moustached tamarins (*Saguinus mystax*). Am J Phys Anthropol 127:449–464
Janson C, van Schaik C (1993) Ecological risk aversion in juvenile primates: slow and steady wins the race. In: Pereira M, Fairbanks L (eds) Juvenile primates: life history, development, and behavior. Oxford University Press, New York, pp 57–74
Jurke M (2002) Reproductive biology of nonhuman primates. Evol Anthropol 11:186–189
Jurke M, Pryce C (1994) Parental and infant behavior during early periods of infant care in Goeldi's monkey, *Callimico goeldii*. Anim Behav 48:1095–1112
Koenig A (1995) Group size, composition, and reproductive success in wild common marmosets (*Callithrix jacchus*). Am J Primatol 35:311–317
Lee P, Bowman J (1995) Influence of ecology and energetics on primate mothers and infants. In: Pryce C, Martin R, Skuse D (eds) Motherhood in human and nonhuman primates. Basel, Karger, pp 47–58
Masataka N (1981) A field study of the social behavior of Goeldi's monkeys (*Callimico goeldii*) in north Bolivia: I Group composition, breeding cycle, and infant development: Kyoto University overseas research reports. Kyoto University, Kyoto, Japan, pp 23–32
Nievergelt C, Digby L, Ramakrishnan U, Woodruff D (2000) Genetic analysis of group composition and breeding system in a wild common marmoset (*Callithrix jacchus*) population. Intl J Primatol 21:1–20
Porter L (2001) *Callimico goeldii* and *Saguinus*: dietary differences between sympatric callitrichines in northern Bolivia. Am J Phys Anthropol 22:961–992
Porter L (2003) Social organization of wild groups of *Callimico goeldii* in northwestern Bolivia. Am J Phys Anthropol 120(S36):170 (abstract)
Porter L, Garber P (2004) Goeldi's monkeys: a primate paradox? Evol Anthropol 13:104–115
Price E (1992) The benefits of helpers: effects of group and litter size on infant care in tamarins (*Saguinus oedipus*). Am J Primatol 26:179–190
Promislow D, Harvey P (1990) Living fast and dying young: a comparative analysis of life-history variation among mammals. J Zool Lond B 220:417–437
Pryce C, Pastorini J, Vasarhelyi K, Christen A (2002) Bio-behavioral description of social and reproductive relationships in captive Goeldi's monkeys. Evol Anthropol 11:190–194
Sanchez S, Pelaes F, Gil-Burmann C, Kaumanns W (1999) Costs of infant-carrying in the cotton-top tamarin. Am J Primatol 48:99–111
Schradin C, Anzenberger G (2001a) Costs of infant carrying in common marmosets, *Callithrix jacchus*: an experimental analysis. Anim Behav 62:289–295
Schradin C, Anzenberger G (2001b) Infant carrying in family groups of Goeldi's monkeys (*Callimico goeldii*). Am J Primatol 53:57–67
Soini P (1982) Ecology and population dynamics of the pygmy marmoset, *Cebuella pygmaea*. Folia Primatol 39:1–21

Steudel K (2000) The physiology and energetics of movement: effects on individuals and groups. In: Boinski S, Garber PA (eds) On the Move: how and why animals travel in groups. The University of Chicago Press, Chicago, pp 9–23

Tardif S (1994) Relative energetic costs of infant care in small-bodied Neotropical primates and its relation to infant-care patterns. Am J Primatol 34:133–143

Tardif S (1995) The bioenergetics of parental behavior and the evolution of alloparental care in marmosets and tamarins. In: Solomon N, French J (eds) Coopertive breeding in mammals. Cambridge University Press, Cambridge, pp 11–33

Tardif S, Bales K (1997) Is infant-carrying a courtship strategy in callitrichid primates. Anim Behav 53:1001–1007

Trivers RL (1971) The evolution of reciprocal altruism. Q Rev Biol 46:35–57

Yamamoto M (1993) From dependence to sexual maturity: the behavioral ontogeny of Callitrichidae. In: Rylands A (ed) Marmosets and tamarins: systematics, behaviour and ecology. Oxford University Press, Oxford, pp 235–254

Yamamoto ME, Arruda M de F, Alencar AI, de Sousa MBC, Araújo A (this volume) Mating systems and female-female competition in the common marmosets, *Callithrix jacchus*. In: Ford SM, Porter LM, Davis LC (eds) The smallest anthropoids: the marmoset/callimico radiation. Springer, New York, pp 119–133

Ziegler T, Snowdon C, Warneke M (1989) Postpartum ovulation and conception in Goeldi's monkey, *Callimico goeldii*. Folia Primatol 52:206–210

Chapter 5
Genetic Structure Within and Among Populations of the Common Marmoset, *Callithrix jacchus*: Implications for Cooperative Breeding

Christopher G. Faulkes, M.F. Arruda, and M.A.O. Monteiro da Cruz

Abstract Cooperative breeding and a reproductive division of labor are characteristic features of callitrichid primates. Of key importance in understanding this kind of social strategy is knowledge of the kin structure of groups, yet until recently the patterns of relatedness within and among social groups have remained unclear, even in the best-studied species such as the common marmoset *Callithrix jacchus*. Of particular interest is the role of kin selection in helping behavior and group dynamics, and the application of molecular genetic techniques to investigate group structure, dynamics, and parentage is now unequivocally addressing such issues. Both microsatellite genotyping and mitochondrial DNA sequence analysis have revealed that while social groups may often be composed of extended family groups, considerable genetic heterogeneity may also arise as a result of inter-group migrations. In one study, a group of nine animals was shown to contain individuals from five matrilines. The occurrence of groups of mixed lineages raises interesting questions about potential reproductive conflicts of interest and the extent of kin-selected altruism in the evolution and maintenance of cooperative behavior in this species.

Resumen La crianza cooperativa y una división reproductiva de labores son rasgos característicos de los primates calitrícidos. El conocimiento de la estructura de parentesco de los grupos es de importancia clave en la comprensión de este tipo de estrategia social; sin embargo, hasta hace poco tiempo los patrones de las relaciones de parentesco dentro y entre los grupos sociales han permanecido poco claros, aún en las especies mejor estudiadas como es el caso de la marmoseta común *Callithrix jacchus*. De particular interés es el papel que juega la selección por parentesco en el comportamiento de ayuda y en las dinámicas de grupo. La aplicación de las técnicas de la genética molecular para investigar la estructura del grupo, las dinámicas y el parentesco claramente se está dirigiendo en esa dirección. Tanto el genotipado de microsatélites como el análisis de secuencias del ADN mitocondrial han revelado que mientras los grupos sociales pueden muchas veces estar compuestos por

C.G. Faulkes (✉)
School of Biological and Chemical Sciences, Queen Mary, University of London, London, UK
e-mail: c.g.faulkes@qmul.ac.uk

S.M. Ford et al. (eds.), *The Smallest Anthropoids*, Developments in Primatology: Progress and Prospects, DOI 10.1007/978-1-4419-0293-1_5,
© Springer Science+Business Media, LLC 2009

grupos de familias completas, considerable heterogeneidad genética podría también presentarse como resultado de migraciones intergrupales. En un estudio, un grupo de nueve animales mostró tener individuos provenientes de cinco matrilíneas. La ocurrencia de grupos con linajes mixtos conlleva preguntas interesantes sobre los potenciales conflictos de interés reproductivos y el grado de altruismo por selección de parentesco en la evolución y mantenimiento de los comportamientos de cooperación en esta especie.

Resumo A reprodução cooperativa com a divisão do cuidado à prole constitui uma característica dos primatas calitriquídeos. O conhecimento da estrutura de parentesco dos grupos é de importância chave para a compreensão deste tipo de estratégia social, mas até recentemente, os padrões de parentesco dentro e entre os grupos sociais não eram conhecidos mesmo em espécies bem estudadas como o sagüi comum, *Callithrix jacchus*. É de particular interesse o papel da seleção de parentesco na participação de ajudantes na dinâmica dos grupos, e a aplicação de técnicas de genética molecular para investigar a estrutura, a dinâmica e o parentesco dentro do grupo está atualmente voltada para estas questões. As análises de genotipagem por microsatélite e seqüenciamento de DNA mitocondrial têm revelado que embora os grupos sejam compostos frequentemente por famílias ampliadas, pode haver uma considerável heterogeneidade genética resultante de migrações entre grupos. Em um estudo, foi encontrado que em um grupo de nove animais, os indivíduos apresentavam cinco linhagens maternas. A ocorrência de grupos com diferentes linhagens levanta questões interessantes sobre conflitos reprodutivos potenciais e os níveis de altruísmo voltado para os animais aparentados, na evolução e manutenção do cuidado cooperativo nesta espécie.

5.1 Cooperative Breeding in the Callitrichidae

In mammals, numerous divergent taxa have evolved cooperative breeding strategies in response to a variety of ecological constraints (for review see Solomon and French 1997). These systems are characterized by group living and a reproductive division of labor (high reproductive skew). Non-reproductive individuals then assist in some way with rearing the offspring of the breeding animals. There may also be overlap of generations within the social group and, in extreme cases, some cooperatively breeding mammals such as African mole-rats have been likened to eusocial insects like bees, ants and wasps (for review see Bennett and Faulkes 2000). Marmosets and tamarins of the family Callitrichidae are unique among primates in exhibiting this kind of social organization. Within the family, cooperative breeding strategies are widespread, and virtually all of the 40 or so species are characterized by small territorial groups of approximately 4–15 individuals, where reproduction is monopolized by one or a small number of dominant individuals of each sex (for review see Tardif 1997; French 1997).

The common marmoset (*Callithrix jacchus*) and its sister species the black-eared marmoset (*Callithrix penicillata*) are two of the most adaptable of the Atlantic forest callitrichid primates, exploiting relatively hostile, seasonal habitats such as the *caatinga* (semi-arid thorn scrub). The trait of tree gouging and gum feeding is most highly developed in these species, and this increased specialization undoubtedly enables them to survive in habitats where fruit may be scarce for long periods of time (Rylands 1984; Ferrari 1993). The social and reproductive system of common marmosets, *Callithrix jacchus*, is typical of the Callitrichidae. Groups contain 3–15 individuals, and usually a single dominant female breeds with a specific breeding male (or males). However, plural breeding among females can occur (Digby and Ferrari 1994; Nievergelt et al. 2000), and mating systems are known to be variable both within and between callitrichid species. Polyandry, polygyny and monogamy have all been reported from behavioral observations (Ferrari and Digby 1996; Ferrari this volume, Chap. 8; Digby and Saltzman this volume, Chap. 7; Yamamoto et al. this volume, Chap. 6).

Irrespective of the mating system, twin offspring are usually produced in callitrichids, and the main cooperative role of non-reproductive helpers in the group is to assist in the care of the breeding female's offspring. This is principally by sharing the burden of carrying the relatively "bulky" twin infants around their arboreal habitat, although cooperation in the form of food provisioning may also occur (Tardif et al. 1993). While breeding animals gain from this behavior, non-breeding helpers forego their own reproduction and also incur energetic costs from helping (Sánchez et al. 1999; Tardif 1997; Bales et al. 2000). Tardif (1997) suggested four main areas where non-breeders may accrue benefits to offset the costs of alloparental behavior. Firstly, by helping, these non-breeders may improve their infant care skill. Later, should they breed themselves, their own offspring survival may be improved as a result of this care-giving experience. This may be true in some cases, and there is evidence that the offspring of primiparous mothers with previous alloparenting experience have increased survival (Tardif 1997).

However, it remains unclear just how many non-breeding animals eventually reproduce, i.e. what the skew in lifetime reproductive success is among callitrichids (ultimately genetic analysis may clarify this issue), and therefore how generally this explanation may apply. Secondly, by staying within a group and helping dominants, rather than dispersing and attempting reproduction elsewhere, non-breeders may increase their opportunities for inheritance of a breeding position. There is some evidence for this as a strategy in the saddle-back tamarin, *Saguinus fuscicollis* (Goldizen et al. 1996). A third suggestion that lacks any clear empirical evidence is that infant carrying functions as a form of courtship, and for potential breeding males this may facilitate increased access to the breeding female (Tardif and Bales 1997). Finally, in a situation where helpers are related to breeders, simple kin selection and inclusive fitness arguments may be invoked to explain cooperation. Thus, determining the genetic structure and patterns of relatedness within and among social groups has become crucial in understanding the role that kin selection may play in the alloparental behavior observed in marmosets and tamarins.

5.2 Genetic Studies of the Callitrichidae

5.2.1 *Background and Difficulties*

As with other species where direct observation of reproductive behavior is difficult (e.g. bats; Burland and Worthington Wilmer 2001), application of molecular genetic techniques to determine unambiguously mating system and group structure would appear to be an ideal approach for the study of small arboreal primates. It is therefore surprising that, to date, very little has been published on the population genetics of the Callitrichidae. This may partly be due to a lack of medium to long-term field studies that can offer sufficient insight into the behavioral correlates of group kin structure, patterns of mating, and dispersal that can be estimated from genetic data. Another problem is the paucity of suitable polymorphic markers. Early attempts to investigate and quantify exactly the mating system in *C. jacchus*, particularly paternity, using minisatellite DNA fingerprinting were hampered by technical difficulties. One problem is that *in utero*, marmoset twins share placental blood, and the subsequent exchange of bone marrow cells produces siblings that are hemopoietic chimeras (Benirschke et al. 1962). As a result, DNA fingerprints generated from DNA derived from tissue rich in blood may be identical (this is not an issue with mitochondrial DNA because it is implicit in the analysis that all offspring share the maternal haplotype). A second problem is that multi-locus minisatellite DNA fingerprints have proved lacking in sufficient variation to be diagnostically useful (Dixson et al. 1992), although recently Signer et al. (2000) have developed new probes that detect increased variability.

Currently the technique of choice for "DNA fingerprinting" is microsatellite genotyping. However, homologous markers or those from closely related species are required, and currently primers have been published only for three callitrichid species: the common marmoset, *C. jacchus* (Nievergelt et al. 1998), the golden lion tamarin, *Leontopithecus rosalia* (Grativol et al. 2001), and the black lion tamarin, *Leontopithecus chrysopygus* (Perez-Sweeney et al. 2005). Microsatellite markers developed for woolly monkeys (*Lagothrix lagotricha*; Di Fiore and Fleischer 2004) and the brown capuchin monkey (*Cebus apella*; Escobar-Páramo 2000) have limited efficacy in the Callitrichidae. In wild populations the power of microsatellite markers may be reduced if allelic variability is generally low. The studies undertaken so far indicate that this may be the case for *C. jacchus* as previously noted by Nievergelt et al. (2000). In captive groups of *C. jacchus* ($n=98$ animals) genotyped at 13 microsatellite loci, the mean ± SEM number of alleles was 6.38 ± 0.93 (range 2–12), compared with data from one field site at Nisia Floresta, NE Brazil, where this value was reduced to 3.09 ± 0.37 ($n=40$, over 11 loci, range 2–6; Nievergelt et al. 1998). A separate study that looked at 78 animals from Nísia Floresta, Tapacurá and Recife in Brazil found on average 2.85 ± 0.69 alleles over 13 loci (range 1–9; CG Faulkes, MF Arruda, and MAO da Cruz unpublished data). The lower allelic diversity observed at the field sites may be due to factors such as founder effects, reduced gene flow between fragmented populations, and the derivation of the captive populations from a wider geographical source.

While a number of molecular phylogenetic analyses of the Callitrichidae have been published (e.g. for the genus *Callithrix*: Tagliaro et al. 1997; Sena et al. 2002; Gonçalves et al. this volume, Chap. 11), little is known at the level of intra-specific population genetics, despite the importance in furthering our understanding of their social behavior and the implications for conservation. The latter has been investigated in two recent studies: Gonçalves et al. (2003, this volume, Chap. 11) used two microsatellite loci developed for *C. jacchus* to examine the effects of habitat fragmentation on the genetic variability of silvery marmosets, *Mico argentatus*, while Grativol et al. (2001) used four species-specific microsatellite loci to investigate genetic variation in fragmented populations of the golden lion tamarin, *L. rosalia*. These were purely conservation genetic studies that did not seek to address questions related to the behavioral ecology of the respective species (but see Gonçalves et al. this volume, Chap. 11). Some new work on wild populations of red-bellied tamarins (*Saguinus labiatus*), again using microsatellite genotyping, has provided some preliminary insights into their reproductive strategies (Suárez et al. 2004). Their initial results support the hypothesis that the mating system, although variable, is most commonly monogamous, and that the mostly male helpers are likely to accrue inclusive fitness benefits due to their relationship with the breeder male or female.

Perhaps the best understood of the Callitrichidae in terms of the proximate control of their reproduction and social behavior is the common marmoset, *C. jacchus*, and a small number of studies have now investigated their population genetics for the first time. The remainder of this chapter will focus on the molecular ecology of this species within and among populations, and what the data obtained so far can tell us about the evolution and maintenance of their cooperative breeding system.

5.2.2 Genetic Structure Among Populations of **Callithrix jacchus**

Due to their increased specializations for gum feeding, common marmosets are able to exploit a variety of habitats, including more marginal biomes such as *caatinga* and this factor may also influence patterns of dispersal, gene flow, and genetic structuring of populations. As a consequence of their adaptability, these species are perhaps less likely than some other callitrichids to suffer from population subdivision as a result of the fragmentation of the primary forest habitat (some *Saguinus* species are also thought to do well in the secondary forest). However, the unusual social system of callitrichids would be predicted to have a profound effect on the spatial and geographic partitioning of genetic variation (for review see Ross 2001).

In an intra-specific study of mitochondrial control region sequence variation in *C. jacchus*, Faulkes et al. (2003) investigated 59 individuals mainly from two geographically distinct wild populations in NE Brazil (Nísia Floresta, Rio Grande do Norte and Tapacurá, Pernambuco). Overall, there was high haplotype diversity but low genetic divergence observed at this locus (approximately 1,100 base pairs were sequenced for each individual). This pattern is consistent with the hypothesis that *C. jacchus* is a recent radiation, the small number of mutations separating the haplotypes indicative of recent common ancestry. Similar low levels of sequence divergence among callitrichid primates have also been reported following analysis of nuclear b_2-microglobin sequences (Canavez et al. 1999). However, despite the small genetic distances observed, a consistent geographic trend in the spatial distribution of mitochondrial control region haplotypes was apparent. Analysis of molecular variation (AMOVA) revealed significant genetic structuring at the population level (see Fig. 5.1).

Phylogenetic trees generated from the sequence data contained shallow branches with allopatric haplotypes (i.e. matrilines are restricted to a particular geographic area), and are consistent with a "Category III" phylogeographic pattern (Avise 2000). This Category III pattern supports a model where although recent gene flow has been limited and resulted in geographic partitioning, the populations were in contact until relatively recently.

5.2.3 Kin Structure Within Social Groups of **Callithrix jacchus**

Initial studies published on the composition and dynamics of common marmoset groups were based on intensive and medium-term field observational studies which lacked a behavioral component. In most cases data were collected using mark-release recapture approaches followed by scan or focal animal sampling to monitor the movements and behaviors of individuals. Again, studies have mainly centered on established field stations in NE Brazil at Nísia Floresta and Tapacurá and originally set out to test the hypothesis that groups were stable, monogamous families with limited inter-group transfer of individuals. The earliest reports from Tapacurá were from studies that were undertaken in the late 1970s and 1980s: Stevenson and Rylands (1988) followed two groups for six months in 1979 and observed no immigration, while two years later Hubrecht (1984) followed two groups (size varied from 3–13 individuals) for five months and recorded the disappearance of one female and the possible immigration of one female. Further fieldwork at Tapacurá in 1986 challenged the monogamous family hypothesis: Scanlon et al. (1988) studied five groups ranging in size from six to 14 individuals and surprisingly found a 50% turnover in group membership over a six-month period, with adult males moving most often. Furthermore, there was also evidence of the presence of two breeding females simultaneously within a group, and the switching of breeding status among adult females within groups. Digby and Barreto (1993) observed three groups with five to 15 individuals at Nísia Floresta field station, all of which contained two breeding females. However, over the study period of 12–18 months, while births, deaths, and emigrations were recorded, there were no immigrations into the groups under investigation. Digby and Barreto concluded from their study that *C. jacchus* live in stable, extended family groups composed of breeders and their offspring, parents and siblings. Follow-up studies by Arruda et al. (2005) confirmed polygyny in two of these groups, while one became monogamous, and immigration into all three groups was also observed (see also Sousa et al. this volume, Chap. 9).

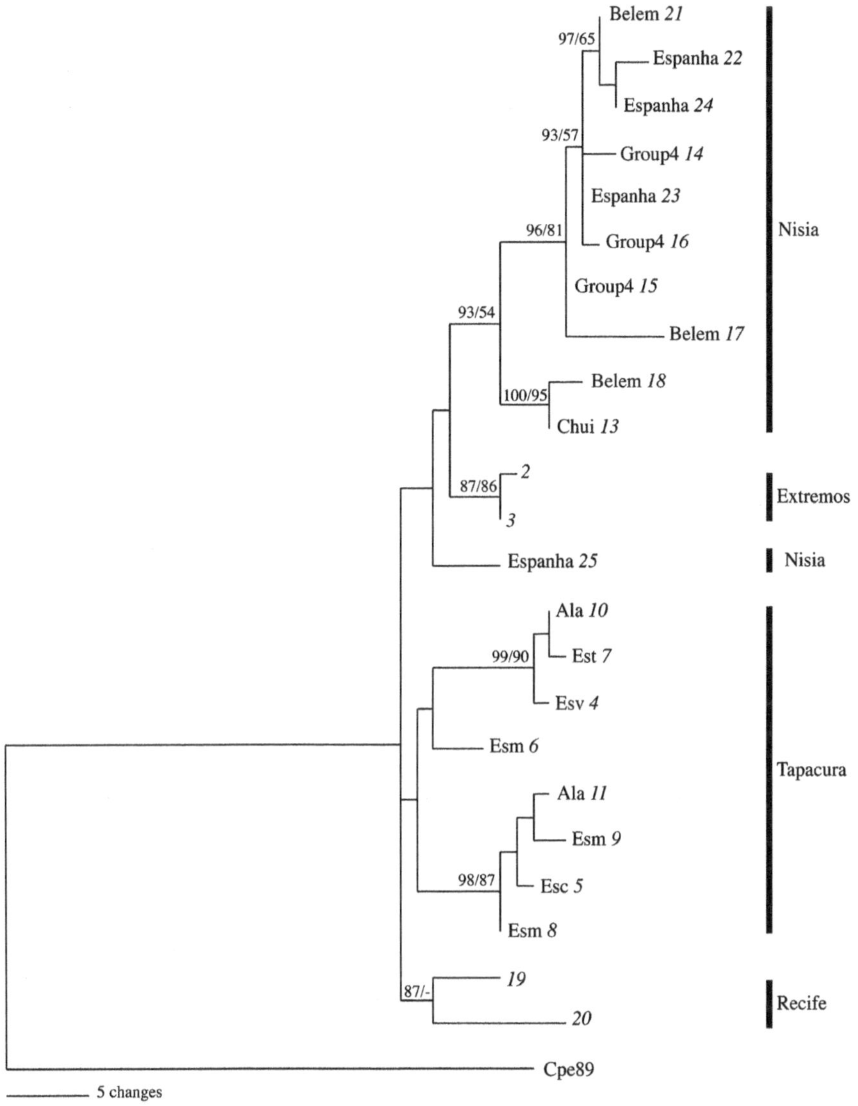

Fig. 5.1 Phylogenetic relationships of 23 *Callithrix jacchus* control region haplotypes and one outgroup species, *C. penicillata* (Cpe 89). Terminal nodes are labeled with the colony name followed by the haplotype, designated by a number in *italics*. Geographical locations are labeled and grouped with vertical bars. The phylogram was generated by the maximum likelihood option in PAUP* using a HKY85+G+I model of sequence evolution (see Faulkes et al. 2003) with molecular clock enforced, tree score=2,259.5. Numbers on nodes are the bootstrap support values for clades, from Bayesian/parsimony analysis, respectively. From Faulkes et al. (2003), used with permission from Blackwell Publishing

The final field study of significance to be published focused on a different field site near Tapacurá, in the Dois Irmãos forest fragment near Recife. Mendes Pontes and Monteiro da Cruz (1995) followed a main study group and four adjacent groups over a three-year period. Group sizes were similar to those reported in previous studies (5–15 individuals), with the main group composed of 10–15 individuals. As with the earlier study of Scanlon et al. (1988), a high rate of inter-group transfers was recorded, and these included adults, sub-adults, and juveniles of both sexes. Altogether, 23 animals left the focal group, and of these, seven transfers to the adjacent study groups were confirmed (emigration as a reproductive strategy is fully discussed in Sousa et al. this volume, Chap. 9). There was evidence for two simultaneously breeding females as well at this field site, confirming that plural breeding of females may be widespread, and that while offspring may most commonly be conceived and reared by a single dominant female in a group, other females are sexually active. This is also evident in a follow-up to the study of Digby and Barreto (1993), in which Digby (1996) elaborates further on the sexual behavior of the three groups studied at Nísia. While within-group sexual behavior was normally restricted to the two breeding females and single dominant male present, out of 101 mounts and copulations that were observed, 24 of these involved the pairing of both breeding and non-breeding group members from different groups. Interestingly, where these inter-group copulations involved breeding females, none occurred at a time when conception was possible, probably as a result of mate guarding from the dominant male of the group. Furthermore, the only reproductive females that copulated were the individuals from polygynous groups, perhaps suggesting that these and non-breeding females involved in extra-group copulations were investigating neighboring groups for future breeding partners (Arruda et al. 2005; Sousa et al. this volume, Chap. 9; Yamamoto et al. this volume, Chap. 6).

Collectively, these behavioral studies indicate a group structure based on a cooperatively breeding extended family with overlapping generations, which may be stable over time, yet subject to rapid turnover at certain periods, for example, after the loss of a breeding female (Lazaro-Perea et al. 2000). Both successful and unsuccessful attempts at plural breeding among females are also clearly documented. Within this dynamic system, a strategy of cooperative breeding with high reproductive skew and altruistic behavior is maintained in both wild and captive populations (see Sect. 5.2.4 below). Understanding how group structure influences social control of reproduction at a proximate level requires a more detailed understanding of group structure than behavioral observations can provide, and quantifying relatedness and parentage using genetic markers. While pedigrees could to some extent be constructed given enough behavioral observations, and maternity confirmed with some certainty (with single breeding females), assignment of paternity with any degree of confidence is almost impossible based exclusively on behavioral observations, and exact relationships among adults are difficult to determine without genetic data.

To date, only two studies have tackled the issue of group composition and relatedness in *C. jacchus* using a more direct genetic approach. Nievergelt et al. (2000) used microsatellite markers to quantify relatedness in the three groups containing two breeding females originally studied at Nísia by Digby and Barreto (1993), together with some individuals from neighboring groups. The results support their behavioral observations suggesting that most group members are close relatives: infants and non-breeding adults were closely related to the breeding female(s), with values of relatedness (R) not significantly different from 0.5 (the mean value for first-order relatives), indicating mother–daughter and sister relationships for these individuals. Relatedness among adult males within groups was significantly lower at approximately $R=0.2$, indicating that immigration had occurred at some point, although no examples were observed by Digby and Barreto (1993) in their behavioral study. This emphasizes the power of genetic analysis in revealing both current and historical events. The relatedness between dominant breeding pairs and between pairs that were observed in other intra-group and extra-group copulations was low (on average R was less than 0.1), confirming that outbreeding was occurring.

The second study by Faulkes et al. (2003) used mitochondrial DNA (mtDNA) control region sequence analysis to examine the distribution of matrilines (corresponding to individual mtDNA haplotypes) in wild populations of *C. jacchus* at Nísia and Tapacurá. At Nísia, two of the groups were the same as those studied by Digby five years earlier, with a few of the original animals still present. At the population level, the mtDNA haplotype data provides a fascinating insight into the dynamics of group structure. Out of seven groups where samples were obtained from the breeding pair, the reproductive males and females were found to have different haplotypes in five cases, indicating that assortative mating for different matrilines (outbreeding) was occurring (Table 5.1), and supporting the study of Nievergelt et al. (2000). There were however, two exceptions: one occurred in Group 4 at Nísia, where the reproductive female (Grecia) had originally been breeding with the male Gandi, who was from a different matriline. However, Gandi was expelled from the group by Goeth, another original male group member having the same haplotype as Grecia, and Goeth then took over as the new breeding male. Although sharing the same matriline, Goeth and Grecia may not necessarily have been close relatives risking inbreeding depression (see below), and they went on to produce over 20 offspring. The second case was found at Tapacurá, in the group Esparancins,

Table 5.1 Genetic structure of 10 wild groups of common marmoset monkeys (*NF* Nísia Floresta field site; *T* Tapacurá; n=group size over period of study from 1995 to 1997). Relatedness estimated from microsatellite genotypes (CG Faulkes, MF Arruda, and MAO da Cruz unpublished data), values in parentheses correspond to estimates of intra-group relatedness from Nievergelt et al.'s (2000) study over the period 1991–1992; mtDNA haplotypes were determined from control region sequences (Faulkes et al. 2003)

Group	Relatedness	mtDNA haplotypes
Paraiba (NF; $n=7$)	0.53	–
Chui (NF; $n=9$)	0.59 (0.29)	2
Espanha (NF; $n=8$)	0.31	4
Group 4 (NF; $n=9$)	0.29	5
Belem (NF; $n=10$)	0.21 (0.47)	2
Esmirradins (T; $n=10$)	0.32	4
Escomprimidins (T; $n=5$)	–	2
Espivitadins (T; $n=8$)	–	2
Esparancins (T; $n=12$)	–	1
Estovadins (T; $n=7$)	–	3

where all individuals carried the same haplotype. Again, some of these may have been more distant relatives. For example, depending on the mutation rate of the mtDNA, it is possible for the same haplotype to be present in cousins once or more removed, second cousins etc. Relationships will be even more diluted in cases where the breeding female has offspring fathered by more than one male.

Out of the nine social groups studied in detail, five contained more than two haplotypes, indicating that in addition to the unrelated breeding pair, other individuals from different matrilines were also present (Table 5.1). One group containing two haplotypes (Escomprimidins) may well have had a third, as the breeding female was not sampled. In the extreme, one group of nine individuals had five haplotypes present at the time of sampling (Group 4 at Nísia). These results suggest that at both Nísia and Tapacurá, the genetic structure of some *C. jacchus* groups may often be more heterogeneous than the extended family model that has previously been suggested (e.g. Digby and Barreto 1993; Nievergelt et al. 2000). The mtDNA data also support the idea that dispersal of both sexes can occur and that individuals immigrate into groups containing other maternal relatives. For example, in the Belém group at Nísia, Bhaskara was an immigrant male and was from the same matriline as the two adult males already in the group (including the breeding male). Bhaskara, although an immigrant, was identified as a frequent helper in the group (MF Arruda unpublished data), behavior that when considered with the genetic data can be explained by kin-selected altruism. Another interesting example of inter-group dynamics was observed in a neighboring group, Chui. In Chui, the haplotype of the original breeding male Chris matched that of adult males in a neighboring group, Belém. When a breeding female "Betty" later joined Chui from Belém (after the two previous breeding females in Chui had died), Chris was replaced as a breeder male by Cazuza, possibly to avoid consanguineous mating: Betty and Cazuza had different haplotypes, while Betty and Chris carried the same. Interestingly, Nievergelt et al. (2000) also sampled and genotyped Chris five years earlier and concluded that he was closely related to their Group B animals (Belém in Faulkes et al.'s 2003 study). With an average relatedness value of 0.41 (first order relatives would have on average 0.5), this is consistent with Faulkes et al.'s mtDNA data, and emphasizes the power and robustness of genetic analysis.

Table 5.1 also compares mtDNA and microsatellite data, and as expected those groups that contain more mtDNA haplotypes have lower mean intra-group relatedness. In two groups that have been resampled (Chui and Belém), changes in mean intra-group relatedness have occurred over time, indicating that inter-group transfers may be constantly challenging the *status quo*, necessitating responses in terms of reproductive control and maintenance of reproductive skew (see Sect. 5.2.4 below).

A final example of genetic relationships offering an explanation for behavioral observations involved two documented cases of group expulsions at Nísia (Faulkes et al. 2003). In one, the breeding male Enrique expelled a juvenile male Eduado (Espanha group), and in the second, breeding male Goeth expelled ex-breeder Gandi (Group 4). In both these cases, the interacting animals were from different matrilines, suggesting competition arising from reproductive conflict may have been the underlying factor in the expulsion.

5.2.4 Group Composition and Its Implication for Social Control of Reproduction in Common Marmosets

In social groups of *C. jacchus*, the common occurrence of non-breeding adults of both sexes from different maternal lineages (Faulkes et al. 2003) and low pairwise relatedness between some group members (Nievergelt et al. 2000) means that in order to maintain a reproductive advantage within groups, the breeding female may need to exert a controlling influence over the adult non-breeding helpers (see Yamamoto et al. this volume, Chap. 6). In the case of a simple family group (e.g. parents and their offspring), incest avoidance alone will result in a reproductive skew and helpers having non-breeding status.

However, as both Nievergelt et al. (2000) and Faulkes et al. (2003) have shown, unrelated adults are frequently present, and there is therefore the potential for out-breeding opportunities among the subordinate helpers within groups. In captivity, reproductive skew in peer groups of unrelated individuals is achieved by a combination of behavioral and pheromonal cues from the dominant female that brings about a hypogonadotrophic state resulting in anovulation in subordinate females (Abbott et al. 1998; Barrett et al. 1993). In family groups studied in captivity, an interesting mix of this dominant control and incest avoidance appears to operate. In this context, 46% of daughters were shown to have ovulated at least once, although none became pregnant because unrelated males were absent and daughters avoided incest (Saltzman et al. 1997). Preliminary data from endocrine monitoring of a wild group of *C. jacchus* have shown that only one female may reproduce in a group, even if subordinate females are undergoing ovulatory cycles (Albuquerque et al. 2001) or show signs of sexual activity or previous pregnancy (Mendes Pontes and Monteiro da Cruz 1995). On the other hand, it is clear from the aforementioned studies of Digby and others that in the wild, while most groups have a single breeding female, plural breeding among females can occur when group composition allows it (i.e. unrelated partners are available). Although the stability and success of such groups in relation to those with a single breeding female remain to be fully quantified, further discussion of female–female competition and its role in the maintenance of reproductive skew are discussed in detail by Yamamoto et al. (this volume, Chap. 6).

5.3 Conclusions and Future Work

It is clear from the data reviewed in this chapter that although kin selection may be an important factor in understanding many aspects of helping behavior and group dynamics in common marmosets, at present there is insufficient knowledge accumulated to say whether relatedness and kin selection can explain all aspects of cooperative breeding behavior. Reproductive conflicts of interest due to the presence of unrelated adults may commonly occur, but it is possible that such heterogeneous groups may be transitory and ultimately stabilize on an extended family group structure. At present the environmental or social triggers for inter-group transfers remain largely unknown, and to date there has also been no attempt to correlate patterns of relatedness (using microsatellite markers) with observations of the occurrence and frequency of altruistic behavior in the wild. In particular, the contribution of kin-selected altruism must be teased apart from other factors that may explain cooperative breeding in mixed-kin groups, such as reciprocity and by-product mutualism.

In a recent paper that studied food-giving in captive groups of cotton-top tamarins (*Saguinus oedipus*), Hauser et al. (2003) showed that unrelated tamarins preferentially give food to others that altruistically give food back. This clearly demonstrates that these callitrichids have the ability to discriminate between altruistic and selfish acts, and have the psychological capacity to carry out acts of reciprocal altruism. Preliminary genotyping studies now need to be extended to include more groups in a population over longer periods of time to address such questions on the relative contribution of kin selection vs. reciprocal altruism, and to understand in detail the possible significance of extra-group copulations.

The power of large scale genetic surveys in revealing hitherto unsuspected mating strategies is exemplified in a recent study of greater horseshoe bats (*Rhinolophus ferrumequinum*): Rossiter et al. (2005) used a detailed microsatellite-based analysis of parentage to show that cooperatively roosting females boost group relatedness by a combination of mate fidelity and intra-lineage polygyny. Clusters of female bats contain overlapping generations of relatives that mate with the same male or males over many years. Furthermore, half sisters may reciprocally mate with their respective fathers, avoiding inbreeding yet at the same time increasing the relatedness between matrilines within a population. The exact benefits to horseshoe bats of this increased relatedness remains unclear at present, but similar social conditions occur for systems like this to operate within populations of marmosets. Genetic studies of callitrichids are still at their early stages, and it is highly likely that in such cognitively advanced mammals there remains hidden complexity in their social and reproductive strategies that will only be revealed by extensive, long-term studies utilizing molecular genetic, behavioral, and ecological data.

Acknowledgements The authors are greatly indebted to Emilia Yamomoto and Hilary Box for all their help and support with the project. Thanks also to IBAMA/EFLEX and the support staff at Nísia Floresta Tapacurá, and to Steve Le Comber for comments on an early version of the manuscript. This work was funded by grants from the Royal Society of London, CnPq (Brazil) and a British Council/CAPES Links Programme, all of whom we gratefully thank.

References

Abbott DH, Saltzman W, Schultz-Darken J, Tannenbaum PL (1998) Adaptations to subordinate status in female marmoset monkeys. Comp Biochem Physiol C 19:261–274

Albuquerque ACSR, Sousa MBC, Santos HM, Ziegler TE (2001) Behavioral and hormonal analysis of social relationships between oldest females in a wild monogamous group of common marmosets (*Callithrix jacchus*). Int J Primatol 22:631–645

Arruda MF, Araújo A, Sousa MBC, Albuquerque FS, Albuquerque ACSR, Yamamoto ME (2005) Two-breeding females within free-living groups may not always indicate polygyny: alternative subordinate female strategies in common marmosets (*Callithrix jacchus*). Folia Primatol 76:10–20

Avise JC (2000) Phylogeography: the history and formation of species. Harvard University Press, Cambridge, Massachusetts

Bales K, Dietz J, Baker A, Miller K, Tardif SD (2000) Effects of allocare-givers on fitness of infants and parents in callitrichid primates. Folia Primatol 71:27–38

Barrett J, Abbott DH, George LM (1993) Sensory cues and suppression of reproduction in subordinate female marmoset monkeys, *Callithrix jacchus*. J Reprod Fert 97:301–310

Benirschke K, Anderson JM, Brownhill LE (1962) Marrow chimerism in marmosets. Science 138:513–515

Bennett NC, Faulkes CG (2000) African mole-rats – ecology and eusociality. Cambridge University Press, Cambridge

Burland TM, Worthington Wilmer J (2001) Seeing in the dark: molecular approaches to the study of bat populations. Biol Rev 76:389–409

Canavez FC, Moreira MAM, Simon F, Parham P, Seuanez HN (1999) Phylogenetic relationships of the Callitrichinae (Platyrrhini, Primates) based on b_2-microglobin DNA sequences. Am J Primatol 48:225–236

Di Fiore A, Fleischer RC (2004) Microsatellite markers for woolly monkeys (*Lagothrix lagotricha*) and their amplification in other New World primates (Primates: Platyrrhini). Mol Ecol Notes 4:246–249

Digby LJ (1996) Sexual behavior and extragroup copulations in a wild population of common marmosets (*Callithrix jacchus*). Folia Primatol 70:136–145

Digby LJ, Barreto CE (1993) Social organization in a wild population of *Callithrix jacchus* I. Group composition and dynamics. Folia Primatol 61:123–134

Digby LJ, Ferrari SF (1994) Multiple breeding females in free-ranging groups of *Callithrix jacchus*. Int J Primatol 15:389–397

Digby LJ, Saltzman W (this volume) Balancing cooperation and competition in callitrichid primates: examining the relative risk of infanticide across species. In: Ford SM, Porter LM, Davis LC (eds) The smallest anthropoids: the marmoset/callimico radiation. Springer, New York, pp 135–153

Dixson AF, Anzenberger G, Monteiro da Cruz MAO, Patel I, Jeffreys AJ (1992) DNA fingerprinting of free-ranging groups of common marmosets (*Callithrix jacchus jacchus*) in N.E. Brazil. In: Martin RD, Dixson AF, Wickings EJ (eds) Paternity in primates: genetic tests and theories. Basel, Karger, pp 192–202

Escobar-Páramo P (2000) Microsatellite primers for the wild brown capuchin monkey *Cebus apella*. Mol Ecol 9:107–118

Faulkes CG, Arruda MF, Monteiro da Cruz MA (2003) Matrilineal genetic structure within and among populations of the cooperatively breeding common marmoset, *Callithrix jacchus*. Mol Ecol 12:1101–1108

Ferrari SF (1993) Ecological differentiation in the Callitrichidae. In: Rylands AB (ed) Marmosets and tamarins: systematics, ecology and behaviour. Oxford University Press, Oxford, pp 314–328

Ferrari SF (this volume) Social hierarchy and dispersal in free-ranging buffy-headed marmosets (*Callithrix flaviceps*). In: Ford SM, Porter LM, Davis LC (eds) The smallest anthropoids: the marmoset/callimico radiation. Springer, New York, pp 155–165

Ferrari SF, Digby LJ (1996) Wild *Callithrix* groups: stable extended families? Am J Primatol 38:19–27

French JA (1997) Singular breeding in callitrichid primates. In: Solomon NG, French JA (eds) Cooperative breeding in mammals. Cambridge University Press, Cambridge, pp 34–75

Goldizen AW, Mendelson J, van Vlaardingen M, Terborgh J (1996) Saddle-back tamarin (*Saguinus fuscicollis*) reproductive strategies: evidence from a thirteen-year study of a marked population. Am J Primatol 38:57–83

Gonçalves EC, Ferrari SF, Silva A, Coutinho PEG, Menezes EV, Schneider MPC (2003) Effects of habitat fragmentation on the genetic variability of silvery marmosets, *Mico argentatus*.

In: Marsh LK (ed) Primates in fragments: ecology and conservation. Kluwer/Plenum, New York, pp 17–28

Gonçalves EC, Ferrari SF, Coutinho PEG, Menezes EV, Silva A, Schneider MPC (this volume) Limited dispersal and genetic structure of silvery marmosets (*Mico argentatus*) in the fragmented landscape of Central Amazonia. In: Ford SM, Porter LM, Davis LC (eds) The smallest anthropoids: the marmoset/callimico radiation. Springer, New York, pp 205–220

Grativol AD, Ballou JD, Fleischer RC (2001) Microsatellite variation within and among recently fragmented populations of the golden lion tamarin (*Leontopithecus rosalia*). Conserv Gen 2:1–9

Hauser MD, Chen MK, Chen F, Chuang E (2003) Give unto others: genetically unrelated tamarins preferentially give food to those that altruistically give food back. Proc R Soc Lond B 270:2363–2370

Hubrecht RC (1984) Field observations on group size and composition of the common marmoset (*Callithrix jacchus jacchus*), at Tapacura, Brazil. Primates 25:13–21

Lazaro-Perea C, Castro CSS, Harrison R, Araújo A, Arruda MF, Snowdon CT (2000) Behavioral and demographic changes following the loss of the breeding female in cooperatively breeding marmosets. Behav Ecol Sociobiol 48:137–146

Mendes Pontes AR, Monteiro da Cruz MAO (1995) Home range, inter-group transfers, and reproductive status of common marmosets *Callithrix jacchus* in a forest fragment in north-eastern, Brazil. Primates 36:335–347

Nievergelt CM, Digby LJ, Ramiakrishnan U, Woodruff DS (2000) Genetic analysis of group composition and breeding system in a wild common marmoset (*Callithrix jacchus*) population. Int J Primatol 21:1–20

Nievergelt CM, Mundy NI, Woodruff DS (1998) Microsatellite primers for genotyping common marmosets (*Callithrix jacchus*) and other callitrichids. Molecular Ecology 7:1431–1439

Perez-Sweeney BM, Valladares-Padua C, Burrell AS, Di Fiore A, Satkoski J, Van Coeverden De Groot PJ, Boag PT, Melnick DJ (2005) Dinucleotide microsatellite primers designed for a critically endangered primate, the black lion tamarin (*Leontopithecus chrysopygus*). Mol Ecol Notes 5:198–201

Ross KG (2001) Molecular ecology of social behaviour: analyses of breeding systems and genetic structure. Mol Ecol 10:265–284

Rossiter SJ, Ransome RD, Faulkes CG, Le Comber SC, Jones G (2005) Mate fidelity and intra-lineage polygyny in greater horseshoe bats. Nature 437:408–411

Rylands AB (1984) Exudate-eating and tree-gouging by marmosets (Callitrichidae, Primates). In: Chadwick AC, Sutton SL (eds) Tropical rain forest: the Leeds symposium. Leeds Philosophical and Literary Society, Leeds, pp 155–168

Saltzman W, Schultz-Darken NJ, Abbott DH (1997) Familial influences on ovulatory function in common marmosets (*Callithrix jacchus*). Am J Primatol 41:159–177

Sánchez S, Peláez F, Gil-Bürmann C, Kaumanns W (1999) Costs of infant carrying in the cotton-top tamarin (*Saguinus oedipus*). Am J Primatol 48:99–111

Scanlon CE, Chalmers NR, Monteiro da Cruz MAO (1988) Changes in the size, composition, and reproductive condition of wild marmoset groups (*Callithrix jacchus jacchus*) in North East Brazil. Primates 29:295–305

Sena L, Vallinoto M, Sampaio I, Schneider H, Ferrari SF, Schneider MPC (2002) Mitochondrial COII gene sequences provide new insights into the phylogeny of marmoset species groups (Callitrichidae, Primates). Folia Primatol 73:240–51

Signer EN, Anzenberger G, Jeffreys AJ (2000) Chimaeric and constitutive DNA fingerprints in the common marmoset (*Callithrix jacchus*). Primates 41:49–61

Solomon NG, French JA (1997) Cooperative breeding in mammals. Cambridge University Press, Cambridge

Sousa MBC de, Rocha Albuquerque ACS da, Yamamoto ME, Araújo A, Fátima Arruda M de (this volume) Emigration as a reproductive strategy of the common marmoset (*Callithrix jacchus*). In: Ford SM, Porter LM, Davis LC (eds) The smallest anthropoids: the marmoset/callimico radiation. Springer, New York, pp 167–182

Stevenson MF, Rylands AB (1988) The marmoset, genus *Callithrix*. In: Mittermeier RA, Rylands AB, Coimbra-Filho AF, da Fonseca GAB (eds) Ecology and behavior of Neotropical primates, vol 2. World Wildlife Fund, Washington DC, pp 131–222

Suárez SS, Böhle UR, Jolly CJ, Disotell TR (2004) Behavioural and genetic data illustrate alternative reproductive strategies in monogamous groups of wild red-bellied tamarins (*Saguinus labiatus*) in northwestern Bolivia. Folia Primatol 75(suppl 1):89

Tagliaro CH, Schneider MPC, Schneider H, Sampaio IC, Stanhope MJ (1997) Marmoset phylogenetics, conservation perspectives, and evolution of the mtDNA control region. Mol Biol Evol 14:674–684

Tardif SD (1997) Parental behavior and evolution of alloparental care. In: Solomon NG, French JA (eds) Cooperative breeding in mammals. Cambridge University Press, Cambridge, pp 11–33

Tardif SD, Bales K (1997) Is infant carrying a courtship strategy in captive callitrichids? Animal Behaviour 53:1001–1007

Tardif SD, Harrison ML, Simek MA (1993) Communal infant care in marmosets and tamarins: relations to energetics, ecology and social organisation. In: Rylands AB (ed) Marmosets and tamarins: systematics, behaviour and ecology. Oxford University Press, Oxford, pp 220–234

Yamamoto ME, Arruda M de F, Alencar AI, de Sousa MBC, Araújo A (this volume) Mating systems and female–female competition in the common marmosets, *Callithrix jacchus*. In: Ford SM, Porter LM, Davis LC (eds) The smallest anthropoids: the marmoset/callimico radiation. Springer, New York, pp 119–133

Chapter 6
Mating Systems and Female–Female Competition in the Common Marmoset, *Callithrix jacchus*

Maria Emilia Yamamoto, Maria de Fátima Arruda, Anuska Irene Alencar, Maria Bernardete Cordeiro de Sousa, and Arrilton Araújo

Abstract Mating systems in callitrichids are flexible, but usually only one female breeds. Reproductive skew models suggest that either dominant individuals can control the incidence of subordinate reproduction, and that when subordinates breed, it represents a concession from the dominant individual (Optimal Skew Model (OSM)), or alternatively, that subordinates breed because dominants are unable to control them (Incomplete Control Model (ICM)). We present data both from a captive study and a long term field study to test which model better explains mating systems and female-female competition in *Callithrix jacchus*. A 10-year study of a wild population allowed us to identify three possible mating systems occurring in groups: monogyny, monogyny with breeding attempts from subordinates (pseudo-polygyny), and polygyny. Comparison of interbirth intervals and time to produce a surviving offspring indicated that polygyny is costly to females, suggesting that females should benefit from monopolization of reproduction. Data from an experimental captive study indicated that dominant females are not always able to physiologically and/or behaviorally suppress subordinate females' reproduction. The literature suggests that helpers are necessary to ensure the survival of dominant females' offspring; however, our field data show that not all adults help equally. A few individuals, mostly adult males, participate significantly more in infant care than other helpers, suggesting that dominant females gain little by retaining subordinate females in their groups. These results indicate that dominant females are not able to control subordinates' reproduction, suggesting that the flexibility of *C. jacchus* mating systems is better explained by ICM than by OSM.

Resumen Los sistemas de apareamiento en los calitrícidos son flexibles, pero usualmente solamente una hembra se reproduce. Los modelos de sesgo reproductivo sugieren que los individuos dominantes pueden ya sea controlar la incidencia de reproducción en individuos subordinados, y que cuando éstos se reproducen

M.E. Yamamoto (✉)
Departamento de Fisiologia, Universidade Federal do Rio Grande do Norte,
Natal, RN, 59072-970, Brazil
e-mail: emiliayamamoto@gmail.com

S.M. Ford et al. (eds.), *The Smallest Anthropoids*, Developments in Primatology: Progress and Prospects, DOI 10.1007/978-1-4419-0293-1_6,
© Springer Science+Business Media, LLC 2009

representa una concesión de los dominantes [Modelo de Sesgo Optimo (OSM)], o bien, que los individuos subordinados se reproducen porque los dominantes no pueden controlarlos [Modelo de Control Incompleto (ISM)]. Presentamos datos sobre un estudio en cautiverio y un estudio de campo a largo plazo para probar cuál modelo explica mejor el sistema de apareamiento y competencia hembra-hembra en *Callithrix jacchus*. Un estudio de 10 años de una población silvestre nos permitió identificar tres posibles sistemas de apareamiento que ocurren los lo grupos: monoginia, monoginia con intentos de reproducción de subordinados (pseudopoliginia) y poliginia. Comparación de intervalos entre nacimientos y tiempo para producir un descendiente que sobrevive indicaron que la poliginia es costosa para las hembras, sugiriendo que las hembras deberían beneficiarse de la monopolización de la reproducción. Datos de un estudio experimental en cautiverio indicaron que las hembras dominantes no siempre son capaces de suprimir la reproducción de las subordinadas, fisiológica o conductualmente. La literatura sugiere que los ayudantes son necesarios para asegurar la supervivencia de la descendencia de la hembra dominante; no obstante, nuestros datos de campo muestran que no todos los adultos ayudan igual. Unos cuantos individuos, principalmente machos adultos, participan significativamente más en el cuidado de los infantes que otros ayudantes, sugiriendo que las hembras dominantes ganan muy poco reteniendo a las hembras subordinadas en sus grupos. Estos resultados indican que las hembras dominantes no son capaces de controlar al reproducción de las subordinadas, sugiriendo que la flexibilidad observada en los sistemas de apareamiento de *C. jacchus* es mejor explicada por ICM que por OSM.

Resumo Os sistemas de acasalamento em calitriquídeos são flexíveis, mas geralmente apenas uma fêmea reproduz. Modelos de viés reprodutivo sugerem que indivíduos dominantes conseguem controlar a reprodução de subordinados, e que sua ocorrência representa uma concessão do dominante (OSM) ou, por outro lado, que subordinados reproduzem porque os dominantes são incapazes de controlá-los (ICM). Nós apresentamos dados de estudos de cativeiro e de campo para testar qual dos modelos melhor explica os sistemas de acasalamento e competição entre fêmeas em *Callithrix jacchus*. Um monitoramento de 10 anos de uma população selvagem nos permitiu identificar três possíveis sistemas de acasalamento: monoginia, monoginia com tentativas de reprodução das subordinadas (pseudo poliginia) e poliginia. A comparação do IBI e do tempo para produzir um filhote sobrevivente indicaram que a poliginia acarreta custos às fêmeas, sugerindo que elas se beneficiariam da monopolização da reprodução. Dados de um estudo experimental em cativeiro indicaram que as fêmeas dominantes nem sempre são capazes de suprimir, seja fisiologicamente ou comportamentalmente, a reprodução de subordinadas. Relatos da literatura sugerem que os ajudantes são necessários para a sobrevivência da prole da fêmea dominante. No entanto, nossos dados de campo mostram que alguns poucos indivíduos, principalmente machos adultos, participam significativamente mais no cuidado do que outros ajudantes, sugerindo que não há motivos para que fêmeas dominantes tentem reter as subordinadas.

Esses resultados indicam que as fêmeas dominantes não são capazes de controlar a reprodução das subordinadas, sugerindo que a flexibilidade observada nos sistemas de acasalamento em *C. jacchus* é melhor explicada por ICM do que por OSM.

6.1 Introduction

The mating system of callitrichids has been suggested to be monogamous, polygynous, and polyandrous. In *Callithrix jacchus*, the common marmoset, groups containing two breeding females as well as one breeding female have been reported in the wild (Scanlon et al. 1988; Digby and Ferrari 1994; Arruda et al. 2005). Reproduction in female common marmosets is usually related to dominance status: the dominant female receives little or no aggression and has complete ovarian function.

Most groups of callitrichids contain only a single breeding female, and the reasons why some groups occasionally deviate from this pattern are controversial. Part of the variation in group composition and mating systems in the Callitrichidae has been attributed to ecological variables, such as the abundance and distribution of foods (Ferrari and Lopes Ferrari 1989). Ferrari and Lopes Ferrari as well as Ferrari and Digby (1996) suggest that marmoset groups are more stable and maintain a higher level of genetic relatedness than tamarin groups, due to the marmoset's ability to exploit stable and aseasonal exudate resources.

C. jacchus are considered to be cooperative breeders, as they present delayed dispersal of offspring from the natal group, reproductive suppression of subordinates, and care of infants that are not genetic relatives (Abbott et al. 1998). Cooperative breeders usually have high reproductive skew, as dominant females produce all or most of the young (Emlen 1995). Emlen (1991) suggests that cooperative breeding groups, such as those observed in *C. jacchus*, will form when there are constraints to reproduction such as when: (1) there are few breeding vacancies for dispersing individuals, or the vacancies available are of poor quality; (2) competition for breeding vacancies is intense, lowering the probability of dispersing youngsters to establish themselves successfully; (3) there are no suitable unrelated mating partners.

6.2 Reproductive Skew Models of Control

Two reproductive-skew models suggest that dominants can control the incidence of subordinate reproduction, with their occurrence representing a concession from the dominant female (Emlen 1995, 1997); or, alternatively, that subordinates breed because dominants are unable to control them (Clutton-Brock 1998; Clutton-Brock et al. 2001).

The Optimal Skew Model (OSM, Emlen 1995, 1997; Emlen et al. 1998) suggests that reproductive sharing within social groups results when dominant females allow reproductive rights to subordinates in return for their cooperation. OSM is based on the idea of a social contract between a dominant individual that benefits from the activities of the same-sex subordinate, and therefore may concede reproduction as an incentive for that subordinate to remain in the group without challenging the dominant. This model assumes that dominants can control reproduction of subordinates without suffering a cost. It also predicts that skew will be high when there are severe constraints to dispersal, there is an asymmetry in dominance (resulting in a low probability that subordinates will successfully challenge the dominant), the dominant and subordinate are closely related, and/or the subordinate's help is important for the survival of the dominant's offspring.

Clutton-Brock (1998) offers a critique of OSM, suggesting the alternative Limited Control Model or Incomplete Control Model (ICM). This model suggests that the low breeding frequency of subordinate females may not be due to an incapacity to breed, but instead is often due to the lack of unrelated males being available. ICM predicts that subordinates are likely to breed when groups contain unrelated males, the dominant's control is reduced, and the benefits of retaining a subordinate exceed the costs of having an additional breeder (Clutton-Brock 1998; Clutton-Brock et al. 2001).

In this chapter, we test the two reproductive skew models on common marmosets using data from long-term studies of wild groups and from an experimental captive study.

6.3 Callitrichid Social Organization and Mating Systems

As in other cooperative breeders, in callitrichids non-reproductive individuals delay their dispersal and help in the care of the dominant female's offspring (Tardif 1997). In captivity, where individuals within groups are highly related, all individuals carry and feed the infants (Feistner and Chamove 1986; Yamamoto and Box 1997). In the wild, kinship varies widely among individuals within groups, and so does the level of allocare individuals provide (Tardif 1997). In *C. jacchus*, wild juvenile animals participate less in care than in captive groups, and adult females are sometimes prevented from carrying infants (Albuquerque 1999; Yamamoto et al. 1996).

Suppression of ovulation in subordinate females was first demonstrated in the common marmoset (Abbott 1984; Abbott et al. 1981) and was later shown to be variable both among marmoset groups and among other taxa. In *Saguinus fuscicollis* (saddle-back tamarins, Epple and Katz 1984) and *S. oedipus* (cotton-top tamarins, Ziegler et al. 1987), subordinate females experience total suppression of ovulation. *Leontopithecus rosalia* (golden lion tamarin) females, on the other hand, show no sign of suppression in the presence of their mothers (French and Stribley 1987). In captive common marmosets, up to half of the daughters escape from this suppression, but only one female will ovulate in peer groups (Abbott 1984; Saltzman et al. 1997a, b), and recent hormonal data from wild common marmoset groups suggest that there is usually more than one ovulating female in all groups (Albuquerque et al. 2004).

Wild and captive *C. jacchus* groups most frequently consist of one breeding female, but polygynous groups have also been reported. Daughters of dominant *C. jacchus* females living in captive families were able to conceive and give birth in their intact family groups (Hubrecht 1989), provided that the daughters were able to copulate with an unrelated male away from the family cage. There have been further reports of polygyny in both captive (Rothe and Koenig 1991) and wild common marmoset groups (Arruda et al. 2005; Digby and Ferrari 1994; Roda and Pontes 1997; Scanlon et al. 1988), and in all these groups close kin relations between the two reproductive females were known or at least suspected.

6.4 Female Competition in a Wild *Callithrix jacchus* Population

6.4.1 Mating Systems

During a 10-year monitoring study of wild *C. jacchus* in Nísia Floresta, northeastern Brazil (Arruda et al. 2005; see also Sousa et al. this volume, Chap. 9), we identified 14 breeding groups that contained either one or two breeding females. Every time the breeding female/females were replaced, for any reason, we considered that a new breeding group was formed. Tenure of the breeding groups varied from 10 to 71 months. Data from these breeding groups suggest three possible mating systems: (1) monogamy or, more specifically, monogyny, as we do not have genetic information on paternity to determine if just one male bred; (2) monogyny with breeding attempts by subordinate females (pseudo-polygyny); and (3) polygyny. Groups were considered as monogynous when only one female bred, a pattern found in almost half of the 14 groups. Pseudo-polygynous groups contained two females that gave birth, a dominant and a subordinate, but there were clear differences in the reproductive output of the females. Dominant females bred repeatedly, their infants had high survivability rates, they were stable members of the group, and within their groups they were the only females to show consistent sexual behavior with the group's dominant male.

Subordinate breeding females, on the other hand, bred only once, their infants did not survive, they did not engage in sexual behavior with their group's breeding male, and they left their groups after losing their infants. All but one of the subordinate females were seen copulating with extragroup males 5–6 months before the birth of their infants. All infants from subordinate females disappeared from unknown causes, except one infant which was observed to be killed by a dominant female. Reproductive output for these subordinate females was always zero in their natal group. Accordingly, this reproductive arrangement is functionally monogynous even though subordinate females give birth to infants, thus our term, pseudo-polygynous. These occurrences are well described by Arruda and colleagues (2005), in which subordinate female pregnancies are attributed to extra-group copulations (EGC).

Finally, true polygynous groups also occurred rarely, in 15% of the groups. In polygynous groups, two females bred repeatedly, had surviving offspring, were stable members of the group, and showed consistent sexual behavior with the same dominant male. Reproductive output of both females was similar. Nevertheless, one of the females could be identified as dominant by the direction of agonistic behavior (Araújo 1996; Arruda et al. 2005). The incidence of polygyny is similar to that described for a wild population of *Leontopithecus rosalia* by Dietz and Baker (1993), suggesting that although this kind of arrangement is rare, it is present across callitrichid taxa.

6.4.2 Costs of Polygyny

Comparison of monogynous (monogynous and pseudo-polygynous) and polygynous groups suggest that polygyny brings some costs to females. First, the duration of the mean interbirth interval (IBI) was significantly higher for polygynous females as compared to monogynous females (Table 6.1). Monogynous females presented an IBI of 5.7 months as compared to the 7.5 months for polygynous females. The difference reflects a higher IBI for both dominant and subordinate females in polygynous groups.

Second, although the percentage of surviving offspring was not significantly different for the three categories of females, monogynous and polygynous dominants and subordinates, the data suggest that with larger sample sizes these differences may be significant (Table 6.1). For example, while the percentage of surviving

Table 6.1 Interbirth interval (IBI), in months, percentage of surviving infants, and time to produce an infant or a surviving infant, in months, in wild monogynous and polygynous common marmoset females from 14 reproductive groups at the Nísia Floresta site, Brazil

		Polygyny ($n=3$)		
Index	Monogyny ($n=11$)	Dom	Sub	Mean (both ♀s)
IBI	5.7[a]	6.8	8.2	7.5[a]
% Surviving infants[b]	69.7	44.8	55.6	49.0
Time to produce an infant[c]	4.0	4.5	4.6	2.3
Time to produce a surviving infant[d]	5.2	9.8	7.0	4.1

[a]Student's t-test: $P<0.005$

[b]Infants that survived to 5 months or older

[c]Number of infants born to each group female divided by the number of months of her tenure

[d]Number of surviving infants born to each group female divided by the number of months of her tenure

infants (that is, the percentage of the total number of infants born to a female that survived until 5 months) from monogynous females varied from 40 to 100%, those of polygynous females, independent of rank, varied from 26.3 to 80%. In the two polygynous groups in this population, females gave birth asynchronously, avoiding competition with one another, which may have increased infant survival rates. In other studies it was observed that when two group females had infants at the same time, infant survival was lower, and infanticide sometimes occurred (Digby 1995; for a review of infanticide in callitrichids see Digby and Saltzmann this volume, Chap. 7).

Finally, we used a measure of the mean time for a female to produce an infant and also to produce a surviving infant (to 5 months old) in both mating systems. This was achieved by dividing the number of infants or the number of surviving infants produced by group females by the number of months of their tenure. Female tenure was considered as the period of time that a female bred in the group. For most groups, this period comprised the time from the female's first breeding to the time of her death, migration or disappearance. In a few cases, the female was already breeding when the group was first identified, and tenure started to be counted from the beginning of the monitoring. This measure of mean time does not reflect the real time to produce an infant, because females usually give birth to twins. But, as the incidence of single births was around 20%, we considered it more accurate to use mean time, as it allows us to compare across all females and litter sizes, and also allowed us to produce a more accurate measure of females' reproductive success. This means that in the case of twin infants (80% of the total), each infant was included in the calculations independently. We found that both the time to produce an infant and to produce a surviving infant were lower for polygynous females (when data from dominant and subordinate females were pooled) than for monogynous females (Table 6.1). This indicates that polygynous groups are more efficient than monogynous groups in the production and survival of infants. Even so, both dominant and subordinate females in polygynous group lose fitness when a second female breeds, as each of them, individually, takes longer to have a surviving infant. The lower fitness probably negatively affects the dominant females most, who would fare better as single breeders, but subordinate females may benefit from such arrangements, as this may represent their only opportunity to breed. Polygynous groups, on the other hand, may increase the fitness of males. If this is true, we should expect males but not females to favor polygyny. Interestingly, polygynous, but not monogynous, breeding females from this population were seen copulating with extra-group males (Digby 1999), indicating that in polygynous groups cuckoldry may decrease males' fitness. Identification of paternity is necessary to provide a measure of loss of fitness for males in polygynous groups that, if sizable, could discourage both males and females from breeding in such a system (see Faulkes et al. this volume, Chap. 5).

These data suggest that females should benefit from monopolization of reproduction, because dominant monogynous females have shorter IBI and take less time to produce a surviving infant. If that is true, then why do dominant females allow subordinates to reproduce? It has been suggested that the occurrence of breeding by subordinates may reflect a strategy by dominants to retain subordinates in their groups if the subordinates improve the dominant's reproductive success (Emlen et al. 1998). Alternatively, a critique of this model suggests that subordinates breed simply because dominants are unable to prevent it (Clutton-Brock 1998). We examined data from an experimental captive study and the role of helpers in common marmoset groups to determine which possibility applies to common marmosets.

6.5 Dominance and Control in Captive Common Marmoset Groups

What factors determine whether groups are monogynous or polygynous ? Some researchers (Digby 1995; Goldizen et al. 1996) suggest that dominant females tolerate polygyny when dominant and subordinate females can give birth far apart, thereby avoiding competition for food and for caregivers. Could there also be different kinds of dominance relationships between callitrichid females that have different consequences for subordinate female fertility?

To answer these questions we assessed ovulatory function and aggressive and submissive behavior exhibited by eight pairs of female common marmosets housed with an unrelated male, a social arrangement meant to trigger female-female competition (Alencar et al. 1995, 2006). Behavioral and hormonal profiles of the pairs of females were different, so we classified them into two groups (after Clarke et al. 1986): "Uncontested" Dominance (UD), in which dominant females showed higher frequencies of dominant behavior during the 15-week period (four pairs), and "Contested" Dominance (CD), those in which the frequencies of dominant behavior were similar between the two females (four pairs).

The relationship between females in CD and UD pairs were distinct. The degree to which dominant females exceeded subordinates in the display of dominant behavior differed among groups, and can be shown as a ratio of dominant/subordinate females' behavior within the two groups (Fig. 6.1). Dominant females exhibited 3–30 times more dominant behavior than subordinates in UD pairs, but equal amounts in CD pairs (see Fig. 6.1).

Moreover, the subordinate/dominant ratio of submissive behavior, measured for two pairs of each group, showed the same asymmetry: UD subordinates displayed 17–53 times more submissive behavior than dominants from their pairs; CD dominant and subordinate females exhibited very low and similar frequencies of submissive behavior (see Fig. 6.2). In summary, subordinates in CD pairs exhibited significantly more agonistic behavior than subordinates in UD pairs, while subordinates in UD pairs displayed significantly more submissive behavior than their dominants.

Female UD and CD pairs also differed in their hormonal profiles when compared using an index of ovulatory function for each female. This index was based on the profile of progesterone concentrations in plasma or feces during the experimental period for each female, and ranges from 3 to 1. An index value of 3 indicates

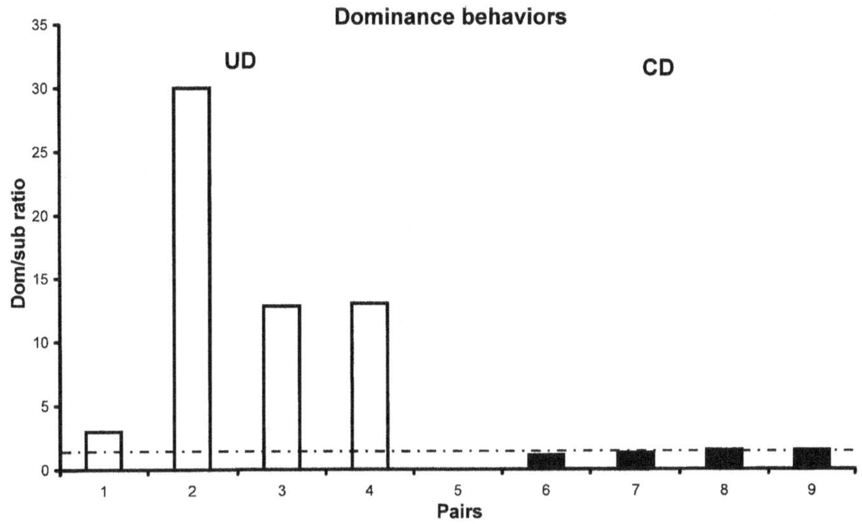

Fig. 6.1 Dominant/subordinate females' ratio of dominance behavior in eight captive *Callithrix jacchus* pairs. The dotted line indicates the point where the ratio is one, meaning that both females present the same frequency of dominance behavior. *UD* uncontested dominance; *CD* contested dominance

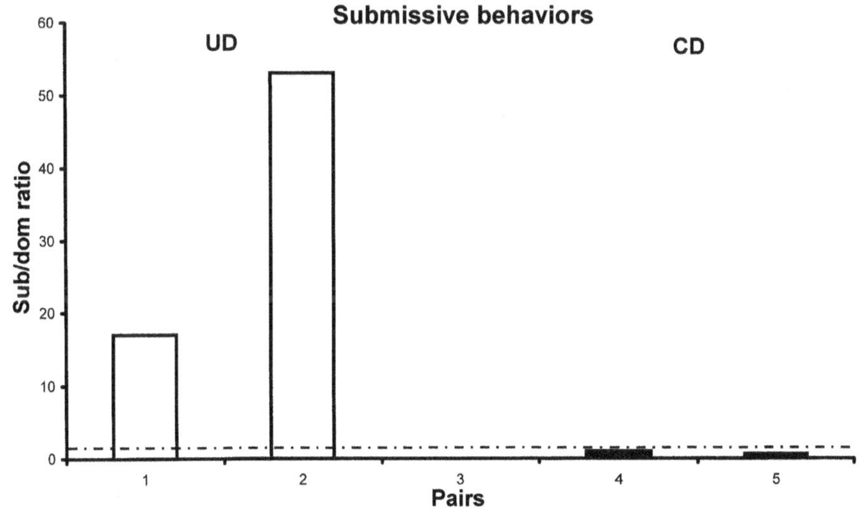

Fig. 6.2 Female subordinate/dominant ratio of submissive behavior in four captive *C. jacchus* pairs. All other conventions as in Fig. 6.1

regular ovulatory cycles with a luteal phase length typical of normal cycles for common marmosets, a value of 2 indicates irregular increases in progesterone concentrations, and a value of 1 indicates a complete absence of progesterone elevations (Alencar et al. 2006). Ovarian functioning ratios between dominant and subordi-

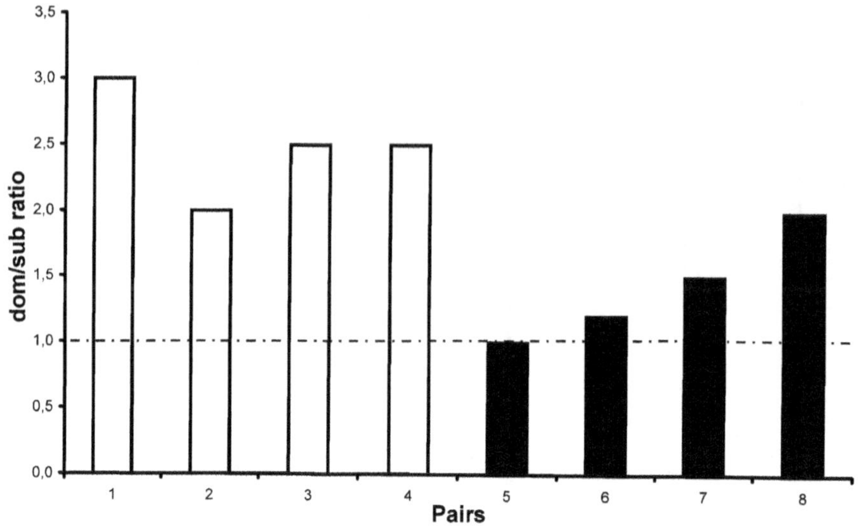

Fig. 6.3 Female dominant/subordinate ratio of ovarian functioning in eight captive *C. jacchus* pairs. The ratio for the ovulatory index between each dominant (D) and subordinate (S) female (determined by progesterone concentrations in plasma or feces) varied from 3 (regular ovulation) to 1 (no ovulation) during three 5-week periods. Other conventions as in Fig. 6.1

nate females varied between 1.0 and 2.0 in CD and between 2.0 and 3.0 in UD (Fig. 6.3). In UD, only the dominant females displayed progesterone levels indicative of ovulatory cycles, while subordinate females showed varying levels of inhibition of ovulatory cycles. In CD, both dominant and subordinate females showed similar progesterone levels during at least part of the observation period.

Our data, together with previously published reports (Arruda et al. 2005; Digby 1995; Digby and Ferrari 1994), suggest that the occurrence of two breeding females in *C. jacchus* groups occurs only when the females are relatives and when they are reproductively and/or behaviorally stimulated by unrelated males. In wild groups, that stimulation could easily arise from males in neighboring groups, since extra-group copulations involving non-breeding females have been reported for *C. jacchus* (Albuquerque et al. 2001; Arruda et al. 2005; Lazaro-Perea et al. 2000). However, while all free-living female marmosets are exposed to adult males, not all free-living *C. jacchus* groups contain two breeding females.

Why may some females be more responsive to unrelated male stimulation than others? Our data regarding the behavior of males suggest that they do not discriminate between dominant and subordinate females as the targets of affiliative and sexual behavior (Alencar et al. 2006). The differential response of subordinate females to the presence of a male results in some subordinate females being less willing to submit to another female without contest, supporting the hypothesis that dominant females are not always able to control the fertility and reproduction of subordinates.

6.6 Do Helpers Help?

Reproduction is energetically very demanding for callitrichid females, as they give birth to twins with high birth weights and may became pregnant again while still nursing the previous set of infants. The father and other non-reproductive individuals may act as helpers, assisting the females in carrying the infants, sharing food with them, being vigilant and defending the territory, and protecting the infants from predators (Snowdon 1996).

Goldizen (1987) proposed that a mating pair of *Saguinus fuscicollis* (particularly the lactating female) would have insufficient energy intake to sustain themselves and carry and feed twin infants. Indeed, wild groups of marmosets or tamarins with only one adult of each sex are extremely rare, although they have been observed (Dietz and Baker 1993; Garber et al. 1993; Koenig 1995).

Many studies indicate that helpers do significantly increase infant survival. For instance, Sussman and Garber (1987) found that the number of surviving offspring in wild *Saguinus midas* (golden-handed tamarin) groups was directly related to the number of adults males in the group. Baker et al. (1993) reported that two-male groups of *Leontopithecus rosalia* reared more offspring than single-male groups. Bales et al. (2000) found a positive correlation between the number of adult males and surviving infants in newly formed groups of the same species, and also between the number of helpers and infant survival in established groups. Koenig (1995), in a review of published data on wild *C. jacchus* from three sites, found that within groups the number of adult males present was related to infant survival, and he concluded that reproductive females benefit from having more males in the group. Santos and Monteiro da Cruz (1997) followed three wild *C. jacchus* groups during 13 months and reported a shorter IBI and no occurrences of infant death or disappearance in the largest of the three groups.

Data from our wild population suggest that neither group size nor the number of adult males affected the percentage of surviving offspring in monogynous groups. However, as all groups, except one, had two or more adult males and at least five animals, excluding infants, they may all have had a minimum necessary number of helpers. Data from Albuquerque (1994) for two groups from this population show that a few non-reproductive adult males generally participated significantly more in infant care than did other helpers. Interestingly, Albuquerque (1999) also found that the breeding females prevented other adult females from carrying infants, suggesting that their help was neither necessary nor wanted. The suggestion that carrying infants is a passport to group membership (Price and McGrew 1991) does not hold for wild *C. jacchus*, as has already been observed in captive *Saguinus oedipus* by Tardif (1997).

In newly formed groups there are two opposing pressures: the need to recruit helpers to assist the breeding female, and competition for the breeding position. The few data from our field site suggest that when a breeding animal dies or disappears, the group becomes unstable and the persistence of the group depends on the establishment of new breeding unities. These were formed usually by the splitting of the old group along sexual lines of males and females, and the two new groups were re-structured with the immigration of two adult females and two adult males respectively (Lazaro-Perea et al. 2000; see Sousa et al. this volume, Chap. 9). These groups emerged after what Emlen (1995) calls a power struggle, during which many of the individuals of the same sex as the new resident breeder left the group, either voluntarily or after suffering aggression.

6.7 Conclusions

The combined results of our field and captive studies indicate that ICM, and not OMS, better explains mating system and subordinate reproduction in common marmoset females. Our data reinforce Saltzman´s (Saltzman 2003; Saltzman et al. 1997b) suggestion that escape from reproductive suppression may be more appropriately interpreted in terms of subordinate females activating their own reproductive function in response to cues from the social environment, particularly those coming from a potential mate. In our groups, these probably came from extra-group males.

In monogynous (strictly and pseudo-polygynous) groups the activation of the reproductive function in subordinate females (Saltzman et al. 1997b) does not result in successful breeding because of several factors (and these are not mutually exclusive): there are no suitable partners, as all males in the group may be related to subordinate females; there are suitable, non-related potential partners, but no opportunities to mate; females do mate while still in their natal group with extra-group males, but there are no adequate conditions for infant survival. Although females may have low reproductive success while in a subordinate position, non-breeding females may inherit the territory and primary breeding role in the event of the death of the breeding female.

In monogynous (pseudo-polygynous) groups, extra-group copulations may constitute an alternative strategy for subordinate females, in that, they can escape harassment by dominant females, they have access to an unrelated male, and they have a chance to scan neighboring groups for breeding vacancies. This tactic proved successful for three wild subordinate females in our field site; they became, eventually, the only breeding female in groups adjacent to their natal group (Arruda et al. in prep.; Lazaro-Perea et al. 2000). We suggest, as also proposed by Emlen (1991), that the availability of breeding positions is unpredictable, and subordinate females should be ready to occupy them whenever they occur.

It is still unclear what the conditions are which give origin to polygynous groups, but one witnessed birth of such a group (Lazaro-Perea et al. 2000) suggests that they may arise from the splitting of well-established groups, after the death of the breeding female.

In conclusion, marmoset mating systems result from the conflicting strategies of various individuals that represent a compromise between males, dominant females, and subordinate females. Moreover, under natural conditions many factors interfere with the breeding success of dominant and subordinate females, and they all probably influence the decisions regarding reproductive strategies for all individuals involved.

Acknowledgements During this project, A.I.A., M.B.C.S. and M.E.Y. were supported by CNPq grants (#521186-97, #301.309-84, and #524409/96, respectively).

References

Abbott DH (1984) Behavioural and physiological suppression of fertility in subordinate marmoset monkeys. Am J Primatol 6:169–186

Abbott DH, McNeilly AS, Lunn SF, Hulme MJ, Burden FJ (1981) Inhibition of ovarian function in subordinate female marmoset monkeys (*Callithrix jacchus jacchus*). J Reprod Fert 63:335–345

Abbott DH, Saltzman W, Schultz-Darken NJ, Tannenbaum PL (1998) Adaptations to subordinate status in female marmoset monkeys. Comp Biochem Physiol 119:261–274

Albuquerque FS (1994) Distribuição do cuidado à prole em grupos de Callithrix jacchus (Callitrichidae: Primates), no ambiente natural. MSc Dissertation, Universidade Federal do Rio Grande do Norte, Natal

Albuquerque FS (1999) Cuidado cooperativo à prole em *Callithrix jacchus*: dinâmica em ambiente natural. PhD Thesis, Universidade de São Paulo, São Paulo

Albuquerque ACSR, Sousa MBC, Santos HM, Ziegler TE (2001) Behavioral and hormonal analysis of social relationship between oldest females in wild monogamous groups of common marmoset, *Callithrix jacchus*. Int J Primatol 22:631–645

Albuquerque ACSR, Nascimento MCL, Santos HM, Sousa MBC (2004) Função ovariana e adrenocortical de fêmeas adultas de sagüi comum (*Callithrix jacchus*) em relação ao comportamento de emigração. In: Mendes SL, Chiarello AG (eds) A primatologia no Brasil, vol 8. IPEMA/SBPr, Vitória, pp 301–316

Alencar AI, Yamamoto ME, Oliveira MS, Lopes FA, Sousa MBC, Silva NG (1995) Behavior and progesterone levels in *Callithrix jacchus* females. Braz J Med Biol Res 28:591–595

Alencar AI, Sousa MBC, Abbott DH, Yamamoto ME (2006) Contested dominance modifies the anovulatory consequences of social subordination in female marmosets. Braz J Med Biol Res 39:647–658

Araújo A (1996) Influence des facteurs écologiques, comportementaux et démographiques sur la dispersion de *Callithrix jacchus*. PhD Thesis, Université Paris-Nord (Paris XIII), Paris

Arruda MF, Araújo A, Sousa MBC, Albuquerque FS, Albuquerque ACSR, Yamamoto ME (2005) Two-breeding females within free-living groups may not always indicate polygyny: alternative subordinate female strategies in common marmosets (*Callithrix jacchus*). Folia Primatol 76:10–20

Baker A, Dietz J, Kleiman D (1993) Behavioural evidence for monopolization of paternity in multi-male group of golden lion tamarins. Anim Behav 46:1091–1103

Bales K, Dietz J, Baker A, Miller K, Tardif S (2000) Effects of allocare-givers on fitness of infant and parents in callitrichid primates. Folia Primatol 71:27–38

Clarke MR, Kaplan JR, Bumsted PT, Koritnik DR (1986) Social dominance and serum testosterone concentration in dyads of male *Macaca fascicularis*. J Med Primatol 15:419–432

Clutton-Brock TH (1998) Reproductive skew, concessions and limited control. TREE 13:288–292

Clutton-Brock T, Brotherton P, O'Riain M, Griffin A, Gaynor D, Kansky R, Sharpe L, McIlrath G (2001) Contributors to cooperative rearing in meerkats. Anim Behav 61:705–710

Dietz JM, Baker AJ (1993) Polygyny and female reproductive success in golden lion tamarins, *Leontopithecus rosalia*. Anim Behav 46:1067–1078

Digby LJ (1995) Infant care, infanticide, and female reproductive strategies in polygynous groups of common marmosets (*Callithrix jacchus*). Behav Ecol Sociobiol 37:51–61

Digby LJ (1999) Sexual behavior and extragroup copulations in a wild population of common marmosets (*Callithrix jacchus*). Folia Primatol 70:136–145

Digby LJ, Ferrari SF (1994) Multiple breeding in free-ranging groups of *Callithrix jacchus*. Int J Primatol 3:389–397

Digby LJ, Saltzman W (this volume) Balancing cooperation and competition in callitrichid primates: examining the relative risk of infanticide across species. In: Ford SM, Porter LM, Davis LC (eds) The smallest anthropoids: the marmoset/callimico radiation. Springer, New York, pp 135–153

Emlen ST (1991) Evolution of cooperative breeding in birds and mammals. In: Krebs JR, Davies NB (eds) Behavioral ecology. Blackwell, Oxford, pp 301–337

Emlen ST (1995) An evolutionary theory of the family. Proc Natl Acad Sci U S A 92:8092–8099

Emlen ST (1997) Predicting family dynamics in social vertebrates. In: Krebs JR, Davies NB (eds) Behavioral ecology. Blackwell, Oxford, pp 228–253

Emlen ST, Reeve HK, Keller L (1998) Reproductive skew: disentangling concessions from control. TREE 13:458–459

Epple G, Katz Y (1984) Social influences on estrogen excretion and ovarian cyclicity in saddle-back tamarins (*Saguinus fuscicollis*). Am J Primatol 6:215–228

Faulkes C, Arruda MF, Monteiro da Cruz MAO (this volume) Genetic structure within and among populations of the common marmoset, *Callithrix jacchus*: implications for cooperative breeding. In: Ford SM, Porter LM, Davis LC (eds) The smallest anthropoids: the marmoset/callimico radiation. Springer, New York, pp 103–117

Feistner A, Chamove A (1986) High motivation toward food increases food-sharing in cotton-top tamarins. Dev Psychobiol 19:439–452

Ferrari SF, Digby LJ (1996) Wild *Callithrix* groups: stable extended families. Am J Primatol 38:19–27

Ferrari S, Lopes Ferrari M (1989) A re-evaluation of the social organisation of the Callitrichidae, with reference to the ecological differences between genera. Folia Primatol 52:132–147

French J, Stribley J (1987) Synchronization of ovarian cycles within and between social groups in golden lion tamarins (*Leontopithecus rosalia*). Am J Primatol 12:469–478

Garber P, Encarnación F, Moya L, Pruetz J (1993) Demographic and reproductive patterns in moustached tamarin monkeys (*Saguinus mystax*): implications for reconstructing platyrrhine mating systems. Am J Primatol 29:235–254

Goldizen A (1987) Facultative polyandry and the role of infant-carrying in wild saddle-back tamarins (*Saguinus fuscicollis*). Behav Ecol Sociobiol 20:99–109

Goldizen AW, Mandelson J, van Vlaadingen M, Terborgh J (1996) Saddle-back tamarins (*Saguinus fuscicolis*) reproductive strategies: evidence from a thirteen-year study of a marked population. Am J Primatol 38:57–83

Hubrecht RC (1989) The fertility of daughters in common marmoset (*Callithrix jacchus jacchus*) family groups. Primates 30:423–432

Koenig A (1995) Group size, composition, and reproductive success in wild common marmosets (*Callithrix jacchus*). Am J Primatol 35:311–317

Lazaro-Perea C, Castro CSS, Harrison R, Araújo A, Arruda MF, Snowdon CT (2000) Behavioral and demographic changes following the loss of the breeding female in cooperatively breeding marmosets. Behav Ecol Sociobiol 48:137–146

Price EC, McGrew WC (1991) Departures from monogamy in colonies of cotton-top tamarins. Folia Primatol 57:16–27

Roda AS, Pontes ARM (1997) Polygyny and infanticide in common marmosets in fragment of the Atlantic forest of Brazil. Folia Primatol 69:372–376

Rothe H, Koenig A (1991) Variability of social organization in captive common marmosets (*Callithrix jacchus*). Folia Primatol 57:28–33

Saltzman W (2003) Reproductive competition among female common marmosets (*Callithrix jacchus*): proximate and ultimate causes. In: Jones CB (ed) Sexual selection and reproductive competition in primates: new perspectives and directions. American Society of Primatologists, Norman, pp 197–229

Saltzman W, Schultz–Darken NJ, Abbott DH (1997a) Familial influences on ovulatory function in common marmoset (*Callithrix jacchus*). Am J Primatol 41:159–177

Saltzman W, Severin JM, Schultz–Darken NJ, Abbott DH (1997b) Behavioral and social correlates of escape from suppression of ovulation in female common marmosets housed with the natal family. Am J Primatol 41:1–21

Santos EM, Monteiro da Cruz MAO (1997) Relação tamanho do grupo X cuidado infantil em *Callithrix jacchus* no ambiente natural. In: Sousa MBC, Menezes AAL (eds) A primatologia no Brasil, vol 6. Ed. Universitária UFRN, Natal, pp 123–137

Scanlon CE, Chalmers NR, Monteiro da Cruz MAO (1988) Changes in the size, composition and reprodution conditions of wild marmoset groups (*Callithrix jacchus jacchus*) in northeast Brazil. Primates 29:295–305

Snowdon CT (1996) Infant care in cooperatively breeding species. In: Rosenblatt JA, Snowdon CT (eds) Parental care: evolution, mechanisms and adaptative significance. Academics, San Diego, pp 643–689

Sousa MBC de, Rocha Albuquerque ACS da, Yamamoto ME, Araújo A, Fátima Arruda M de (this volume) Emigration as a reproductive strategy of the common marmoset (*Callithrix jacchus*). In: Ford SM, Porter LM, Davis LC (eds) The smallest anthropoids: the marmoset/callimico radiation. Springer, New York, pp 167–182

Sussman R, Garber P (1987) A new interpretation of the social organization and mating system of the Callitrichidae. Int J Primatol 8:73–92

Tardif S (1997) The bioenergetics of parental behavior and the evolution of alloparental care in marmosets and tamarins. In: Solomon N, French J (eds) Cooperative breeding in mammals. Cambridge University Press, Cambridge, pp 11–33

Yamamoto ME, Box H (1997) The role of non-reproductive helpers in infant care in captive *Callithrix jacchus*. Ethology 103:1–12

Yamamoto ME, Box H, Albuquerque FA, Miranda MFA (1996) Carrying behaviour in captive and wild marmosets (*Callithrix jacchus*): a comparison between two colonies and a field site. Primates 37:295–302

Ziegler T, Savage A, Scheffler G, Snowdon C (1987) The endocrinology of puberty and reproductive functioning in female cotton-top tamarins (*Saguinus oedipus*) under varying social conditions. Biol Reprod 37:618–627

Chapter 7
Balancing Cooperation and Competition in Callitrichid Primates: Examining the Relative Risk of Infanticide Across Species

Leslie Digby and Wendy Saltzman

Abstract At least eight cases of infanticide by females other than the mother have been observed in wild groups of common marmosets (*Callithrix jacchus*), with several more cases described for captive groups. Infanticide by females other than the mother has not, however, been documented for wild groups of other callitrichid species. Why might such overt aggression toward infants be more likely in one species than in others? In the common marmoset, a variety of social, reproductive and ecological characteristics – including short inter-birth intervals (and the resulting potential for overlap of pregnancies and births), habitat saturation, small home ranges, and low cost of infant care (including decreased travel costs and short dependency periods compared to other callitrichids) – may contribute to an increased likelihood of two breeding females being present in a group, which in turn may give rise to the potential for competition between breeding females and, ultimately, to infanticide. These conditions are less common in wild groups of most other callitrichid species. All callitrichids balance the need for cooperative care of young with the reproductive competition that results from limited reproductive opportunities; however, ecological and social conditions appear to tip the balance toward infanticide more frequently in common marmosets than in other callitrichid species.

Resumen Al menos ocho casos de infanticidio por hembras distintas a la madre han sido observados en grupos silvestres de marmosetas comunes (*Callithrix jacchus*), con muchos más casos descritos para grupos cautivos. Sin embargo, infanticidio por hembras distintas a la madre no ha sido documentado en grupos silvestres de otras especies de calitrícidos ¿Porqué podría la agresión tan evidente hacia infantes ser más probable en una especie que en otras? En las marmosetas comunes, una variedad de características sociales, reproductivas y ecológicas – incluyendo intervalos cortos entre nacimientos (y el resultado potencial de traslape de

L. Digby (✉)
Department of Evolutionary Anthropology, Duke University, Durham, NC, USA
e-mail: ldigby@duke.edu

S.M. Ford et al. (eds.), *The Smallest Anthropoids*, Developments in Primatology: Progress and Prospects, DOI 10.1007/978-1-4419-0293-1_7,
© Springer Science+Business Media, LLC 2009

embarazos y nacimientos), la saturación del hábitat, ámbitos hogareños pequeños y bajo costo de cuidado infantil (incluyendo disminución de costos de viaje y períodos cortos de dependencia comparados con otros calitrícidos) – pueden contribuir al aumento de la posibilidad de que dos hembras reproductoras se presenten en un grupo, lo cual aumenta el potencial para competencia entre la hembras reproductoras y, ultimadamente, a cometer infanticidio. Dichas condiciones son menos comunes en los grupos silvestres de otras especies calitrícidos. Todos los calitrícidos equilibran la necesidad de cuidado cooperativo de los jóvenes con la competencia reproductiva que resulta de oportunidades reproductivas limitadas; no obstante, las condiciones ecológicas y sociales parecen inclinar la balanza hacia el infanticidio con más frecuencia en las marmosetas comunes que en otras especies de calitrícidos.

Resumo Pelo menos oito casos de infanticídio realizados por fêmeas que não as mães foram observados em grupos de sagüi comum selvagens (*Callithrix jacchus*), com vários outros casos sendo descritos para grupos vivendo em cativeiro. Entretanto, infanticídios por fêmeas que não as mães não têm sido bem documentados em grupos selvagens de outras espécies de calitriquídeos. Por que esta agressão em relação aos infantes é mais comum em uma espécie do que em outras? Em sagüi comum, uma variedade de características sociais, reprodutivas e ecológicas – incluindo o intervalo curto entre nascimentos (e o resultante potencial de sobreposição de gestações e nascimentos) saturação de habitat, áreas de uso pequenas e baixo custo do cuidado parental (incluindo diminuição no custo do deslocamento e o período de dependência curto comparado com outros calitriquídeos) – podem contribuir para o aumento da probabilidade de duas fêmeas reprodutoras estarem presentes em um mesmo grupo, que por sua vez favorece o surgimento de competição potencial entre as fêmeas reprodutoras e, eventualmente, o infanticídio. Estas condições são menos comuns em grupos selvagens da maioria das demais espécies de calitriquídeos. Todos os calitriquídeos fazem o balanço entre a necessidade de cuidado cooperativo dos infantes e a competição reprodutiva que resulta das oportunidades limitadas de reprodução; entretanto, as condições ecológicas e sociais parecem inclinar a balança em relação ao infanticídio mais frequentemente em sagüis comuns do que em outras espécies de calitriquídeos.

7.1 Introduction

Female–female competition plays an important role in the reproductive strategies of female primates (e.g., Hrdy 1981; Altmann 1997; Jones 2003) and is likely to play a critical role in the evolution of mammalian social systems (Gowaty 1997; Digby 2000). In recent years an increasing number of studies have documented the roles of social dominance, aggression, and interference by females in controlling and/or limiting the ability of other females to breed and raise young successfully (e.g., Pusey et al. 1997; Gowaty 1997; Saltzman and Abbott 2005). This type of competition is carried to an extreme in cases where one female kills another female's young (e.g., Sherman 1981; Hoogland 1995, reviewed in Digby 2000).

At first glance, the cooperatively breeding callitrichids may not appear to be a likely example of such extreme reproductive competition. This group is better known for shared infant care, flexible mating strategies, and a relatively low rate of aggression (Tardif et al. 1993, 2003; Garber 1997; Digby et al. 2007, Porter and Garber this volume, Chap. 4; Ferrari this volume, Chap. 8). However, cooperatively breeding species are also known for restricting reproduction to a small subgroup of adults within each group (Abbott et al. 1993; French 1997). Such restriction of reproduction will naturally lead to competition over which animals lay claim to reproductive sovereignty (Saltzman 2003; Digby et al. 2007, Yamamoto et al. this volume, Chap. 6). Indeed, it is likely that the delicate balance between the need for cooperation and the conditions that give rise to competition have ultimately shaped many aspects of callitrichid social organization and reproductive tactics.

Callitrichid females may influence the breeding success of other females in a variety of ways, including physiological suppression of ovulation, inhibition of sexual behavior (including interference during copulation), and aggression toward females attempting to join a group (reviewed in Saltzman 2003; note that subordinate females also likely play an active role in restricting their own reproduction, thus avoiding wasted reproductive effort; Wasser and Barash 1983; Saltzman 2003; Abbott et al. 2009; Saltzman et al. 2009 Yamamoto et al. this volume, Chap. 6 2009). Perhaps the most dramatic means by which one callitrichid female can influence the reproductive success of another, however, is via infanticide. Eight cases of infanticide by females other than the mother have been observed among wild callitrichids (all in the common marmoset, *Callithrix jacchus*), with another four observed and several inferred under similar circumstances, in captivity (see Table 7.1; Saltzman 2003; Abbott et al. 2009).

Across mammals, infanticide has been observed under a variety of circumstances, with perpetrators being male or female, related or unrelated to the victim. Of the five hypotheses put forth to explain infanticide (Hrdy 1979), the sexual selection hypothesis is the one most commonly invoked to explain infanticide in primates (van Schaik and Janson 2000). This hypothesis postulates that one individual, typically a male, will gain access to a potential mate more quickly by killing that individual's dependent young. For most callitrichid species, however, the killing of dependent young would have minimal influence on the mother's resumption of fertility, because lactation has little or no inhibitory effect on ovulation (Abbott et al. 1993; French 1997; Digby et al. 2007). Correspondingly, infanticide by males has never, to our knowledge, been reported in wild callitrichid groups (see Table 7.1).

Only one occurrence of maternal infanticide has been reported in wild callitrichids: a saddle-back tamarin (*Saguinus fuscicollis*) infant was cannibalized after falling off

Table 7.1 Reported cases of infanticide in callitrichids

Species	Field/ lab	Observed/ inferred	Perpetrator	Relationship to victim	Comments	References
Callithix jacchus	Field	1 Obs.	Other (subordinate) breeding female, but subsequently became dominant	Unknown	Perpetrator attacked 2-week-old infant on two occasions; gave birth approximately 1 month later	Roda and Mendes Pontes (1998)
C. jacchus	Field	1 Obs.	Probably other (dominant) breeding female	Not closely related based on genetic tests ($R<0.14$)	Perpetrator gave birth 2 days later	Digby (1995) and Nievergelt et al. (2000)
C. jacchus	Field	3 Obs.	Other (dominant) breeding female	Grandmother in one case; others unknown		Yamamoto et al. (1996) and Arruda et al. (2005)
C. jacchus	Field	1 Obs.	Other (dominant) breeding female	Unknown	Perpetrator gave birth 1 month after infanticide	Lazaro-Perea et al. (2000)
C. jacchus	Field	1 Obs.	Dominant female from neighboring group	Unknown	Involved incursions into a recently vacated home range	Melo et al. (2003)
C. jacchus	Field	1 Obs.	Other (subordinate) breeding female, but subsequently became dominant	Unknown	Perpetrator attacked and cannibalized infant after it had apparently been wounded by birds; gave birth 2 months later	Bezerra et al. (2007)
C. jacchus	Lab	2 Obs.	Eldest daughter of breeding female	Full- or half-sister	Perpetrator subsequently became dominant female	Alonso (1986)
C. jacchus	Lab	1 Obs./1 inf.	Other breeding female/unknown	Grandmother	Perpetrator gave birth 1 week later/ other breeding female gave birth 4 days later	Kirkpatrick-Tanner et al. (1996)
C. jacchus	Lab	1 Obs./ several inferred	Other (transiently subordinate) breeding female	Grandmother	Perpetrator gave birth 1 week later	Saltzman (2003) and Saltzman et al. (2008)
Saguinus fuscicollis	Field	1 Obs.	Mother of infant	Mother	Infant fell repeatedly immediately before infanticide; other adult female gave birth 2.5 months later	Herrera et al., (2000)

its carrier several times (Herrera et al. 2000). Two hypotheses may apply to this case: exploitation (e.g., cannibalism or "conspecific predation"; Hrdy 1979; Ebensperger 1998; note that cannibalism is unlikely to have a major fitness impact on the perpetrator in terms of nutrition, see Digby 2000) and maternal manipulation, in which a mother kills an infant in the hope of improving her future reproductive success (Hrdy 1979). Maternal infanticide is very rare in free-ranging populations of non-human primates (Hrdy 1999), so this case is of great interest in itself. The majority of infanticides in callitrichids, however, involve the killing of infants by breeding females other than the mother. These cases will be the focus of the remainder of this chapter.

Most cases of infanticide by females other than the mother can best be explained by the resource competition hypothesis (Hrdy and Hausfater 1984; Digby 2000). This model predicts that by killing unrelated infants, a female will gain increased access to resources for herself and her young, both immediately and in the future (Hrdy 1979; Sherman 1981; Hoogland 1995). While typically construed in terms of access to food or sleeping sites, the resource competition model can also apply to other limited resources such as breeding opportunities or access to helpers. As such, the model is compatible with the restrictive breeding patterns observed in most callitrichid species. Indeed, in most observed cases (Table 7.1), the socially dominant breeding female has killed infants born to a subordinate female, thereby allowing the dominant to maintain reproductive sovereignty (e.g., Digby 1995; Kirkpatrick-Tanner et al. 1996; Roda and Mendes Pontes 1998; see Fig. 7.1).

Fig. 7.1 Common marmoset infant killed by fellow group member (most likely the dominant female). The 24-day-old infant exhibits puncture wounds over the body and skull, a broken jaw, and torn skin around the genitalia. The infanticide was directly observed, but identification of the perpetrator could only be narrowed down to three group members: the dominant female (who gave birth 2 days later), the subordinate adult male that had been carrying the infant (and stayed with the infant for close to an hour after the attack, trying to pick it up), and a subadult male who, based on genetic analysis, was likely a maternal half-sibling of the victim. The mother of the victim went on to both carry and occasionally nurse the infants of the dominant female (Digby 1995; Nievergelt et al. 2000; photo by L. Digby)

Resource limitations and competition for breeding opportunities are likely to occur across the callitrichids. In addition, it is clear that breeding is not always restricted to a single female, and that infants born to secondary females in common marmosets, as well as in a handful of other callitrichid species, do sometimes survive. Why, then, do common marmoset infants appear to be more vulnerable to infanticide than infants of other species? To address this question, we will examine the delicate balance between the reproductive strategies of subordinate females (e.g., when do the benefits of breeding as a secondary female outweigh the costs of potential infant loss?) and those of dominant females (e.g., when secondary females do breed, when is the cost of sharing resources high enough to trigger infanticide?). We focus here on the idea that infanticide occurs when the reduced costs that favor plural breeding change just enough to push dominant breeding females over a threshold, such that the once-tolerated plural breeding becomes too costly and intense reproductive competition results. We examine species differences in breeding patterns, group composition, travel costs, infant care, population density, dominance relationships, and reproductive physiology in order to develop a preliminary framework to explain why common marmosets appear to be more susceptible to shifts in this balance between single and plural breeding and, therefore, more vulnerable to infanticide than the other callitrichid species.

7.2 Conditions that may Give Rise to Infanticide in Callitrichids

7.2.1 Multiple Breeding Females

In all but one case of infanticide observed in callitrichids under natural conditions, the groups contained two breeding females (the exception occurred when a female common marmoset from one group attacked an infant from another group: Melo et al. 2003). This is notable in a taxon known for restricting breeding to a single female in each group (reviewed in Abbott et al. 1993; Tardif et al. 2003) and suggests that when the mechanisms that usually restrict breeding fail, females may resort to infanticide to regain reproductive sovereignty. Although non-breeding females could potentially kill infants, such cases are rare among mammals (Digby 2000).

Groups containing multiple breeding females have been documented in 11 species of callitrichids (see Table 7.2; see also French 1997). More cases of plural breeding have been reported for common marmosets (11+ cases; see Table 7.2; up to 50% of surveyed groups ($n=6$); Digby personal observation), golden lion tamarins (*Leontopithecus rosalia*; 10% of group-years: Dietz and Baker 1993), and saddle-back tamarins (12% of group-years: Goldizen et al. 1996) than for the other callitrichids. It should be noted, however, that systematic data on the rates of plural breeding are not available for other species; there are only some case

Table 7.2 Plural breeding and infant outcomes in wild callitrichids

Species	Number of observed cases of multiple breeding females	Number of infants or pregnancies lost/total observed in plural breeding groups	(Survival of infants born to subordinate vs. dominant females); context/comments	References
Callithrix jacchus				
Nisia Floresta – 1	3	11/23 Infants	(33% vs. 62%); subordinate females sole carriers of own infants for at least 10 days; in contrast dominant females allow helpers to carry immediately after birth	Digby and Barreto (1993) and Digby (1995)
Nisia Floresta – 2	7+	26/68 Infants	(0% vs. 76%)[a]	Arruda et al. (2005)
Santo Antonio Ranch	1	8/13 Infants	(66% vs. 75%; latter female became dominant during the study)	Roda and Mendes Pontes (1998)
Callithrix aurita	1	2/7 Infants	(67% vs. 75%) both breeding females carried own infants exclusively for relatively long periods (10 days to 4 weeks)	Coutinho and Corrêa (1995)
Callithrix flaviceps	1	7/9 Infants	(50% vs. 100%); subordinate females sole carriers of own infants for 15 days; dominant female allowed allomaternal care at 2 days	Guimarães (1998)
Cebuella pygmaea	1[b]			de la Torre et al. (2000)
Callimico goeldi	4[c]		Mothers tolerated allomaternal care when infants less than 10 days old	Christen (1999), Porter (2001), Masataka (1981) as cited in Porter (2001), and Rehg (pers. comm.)
Callibella humillis	"Several"		Singleton births?; lack of allomaternal care; parking of infants	van Roosmalen and van Roosmalen (2003)
Saguinus fuscicollis				
Acre	1	0/4 Infants	(100% vs. 100%); all survived for 4+ months	Calegaro-Marques et al. (1995)
Cocha Cashu	5 (10.6% of group-years)	1/5 Pregnancies	Dominance not specified; at least one breeding daughter lost infant immediately after birth (breeder's mother was in last month of pregnancy); surviving infants born at least 3 months apart	Goldizen et al. (1996)

(continued)

Table 7.2 (continued)

Species	Number of observed cases of multiple breeding females	Number of infants or pregnancies lost/total observed in plurally breeding groups	(Survival of infants born to subordinate vs. dominant females); context/comments	References
Quebrada Blanco	1	1/3 Infants	(100% vs. 0%). Infants born to subordinate after maternal infanticide. Survived at least 2 months	Herrera et al. (2000)
Saguinus mystax				
Padre Isla	3[d]	1 female per group had surviving offspring	Dominance status not noted; plural breeding groups were among the largest in population, with an average of 10 individuals	Garber et al. (1993)
Quebrada Blanco	1	1/3 Infants	Infant died on day of birth; other breeding female gave birth next day; dominance status not noted	Smith et al. (2001) and Löttker et al. (2004)
Saguinus oedipus	3[e]	3/4 Infants	New males had immigrated into group; dominance status not noted	Savage et al. (1996, 1997)
Leontopithecus rosalia	11 (in 6 groups)/10% of group samples[f]	6 Infants; 7–8% of group-breeding seasons	(33% vs. 68%) Subordinates/daughters more likely to lose infants; some groups had unrelated males, others did not. Incestuous matings never resulted in successful young	Dietz and Baker (1993) and Baker et al. (2002)
Leontopithecus chrysomelas	1?			Raboy and Dietz (2004)

[a]Based on groups with two breeding females that, however, maintained a monogamous mating system (i.e., using extra-group copulations)

[b]Based on short interbirth interval

[c]Based on presence of two infants

[d]Based on presence of two pregnant females or one pregnant and one lactating female

[e]Based on presence of two pregnant females

[f]Dietz and Baker (1993) assume polygyny as long as the two females that have bred in the past remain in the group, even if only one female continues to breed

studies. Nevertheless, it seems likely that species differ in their propensities toward plural breeding.

Why some groups contain a second breeding female and others do not is not well understood, but several determinants of plural breeding have been proposed, including group composition, infant-care costs, timing of births, reduced opportunities for dispersal, and dominance relationships (see reviews by French 1997; Smith et al. 2001, Yamamoto et al. this volume, Chap. 6). A thorough examination of this issue is beyond the scope of this chapter, but it is clear that a propensity toward plural breeding is an important factor in determining whether or not infanticide is likely to occur in a given species. We cover some of the basic costs of these variables and their possible relationship to the threat of infanticide, below.

7.2.2 *Infant Care Costs, Ecology, and the Propensity for Multiple Breeding Females*

7.2.2.1 Group Composition

The degree of tolerance for a second breeding female and her infants may be determined by group size and composition. Some studies have indicated that infant survivorship is significantly correlated with the number of adults in a group (e.g., common marmoset: Koenig 1995; pygmy marmoset, *Cebuella*: Heymann and Soini 1999, reviewed in Yamamoto et al. this volume, Chap. 6), and one might assume that in a large group, the large number of potential helpers would enable sharing of infant care costs to a degree that two breeding females could successfully raise young. Nevertheless, an association has not been found between group size and plural breeding (Dietz and Baker 1993; Digby 1995; Goldizen et al. 1996; Table 7.3).

Instead, the actual composition of the group, especially relatedness among group members, may be more important than group size. Infanticide might be predicted to be less likely to occur in groups with closely related breeding females (mother and daughter, sisters) compared to those with unrelated females, because of inclusive fitness costs. The data do not currently support this prediction, however, as infanticide has been observed in groups with both unrelated (e.g., Digby 1995) and related (e.g., Yamamoto et al. 1996) breeding females (Table 7.1). The fact that females have killed related infants may indicate that these females would have incurred very high costs by tolerating the presence of other females' infants in the group (e.g., Hager and Johnstone 2004).

The presence of an unrelated male may also increase both the likelihood of a group breeding plurally (Saltzman et al. 2004, 2008) and the survivorship of young (e.g., no infants survived from father–daughter inbreeding in golden lion tamarins: Dietz and Baker 1993). Notably, in some plurally breeding groups of common marmosets, females may avoid inbreeding by mating with extra-group males (Arruda et al. 2005). Species or populations may differ in females' access to extra-group

Table 7.3 Group size, home range and daily path length in wild callitrichids

Species	Group size (average)	Home range size (ha)	Daily path length (m/day)	References
Callithrix jacchus	3–16 (8.7)	2–5.2	912–1,243	Digby and Barreto (1996), Lazaro-Perea et al. (2000), and Koenig (1995)
Callithrix flaviceps	5–20 (13)	33.9–35.5	884–1,223	Ferrari and Diego (1992), Ferrari and Rylands (1994), and Guimarães (1998)
Callithrix aurita	4–11	16.5–35.3	959–986	Ferrari et al. (1996) and Martins (1998)
Mico intermedius	9–15 (12)	22.1	772–2,115	Rylands (1986a, b) and Ferrari and Rylands (1994)
Mico argentatus	6–10 (7.3)	4–35		Albernaz and Magnusson (1999) and Tavares and Ferrari (2002)
Cebuella pygmaea	2–9 (5.0)	0.1–1.09	280–300	Heymann and Soini (1999) and de la Torre et al. (2000)
Callimico goeldii	4–12 (7.7)	30–150	2,000	Porter (2001)
Callibella humilis	up to 30 (6–8)	–	–	van Roosmalen and van Roosmalen (2003)
Saguinus fuscicollis	2–10 (5.3)	30–149	1,150–2,700	Goldizen (2003) and Heymann (2000, 2001)
Saguinus mystax	3–11 (5.3)	40–45	1,500–1,720	Garber (1988) and Heymann (2000)
Saguinus tripartitus	2–9 (5.1)	16–21	500–2,300	Kostrub (2003)
Leontopithecus rosalia	2–11 (5.4)	21.7–229	955–2,405	Baker et al. (2002), Dietz et al. (1997) and Kierulff et al. 2002
Leontopithecus chrysomelas	3–10 (6.7)	36–130	1,410–2,044	Baker et al. (2002) and Raboy and Dietz (2004)

males due to differences in the degree of overlap in home ranges and the frequency of encounters with neighboring groups. It is unclear, however, whether or not the presence of an unrelated male influences vulnerability to infanticide other than by increasing the likelihood of breeding by a secondary female.

7.2.2.2 Travel Costs

The ability of a group to support two breeding females, and the threshold at which one female will no longer tolerate the presence of another female's young, are likely also tied to costs of infant care. Though direct measures of energy requirements are difficult to obtain under natural conditions, the distance that a parent or helper must carry infants can be a useful indirect measure of costs (e.g., Tardif 1997).

Digby and Barreto (1996) noted that common marmosets that were carrying infants traveled significantly less than the group as a whole traveled when no young infants were present. Qualitative observations indicated that infant carriers often stayed in central areas of the group's relatively small home range (see Table 7.3), while other group members foraged in a larger circuit around them. This reduction in travel may reduce infant-care costs to the point where a second set of twins could be raised without energetically over-taxing helpers (Digby and Barreto 1996). Such a strategy would be difficult for species with much larger home ranges, as fewer active infant carriers would be left behind.

The cost of carrying infants should also vary with the length of time that infants take to achieve locomotor independence, which differs considerably among the callitrichid species. Under captive conditions, for example, in the first 8 weeks after birth, common marmosets are carried less than 20% of the time, while saddle-back tamarins are carried more than 60% and golden lion tamarins about 30% of the time (Tardif et al. 1993, 2003). Shorter infant dependency periods may allow for a greater initial tolerance of two sets of infants, but can also leave the offspring of secondary breeding females vulnerable to infanticide, if other costs, such as competition over access to helpers, become too great.

7.2.2.3 Infant Care and Overlapping Dependency Periods

While both small home range size and shorter dependency periods should reduce the costs of infant care and, therefore, favor plural breeding in common marmosets, they may not be enough to counter the cost of two sets of infants born in close succession, resulting in overlapping dependency periods and more intense competition for limited resources (e.g., helpers). Infants born to subordinate female common marmosets were unlikely to survive if they were born less than a month before or after the infants of the dominant female (Digby 1995; Saltzman et al. 2008). Similarly, for saddle-back tamarins, Goldizen et al. (1996) suggested that a 3-month interval before or after the birth of the infants of the primary breeding female is necessary for the successful rearing of infants by a secondary female. Breeding asynchronously may be more difficult in the *Callithrix* species than in other species, as the *Callithrix* species usually breed twice per year, with births tending to cluster around two birth peaks (reviewed in Digby et al. 2007). Golden lion tamarins have one of the more pronounced birth peaks among the callitrichids, with most births occurring in a 3-month window (French et al. 2002); thus, plurally breeding groups in this species may frequently face periods in which two sets of infants need to be cared for simultaneously. *Saguinus* species, on the other hand, exhibit only weak seasonality and annual births (reviewed in French 1997; Digby et al. 2007), making them less vulnerable to overlapping periods of infant dependency.

7.2.2.4 Population Density/Habitat Saturation

Habitat saturation has also been put forth as a possible correlate of plural breeding in callitrichids. In Poço das Antas, Brazil (golden lion tamarins), Cocha Cashu, Peru (saddle-back tamarins) and Santo Antonio, Brazil (common marmosets), high population densities appear to make it difficult for an adult female to find a breeding position outside of her natal group (Dietz and Baker 1993; Goldizen et al. 1996; Roda and Mendes Pontes 1998). Based on the hypothesis that emigration is risky, Dietz and Baker (1993) modeled the costs and benefits to dominant female golden lion tamarins, of tolerating a breeding daughter, and found that mothers gain an overall benefit by allowing daughters to remain and breed in the natal group. When the costs and benefits to the primary breeding female shift, however – for example, if breeding within the natal group results in inbreeding – then eviction of the daughter from the natal group is predicted (Dietz and Baker 1993; Dietz et al. 2000; Baker et al. 2002). Notably, when female golden lion tamarins do breed with related males, there are much higher mortality rates (Dietz et al. 2000). Such a pattern may indicate that secondary females are willing to risk reproduction within their natal group if the risk associated with emigration is sufficiently high. Infants born to these females would very likely be all the more vulnerable to infanticide.

7.2.3 Dominance Relationships and the Propensity for Infanticide

For infanticide to occur, a perpetrator must gain access to the intended victim. Among primates, which typically carry young infants, this usually means that the perpetrator must have either greater physical strength or greater social power than the infant's caretaker (Hrdy 1976; Maestripieri and Carroll 1998; Treves 2000). Callitrichids typically share the care of young, sometimes even from the first day following birth; however, subordinate breeding female common marmosets appear to be more cautious than dominant females, not allowing others to carry their infants for up to 10 days postpartum (Digby 1995). It is unclear if similar patterns occur in plurally breeding groups of other callitrichid species.

Most cases of infanticide in common marmosets appear to be perpetrated by females that are already behaviorally dominant over the victim's mother or that will become dominant following the infanticide (see Table 7.1; it is to be noted, however, that this is not necessarily the case under captive conditions: Saltzman 2003; Saltzman et al. 2008). In golden lion tamarins, daughters (which are usually, but not always, subordinate to their mothers) typically suffer higher infant mortality than their mothers when two females breed within the same group (Dietz and Baker 1993); however, the contribution of infanticide is unclear. Further research on potential differences in the intensity and/or nature of dominance relationships across the callitrichid species is necessary to elucidate the role they may play in the relative vulnerability of their young.

7.2.4 Reproductive Physiology

7.2.4.1 Suppression of Reproduction

Callitrichid species differ markedly, both in the degree to which reproduction is suppressed in subordinate females and in the mechanisms of suppression. As described above, laboratory studies indicate that mechanisms of reproductive failure range from inhibition of sexual behavior in fully fertile subordinate females (e.g., golden lion tamarins: French et al. 2002) to inhibition of fertility as a consequence of ovulation suppression (e.g., common marmosets: Abbott 1984; cotton-top tamarins, *S. oedipus*: Snowdon et al. 1993). Among species exhibiting suppression of reproductive physiology, the consistency of this suppression varies considerably, at least in captivity, from anovulation and reproductive failure in virtually all subordinate females (e.g., cotton-top tamarins: Snowdon et al. 1993) to the occurrence of ovulatory cyclicity and even conceptions in a sizeable minority of subordinates (e.g., common marmoset: Saltzman et al. 1997, 2004).

Although these patterns have not always been borne out in studies of free-living animals (reviewed in Digby et al. 2007), it is tempting to speculate that interspecific differences in the stringency of reproductive suppression may be associated with differences in the ability of groups to support the infants of more than a single breeding female. For example, the absence of strict suppression in callimico (*Callimico goeldii*; Dettling and Price 1999), in which secondary breeding females are routinely tolerated (Porter 2001), may reflect the low costs of raising singleton infants when compared to the costs of raising twins, typical in the other callitrichid species. The more context-dependent suppression (e.g., varying with age: Baker et al. 2002; group composition: Abbott 1984; Saltzman et al. 1997, 2004) seen in golden lion tamarins and common marmosets may suggest that ecological and/or social conditions change frequently enough to make it cost-effective for females to employ a more flexible strategy, attempting to breed as subordinates under some circumstances, but forgoing breeding attempts under other circumstances (Yamomoto et al. this volume, Chap. 6).

Differences in the extent of suppression may be associated with differences in vulnerability to infanticide. Strict forms of reproductive suppression may have evolved in subordinate breeding females in response to a high risk of reproductive failure; this may enable the females to avoid wasting reproductive effort (Wasser and Barash 1983; Abbott et al. 2009; Saltzman et al. 2009). It also follows that this stricter reproductive suppression is expected to reduce the frequency of plural breeding and, consequently, the likelihood of infanticide. In contrast, more flexible forms of suppression may reflect an increased likelihood that subordinate females might, at least occasionally, breed successfully (depending on timing of births, etc.). By enabling subordinate females to breed, these flexible forms of suppression may increase both the rate of plural breeding and, consequently, the likelihood of infanticide. Further research into species differences in the extent and mechanisms of reproductive suppression might therefore shed light on patterns of infanticide as well.

7.2.4.2 Proximate Mechanisms of Female Aggression Toward Infants

When two females breed concurrently within the same social group, the relative timing of births may be a critical determinant of infanticide. Infanticide was committed by female common marmosets in the late stages (i.e., final 1–2 months) of pregnancy in most known cases from wild populations (in which pregnancy status was reported) and in many captive cases as well (Saltzman 2003). Correspondingly, in a recent study of captive common marmosets, multiparous females exhibited a markedly reduced attraction to, as well as tolerance of, unfamiliar infants during late pregnancy, as compared to early pregnancy and the early post-partum period (Saltzman and Abbott 2005). These findings suggest that the hormonal milieu of late pregnancy may inhibit maternal responsiveness and promote infanticide in this species. Such a pattern would contrast strikingly with that in other mammals, in which hormonal changes in late pregnancy typically enhance maternal responsiveness (Numan and Insel 2003). Assessment of variation in this tendency in other callitrichid species awaits further study.

7.3 Discussion: Balancing Cooperation and Competition in Callitrichids

Callitrichids exhibit numerous behavioral, ecological, and reproductive differences among species that may render the infants of some species more likely to fall victim to infanticide. Strikingly, almost all the reported cases of infanticide in wild callitrichids have occurred in plural breeding groups, with one breeding female killing the offspring of another. Perhaps the most important determinant of the likelihood of infanticide, therefore, is the propensity for groups to contain a second breeding female. With the exception of callimico and, perhaps, dwarf marmosets (*Callibella*), plural breeding in callitrichid groups appears to be the exception rather than the rule (French 1997). In species such as golden lion tamarins and saddle-back tamarins, ecological and social factors such as high population density and/or the presence of unrelated males in the groups may encourage secondary females to attempt breeding in the presence of a primary breeding female. In common marmosets, saturated habitats, the ability to live in small, highly overlapping home ranges, and access by females to unrelated males in either their own or a neighboring group may be key factors giving rise to plural breeding.

Once multiple females breed within a group, these females may still attempt to limit each other's reproductive success. Overlapping periods of infant dependency, births occurring during the final stages of another female's pregnancy, relatedness and relative social status of breeding females may all contribute to the fitness costs associated with tolerating a second set of infants within a group. When the costs exceed a given threshold and become too high, females may resort to infanticide as a means of decreasing resource competition and increasing their own overall reproductive success. The ecology, social organization and reproductive biology of common marmosets appear to put this species at greater risk for infanticide than other callitrichids.

Could infanticide be occurring in other callitrichid species? Importantly, in most of these species, when groups are observed to contain two pregnant females, most secondary females are not successful in raising their young (i.e., one or both of the pregnancies are not carried to term or neonates from one or both females are never found; see Table 7.2). In golden lion tamarins, for example, infant survival rates in polygynous groups were roughly half of those in monogamous groups (47% vs. 83%: Dietz and Baker 1993; Baker et al. 2002), and the infant survival rate for subordinate females was significantly lower than that for dominant females (33% vs. 68%: Dietz and Baker 1993). A similar pattern arises in cotton-top tamarins, with 75% ($n = 4$ pregnancies) of plural breeding females being unsuccessful (infants not seen) compared to only 8% loss ($n = 13$ pregnancies) in groups with a single breeding females (excluding losses that occurred in a drought year; Savage et al. 1996). Thus, females in several callitrichid species lose infants under conditions that are consistent with the resource competition model. It remains to be determined whether infant losses under these conditions result from infanticide or from other sources of mortality.

Why have no infanticides been reported for these species? Perhaps it is not surprising that infanticide has not been observed, given the probable rarity of the behavior combined with the timing of most births (occurring at night or out of sight of the observer). In fact, in view of these constraints, the fact that infanticide has been observed repeatedly in free-living common marmosets is quite striking. Importantly, the disparity between common marmosets and at least some other callitrichid species is unlikely to be caused by differences in the number of observation hours. Golden lion tamarins and saddle-back tamarins, for example, are both subjects of long-term field studies. Another possibility is that some species employ an earlier form of infant loss, in which unfavorable social or ecological conditions promote fetal reabsorption or spontaneous abortion. Only with additional field studies and more long-term data will we know whether these disappearances are part of a typical pattern of spontaneous mortality (reflecting, for example, differences in infant viability or maternal care between experienced and primiparous breeding females, or inbreeding by subordinate females) or whether they reflect a greater risk of neonatal infanticide in plural breeding groups.

At this point, we can only speculate about the relative importance of the ecological, behavioral, and physiological parameters discussed here, in determining patterns of infanticide. Future studies on species differences in the propensity for infanticide will need to focus observations on females in plural breeding groups, especially during the period immediately following birth. In addition, as more data become available, detailed analyses of the costs and benefits to primary breeding females of tolerating a second breeding female and her young, as performed for golden lion tamarins by Dietz and Baker (1993; Baker et al. 2002),will need to be carried out for more species. Ultimately, we believe that such studies will point to a delicate balance between the need for cooperative care of the young and the inherent competition that results from the restriction of reproduction to a small minority of adults. Thus, perhaps ironically, it is the cooperative nature of callitrichid reproductive strategies that ultimately gives rise to the conditions driving some females to infanticide.

References

Abbott DH (1984) Behavioral and physiological suppression of fertility in subordinate marmoset monkeys. Am J Primatol 6:69–186

Abbott DH, Barrett J, George LM (1993) Comparative aspects of the social suppression of reproduction in female marmosets and tamarins. In: Rylands AB (ed) Marmosets and tamarins: systematics, behavior, and ecology. Oxford Science Publications, Oxford, pp 152–162

Abbott DH, Digby L, Saltzman W (2009) Reproductive skew in female common marmosets: contributions on infanticide and subordinate self-restraint. In: Jones C, Hager R (eds) Reproductive skew in vertebrates: proximate ultimate causes. Cambridge University Press, Cambridge

Albernaz AL, Magnusson WE (1999) Home-range size of the bare-ear marmoset (*Callithrix argentata*) at Alter do Chão, Central Amazonia, Brazil. Int J Primatol 20(5):665–677

Alonso C (1986) Fracasso no inibição da reprodução du uma fêmea subordinada e troca de hierarquia em um grupo familiar de *Callithrix jacchus jacchus*. In: Thiago de Mello M (ed) A primatologia no Brasil – 2. Sociedade Brasileira de Primatolgia, Campinas, p 203

Altmann J (1997) Mate choice and intrasexual reproductive competition: contributions to reproduction that go beyond acquiring more mates. In: Gowaty PA (ed) Feminism and evolutionary biology: boundaries, intersections and frontiers. Chapman & Hall, New York, pp 320–333

Arruda MF, Araujo A, Sousa MBC, Albuquerque FS, Albuquerque ACSR, Yamamoto ME (2005) Two breeding females within free-living groups may not always indicate polygyny: alternative subordinate female strategies in common marmosets (*Callithrix jacchus*). Folia Primatol 76:10–20

Baker AJ, Bales K, Dietz JM (2002) Mating system and group dynamics in lion tamarins. In: Kleiman DG, Rylands AB (eds) Lion tamarins: biology and conservation. Smithsonian Institution Press, Washington, DC, pp 188–212

Bezerra BM, Souto AS, Schiel N (2007) Infanticide and cannibalism in a free-ranging plurally breeding group of common marmosets (*Callithrx jacchus*). Am J Primatol 69:945–952

Calegaro-Marques C, Bicca-Marques JC, Azevado MAO (1995) Two breeding females in a *Saguinus fuscicollis weddelli* group. Neotrop Primates 3(4):183

Christen A (1999) Survey of Goeldi's monkeys (*Callimico goeldii*) in northern Bolivia. Folia Primatol 70(2):107–111

Coutinho PEG, Corrêa HKM (1995) Polygyny in a free-ranging group of buffy-tufted-ear marmosets, *Callithrix aurita*. Folia Primatol 65:25–29

de la Torre S, Snowdon CT, Bejarana M (2000) Effects of human activities on wild pygmy marmosets in Ecuadorian Amazonia. Biol Conserv 94(2):153–163

Dettling A, Price CR (1999) Hormonal monitoring of age at sexual maturation in female Goeldi's monkey (*Callimico goeldii*) in their family groups. Am J Primatol 48:77–83

Dietz JM, Baker AJ (1993) Polygyny and female reproductive success in golden lion tamarins, *Leontopithecus rosalia*. Anim Behav 46:1067–1078

Dietz JM, Peres CA, Pinder, L (1997) Foraging ecology and use of space in wild golden lion tamarins (*Leontopithecus rosalia*). Amer J Primatol 41(4): 289–305

Dietz JM, Baker AJ, Ballou JD (2000) Demographic evidence of inbreeding depression in wild golden lion tamarins. In: Young AG, Clarke GM (eds) Genetics. Demography and Viability of Fragmented Populations. Cambridge University Press, Edinburgh, pp 203–211

Digby L (1995) Infant care, infanticide, and female reproductive strategies in polygynous groups of common marmosets (*Callithrix jacchus*). Behav Ecol Sociobiol 37:51–61

Digby LJ (2000) Infanticide by female mammals: implications for the evolution of social systems. In: van Schaik CP, Janson CH (eds) Infanticide by males and its implications. Cambridge University Press, Cambridge, pp 423–446

Digby LJ, Barreto CE (1996) Activity and ranging patterns in common marmosets (*Callithrix jacchus*): implications for reproductive strategies. In: Norconk M, Rosenberger A, Garber PA (eds) Adaptive radiations of Neotropical Primates. Plenum, New York, pp 173–185

Digby, L.J., Barreto CE (1993) Social organization in a wild population of *Callithrix jacchus*. Part I: group composition and dynamics. Folia Primatologica 61:123–134

Digby LJ, Ferrari SF, Saltzman W (2007) Callitrichines: the role of competition in a cooperatively breeding species. In: Campbell C, Fuentes A, MacKinnon KC, Panger M, Bearder S (eds) Primates in perspective. Oxford, Oxford University Press, pp 85–105

Ebensperger LA (1998) Strategies and counterstrategies to infanticide in mammals. Biol Rev 73:321–326

Ferrari SF (this volume) Social hierarchy and dispersal in free-ranging buffy-headed marmosets (*Callithrix flaviceps*). In: Ford SM, Porter LM, Davis LC (eds) The smallest anthropoids: the marmoset/callimico radiation. Springer, New York, pp 155–165

Ferrari SF, Diego VH (1992) Long-term changes in a wild marmoset group. Folia Primatol 58(4):215–218

Ferrari SF, Rylands AB (1994) Activity budgets and differential visibility in three species of marmosets. Folia Primatol 63(2):78–83

Ferrari SF, Correa HKM, Coutinho PEG (1996) Ecology of the "southern" marmosets (*Callithrix aurita* and *Callithrix flaviceps*). In: Norconk MA, Rosenberger AL, Garber PA (eds) Adaptive radiations of Neotropical Primates. Plenum, New York, pp 157–171

French JA (1997) Proximate regulation of singular breeding in callitrichid primates. In: Soloman NG, French JA (eds) Cooperative breeding in mammals. Cambridge University Press, Cambridge, pp 34–75

French JA, de Vleeschouwer K, Bales K, Heistermann M (2002) Lion tamarin reproductive biology. In: Kleiman DG, Rylands AB (eds) Lion tamarins: biology and conservation. Smithsonian Institution Press, Washington, DC, pp 133–156

Garber PA (1988) Diet, foraging patterns, and resource defense in a mixed species troop of *Saguinus mystax* and *Saguinus fuscicollis* in Amazonian Peru. Behaviour 105:18–34

Garber PA (1997) One for all and breeding for one: cooperation and competition as a tamarin reproductive strategy. Evol Anthropol 5:187–199

Garber PA, Encarnacion F, Moya L, Pruetz JD (1993) Demographic and reproductive patterns in moustached tamarin monkeys (*Saguinus mystax*): implications for reconstructing platyrrhine mating systems. Am J Primatol 29(4):235–254

Goldizen AW (2003) Social monogamy and its variations in callitrichids: do these relate to the costs of infant care? In: Reichard UH, Boesch C (eds) Monogamy: mating strategies and partnerships in birds, humans, and other mammals. Cambridge University Press, Cambridge, pp 232–247

Goldizen AW, Mendelson J, van Vlaardingen M, Terborgh J (1996) Saddle-back tamarin (*Saguinus fuscicollis*) reproductive strategies: evidence from a thirteen-year study of a marked population. Am J Primatol 38(1):57–84

Gowaty P (1997) Sexual dialectics, sexual selection, and variation in female reproductive behavior. In: Gowaty PA (ed) Feminism and evolutionary biology: boundaries, intersections and frontiers. Chapman & Hall, New York, pp 351–84

Guimarães A (1998) Ecology and social behavior of buffy-headed marmosets, *Callithrix flaviceps*. Neotrop Primates 6(2):51–52

Hager R, Johnstone RA (2004) Infanticide and control of reproduction in cooperative and communal breeders. Anim Behav 67:941–949

Herrera ERT, Knogge C, Heymann EW (2000) Infanticide in a group of wild saddle-back tamarins, *Saguinus fuscicollis*. Am J Primatol 50:153–157

Heymann EW (2000) The number of adult males in callitrichine groups and its implications for callitrichine social evolution. In: Kappeler PM (ed) Primate males: causes and consequences of variation in group composition. Cambridge University Press, Cambridge, pp 64–71

Heymann EW (2001) Interspecific variation of scent-marking behaviour in wild tamarins, *Saguinus mystax* and *Saguinus fuscicollis*. Folia Primatol 72:253–267

Heymann EW, Soini P (1999) Offspring number in pygmy marmosets, *Cebuella pygmaea*, in relation to group size and number of adult males. Behav Ecol Sociobiol 46:400–404

Hoogland JL (1995) The black-tailed prairie dog: social life of a burrowing mammal. University of Chicago Press, Chicago
Hrdy SB (1976) Care and exploitation of nonhuman primate infants by conspecifics other than the mother. In: Rosenblatt L, Hinde RA, Shaw R, Beer C (eds) Advances in the study of behavior, vol 6. Academic, New York, pp 101–58
Hrdy SB (1979) Infanticide among animals: a review, classification, and examination of the implications for the reproductive strategies of females. Ethol Sociobiol 1:13–40
Hrdy SB (1981) The woman that never evolved. Harvard University Press, Cambridge, MA
Hrdy SB (1999) Mother nature: maternal instincts and how they shape the human species. Ballantine Books, New York
Hrdy SB, Hausfater G (1984) Comparative and evolutionary perspectives on infanticide: introduction and overview. In: Hausfater G, Hrdy SB (eds) Infanticide: comparative and evolutionary perspectives. Aldine de Gruyter, New York, pp xiii–xxxv
Jones C (2003) Sexual selection and reproductive competition in Primates: new perspectives and directions. American Society of Primatologists, Norman
Kirkpatrick-Tanner M, Aeschlimann C, Anzenberger G (1996) Occurrence of an infanticide within captive polygynous group of common marmosets, *Callithrix jacchus*. Folia Primatol 67(1):52–58
Koenig A (1995) Group size, composition, and reproductive success in wild common marmosets (*Callithrix jacchus*). Am J Primatol 35:311–17
Kostrub CE (2003) The Social Organization and Behavior of Golden Mantled Tamarins, *Saguinus tripartitus*, in eastern Ecuador [PhD Thesis]. University of California Davis
Lazaro-Perea C, Castro CSS, Harrison R, Araujo A, Arruda MF, Snowdon CT (2000) Behavioral and demographic changes following the loss of the breeding female in cooperatively breeding marmosets. Behav Ecol Sociobiol 48:137–146
Löttker P, Huck M, Heymann EW (2004) Demographic parameters and events in wild moustached tamarins (*Saguinus mystax*). Am J Primatol 64:425–449
Maestripieri D, Carroll KA (1998) Risk factors for infant abuse and neglect in group-living rhesus monkeys. Psychol Sci 9:143–5
Martins MM (1998) Feeding ecology of *Callithrix aurita* in a forest fragment of Minas Gerais. Neotrop Primates 6(4):126–127
Melo L, Mendes Pontes AR, Monteiro da Cruz MAO (2003) Infanticide and cannibalism in wild common marmosets. Folia Primatol 74(1):48–50
Nievergelt CM, Digby LJ, Ramakrishnan U, Woodruff DS (2000) Genetic analysis of group composition and breeding system in a wild common marmoset (*Callithrix jacchus*) population. Int J Primatol 21(1):1–20
Numan M, Insel TR (2003) The neurobiology of parental behavior. Springer, New York
Porter LM (2001) Social organization, reproduction and rearing strategies of *Callimico goeldii*: new clues from the wild. Folia Primatol 72(2):69–79
Porter LM, Garber PG (this volume) Social behavior of callimicos: mating strategies and infant care. In: Ford SM, Porter LM, Davis LC (eds) The smallest anthropoids: the marmoset/callimico radiation. Springer, New York, pp 87–101
Pusey A, Williams J, Goodall J (1997) The influence of dominance rank on the reproductive success of female chimpanzees. Science 277:828–831
Raboy BE, Dietz JM (2004) Diet, foraging, and use of space in wild golden-headed lion tamarins. Am J Primatol 63:1–15
Roda SA, Mendes Pontes AR (1998) Polygyny and infanticide in common marmosets in a fragment of Atlantic Foreset of Brazil. Folia Primatol 69(6):372–376
Rylands AB (1986a) Ranging behaviour and habitat preference of a wild marmoset group, *Callithrix humeralifer* (Callitrichidae, Primates) J. Zool Lond (A) 210:489–514
Rylands AB (1986b) Infant-carrying in a wild marmoset group, *Callithrix humeralifer*: evidence for a polyandrous mating system. In: Thiago de Mello M (ed) A primatologia No Brasil – 2. Sociedade Brasileira de Primatologia, Brasilia, pp 131–144
Saltzman W (2003) Reproductive competition among female common marmosets (*Callithrix jacchus*): proximate and ultimate causes. In: Jones C (ed) Sexual selection and reproductive

competition in primates: new perspectives and directions. American Society of Primatologists, Norman, pp 197–229

Saltzman W, Abbott DH (2005) Diminished maternal responsiveness during pregnancy in multiparous female common marmosets. Horm Behav 47:151–163

Saltzman W, Schultz-Darken NJ, Abbott DH (1997) Familial influences on ovulatory function in common marmosets (*Callithrix jacchus*). Am J Primatol 41(3):159–177

Saltzman W, Pick RR, Salper OJ, Liedl KJ, Abbott DH (2004) Onset of plural cooperative breeding in common marmoset families following replacement of the breeding male. Anim Behav 68:59–73

Saltzman W, Liedl KJ, Salper OJ, Pick RR, Abbott DH (2008) Post-conception reproductive competition in cooperatively breeding common marmosets. Horm Behav 53:274–286

Saltzman W, Digby LH, Abbott DH (2009) Reproductive skew in female common marmosets: what can proximate mechanisms tell us about ultimate causes? Proc Biol Sci 276:389–399

Savage A, Giraldo LH, Soto LH, Snowdon CT (1996) Demography, group composition, and dispersal in wild cotton-top tamarins (*Saguinus oedipus*) groups. Am J Primatol 38(1):85–100

Savage A, Shideler SE, Soto LH, Causado J, Giraldo LH, Lasley BL, Snowdon CT (1997) Reproductive events of wild cotton-top tamarins (*Saguinus Oedipus*) in Colombia. Am J Primatol 43:329–337

Sherman PW (1981) Reproductive competition and infanticide in Belding's ground squirrels and other animals. In: Alexander RD, Tinkle D (eds) Natural selection and social behavior: recent research and theory. Chiron Press, New York, pp 311–331

Smith AC, Tirado Herrara ER, Buchanan-Smith HM (2001) Multiple breeding females and allonursing in a wild group of moustached tamarins (*Saguinus mystax*). Neotrop Primates 9(2):67–69

Snowdon CT, Ziegler TE, Widowski TM (1993) Further hormonal suppression of eldest daughter cotton-top tamarins following birth of infants. Am J Primatol 31(1):11–21

Tardif SD (1997) The energetics of parental behavior and the evolution of alloparental care in marmosets and tamarins. In: Soloman NG, French JA (eds) Cooperative breeding in mammals. Cambridge University Press, Cambridge, pp 11–33

Tardif SD, Harrison ML, Simek MA (1993) Communal infant care in marmosets and tamarins: relation to energetics, ecology, and social organization. In: Rylands AB (ed) Marmosets and tamarins: systematics, behaviour, and ecology. Oxford Science Publications, Oxford, pp 220–234

Tardif SD, Smucny DA, Abbott DH, Mansfield K, Schultz-Darken N, Yamamoto ME (2003) Reproduction in captive common marmosets (*Callithrix jacchus*). Comp Med 53:364–368

Tavares LI, Ferrari SF (2002) Diet of the silvery marmoset (*Callithrix argentata*) at CFPn: seasonal and longitudinal variation. In: Lisboar PLB (ed) Caxiuanã, Populações Tradicionais. Meio Físico e Diversidade Biológica. CNPq/MCT, Belèm, pp 707–719

Treves A (2000) Prevention of infanticide: the perspective of infant primates. In: van Schaik CP, Janson CH (eds) Infanticide by males and its implications. Cambridge University Press, Cambridge, pp 223–238

van Roosmalen MGM, van Roosmalen T (2003) The description of a new marmoset genus, *Callibella*, (Callitrichinae, Primates), including its molecular phylogenetic status. Neotrop Primates 11(1):1–10

van Schaik CP, Janson CH (2000) Infanticide by males and its implications. Cambridge University Press, Cambridge

Wasser SK, Barash DP (1983) Reproductive suppression among female mammals: implications for biomedicine and sexual selection theory. Q Rev Biol 58:513–538

Yamamoto ME, Arruda MF, de Sousa MBC, Alencar AI (1996) Mating systems and reproductive strategies in *Callithrix jacchus* females. Abstracts XVI Congress of the International Primatological Society and XIX Conferences American Society of Primatologists, p 56

Yamamoto ME, Arruda MF, Alencar AI, de Sousa MBC, Araújo A (this volume) Mating systems and female–female competition in the common marmosets, *Callithrix jacchus*. In: Ford SM, Porter LM, Davis LC (eds) The smallest anthropoids: the marmoset/callimico radiation. Springer, New York, pp 119–133

Chapter 8
Social Hierarchy and Dispersal in Free-Ranging Buffy-Headed Marmosets (*Callithrix flaviceps*)

Stephen F. Ferrari

Abstract Behavior patterns in a large, free-ranging social group of buffy-headed marmosets (*Callithrix flaviceps*) were monitored over a 7-month period during which four adults – one male and three females – emigrated in two separate events. Social interactions such as play and allogrooming were relatively frequent, but agonistic interactions were rare, being observed, on average, less than twice per observation day for a group with between seven and ten adult members at any given time. With the exception of events involving the breeding female, intra-sexual agonism was almost non-existent between adults, and male→female aggression was five times more frequent than female→male. Absent between males and rare in females, submissive behavior was almost invariably directed by non-breeding females towards males and the breeding female. Taken together, these interactions point to a three-tiered social hierarchy within the group, with the breeding female in the top tier, followed by males in the second, and finally, non-breeding females. However, no one male was more dominant socially than any other, nor was any non-breeding female more subordinate. Male and female group members dispersed under different circumstances, but there is little evidence in either case to suggest that emigrations were a consequence of intra-group agonism, related to social rank or to competition for resources.

Resumen Patrones de comportamiento en un grupo de *Callithrix flaviceps* fueron monitoreados a lo largo de un período de siete meses durante el cual cuatro adultos –un macho y tres hembras- emigraron en dos eventos distintos. Las interacciones sociales como un comportamiento lúdico y acicalamiento fueron relativamente frecuentes, pero las interacciones agonísticas fueron raras, siendo observadas, en promedio, menos de dos veces por día de observación para un grupo de entre siete y diez miembros adultos en un momento dado. Con la excepción de los eventos que involucraron a la hembra en estado de reproducción, el agonismo intrasexual estuvo prácticamente ausente entre los adultos, y la agresión macho→hembra fue

S.F. Ferrari (✉)
Department of Biology, Universidade Federal de Sergipe, São Cristóvão, SE, Brazil
e-mail: ferrari@pq.cnpq.br

S.M. Ford et al. (eds.), *The Smallest Anthropoids*, Developments in Primatology: Progress and Prospects, DOI 10.1007/978-1-4419-0293-1_8,
© Springer Science+Business Media, LLC 2009

cinco veces más frecuente que entre hembra→macho. Ausente entre los machos, y raro entre las hembras, el comportamiento sumiso fue casi siempre dirigido hacia machos y a hembras reproductoras por hembras no reproductoras. Analizadas en conjunto, las interacciones indican que una jerarquia social de tres camaradas dentro de un grupo: hembra reproductora→macho→hembra no reproductora. Sin embargo, ningún macho fue más dominante socialmente que cualquier otro, y ninguna hembra fue más subordinada. Los miembros de los dos sexos se dispersaron bajo diferentes circunstancias, pero existe poca evidencia de que las emigraciones estuvieran relacionadas de alguna forma con agonismo dentro del grupo, sea relacionado con la posición social o a la competición por recursos.

Resumo Padrões comportamentais em um grupo silvestre de sagüis-da-serra (*Callithrix flaviceps*) foram monitorados ao longo de um período de sete meses durante o qual quatro adultos – um macho e três fêmeas – emigraram em dois eventos distintos. Interações sociais como o comportamento lúdico, e a alocatação foram relativamente freqüentes, embora interações agonísticas foram raras. Agonismo foi observado menos do que duas vezes por dia, em média, neste grupo, que continha de sete a dez membros adultos, em um dado momento. Com a exceção de eventos que envolveram a fêmea reprodutora, o agonismo intrasexual foi praticamente ausente entre adultos, e a agressão macho→fêmea foi cinco vezes mais freqüente que fêmea→macho. Ausente entre machos, e raro em fêmeas, o comportamento submissivo foi quase sempre direcionado a machos e a fêmea reprodutora por fêmeas não reprodutivas. Analisadas em conjunto, as interações indicam uma hierarquia social de três camadas dentro do grupo, com a fêmea reprodutora na primeira camada, seguido pelos machos na segunda, e finalmente as fêmeas não reprodutivas. Entretanto, nenhum macho foi mais dominante socialmente do que qualquer outro, e nenhuma fêmea foi mais subordinada. Os emigrantes dos dois sexos dispersaram sob circunstâncias diferentes, mas existe pouca evidência de que as emigrações foram relacionadas de alguma forma com agonismo dentro do grupo, relacionado a posição social ou a competição por recursos.

8.1 Introduction

Social behavior in groups of marmosets and tamarins (the Callitrichidae) is characterized by the dominance of a single reproductive female or breeding pair (Abbott et al. 1993; French 1997), even when polygynous breeding is the norm (e.g., Dietz and Baker 1993; Digby 1995a; Coutinho and Corrêa 1996). The cooperative breeding system mediates tolerance of mature, non-reproductive helpers by breeding animals, but it also creates the potential for increased competition for both resources and breeding positions. However, intra-group agonism is normally rare in free-ranging callitrichids (Goldizen 1989; Digby 1995a), and in captivity, it generally occurs between members of the same sex (e.g., *Callithrix jacchus*: Epple 1975; Rothe 1978; Evans 1983; Abbott 1984; Stevenson and Rylands 1988).

As only one or two helpers may make a significant contribution to a breeding female's reproductive success (Goldizen 1987; Sánchez et al. 1999), it would seem reasonable to expect declining tolerance in groups with more than four adult members (Ferrari and Digby 1996). Many free-ranging marmoset (*Callithrix*) groups contain five or more adults, however, which suggests that such tolerance may depend on additional factors. The marmosets' unique set of specializations for the dietary exploitation of plant exudates, which provides a stable resource base, and the relatedness of group members and the risks of dispersal is likely to help explain tolerance within large groups (Ferrari and Lopes 1989), as are the relatedness of group members and the risks of dispersal (Goldizen and Terborgh 1989; Ferrari and Digby 1996).

In the present study, agonistic behavior in a free-ranging group of buffy-headed marmosets, *Callithrix flaviceps*, is analyzed during a period when the group reached its maximum size (15 members) and four adult members emigrated. The results of the study point to a three-tiered, gender-based social hierarchy, but no evidence of any increase in agonism was found with decreasing resource availability, or in relation to the emigration of group members. Dispersal thus appeared to be triggered by external (availability of potential mates) rather than internal factors.

8.2 Methods

8.2.1 Study Site and Animals

The *C. flaviceps* study group was monitored at the Fazenda Montes Claros (19°50′S, 41°50′W) in the municipality of Caratinga, in Minas Gerais, Brazil, between 1985 and 1991 (see Ferrari 1988; Ferrari and Diego 1992). The observational records analyzed here cover the 7-month period between January and July, 1986, during which a pair of twins were born and four adult group members emigrated. During this period, group size varied between a maximum of 15 members (following the birth) and a minimum of 11, following emigrations (Table 8.1). The study subjects were fully habituated to the presence of human observers (observer-subject distances were often less than 1 m) and individually identifiable through differences in pelage markings.

8.2.2 Behavioral Data

Quantitative behavioral records were collected using instantaneous scan sampling (Altmann 1974), with a scan of 1-min duration carried out at 5-min intervals throughout the daily activity period of the study group (sleep-tree to sleep-tree) during 10 days each month. The behavior of each group member located during the scan was recorded, allowing the compilation of activity budgets (Ferrari 1988, 1992). Behavior sampling (Martin and Bateson 1993) was also used for the collection of records of social behavior. In all cases, the identity of the animals involved and the sequence of events were recorded whenever possible. For the purposes of

Table 8.1 Composition of the *Callithrix flaviceps* study group during the study period

Identity	Age/sex class	Observations
Maggie	Adult female	Reproductive female
Spock	Adult female	Emigrated in 06/86
Cuba	Adult female	Emigrated in 06/86
Dida	Adult female	Emigrated in 06/86
Máscara	Adult female	
Jocky	Subadult female[1]	Daughter of Maggie (b. 01/85), twin sister of Jimmy
Bob	Adult male	Reproductive male (?)
Simão	Adult male	Emigrated between 05/86 and 06/86 observation periods
Gordão	Adult male	
Smell	Adult male	
Jimmy	Subadult male[a]	Son of Maggie (b. 01/85), twin brother of Jocky
Pablo	Juvenile male	Son of Maggie (b. 07/85), twin brother of Paxo
Paxo	Juvenile male	Son of Maggie (b. 07/85), twin brother of Pablo
Oz1	Infant	Offspring of Maggie (b. 01/86), twin of Oz2
Oz2	Infant	Offspring of Maggie (b. 01/86), twin of Oz1

[a]Reclassified as adult in June, 1986, at 18 months of age

Table 8.2 Categories of social behavior used in the present study, adapted from Stevenson and Poole (1976), Ferrari (1988), and Digby (1995a)

Category	Description
Agonistic	
Advance	Advancing rapidly on other individual(s) without physical contact;
Arch-back walk	Strutting with arched back, normally pilo-erected;
Attack	Engaging other individuals in physical contact (grappling, biting, hitting);
Chase	Moving rapidly behind retreating individual;
Threat	Facial (open-mouth stare) or postural (arch-back) threat.
Submissive	
Avoid	Moving out of the path of an approaching individual or moving rapidly away from an advancing or attacking individual;
Grimace	Squinting, open-mouth expression, usually with head lowered;
Submit	Lowering upper body towards substrate ("cringing") and withdrawing body while looking towards aggressor.

the present study, only records of agonistic (dominant) and submissive (subordinate) behavior categories (Table 8.2) were analyzed. The dominant member of a dyad was defined according to the ratio of agonistic to submissive behavior. The characteristics of the data proscribed the use of a more systematic, quantitative measure, such as Nishida's (1988) index of reciprocity. Resource availability was monitored throughout the main study period (for details, see Ferrari 1988).

As the composition of the group changed during the study period, comparisons between sexes and across months are based on the rate of a given behavior category per individual per standard observation month (10 days) rather than absolute values. Group composition only changed mid-month in June, at the end of the fifth day of

Table 8.3 Records of agonistic behaviour categories among *Callithrix flaviceps* adults by sex and month

Month	Adult males: adult females in group	Records of agnostic behaviour (involving the breeding female):			
		Male → male	Male → female	Female → male	Female → female
January	4:5	1	16 (0)	1 (1)	8 (7)[a]
February	4:5	0	7 (1)	2 (1)	1 (1)
March	4:5	0	22 (1)	6 (2)	3 (3)
April	4:5	0	17 (0)	4 (1)	3 (3)
May	4:5	0	5 (0)	5 (1)	1 (1)
June 1	4:6	0	8 (0)	2 (0)	0
June 2	4:3	0	7 (0)	2 (0)	0
July	4:3	0	13 (0)	6 (1)	0
Total		1	95 (2)	28 (7)	16 (15)

[a]All records in parentheses in this column refer to the breeding female as aggressor

Table 8.4 Records of submissive behavior categories among *Callithrix flaviceps* adults by sex and month

Month	Adult males: adult females in group	Records of submissive behaviour (involving the breeding female):			
		Male → male	Male → female	Female → male	Female → female
January	4:5	0	1 (1)	30 (0)	4 (4)[a]
February	4:5	0	1 (1)	9 (0)	5 (2)
March	4:5	0	2 (2)	25 (0)	2 (1)
April	4:5	0	0	27 (0)	1 (0)
May	4:5	0	0	20 (0)	1 (1)
June 1	4:6	0	1 (0)	13 (0)	0
June 2	4:3	0	0	23 (0)	0
July	4:3	0	2 (2)	14 (0)	0
Total		0	7 (6)	170 (0)	13 (8)

[a]All records in parentheses in this column refer to the breeding female as recipient

observation. The data for this month are thus presented separately, with June 1 referring to the period preceding the emigration, and June 2 to that following the change in group composition. The distribution of events among individuals was tested using χ^2.

Standardized rates were calculated for each sex (Tables 8.3 and 8.4), and for the group as a whole (Fig. 8.1). For the χ^2 analyses, the individual values (number of events) were used, and compared to a homogeneous distribution.

8.3 Results

Infant care-giving behavior and changes in the composition of the *C. flaviceps* group during the present study period have been described elsewhere (Ferrari 1992; Ferrari and Diego 1992). Overall, there was little difference in the contribution of

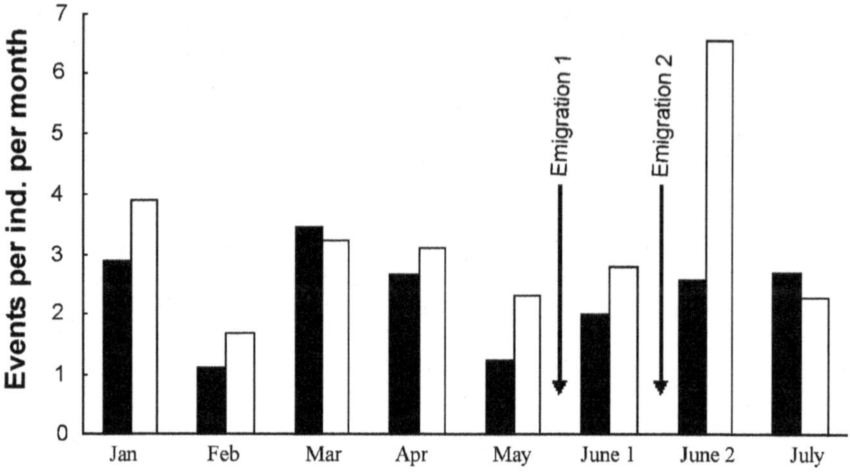

Fig. 8.1 Standardised monthly rates of agonistic (*shaded*) and submissive (*unshaded*) behaviour recorded in the *Callithrix flaviceps* study group. *Emigration 1* Simão left group; *Emigration 2* Cuba, Dida and Spock left group

males and non-breeding females to the rearing of infants (carrying and food sharing). The adult male Simão left the study group at the end of May to join a neighboring group containing a single adult female and two adult males. In mid-June, the non-breeding females Cuba, Dida, and Spock left the study group to join the two males from the group into which Simão had immigrated for the formation of a new social unit of five adults; they were subsequently observed within the home range of the main study group on a regular basis.

During the 7 months considered here, social behavior, predominantly play, allo-grooming, and scent marking, accounted for 10.4% of scan sample records ($n=40713$). Agonistic behavior accounted for no more than 0.2% of records in any one month. Even when behavioral sampling is included, agonism was observed at a maximum rate of 3.4 records per adult in any 1 month (March), and physical contact was extremely rare. The maximum monthly rate for submissive categories was similar (3.8), as might be expected given the complementary nature of these categories.

Agonistic interactions between males were virtually nonexistent, with a single record collected in January (Table 8.3). Interactions between females were more frequent, but account for a little over one tenth of the records, and almost invariably involved the breeding female (Maggie). Maggie was aggressor in all but one record of intra-female agonism, and received two-thirds of intra-female submissions (Table 8.4).

Intra-female agonism was concentrated in the first 3 months of the study period, when dependent infants were present in the group, and basically resulted from the breeding female's monitoring of caregivers. When infant distress vocalizations were insistent, the breeding female normally reacted by approaching, threatening, and even attacking adult caregivers (both male and female). This occurred in the context of both infant carrying and the solicitation of food items (see Ferrari 1987).

By contrast, no agonism between females was recorded in June, the month during which Cuba, Dida, and Spock emigrated.

Most of the inter-sexual agonism involving the breeding female was also related to the monitoring of caregivers. While males were aggressive towards Maggie on two occasions, she was never observed exhibiting submissive behavior towards individuals of either sex, which would appear to confirm her dominant position within the group. Similarly, a male was submissive towards a non-breeding female on only one occasion, in contrast with 170 records of non-breeding female→male submission. Males were also five times more aggressive towards non-breeding females than vice versa. Overall, then, the data indicate that males were socially dominant over non-breeding females.

While males as a whole were clearly dominant over non-breeding females, the almost total lack of intra-gender agonism obscures any possible linear hierarchy. There was also no clear tendency for a given male to be more aggressive than others (January to May: $\chi^2=4.96$, df=3, $p>0.20$), nor for a given non-breeding female to be more submissive ($\chi^2=6.22$, df=3, $p>0.05$).

Despite being able to exploit plant exudates systematically as a substitute source of carbohydrates during periods of fruit scarcity, marmosets are vulnerable to seasonal fluctuations in the abundance of arthropod prey, which declined progressively during the present study period, accompanying the transition from the wet to the dry season (Ferrari 1988). Over the year as a whole, feeding on animal material declined significantly with decreasing arthropod abundance, as did foraging efficiency and prey selectivity. However, the relative contribution of animal material to the *C. flaviceps* diet varied little between January (17.4% of feeding records) and June (19.8%), whereas it fell to 9.3% in July, that is, after the emigrations.

The exact consequences of such seasonal changes are unclear, but the fact that Maggie gave birth to a set of healthy twins towards the end of September appears to support the idea that nutritional stress was not a factor. While the observed changes do suggest that intra-specific competition increased during the course of the study period, there is no evidence to suggest that this resulted in any change in the relationships between group members (see Fig. 8.1), in particular with regard to the possibility that emigrations were stimulated by an increase in intra-group agonism. Inter-gender aggression was in fact recorded less frequently than expected (homogeneous distribution) in May ($\chi^2=3.83$, df=1, $p>0.05$) prior to the emigration of Simão, and barely more than average in June 1 ($\chi^2=0.18$, df=1, $p>0.70$).

8.4 Discussion

The social dominance of a single breeding female or male/female pair is a well-documented characteristic of callitrichid social organization, one which has a well-defined role in the reproductive biology of these primates (Abbott et al. 1993;

French 1997). In general, relationships between other group members have received less attention, although tolerance is the norm (at least in the wild), even between unrelated group members or potentially reproductive adult males. Captivity often accentuates intra-group agonism (Epple 1975; Rothe 1978; Box and Morris 1980; Abbott 1984; Snowdon and Soini 1988; Moura 2003), which may result in the formation of a distinct, intra-sexual hierarchy (Rothe 1978; Fuchs et al. 1991). But up to now, little evidence has been found of social dominance between the sexes (Evans and Poole 1984; Sutcliffe and Poole 1984; Digby 1995a), as might be expected for such a sexually monomorphic mammal (Kleiman 1977).

The gender-based social hierarchy described here thus represents a somewhat novel interpretation of the social organization of a callitrichid species. Unfortunately, with the exception of data collected at the present study site (Ferrari and Diego 1992; Guimarães 1998), virtually nothing is known of the behavior of *C. flaviceps*, so it remains unclear whether the patterns reported here are typical of the species, and even less so, to what extent similar patterns might be expected in other members of the genus. In fact, with one exception (Digby 1995a), very little is known of intra-group relations in *Callithrix* species. Digby found some evidence of an age-based hierarchy in the common marmoset, *Callithrix jacchus*, although the social context was different from that in *C. flaviceps*, in particular the presence of two breeding females in each of the three study groups.

What does seem most likely, from these scant data, is that social structure and intra-group relations in marmoset species are as variable as their mating patterns (Ferrari and Digby 1996; Ferrari et al. 1996). As for mating patterns, group composition may play at least as important a role as ecological variables. Except for recruitment through births, the composition of the *C. flaviceps* study group remained unchanged for at least 18 months prior to the emigrations reported here, and it seems likely that most if not all group members were closely related (Ferrari and Diego 1992). All three *C. jacchus* groups appeared to be less stable, by contrast, suffering losses of both adult and immature members (Digby and Barreto 1993), in addition to at least one infanticide (Digby 1995b). The latter, in particular, appears to reflect more intense competition for reproductive success, as might be expected from the presence of two breeding females in each group.

The relatedness of group members may be a key factor in intra-group relations, and may account for much of the difference between captive and field data, given that the stable resource base provided by the marmosets' gum-feeding specializations is common to all species (Soini 1982; Ferrari and Digby 1996; Ferrari et al. 1996; Passamani 1996; Veracini 1997, this volume Chapt. 12). Ideally, it would be valuable to compare systematically the social relations in groups of different composition, although up to now, there has been only one field study in which the genetic relatedness of all group members is known (see Faulkes et al. Chapt. 5 this volume). Systematic differences between marmosets and tamarins (*Leontopithecus* and *Saguinus*) might also be expected, given the ecological and demographic differences between these two groups (Ferrari and Lopes 1989).

The present study also reinforces the highly systematic nature of dispersal and group formation in this species, at least. The only pattern observed in the present study was an apparent increase in agonistic interactions between the breeding female and non-breeding females at the time of the birth. Certainly, there is no evidence to suggest that intra-group agonism was influenced in any way by resource abundance or intra-group competition, nor that emigrations were related to either factor. It thus seems more likely that the primary determinant of both emigrations was the availability of potential mates in neighboring groups. Timing may also have been important, given that Simão's emigration coincided with the period during which Maggie would have conceived the litter born in late September.

Acknowledgements This study was supported by the Brazilian National Research Council (CNPq, process no. 307506/2003-7), the Medical Research Council of Great Britain, the A.H. Schultz-Stiftung, London University Central Research Fund, the Boise Fund and the Leakey Trust. I am especially grateful to Bob Martin, Cida Lopes, and Vânia Diego. An early version of this paper was presented at the 15th IPS Congress, in the symposium on "New World primates with parental care" organized by Hilary Box and Gisela Epple.

References

Abbott DH (1984) Behavioral and physiological suppression of fertility in subordinate marmoset monkeys. Am J Primatol 6:169–186

Abbott DH, Barrett J, George LM (1993) Comparative aspects of the social suppression of reproduction in female marmosets and tamarins. In: Rylands AB (ed) Marmosets and tamarins: systematics, behaviour, and ecology. Oxford University Press, Oxford, pp 152–163

Altmann J (1974) Observational study of behavior: sampling methods. Behaviour 48:1–41

Box HO, Morris JM (1980) Behavioural observations on captive pairs of wild caught tamarins (*Saguinus mystax*). Primates 21:53–65

Coutinho PEG, Corrêa HKM (1996) Polygyny in a free-ranging group of buffy-tufted-ear marmosets, *Callithrix aurita*. Folia Primatol 65:25–29

Dietz JM, Baker AJ (1993) Polygyny and female reproductive success in golden lion tamarins, *Leontopithecus rosalia*. Anim Behav 46:1067–1078

Digby LJ (1995a) Social organization in a wild population of *Callithrix jacchus*. II. Intragroup social behavior. Primates 36:361–375

Digby LJ (1995b) Infanticide, infant care, and female reproductive strategies in a wild population of common marmosets. Behav Ecol Sociobiol 37:51–61

Digby LJ, Barreto CE (1993) Social organization in a wild population of *Callithrix jacchus*. Part I: group composition and dynamics. Folia Primatol 6:123–134

Epple G (1975) The behavior of marmoset monkeys (Callitrichidae). In: Rosenblum LA (ed) Primate Behavior, vol 4. Academic, New York, pp 195–239

Evans S (1983) The pair-bond of the common marmoset, *Callithrix jacchus jacchus*: an experimental investigation. Anim Behav 31:651–658

Evans S, Poole TB (1984) Long-term changes and maintenance of the pair-bond in common marmosets, *Callithrix jacchus jacchus*. Folia Primatol 42:33–41

Faulkes C, Arruda MF, Monteiro da Cruz MAO (this volume) Genetic structure within and among populations of the common marmoset, *Callithrix jacchus*: implications for cooperative breeding. In: Ford SM, Porter LM, Davis LC (eds) The smallest anthropoids: the marmoset/callimico radiation. Springer, New York, pp 103–117

Ferrari SF (1987) Food transfer in a wild marmoset group. Folia Primatol 48:203–206

Ferrari SF (1988) The behaviour and ecology of the buffy-headed marmoset, *Callithrix flaviceps* (O. Thomas, 1903). Ph.D thesis, University of London.

Ferrari SF (1992) The care of infants in a wild marmoset, *Callithrix flaviceps*, group. Am J Primatol 26:109–118

Ferrari SF, Diego VH (1992) Long-term changes in a wild marmoset group. Folia Primatol 58:215–218

Ferrari SF, Digby LJ (1996) Wild *Callithrix* groups: stable extended families? Am J Primatol 38:19–27

Ferrari SF, Lopes MA (1989) A re-evaluation of the social organisation of the Callitrichidae, with reference to the ecological differences between genera. Folia Primatol 52:132–147

Ferrari SF, Corrêa HKM, Coutinho PEG (1996) Ecology of the "southern" marmosets (*Callithrix aurita* and *Callithrix flaviceps*): how different, how similar? In: Norconk MA, Rosenberger AL, Garber PA (eds) Adaptive Radiations of Neotropical Primates. Plenum, New York, pp 157–171

French JA (1997) Proximate regulation of singular breeding in callitrichid primates. In: Solomon RG, French JA (eds) Cooperative Breeding in Mammals. Cambridge University Press, Cambridge, pp 34–75

Fuchs E, Rosenbusch J, Anzenberger G (1991) Urinary protein pattern reflects social rank in male common marmosets (*Callithrix jacchus*). Folia Primatol 57:177–180

Goldizen AW (1987) Facultative polyandry and the role of infant carrying in wild saddle-back tamarins (*Saguinus fuscicollis*). Behav Ecol Sociobiol 20:99–109

Goldizen AW (1989) Social relationships in a cooperatively polyandrous group of tamarins (*Saguinus fuscicollis*). Behav Ecol Sociobiol 24:79–89

Goldizen AW, Terborgh J (1989) Demography and dispersal patterns of a tamarin population: possible causes of delayed breeding. Am Nat 134:208–224

Guimarães A (1998) Ecology and social behaviour of buffy-headed marmosets, *Callithrix flaviceps*. Neotrop Primates 6:51–52

Kleiman DG (1977) Monogamy in mammals. Q Rev Biol 52:39–69

Martin P, Bateson P (1993) Measuring Behaviour: an introductory Guide. Cambridge University Press, Cambridge

Moura ACA (2003) Sibling age and intragroup aggression in captive *Saguinus midas midas*. Int J Primatol 24:639–652

Nishida T (1988) Development of social grooming between mother and offspring in wild chimpanzees. Folia Primatol 50:109–123

Passamani M (1996) Ecologia e Comportamento de um Grupo de Sagüi-da-Cara-branca (*Callithrix geoffroyi*) em um Fragmento de Mata Atlântica no Espírito Santo. Universidade Federal de Minas Gerais, Belo Horizonte, M.Sc dissertation

Rothe H (1978) Sub-grouping behaviour in captive *Callithrix jacchus* families: a preliminary investigation. In: Rothe H, Wolters HJ, Hearn JP (eds) Biology and Behaviour of Marmosets. Eigenverlag H. Rothe, Göttingen, pp 233–257

Sánchez S, Peláez F, Gil-Burmann C, Kaumanns W (1999) Costs of infant-carrying in the cotton-top tamarin (*Saguinus oedipus*). Am J Primatol 48:99–111

Snowdon CT, Soini P (1988) The tamarins, genus *Saguinus*. In: Mittermeier RA, Rylands AB, Coimbra-Filho AF, da Fonseca GAB (eds) Ecology and behavior of neotropical primates, vol. 2. World Wildlife Fund, Washington, pp 223–298

Soini P (1982) Ecology and population dynamics of the pygmy marmoset, *Cebuella pygmaea*. Folia Primatol 39:1–21

Stevenson MF, Poole TB (1976) Ethogram of the common marmoset (*Callithrix jacchus jacchus*): general behavioural respertoire. Anim Behav 24:428–451

Stevenson MF, Rylands AB (1988) The marmosets, genus *Callithrix*. In: Mittermeier RA, Rylands AB, Coimbra-Filho AF, da Fonseca GAB (eds) Ecology and Behavior of Neotropical Primates, vol 2. World Wildlife Fund, Washington, pp 131–222

Sutcliffe MG, Poole TB (1984) Intragroup agonistic behavior in captive groups of the common marmoset (*Callithrix jacchus jacchus*). Int J Primatol 5:473–489

Veracini C (1997) O comportamento alimentar de *Callithrix argentata* (Linnaeus 1771) (Primata, Callitrichinae). In: Lisboa PLB (ed) Caxiuanã. MCT/CNPq, Belém, pp 437–446

Veracini C (this volume) Habitat use and ranging behavior of the silvery marmoset (*Mico argentatus*) at Caxiuanã National Forest (eastern Brazilian Amazonia). In: Ford SM, Porter LM, Davis LC (eds) The smallest anthropoids: the marmoset/callimico radiation. Springer, New York, pp 221–240

Chapter 9
Emigration as a Reproductive Strategy of the Common Marmoset (*Callithrix jacchus*)

Maria Bernardete Cordeiro de Sousa, Ana Claudia Sales da Rocha Albuquerque, Maria Emilia Yamamoto, Arrilton Araújo, and Maria de Fátima Arruda

Abstract Demographic studies provide critical data for understanding the evolution, ecology, social organization and mating systems of primates. One influence on the demographic structure of a population is the pattern of emigration of individuals from their natal groups. In cooperative breeders, the offspring may stay in their natal group even after maturity, and many researchers are investigating the possible ecological and social factors responsible for this pattern and the resulting reproductive skew that occurs within these populations. In *Callithrix jacchus*, the common marmoset individuals disperse from their natal groups as adults, females disperse more frequently than males, and dispersals are more frequent during the rainy season than during the dry season. However, neither group size nor tertiary sex ratio is correlated with dispersal. Captive animals that undergo experimental procedures that simulate natural conditions after emigration (isolation from a group followed by movement into a new environment) trigger different physiological and behavioral responses in each sex. Males increased significantly in cortisol levels after separation, and upon placement in a new environment together with other males, they intensified their affiliative relationships and cooperated in tasks to acquire food. Females were less affected by isolation (as measured by changes in cortisol levels) and significantly increased piloerection and scent-marking behaviors when they were moved (as dyads) to a new environment. These differences suggest that males and females have evolved different strategies for dispersal and breeding success, and the dynamics of social relationships between females seem to be the main determinant of the demographic profile of common marmoset populations. Moreover, besides adaptations related to reproduction, females show additional behavioral and physiological mechanisms that may be selected to cope with social isolation following emigration.

M.B.C. de Sousa (✉)
Departamento de Fisiologia, Universidade Federal do Rio Grande do Norte,
Caixa Postal 1511, Natal, RN, 59072-970, Brazil
e-mail: mdesousa@cb.ufrn.br

S.M. Ford et al. (eds.), *The Smallest Anthropoids*, Developments in Primatology: Progress and Prospects, DOI 10.1007/978-1-4419-0293-1_9,
© Springer Science+Business Media, LLC 2009

Resumen Los estudios demográficos proveen datos críticos para entender la evolución, ecología, organización social y sistemas de apareamiento de los primates. La permanencia o emigración del grupo natal de un miembro influye las estructuras demográficas de la población. En las especies de reproducción cooperativa los descendientes pueden permanecer en un grupo natal aún después de alcanzar la madurez y muchos investigadores buscan identificar los posibles factores sociales y ecológicos responsables de estos patrones y de la asimetría reproductiva que se produce en estas poblaciones. En *Callithix jacchus*, los individuos se dispersan de sus grupos natales como adultos, las hembras se dispersan con mayor frecuencia que los machos y las dispersiones son más frecuentes en la estación húmeda que en la seca. Sin embargo, ni el tamaño de grupo ni la tasa sexual terciaria están correlacionados con la dispersión. Animales cautivos que enfrentaron procedimientos experimentales que simulan condiciones naturales después de la emigración (aislamiento de un grupo seguido de movimiento hacia un ambiente nuevo) mostraron diferentes respuestas fisiológicas y conductuales para cada sexo. Los machos incrementaron significativamente los niveles de cortisol luego de la separación y, cuando fueron ubicados dentro de un nuevo ambiente con otros machos, intensificaron las relaciones afiliativas y cooperaron en las tareas para adquirir alimento. Las hembras fueron menos sensibles al aislamiento (medido por los cambios en niveles de cortisol) e incrementaron significativamente la piloerección y las conductas de marcaje por olor cuando fueron movidas en diadas a un ambiente nuevo. Estas diferencias sexuales sugieren que machos y hembras han evolucionado estrategias distintas para la dispersión y el éxito reproductivo, y las dinámicas de relaciones sociales entre hembras parecen ser el determinante principal del perfil demográfico de las poblaciones de marmosetas comunes. Aunado a las adaptaciones relacionadas a la reproducción, las hembras muestran mecanismos conductuales y fisiológicos adicionales que pueden estar seleccionados para enfrentar el aislamiento social que conlleva la emigración.

Resumo Os dados demográficos de uma população fornecem informações críticas para a compreensão sobre evolução, ecologia, organização social e sistemas de acasalamento de primatas. Um dos fatores que influenciam a estrutura demográfica de uma população é o padrão de emigração dos indivíduos do seu grupo natal. Nas espécies de reprodução cooperativa os filhos podem permanecer no grupo natal após a maturidade reprodutiva e várias pesquisas buscam identificar os fatores sociais e ecológicos responsáveis pela assimetria que se instala na reprodução. Em sagüi comum, *Callithrix jacchus*, os indivíduos dispersam do seu grupo natal como adultos sendo mais freqüente nas fêmeas durante a estação chuvosa. Entretanto nem o tamanho do grupo nem a razão sexual terciária se correlacionam com a dispersão. Procedimentos experimentais simulando situações naturais com as quais os animais se confrontam após emigração (isolamento após a saída do grupo e mudança para um novo ambiente) desencadeiam respostas fisiológicas e comportamentais diferentes para ambos os sexos. Os machos elevam significativamente mais o cortisol quando isolados e intensificam as interações afiliativas com outros machos e outros experimentos mostram que eles cooperam em tarefas para obtenção de alimento. As fêmeas são menos afetadas pelo isolamento (não elevam significativamente o cortisol) e aumentam a piloereção e marcação de cheiro quando expostas em duplas a um ambiente novo. Esses resultados sugerem que machos e fêmeas desenvolveram diferentes estratégias para dispersarem e obterem o sucesso reprodutivo e que a dinâmica das relações entre fêmeas parece ser o principal determinante do perfil demográfico de uma população de *C. jacchus*. Adicionalmente, além das adaptações relacionadas à reprodução, as fêmeas apresentam mecanismos fisiológicos e comportamentais que podem ter sido selecionados para enfrentar o isolamento social após a emigração.

9.1 Demographic Data for Primate Populations

The demographic profile of an animal population includes the rate of births, deaths and migration (emigration and immigration) of individuals among groups, as well as the composition of the population in terms of age and sex classes (Dunbar 1988a, b). In this chapter, we will focus on data collected on the common marmoset (*Callithrix jacchus*) with regard to migration patterns of males and females. We will use the terms dispersal and emigration interchangeably to refer to individuals who departed from their natal groups and were later seen alone or integrating into another social group.

Mammals and birds are the most studied taxa in relation to dispersal, and different patterns are seen in each. In the majority of mammals, males show higher rates of dispersal than females do, whereas in birds, females are more likely to leave their groups than males (Shields 1987).

In mammals, emigration is the most important process regulating local population size and occurs as a natural process for both immature and mature individuals. The proximate mechanisms that influence primate dispersal are ecological (food availability, predation risk), demographic (operational sex ratio, group size, group composition) and social (hierarchy relationship, helper availability, risk of infanticide). The ultimate mechanism responsible for dispersal seems to be inbreeding avoidance: animals leave their natal group to avoid breeding with closely related individuals.

Evidence from the primate literature has shown that among Old World primates males typically disperse from their natal group (Moore 1984), whereas females do not, a pattern called female philopatry. In prosimians, male-biased emigration is also found, although females are less gregarious than in Old World Monkeys (Moore 1984). For New World primates no systematic data are available, and emigration of species appears to vary, and be determined by whether dispersal is female or male biased. The demographic structure of wild callitrichid populations, as for other primates, is thought to be less heavily influenced by migration than by births and deaths (Dunbar 1988a). Baker and Dietz (1996), studying 17 territorial groups of golden lion tamarins (*Leontopithecus rosalia*) for a period of 10–76 months, found that emigration is male-biased (85% of emigrants), and only 0.48 individuals per group, per year left their groups, generally to replace breeding individuals in neighboring groups. In a 5-year study of cotton-top tamarins (*Saguinus oedipus*), Savage et al. (1996), did not find sex-biased dispersal, but discovered that males were more likely to emigrate to a group to replace a resident male, whereas females were apparently tolerant with immigrating females although they defend their breeding position during the estrus period.

Common marmoset (*C. jacchus*) dispersal occurs at low levels as demonstrated by Monteiro da Cruz (1998), who monitored five free-ranging groups for 36 months (from 1994 to 1996) in Tapacurá, state of Pernambuco, a field station in northeastern Brazil. In this population, dispersal is female-biased, and of 22 episodes of emigration recorded, 15 were from females. Field studies at Nisia Floresta, another field station in the same region (Araújo 1996) recorded 37 emigration episodes during approximately 10 years of long-term monitoring; 63.4% of these emigrants were females and, although age could not be precisely established for 2/3 of the individuals, all emigrants of known ages were adults between 17 and 48 months old.

9.2 Hormonal Aspects of Male and Female Common Marmoset Dominance

During the last four decades, studies of the behavioral ecology of common marmosets (Hubretch 1984; Stevenson and Rylands 1988; Alonso and Langguth 1989; Digby and Barreto 1993; Digby 1995a, b; Digby and Ferrari 1994; Monteiro da Cruz 1998; Lazaro-Perea 2001; Lazaro-Perea et al. 2000, 2004; Arruda et al. 2005) have provided a body of information on their social organization, mating systems, and reproductive strategies. These studies show that common marmoset group sizes vary from 3 to 19 individuals, and that they inhabit gallery and secondary forests and small patches in central and northeast Brazil (Stevenson and Rylands 1988; Santee and Arruda 1994). Group home range sizes are small (0.5–5.0 ha) as a result of their extensive exudativory feeding habits that allow them to exploit a few trees intensively in a small area of forest (Stevenson and Rylands 1988; Alonso and Langguth 1989) when compared to the more frugivorous species such as *Mico humeralifer* (the tassel-eared or Santarem marmoset, Rylands 1986) and *Callithrix flaviceps* (the buffy-headed marmoset, Ferrari 1988). The average estimated density of marmosets is around eight individuals per hectare, and this seems be dependent on the availability of food and gum trees (Stevenson and Rylands 1988). Group density does not appear to be dependent on sleeping site availability, as Sousa et al. (2002a) found that two groups followed for 14 and 16 months used only a few sleeping trees (five and six) during the study and did not require trees with special characteristics, as *Leontopithecus* do.

Common marmosets are considered to live as cooperative breeders, that is, a system of breeding in which other adults give significant care to infants that are not their genetic offspring (Emlen 1997), and older siblings help their parents to rear younger siblings (Snowdon 1996). According to family definitions proposed by

Emlen (1997), common marmosets form extended biparental family units in which unrelated individuals, other than offspring of the breeding pair, can also be part of the group (Nievergelt et al. 2000).

Emlen (1997) used reproductive skew to explain why offspring remain in their natal group instead of dispersing and trying to breed independently. He categorized families into either simple families (monopolization of reproduction by a few dominant individuals, skew = 1) or extended families (shared reproduction, skew <1) and predicted that subordinate females would attempt different strategies for reproductive success within these family types. Subordinate females may follow several alternative strategies: first, they may replace the dominant female in a group by displacement; second, they may assume the breeding position in their natal group after the loss of a parent; third, they may become a breeder in their natal group by sharing reproduction with the dominant female; fourth, they may remain in their natal group as a nonreproductive helper and invest in the offspring of the related dominant female; fifth, they may disperse to a different group where they can breed independently.

Our understandings of the social organization of common marmosets are derived from investigations of the social dynamics of family groups and newly formed heterosexual pairs in captivity (Box 1975; Rothe 1975; Evans and Poole 1983; Kendrick and Dixson 1983). These studies revealed that a dominance hierarchy is established within both the female and male sexes in a social group, but the hierarchy differs between the sexes (Abbott 1984), with hierarchy within females being characterized by higher agonism levels. Furthermore, although dominant females can sometimes inhibit breeding among subordinate females in their group (Abbott 1993), the extent of this inhibition depends on whether unrelated males are present in the group (Saltzman et al. 1997a, b) and whether a clear dominance relationship exists between females (Alencar et al. 1995, 2006; Saltzman et al. 2004). In addition to reproductive competition among females, males also compete for breeding positions. Data indicate that males primarily use mechanisms of behavioral inhibition to suppress reproduction amongst each other (Abbott 1993; Baker et al. 1999); however, there is some evidence that testosterone levels might also be lower in subordinate males (Abbott 1993).

The development of techniques to allow the hormonal analysis of feces (Ziegler et al. 1996; Sousa and Ziegler 1998) has enabled the study of reproductive success and physiology in wild common marmosets. In a monogamous free-ranging group of common marmosets (Albuquerque et al. 2001), the subordinate female, probably a daughter of the dominant female, was able to ovulate during the pregnancy of the dominant female. Another study using long-term data from three monogamous groups (Sousa et al. 2005) also indicates that reproductive inhibition of subordinate females by dominant females occurs primarily through behavioral, rather than hormonal, inhibition. In all three groups monitored by Sousa et al. (2005), subordinate females showed ovarian cyclicity at least once during the study and, in two groups, the subordinate female gave birth, although the offspring did not survive. Subordinate females tried to reproduce in the natal group (probably impregnated by extra-group males) and dispersed following their unsuccessful attempts (see Yamamoto et al. Chapt. 6 this volume). Therefore, emigration seems to be an important component of the reproductive strategies of subordinate common marmoset females, particularly if attempts to breed in their natal groups have failed.

9.3 Long-Term Data About Emigration in a Population of Common Marmosets

9.3.1 General Emigration Data

Long-term data on emigration were collected from free ranging groups of common marmosets at the Nísia Floresta field station (see Santee and Arruda 1994, for full description of the area) from January 1991 to April 2001. During this time, eight free-ranging groups (Plantação, Chuí, Belém, Marajó, Espanha, Teruel, PB and G4) were monitored between 15 and 71 months, depending on the group. Partial information about emigration episodes, excluding the Marajó group, was collected in previous studies (Digby 1995b; Araújo 1996; Lazaro-Perea 2001; Lazaro-Perea et al. 2000; Arruda et al. 2005; Sousa et al. 2005).

In total, 37 emigration events were observed. In our observations, single individuals emigrated from a group; however, in a different study, pairs of emigrant females were observed (Monteiro da Cruz 1998).

As pointed out earlier, common marmoset dispersal is female-biased (23 of 37 episodes) and the annual number of emigrations ranged from a minimum of 0.42 to a maximum of 3.44 individuals per group during the 10 years. In all the groups, both sexes emigrated as adults: male age at emigration varied between 17 and 48 months (25 ± 9.8 months), and female age at emigration varied from 17 to 36 months (28 ± 6.8 months). Group size varied between 5 and 19 individuals when emigration was recorded, and group size was not correlated with female ($r_s = 0.34$, $p = 0.34$) or male ($r_s = 0.1$, $p = 0.84$) emigration episodes (Table 9.1). Tertiary sex ratio (TSR), or the sex ratio among the adults of a population (Dunbar 1988b) before dispersal, ranged from a minimum of 0.2 (male/female) to a maximum of 3.0. As for group size, no correlation was found between the TSR and male ($r_s = 0.44$; $p = 0.10$) or female ($r_s = 0.43$, $p = 0.70$) dispersal, and emigration events occurred (7/14 of males and 17/23 of females) with TSR ratios ranging from 0.2 and 1.0, suggesting that other factors besides TSR lead individuals to emigrate from their natal groups (Table 9.1). In both sexes, emigration events that were voluntary and involuntary were observed. During involuntary expulsion, emigrants remained on the periphery of the group for a number of days before dispersal (Arruda personal communication).

Concerning emigration seasonality, the results obtained by Araújo and Arruda (in prep) were similar to those of Monteiro da Cruz (1998). In both studies, significant differences were recorded between the dry season (this study, September

Table 9.1 Demographic data of eight groups of common marmosets living in a population at Nísia Floresta field station from 1991 to 2001

	Emigration		
Variables	Males	Females	Total
Group size			
0–5	1	1	2
6–10	8	8	16
11–15	4	9	13
16–19	1	5	6
Total	14	23	37
Tertiary male/female ratio			
0.2–1.0	7	17	24
1.1–2.0	4	6	10
2.1–3.0	3	0	3
Total	14	23	37

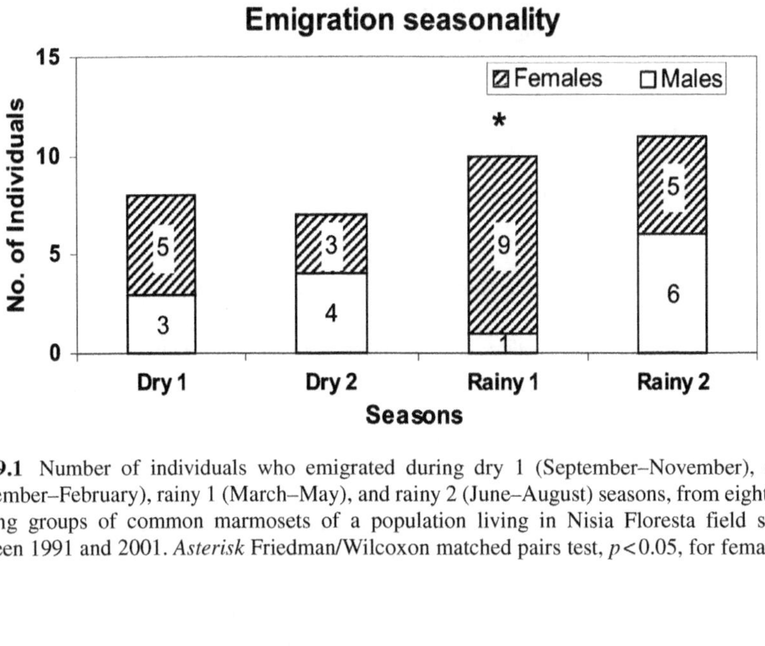

Fig. 9.1 Number of individuals who emigrated during dry 1 (September–November), dry 2 (December–February), rainy 1 (March–May), and rainy 2 (June–August) seasons, from eight free-ranging groups of common marmosets of a population living in Nisia Floresta field station between 1991 and 2001. *Asterisk* Friedman/Wilcoxon matched pairs test, $p<0.05$, for females

to February; in Monteiro da Cruz 1998, September to March) and the rainy season (this study, March to August; Monteiro da Cruz 1998, April to August) for female dispersal, but not for the dispersal of males. Sixty-four percent of the females emigrated during the first 3 months of the rainy season (referred to as rainy season 1, see Fig. 9.1).

9.3.2 Hormonal Data of Emigrating Females

From 1997 to 2001 we hormonally monitored four dominant and seven subordinate females living in three monogamous and one polygynous group (Albuquerque 2003; Albuquerque et al. 2004; Sousa et al. 2005). Three of the females we monitored are included in the 23 dispersal records shown in Table 9.1. Fecal collection lasted for periods varying from 10 to 17 months. Monitoring was carried out in the G4 group, in which data were collected from the dominant and four subordinate females on two occasions (phase 1: subordinate females GRA, GIO and phase 2: subordinate females GER and GIS). During phase 1, GRA ovulated but did not emigrate, whereas GIO disappeared while she was not cycling and when she showed low cortisol levels (Albuquerque et al. 2004). In phase 2, both the dominant and subordinate (GER) females gave birth to twin offspring but GER's offspring were killed between the fourth and ninth day after birth. Subordinate (GER) emigrated during ovarian cycling and 3 months after the infanticide event, when she was showing high cortisol levels. The second female monitored in this phase (GIS) was not ovulating prior to her disappearance and had low cortisol levels (Table 9.2). In the G4 group another episode of emigration of one subordinate pregnant female (not included in Table 9.2) was also witnessed, but her hormones were not monitored. As she was pregnant when she emigrated, this observation suggests that subordinate females might be copulating and successfully becoming impregnated with extra-group males (Digby 1999; Arruda et al. 2005).

Evidence from long-term studies show that subordinate females can become the reproductive female in a neighboring group, indicating that they might have had prior contact with that group (Arruda et al. 2005). In another monogamous group

Table 9.2 Hormonal profiles before dispersing or disappearance of common marmoset subordinate females

Groups	Females	Female's status	Reproductive condition before emigration (E) or disappearance (D)	Fecal cortisol levels
Monogamous				
G4 phase 1	GRE	Dom	Regular cycling and pregnancy	–
	GRA	S1	No cycling[a]	Low
	GIO	S2	No cycling (D)	Low
G4 phase 2	GER	S1	Cycling after reproductive attempt (E)	High
	GIS	S2	No cycling (E)	Low
Teruel	TER	Dom	Regular cycling and pregnancy	–
	TIN	S	No cycling after reproductive attempt (D)	Low
Pbf	PAT	Dom	Regular cycling and pregnancy	–
	PAL	S	Cycling after reproductive attempt (E)	High
Polygynous				
PBm	POL	Dom	Regular cycling and pregnancy	–
	PER	S	No cycling (D)	Low

Dom dominant females; *S1* and *S2* subordinate females; *D* disappearance; *E* emigration
[a]Stayed in the group until the end of the hormonal monitoring

(Teruel), before emigration, one subordinate female gave birth to twins, but her infants disappeared 6 days after birth. She remained in the group with ovarian suppression and emigrated, without apparent conflict, 6 months after her infants disappeared (Table 9.2). In the third monogamous group (PBf), a subordinate female reproduced, but her infants were killed by the dominant female (infanticide), and the subordinate emigrated while ovulating and while showing elevated cortisol levels. After the death of the dominant female, she returned and reproduced in her natal group. In the polygynous group (PBm), the subordinate female disappeared without signs of ovulation and while showing low fecal cortisol levels (Table 9.2).

9.4 Captive Studies and the Response of Males and Females to Social and Environmental Changes

Aggression among primates is common in highly kin-based societies (Walters and Seyfarth 1987). However, in common marmoset family groups, no overt aggression between dominant and subordinate males or between females was recorded (Abbott 1984). Agonistic interaction between same-sex adults in wild groups was rare (Lazaro-Perea et al. 2004), but when peer groups were formed in captivity, aggression between same-sexed individuals was observed (Abbott 1984). Dominance of one sex over the other has not been recorded in common marmosets (but see Ferrari Chapt. 8 this volume).

Mendoza and Mason (1986) showed that species with different social organization responded differently both behaviorally and physiologically, to artificial challenges during captive tests. According to the theoretical model developed by Troisi (2001), it can be speculated that within a species, males and females have evolved different mechanisms for coping with social stress, according to which social behaviors lead to the greatest reproductive success.

Based on this model, our laboratory is interested in using captive groups in which social deprivation and environmental changes can be induced, in order to trigger the physiological and behavioral responsiveness that would occur during emigration. This would lead to a better understanding of how male and female social strategies differ. In one of these studies, performed by Silva et al. (2008), the hormonal and behavioral response to social isolation of common marmoset males and females was investigated using two types of groups: (1) eight adult heterosexual pairs, and (2) nine subadult same-sex twin pairs (male–male = 5; female–female = 4). Observation sessions and two fecal collections for hormonal assays were performed each week during the baseline portion of the study (4 weeks), and daily, during the separation phase of the study (1 week). Separation for the heterosexual pairs was between males and females, whereas for the twins, separation involved removing the twins together from their natal group. The results of these studies demonstrate that adult males and females showed different profiles in relation to cortisol changes, with seven of eight males showing approximately 96% increase in fecal cortisol, and three of eight females, a 30% increase in fecal cortisol during the first 2 days after separation. The percentage increase in subadult twins showed a similar pattern, with male twins showing greater cortisol increases after separation from the natal group than female twins (Table 9.3). Behavioral changes recorded after separation included a significant decrease in scent-marking behavior for adult females during separation from their mates, but scent-marking behavior significantly increased in both male–male and female–female twin pairs when isolated from the family group (Sousa et al. 2002b) (Table 9.3). Subadult twin males showed a significantly higher frequency of affiliative interactions than did twin females at the same age, as evaluated by the frequency of social grooming and huddling (Silva 2003) after separation from the family group.

Increased huddling behaviour was also observed when same-sex adult related dyads were exposed to a new cage. On the other hand, same-sex adult related females increased piloerection and significantly decreased allogrooming behavior when they were moved to a new cage, thus showing an increase in agonistic interactions rather than affiliative relationships (Silva et al. 2008) (Table 9.3).

These data indicate that even when the females are relatives, changes in social and physical environment increase competition among them, whereas they increase affiliative behaviors among males. These findings indicate general and widespread differences between male and female sexual strategies, since, as previously shown, females disperse more frequently than males. The high responsiveness in males to separation, suggested by high cortisol levels, might reflect a more natural tendency toward philopatry, as evidenced by field studies. Cortisol increases in subordinate females is associated with competition among females before emigration and not with social deprivation after isolation. These results verify that common marmoset females have developed physiological adaptations to cope with the stress of dispersal, demonstrated by low cortisol changes after separation from their family groups or from their mates (Silva et al. 2008).

Faulkes et al. (2003, Chapt. 5 this volume), studying genetic relationships between two populations of common marmosets living in Tapacurá ($n=6$ groups) and Nísia Floresta ($n=4$ groups, all monitored for collection of emigration data as presented earlier in this chapter) in northeastern Brazil, found five haplotypes in one group (G4) of nine animals. That is, different maternal lineages were present in the group at the same time of sampling (see Faulkes et al. Chapt. 5 this volume). Although subordinate females were frequently seen with infants in both natural monogamous and polygynous groups (Arruda et al. 2005), the immigration of new members into groups, as well as the occurrence of multiple breeding females in a group, suggest a lower degree of genetic relatedness among group members than reported previously (Faulkes et al. 2003). As shown by Yamamoto et al. (Chapt. 6 this volume), polygynous groups are reproductively more successful than monogamous groups (where subordinates were probably reproducing with extra-group males), but dominant monogamous females are individually more successful than both dominant and subordinate polygynous females.

In addition to monogamy and polygyny, some groups may adopt polyandrous mating systems. Polyandry could become an evolutionary stable strategy if males have limited dispersal options, thereby making the sharing of one mate better than

Table 9.3 Hormonal and behavioral consequences of social and environmental changes for captive immature and mature male and female common marmosets

Groups (*n*)	Study period	Significant behavioral changes	Percentage of cortisol increase: Males	Percentage of cortisol increase: Females
Subadult twins				
Male: $n=5$,	After separation from family group			
Female: $n=4$	After exposure together to a new environment	Males: increase huddling, social grooming Males and females: increase scent marking	+44.9*-	+10.2#-
Heterosexual adult pairs $n=8$	After separation from the mate	Females: decrease scent marking	–	–
Same-sex adult related dyads		Males: increase huddling and the frequency of approximation	+95.8*	+29.6#
Male: $n=4$, Female: $n=4$	After exposure together to a new environment	Females: increase piloerection and scent marking; decrease allogrooming	–	–
Same-sex adults unrelated dyads	After exposure together to a new		+37.5#	+10#
Male: $n=3$	environment	Males: increase social grooming	–50%*	NA

Statistical differences between before and after separation: **t* test, $p<0.05$; #*t* test NS. + increase; – decrease
NA not available

having no mate at all. In addition, polyandry can be beneficial for females who gain additional help from group males in caring for infants (Rylands 1996). Evidence from intruder studies (Anzenberger 1985; Araújo and Yamamoto 1993) and from an experimental study where all adult male–male dyads cooperated successfully in the instrumental task, regardless of the amount of food available (Werdenich and Huber 2002) reinforce the speculation that males will, willingly, cooperate with one another. Indeed, data showing that males delay dispersal, and that even unrelated males engage in affiliative behaviors more than aggressive behaviors (Silva et al. 2008), suggest that males have evolved cooperative, rather than competitive, social strategies. However, so far, no genetic or behavioral evidence is available to confirm or rule out the occurrence of polyandry in common marmosets.

Additional evidence, based on observations of a wild group recorded by Lazaro-Perea et al. (2000) occurred, when a group (PB) fissioned along sexual lines of males and females, and the two groups were restructured with the immigration, respectively, of two adult females and two adult males. Following fission, the male group was more successful in terms of recruiting mates, reproducing, and maintaining group stability than the female group. Therefore, for males, it appears that their strategy is to remain within their natal group, build affiliate bonds with other males, and help in infant care, thereby benefiting to some degree from inclusive fitness. On the other hand, the female strategy seems to be more varied, probably due to the high costs of rearing twin offspring, and the ability to undergo postpartum ovulation and gestation, making helpers crucial to their reproductive success (Snowdon 1996).

9.5 Why Do Adult Common Marmoset Females Disperse?

In general, the risk of predation and the availability and dispersion of feeding resources are the two main factors responsible for the regulation of group size in primates, as well as for tension in the social group (van Schaik 1983; Dunbar 1988a).

During the 3 years of monitoring wild common marmoset groups, we found predation risk to be minimal (only one confirmed predation event). On the other hand, food availability and intergroup tension appear to play major roles in group dynamics (Araújo 1996). Common marmosets primarily eat exudates and insects. However, although gum is available all year long, its nutritional content is not adequate enough for marmosets to rely exclusively on this food item, thus, a large variety of other foods are included in their diet (Hubretch 1984; Stevenson and Rylands 1988; Alonso and Langguth 1989). During the wet season, the amount of fruit and insects increases (Scalon et al. 1991); this may explain why the dispersal of females is significantly higher during the rainy season –the dispersal conditions are more favorable. As noted previously, the tertiary male/female ratio was not correlated with dispersal, and females emigrated even when TSR was as low as 0.2 and 0.6 for males. Therefore, the level of conflict between dominant and subordinate females seems to be the mechanism triggering the dispersal of subordinates, and this may be food related. Although emigration might appear to be voluntary rather than forced, our hormonal data indicate that in two out of three occasions, emigrant females exhibited high fecal cortisol levels. These high cortisol levels likely indicate high levels of competition within the social group.

This behavior in common marmosets probably involves multiple and complementary factors that could be grouped as proximate and ultimate causes, similar to what was proposed by Holecamp and Sherman (1989) regarding male ground squirrel dispersal. Proximate causes include aspects of ontogeny and underlying related physiological mechanisms. In this perspective, evidence collected by Johnson et al. (1996) and Raminelli et al. (2001) has shown that females have higher basal levels of cortisol than males do. Complementary data show that increases in cortisol levels after separation or after moving to a new environment are lower in females than in males (Sousa et al. 2003; Silva et al. 2008) and indicate that common marmoset females have different adaptations to deal with social changes after dispersal. Indeed, the reproductive suppression of subordinates appears to depend on the type of behavioral relationship between dominant and subordinate females (Alencar et al. 2006; Saltzman et al. 2004). Therefore, the main determinant for common marmoset female dispersal seems to be intra-sexual competition between dominant and subordinate females, and the ultimate result is inbreeding avoidance.

Understanding the genetic relatedness between group females is important for further understanding the factors that influence female reproductive and dispersal patterns. However, at the moment there is insufficient genetic data available to fully assess genetic relationships within social common marmoset groups (but see Faulkes et al. Chapt. 5 this volume). Indeed, the costs of subordinates that attempt to breed (lost competitive interactions, unsuccessful attempts to breed, little assistance with infant care, and infanticide) (Clutton-Brock 1998) must be considered in order to understand the social dynamics between dominant and subordinate common marmoset females. Dominant females can use at least four different strategies to decrease the reproductive success of the subordinate females: ovulation inhibition, direct aggression, infanticide, and competition for access to helpers. However, despite these daunting challenges, subordinate females can, occasionally, become successful breeders in their natal group. Common marmosets, therefore, have evolved a system by which males and females have different physiological and behavioral mechanisms for cooperating and competing in order to maximize their reproductive success.

References

Abbott DH (1984) Behavioral and physiological suppression of fertility in subordinate marmoset monkeys. Am J Primatol 6:169–186

Abbott DH (1993) Social conflict and reproductive suppression. In: Mason WA, Mendoza SP (eds) Primate Social Conflict. State University of New York Press, Albany, pp 331–372

Albuquerque ACSR (2003) Estratégias reprodutivas de fêmeas de *Callithrix jacchus* em grupos poligínicos e monogâmicos no ambiente natural. PhD Thesis, Natal:UFRN 138 pp

Albuquerque ACSR, Sousa MBC, Santos HM, Ziegler TE (2001) Behavioral and hormonal analysis of social relationship between oldest (reproductive and non-reproductive) females in a wild monogamous group of common marmoset, *Callithrix jacchus*. Int J Primatol 22:631–645

Albuquerque ACSR, Nascimento MCL, Santos HM, Sousa MBC (2004) Função ovariana e adrenocortical de fêmeas adultas de sagüi comum (*Callithrix jacchus*) em relação ao comportamento de emigração. In: Mendes SL, Chiarello AG (org), vol 8, IPEMA/SBPr, Vitória, pp 301-316

Alencar AI, Yamamoto ME, Oliveira MS, Lopes FA, Sousa MBC, Silva NG (1995) Behavior and progesterone levels in *Callithrix jacchus* females. Braz J Med Bio Res 28:591–595

Alencar AI, Sousa MBC, Abbott DH, Yamamto ME (2006) Contested dominance modifies the anovulatory consequences of social subordination in female marmosets. Braz J Med Biol Res 39:647–658

Alonso C, Langguth A (1989) Ecologia e comportamento de *Callithrix jacchus* (Primates: Callitrichidae) numa ilha de floresta atlântica. Rev Nordestina Biol 6:105–137

Anzenberger G (1985) How stranger encounters of common marmosets (*Callithrix jacchus jacchus*) are influenced by family members: the quality of behavior. Folia Primatol 45:204–224

Araújo A (1996) Influence des facteurs écologiques, comportementaux et démografhiques sur la dispersion de *Callithrix jacchus*. Thesis. Paris: Université Paris-Nord (Paris XIII). 206 pp

Araújo A, Yamamoto ME (1993) Reação a intrusos da mesma espécie em *Callithrix jacchus*: influência do status social. In: Yamamoto ME, Sousa MBC (org), A Primatologia no Brasil. EDFURN, Natal, pp 15-34

Arruda MF, Araújo A, Sousa MBC, Albuquerque FS, Albuquerque ACSR, Yamamoto ME (2005) Two-breeding females within free-living groups may not always indicate polygenic: alternative subordinate female strategies in common marmosets (*Callithrix jacchus*). Folia Primatol 76:10–20

Baker AJ, Dietz JM (1996) Immigration in wild groups of golden lion tamarins (*Leontopithecus rosalia*). Am J Primatol 38:47–56

Baker JV, Abbott DH, Saltzman W (1999) Socal determinants of reproductive failure in male common marmosets housed with their natal family. Anim Behav 58:501–513

Box HO (1975) Quantitative studies of behaviour within captive groups of marmosets monkeys (*Callithrix jacchus*). Primates 16:155–174

Clutton-Brock TH (1998) Reproductive skew, concessions and limited control. Trends Ecol Evol 13:288–292

Digby LJ (1995a) Infant care, infanticide and female reproductive strategies in polygynous groups of common marmosets (*Callithrix jacchus*). Behav Ecol Sociobiol 37:51–61

Digby LJ (1995b) Social organization in a wild population of *Callithrix jacchus*: II. Intragroup social behavior. Primates 36:361–375

Digby LJ (1999) Sexual behavior and extragroup copulations in a wild population of common marmosets (*Callithrix jacchus*). Folia Primatol 70:136–145

Digby LJ, Barreto CE (1993) Social organization in a wild population of *Callithrix jacchus*. Folia Primatol 61:123–134

Digby L, Ferrari SF (1994) Multiple breeding females in free-ranging groups of *Callithrix jacchus*. Int J Primatol 15:389–397

Dunbar RIM (1988a) Primate social systems. Demography processes: 1. Life history variables. Croom Helm, London, pp 55–75

Dunbar RIM (1988b) Primate social systems. Demography processes: 2. Population parameters. Croom Helm, London, pp 76–89

Emlen ST (1997) Predicting family dynamics in social vertebrates. In: Krebs JR, Davies NB (eds) Behavioural ecology: an evolutionary approach, 4th edn. Blackwell, Oxford, pp 228–253

Evans S, Poole TB (1983) Pair-bond formation and breeding success in the common marmoset *Callithrix jacchus jacchus*. Int J Primatol 4:83–96

Faulkes C, Arruda MF, Monteiro da Cruz MAO (2003) Matrilineal genetic-structure within and among populations of the cooperatively breeding common marmoset, *Callithrix jacchus*. Mol Ecol 12:1101–1108

Faulkes C, Arruda MF, Monteiro da Cruz MAO (this volume) Genetic structure within and among populations of the common marmoset, *Callithrix jacchus*: implications for cooperative breeding. In: Ford SM, Porter LM, Davis LC (eds) The smallest anthropoids: The marmoset/callimico radiation. Springer Press, New York, pp 103–117

Ferrari SF (1988) The behaviour and ecology of the buffy-headed marmoset, *Callithrix flaviceps* (O. Thomas, 1903). PhD Thesis, London, University College, 278 pp

Ferrari SF (this volume) Social hierarchy and dispersal in free-ranging buffy-headed marmosets (*Callithrix flaviceps*). In: Ford SM, Porter LM, Davis LC (eds) The smallest anthropoids: The marmoset/callimico radiation. Springer Press, New York, pp 155–165

Holecamp KE, Sherman PW (1989) Why male ground squirrel disperse? In: Sherman PW, Alcock J (eds) Exploring animal behavior: readings from American scientist. Sinauer Associates, Sunderland, pp 41–48

Hubretch RC (1984) Field observations on group size and composition of the common marmoset (*Callithrix jacchus jacchus*), at Tapacura, Brazil. Primates 25:13–21

Johnson EO, Kamilaris TC, Carter CS, Calogero AE, Gold PW, Chrousos GP (1996) The behavioral consequences of psychogenic stress in a small, social primate (*Callithrix jacchus jacchus*). Biol Psychiatry 40:317–337

Kendrick KM, Dixson AF (1983) The effect of the ovarian cycle on the behavior of common marmoset (*Callithrix jacchus*). Physiol Behav 30:735–742

Lazaro-Perea C (2001) Intergroup interactions in wild common marmosets, *Callithrix jacchus*: territorial defense and assessment of neighbours. Anim Behav 62:11–21

Lazaro-Perea C, Castro CCS, Harisson R, Araújo A, Arruda MF, Snowdon T (2000) Behavioral and demographic changes following the loss of the breeding female in cooperatively breeding marmosets. Behav Ecol Sociobiol 48:137–146

Lazaro-Perea C, Arruda MF, Snowdon CT (2004) Grooming as a reward? Social function of grooming between females in cooperatively breeding marmosets. Anim Behav 67: 627–636

Mason WA (1974) Comparative studies of social behavior in *Callicebus* and *Saimiri:* behavior of male-females pairs. Folia Primatol 22:1–8

Mendoza SP, Mason WA (1986) Contrasting responses to intruders and to involuntary separation by monogamous and polygynous New World monkeys. Physiol Behav 38:795–801

Monteiro da Cruz MAO (1998) Dinâmica reprodutiva de uma população de saguis-do-nordeste (*Callithrix jacchus*) na estação ecológica de Tapacurá, PE. PhD Thesis, São Paulo, 191 pp

Moore J (1984) Female transfer in primates. Int J Primatol 5:537–589

Nievergelt CM, Digby L, Ramakrishnan U, Woodruff DS (2000) Genetic analysis of group composition and breeding system in a wild common marmoset (*Callithrix jacchus*) population. Int J Primatol 21:1–20

Raminelli JLF, Sousa MBC, Cunha MS, Barbosa MFV (2001) Morning and afternoon patterns of fecal excretion of cortisol among reproductive and non-reproductive male and female common marmosets, *Callithrix jacchus*. Biol Rhythm Res 32:159–167

Reeve HK, Emlen ST, Keler L (1998) Reproductive sharing in animal societies: reproductive incentives or incomplete control by dominant breeders? Behav Ecol 9:267–278

Rothe H (1975) Some aspects of sexuality and reproduction in groups of captive marmosets (*Callithrix jacchus*). Zietsch Tierp 35:253–273

Rylands AB (1986) Ranging behaviour and habitat of a wild marmoset group, *Callithrix humeralifer* (Callitrichidae, Primates). J Zool London 210:489–514

Rylands AB (1996) Habitat and the evolution of social and reproductive behavior in Callitrichidae. Am J Primatol 38:5–18

Saltzman W, Schultz-Darken N, Abbott DH (1997a) Behavioral and social correlates of escape from suppression of ovulation in female common marmosets housed with natal family. Am J Primatol 41:1–21

Saltzman W, Schultz-Darken NJ, Abbott DH (1997b) Familial influences on ovulation function in common marmosets (*Callithrix jacchus*). Am J Primatol 41:159–177

Saltzman W, Pick RR, Salper OJ, Liedl KJ, Abbott DH (2004) Onset of plural cooperative breeding in common marmoset families following replacement of the breeding male. Anim Behav 68:59–73

Santee D, Arruda MF (1994) The Nisia Floresta common marmoset research station. Neotrop Primates 2:8–11

Savage A, Giraldo LH, Soto LH, Snowdon CT (1996) Demography, group composition, and dispersal in wild cotton-top tamarins (*Saguinus oedipus*) groups. Am J Primatol 38:85–100

Scalon CE, Monteiro da Cruz MAO, Rylands AB (1991) Exploração de exudatos vegetais pelo sagüi-comum, *Callithrix jacchus*. In: Rylands AB, Bernardes AT (eds) A Primatologia no Brasil, vol 3. Fundação Biodiversitas para a Conservação da Diversidade Biológica, Belo Horizonte, MG, pp 197–205

Shields WM (1987) Dispersal and mating system: investigationing their causal connections. In: Chepko-Sade BD, Halpin ZT (eds) Mammalian dispersal patterns. The Chicago University Press, Chicago, pp 3–24

Silva HPA (2003) Variações comportamentais e hormonais entre gêmeos de Callithrix jacchus e sua relação com a aquisição de posto social. PhD Thesis, USP, São Paulo, 168 pp

Silva HPA, Leão AC, Sousa MBC (2008) Sexual differences on behavioral and hormonal response to social change in adult and subadult common marmosets, *Callithrix jacchus*. In: Ferrari S, Rímoli J (eds) A Primatologia no Brasil, vol 9, SBPr, Aracaju, SE, pp 154–168

Snowdon CT (1996) Infant care in cooperatively breeding species. In: Rosenblatt JS, Snowdon CT (eds) Parental care: evolution, mechanisms, and adaptative significance. Academic Press, San Diego, pp 643–689

Sousa MBC, Ziegler TE (1998) Diurnal variation on the excretion patterns of steroids in common marmoset (*Callithrix jacchus*) females. Am J Primatol 46:105–117

Sousa MBC, Nascimento MCL, Albuquerque ACSR (2002a) Fecal collection in free-ranging common marmosets, *Callithrix jacchus*. Neotrop Primates 10:21–24

Sousa MBC, Silva HPA, Leão AC (2002b) Sexual differences on behavior and fecal cortisol using the separation paradigm in common marmosets, *Callithrix jacchus*. Abstracts, IPS Conference, Beijing, 2002, p 94

Sousa MBC, Albuquerque ACSR, Albuquerque FS, Araújo A, Yamamoto ME, Arruda MF (2005) Behavioral strategies and hormonal profile of dominant and subordinate common marmoset (*Callithrix jacchus*) females in wild monogamous groups. Am J Primatol 67:37–50

Stevenson MF, Rylands AB (1988) The marmosets, genus *Callithrix*. In: Mittermeier RA, Rylands AB, Coimbra-Filho A, Fonseca GAB (eds) Ecology and behavior of neotropical primates, vol 2. Littera Maciel, Contagem, pp 131–222

Troisi A (2001) Gender differences in vulnerability to social stress: a Darwinian perspective. Physiol Behav 73:443–449

van Schaik CP (1983) Why are diurnal primates living in groups? Behaviour 87:120–144

Walters JR, Seyfarth RM (1987) Conflict and cooperation. In: Smuts BB, Cheney DL, Seyfarth RM, Wrangham RW, Strusaker TT (eds) Primate Societies. University of Chicago Press, Chicago, pp 306–317

Werdenich D, Huber L (2002) Social factors determine cooperation in marmosets. Anim Behav 64:771–781

Yamamoto ME, Arruda M de F, Alencar AI, de Sousa MBC, Araújo A (this volume) Mating systems and female-female competition in the common marmosets, *Callithrix jacchus*. In: Ford SM, Porter LM, Davis LC (eds) The smallest anthropoids: The marmoset/callimico radiation. Springer Press, New York, pp 119–133

Ziegler TE, Scheffler G, Wittwer DJ, Schultz-Darken NJ, Snowdon CT, Abbott DH (1996) Metabolism of reproductive steroids during the ovarian cycle in two species of callithrichids, *Saguinus oedipus* and *Callithrix jacchus*, and estimation of the ovulatory period from fecal steroids. Biol Reprod 54:91–99

Chapter 10
Social and Physical Cognition in Marmosets and Tamarins

Ludwig Huber and Bernhard Voelkl

Abstract In this paper, we summarize and discuss recent research on the cognitive abilities of marmosets and tamarins, and compare its results with findings from other primates. The focus animal in this chapter will be the common marmoset (*Callithrix jacchus*), which we have studied extensively in our lab in Vienna. We present accumulated evidence for callitrichids being likely to: (1) locate food by using some sort of cognitive map, (2) represent objects and their movements in an abstract manner, (3) benefit from social influences that aid in learning about new food by motivational and perceptual factors, and (4) learn new foraging techniques imitatively by observing skillful conspecifics. Together, these findings provide evidence for quite a surprising level of understanding of the physical and social world of these monkeys and shake the long-held belief that callitrichids have low intelligence. These new insights suggest a reevaluation of the traditional monkey-to-ape shift in primate cognition.

Resumen En el presente estudio, resumimos y discutimos investigación reciente sobre las habilidades cognitivas de marmosetas y tamarinos y comparamos sus resultados con hallazgos en otros primates. El animal foco de atención en el presente estudio es la marmoseta común (*Callithrix jacchus*), el cual hemos estudiado extensivamente en nuestro laboratorio en Viena. Presentamos evidencia acumulada de sugiere que los calitrícidos: (1) localizan el alimento usando algún tipo de mapa cognitivo, (2) representan objetos y sus movimientos de una manera abstracta, (3) se benefician de influencias sociales que ayudan en el aprendizaje sobre nuevos alimentos por factores motivacionales y de percepción y (4) aprenden nuevas técnicas de forrajeo por imitación, observando conespecíficos habilidosos. Juntos, estos hallazgos proveen evidencia de un nivel de entendimiento bastante sorprendente del mundo físico y social en estos monos, y debilita la creencia de que los calitrícidos

L. Huber (✉)
Department for Behavior, Neurobiology and Cognition,
University of Vienna, Althanstrasse 14, 1090, Vienna, Austria
e-mail: ludwig.huber@univie.ac.at

S.M. Ford et al. (eds.), *The Smallest Anthropoids*, Developments in Primatology: Progress and Prospects, DOI 10.1007/978-1-4419-0293-1_10,

© Springer Science+Business Media, LLC 2009

poseen una inteligencia baja. Estos nuevos elementos sugieren una reevaluación del cambio tradicional de mono a simio en la cognición en primates.

Resumo Neste estudo nós resumimos e discutimos pesquisas recentes sobre as habilidades cognitivas de sagüis e micos e comparamos os resultados com achados de outros primatas. O animal foco deste capítulo é o sagüi comum, *Callithrix jacchus* o qual vem sendo estudado extensivamente no nosso Laboratório em Viena. Nós apresentamos dados cumulativos para calitriquídeos que se mostraram aptos para: (1) localizar comida usando algum tipo de mapa cognitivo, (2) representar objetos e seus movimentos de maneira abstrata, (3) se beneficiar de influências sociais que ajudam no aprendizado sobre novos tipos de alimentos por meio de fatores motivacionais e perceptuais e (4) aprender novas técnicas de forrageio por meio da imitação, observando as habilidades de co-específicos. Juntos, estes resultados fornecem evidências sobre os níveis surpreendentes de compreensão do mundo físico e social nestes macacos e questiona as crenças de que os calitriquídeos possuem baixa inteligência. Estas novas evidências sugerem uma re-avaliação do tradicional desvio na passagem macacos-para-pongídeos na cognição de primatas.

10.1 Introduction

Amazing progress has been made on the study of primate cognition in the past few years. Thanks to the combined efforts of ethologists, comparative psychologists, anthropologists, primatologists and neuroscientists we have available an impressive body of knowledge about the problem solving abilities and mental capacities of our nearest relatives, the great apes (see Parker 1990, and Tomasello and Call 1997, for reviews of the scientific development of primate cognition). Furthermore, studies such as those of Kummer (1968) on tripartite relations and social knowledge in hamadryas baboons (*Papio hamadryas*) and of Seyfarth et al. (1980) on reference and mental representation in the communication of vervet monkeys (*Cercopithecus aethiops*) have revealed considerable cognitive capacities in selected species of Old World monkeys. However, the respective database for New World monkeys is comparably modest (the notable exception is the genus *Cebus*, see Fragaszy et al. 2004b).

One reason for this difference in research effort may be that New World monkeys are less attractive than catarrhines for cognitive scientists seeking human-like features of cognition. New World monkeys separated from the catarrhines about 30–40 million years ago, when they became isolated from other primates because of their introduction to the then island-continent of South America (Fleagle 1999); thus, they are more distant relatives to humans than the African apes and monkeys (Martin 1992). A further reason might be the relatively recent flourishing of fieldwork done in Central and South America, as compared to the fieldwork on Old World monkeys and great apes in Africa and Asia (Parker 1990). These two developments may have contributed to the lack of understanding of the cognitive potentials in New World monkeys and the general attitude that New World monkeys are inferior in the flexibility of their behavior as compared to the Old World monkeys and apes.

Callitrichid monkeys are small New World monkeys, once thought to have retained many primitive primate characters and to be rather simple-minded (Hershkovitz 1977). Therefore, marmosets and tamarins would not seem likely candidates for studies of complex cognition. In time, this evaluation has changed, and it is currently accepted that they have developed a number of remarkably original adaptations for their unusual lifestyle (Snowdon 2001). However, these monkeys do have relatively small brains among primates, not only in terms of absolute brain volume, but also in terms of the absolute and relative size of the executive brain. For instance, the respective measures for common marmosets (*Callithrix jacchus*) are 7,241 mm^3 total brain volume; 4,371 mm^3 neocortex volume; and 7.74 neocortex/brainstem ratio (Reader and MacDonald 2003). The literature also suggests that marmosets have quite low frequencies of social learning, innovation, and tool-use, representing three ecologically relevant measures of cognitive ability (Reader and Laland 2002). Most importantly, although to date there is no direct, unequivocal support for a link between brain size and general behavioral flexibility, Reader and Laland (2002) found a positive correlation between executive brain volume and the three above mentioned measures of cognitive flexibility using data from more than 100 species of primates.

Nevertheless, only the combined efforts of observation of natural behavior and of rigorous testing under controlled conditions can reveal the true repertoire and potential of the cognitive behavior of a species. The aim of this review is to provide those interested in the biology of the smallest anthropoids with information on the cognitive abilities of marmosets and tamarins emerging from current research, with references to related findings in other primates. The focus animal in this chapter will be the common marmoset (*Callithrix jacchus*), which we have studied extensively over the last decade. We hope to convince the reader that members of this species display considerable cognitive abilities in both the social and the non-social domain. Our findings provide strong evidence that marmosets and tamarins are remarkably sensitive and responsive to cues from other social companions as well as responsive to the physical environment, and they are supported by results from many other studies which were reviewed recently by Snowdon (2001).

Many studies have revealed that primates use a variety of cognitive mechanisms in understanding such things as space, tools, object categories, quantities, causality, and the behavior and possible mental states of conspecifics in interactions involving cooperation, competition, communication, and social learning (Tomasello and Call 1997; Tomasello 2000). Early primatologists were trained ethologists or physical anthropologists who often viewed cognitive issues with suspicion considering them as anthropomorphic impositions, but current students of primate behavior do not hesitate to apply and test models of human cognition. In the last few decades, such models have become less symbolic in nature and less explicitly focused on language, and primatologists have intensified their communication with cognitive scientists (Tomasello 2000). Mainstream cognitive science has been equally helpful in developing increased interest in understanding the importance of early human cognitive development and the adaptive role of the human mind.

Despite the recent interest in human cognitive science, the major concerns of primatologists have remained issues of survival, adaptation, and reproductive success. Cognitive mechanisms are considered important as behavioral adaptations that work by empowering individual organisms to make flexibly appropriate decisions based on individually acquired, mentally represented information. These decisions can only be evaluated in view of the permanent needs of the individual embedded in a complex network of demands from the social and asocial environment. Both ecological and behavioral studies can inform students of primate cognition about the problems primates have to cope with in nature.

Primates evolved grasping hands and sophisticated visual and tactile systems, which have allowed them to discover and exploit a huge variety of food sources. Grasping hands, one of the most distinctive morphological adaptations of primates, seem to have evolved for arboreal life and especially for obtaining and processing food items (Passingham 1982). The view that foraging creates the main selective pressure for morphological, sensual, and cognitive adaptations, resulting in the emergence of large brains among primate species, continues to be a matter of debate (Milton 1981). Obviously, foraging presents special challenges for primates relative to non-arboreal mammals because the spatio-temporal distribution of food is much less uniform and predictable than in most non-arboreal environments (Garber 1989). Fruits, in particular, occur only intermittently throughout tropical rainforests, ripening only in some seasons of the year, and are difficult to find without advanced memories encoding their distribution in time and space (Garber 1989; Garber and Hannon 1993).

Parker and Gibson (1977) noted that concealed and hard-to-extract foods create special problems to animals that lack specialized anatomy for extraction. They hypothesized that when the need to forage extractively arises seasonally and over a wide range of foods, advanced forms of sensorimotor intelligence may evolve to improve foraging. More recently, this hypothesis has been narrowed down from the evolution of primate cognition in general, to apply only to the origin of the ape/monkey grade-shift in intelligence (Byrne 1997). The term "Technical intelligence hypothesis" (Byrne 1997) is used as a banner for the collection of several related hypotheses put forward to argue that the great apes evolved a greater efficiency in foraging behavior compared to that seen in monkeys, because of their large size, the difficulties of brachiation for distance travel, and their reliance on a high quality diet.

The "Social function of intellect hypothesis" is an alternate but complementary hypothesis suggesting that problem solving capacity in primates is a correlate of having a complex social system and a long life (Jolly 1966; Humphrey 1976). Although this hypothesis, now widely known as the "Machiavellian intelligence hypothesis" (Byrne and Whiten 1988), has been used as an explanation for the cognitive evolution of many non-primate species, most primates appear especially gifted in solving the demands of a complex social life with intricate relationships and elaborate interactions, in which others are singled out as agents and subjects of behavior. Compared to other mammals, primate societies are especially complex in terms of the number and nature of social relationships formed and the range of communicative mechanisms utilized (Cheney and Seyfarth 1990). A cognitive "arms race" likely occurred as primate individuals faced ever-increasing demands to compete with members of their own group; at the same time, primate individuals may also have gained equal adaptive advantage by cooperating with selected friends and learning important things from one another, socially (Tomasello 2000).

For many animal species the most pressing problems arise in locating and obtaining food. This fact implies that many important cognitive skills for many animal species evolved in the context of foraging, and the latter was likely a major influence on the evolution of New World monkeys.

Following the scheme proposed by Tomasello and Call (1997), foraging can be divided into three categories based on the need to: (1) find food, (2) obtain or process food, and (3) identify and categorize food. Locating food requires an understanding of large-scale foraging space and the movements of objects from one location to another in small-scale manipulative space. The necessary cognitive mechanisms of spatial cognition and object representation found in marmosets and tamarins are dealt with in the following two sections of this review. Then we will discuss socially influenced learning of the location, quality, and manipulation of food. We will conclude with a comparison of the cognitive abilities of callitrichids with those of other nonhuman primates.

10.2 Spatial Cognition

Finding food efficiently in large-scale space (i.e., an area that cannot be seen entirely from one single vantage point) requires having an adequate mental representation of the space and the kinds of objects that it includes. It has long been suggested that primates have a quite precise knowledge of where and how they are searching for food (reviewed in Clutton-Brock 1977; Cheney and Seyfarth 1990). Systematic field studies of the highly seasonal nectar feeding in moustached tamarins (*Saguinus mystax*) and saddleback tamarins (*Saguinus fuscicollis*) in the Amazon forests of Northeastern Peru have generated results that support this assumption (Garber 1989). The routes taken by these primates are direct and straight, from one rich nectar source to the next, bypassing food trees with lesser reward. Comparison of foraging routes used by these tamarins with various computer models revealed that the most likely explanation for their efficient foraging is the use of a cognitive map that encodes the location of hundreds of feeding trees in their home range (Garber and Hannon 1993). Thus, these monkeys will ignore a nearby resource of food that is fully visible, but poor quality in order to go to a more distant, currently invisible, but much richer resource they have found before. Such selective choice of rich, abundant resources requires having a detailed mental picture of the distribution of relevant objects and feeding sites in a complex environment.

Saddleback tamarins (*Saguinus fuscicollis*) have also been shown to meet this expectation (Menzel and Menzel 1979; Menzel and Juno 1982). These monkeys look for fruits that grow in small quantities in a varied number of locations and ripen at different times, even within the same bush or tree. This means that a given location is not permanently exhausted of food, and that it will pay to return to it in the near future. Rather than sampling their home range arbitrarily, trusting that they will find enough food each day by luck, saddle-backs appear to be able to remember where they have found food before and re-visit sites appropriately. When a group was tested to see if they could remember which of the 30 objects presented in an artificial search room were baited the day before, saddleback tamarins chose the correct objects in 75% of the occasions, after one trial (Menzel and Menzel 1979). Moreover, the tamarins not only recognized the previously baited object, but they also remembered its position. If the originally baited object was moved to a new location in the next trial, they went to the now empty location where it had been before. If there was now a new object there or one that had not previously been baited, they went to the food-associated object in its new location in 75% of trials, suggesting that their search was guided by some representation of both the appearance and location of an object. Further tests revealed that the tamarins discriminated between similar objects very accurately by not only remembering categorical differences between them, but also by identifying individual objects within the same category (Menzel and Menzel 1979). These primates, therefore, appear to be able to build very detailed representations of their foraging environments, both in terms of spatial locations and of the individual objects situated in different locations.

10.3 Object Permanence

Locating objects precisely in the complex arboreal environment of New World primates requires not only knowledge about stable large-scale relationships between objects, but also the ability to track objects if they move or if they change their position relative to the foraging animal. Such moving objects could be mates, predators, or prey. In everyday life, objects appear, disappear, and reappear, requiring that monkeys have the capacity to recognize the important object through its repeated appearances over time.

In order to maintain interaction, the animal must use functional invariants; that is, if an object has disappeared, leaving no traces or perceptual cues to rely on, the search problem requires a solution on an abstract level (a representation of the absent target object in memory). Are New World primates, such as marmosets and tamarins, able to locate objects hidden from view if they have witnessed the hiding? Can they infer the possible new location of an object that continues to move out of view? Although it is known that a wide variety of animals can maintain a representation of an object that has disappeared from view, psychologists' understanding of which information is extracted from this event and the knowledge of which animals use which information to keep track of the object are far from complete.

The ability to use a functional invariant necessary to maintain interaction with moving objects is called object permanence. According to Piaget (1937), this last perceptual and first conceptual invariant enables the animal to unify objects across the spatial and temporal discontinuities of their perceptual appearance. It allows a subject to understand that objects continue to exist, even when they are no longer available for immediate perception. Piaget's experiments with his own children have been repeatedly used to test animals on their skills in finding objects that the experimenter has hidden in various ways. Although a large number of studies on object permanence have been conducted on non-human primates (for a review, see Doré and Dumas 1987; Tomasello and Call 1997; Parker and McKinney 1999), only a small fraction has been conducted on New World monkeys, including cotton-top tamarins (*Saguinus oedipus* – Santos et al. 1999; Hauser et al. 2001), tufted (white-throated) capuchins (*Cebus capucinus* – Mathieu et al. 1976; Snyder et al. 1978), brown capuchins (*Cebus apella* – Dumas and Brunet 1994; Schino et al. 1990; Snyder et al. 1978; Spinozzi 1989), yellow-tailed woolly monkeys (*Lagothrix* (=*Oreonax*) *flavicauda* – Mathieu et al. 1976), and squirrel monkeys (*Saimiri sciureus* – de Blois et al. 1998; Vaughter et al. 1972). Any general hypotheses about the object permanence skills of primates would be premature without some understanding of the abilities of marmosets and tamarins, which is currently lacking.

Both New World and Old World monkeys are able to solve visible displacement problems, in which the subject observes the single or repeated hiding of the target object, and then, is given the opportunity to find it. However, whether both groups are also able to solve invisible displacement problems, in which the target object is covered by a container during the hiding event, is an issue of current debate. The latter task requires not only the mental representation of an object, but also of the movement of an object in small-scale space. From experiments with orangutans and squirrel monkeys, de Blois et al. (1998) concluded that the ability to mentally represent unperceived events is shared by humans and great apes but not by monkeys.

The first, and to our knowledge only, object permanence tests with marmosets have been conducted in our Viennese laboratory (Mendes and Huber 2004). Using a battery of standard Piagetian object-permanence tasks, these experiments assessed whether marmosets use object permanence in their search strategies and also whether they display alternate search strategies. Six male and five female common marmosets, members of four separated family groups and varying in age between 2 and 16 years, were tested. The marmosets were allowed to search for a food reward, which was hidden behind one of three plastic boxes prior to testing; the marmosets were able to observe the placement. The hiding procedure was systematically varied across five visible and four invisible displacement tests, all of which had been used in at least similar form in previous experiments with the animals. In the visible displacement tests the experimenter placed the food under one of three plastic boxes before allowing the animal to search for it; in the invisible displacement tests, the experimenter placed the food in a cup and moved the cup underneath one of the boxes. These tests were administered in different sequences and counterbalanced across subjects to control for effects from previous tests.

As determined on the group level, the subjects solved eight of the nine tests (Mendes and Huber 2004). They only failed in one sequential invisible test, in which the experimenters altered the hiding box. In all the other tests, marmosets found hidden objects with greater success than would be predicted by pure chance. The test performances of the marmosets were fairly constant across all the trials of each test (maximum trial number: 24), ruling out that the animals were adapting to the task or learning the rules of thumb. However, several search strategies employed by the marmosets prevented their consistently finding the food rewards. These alternate strategies included preferences for specific locations or the win-stay strategy (always try the response that was last rewarded). In addition, subjects did not always pay attention when the experimenter was hiding the objects, which hampered the subjects' ability to figure out the location of the food and to separate relevant from irrelevant objects and events. This was particularly true in successive displacement tasks, in which up to three boxes were manipulated by the experimenter before the target was entirely displaced. All those factors led to considerable inter-individual differences in performance and inconsistencies across tests.

Nevertheless, the results showed that marmosets are able to find food that was hidden behind a different screen on every trial or behind various screens within the same trial (this ability corresponds to Piagetian Stage 5 of object permanence development; see de Blois et al. 1998). Moreover, the excellent performance of two subjects that also achieved very high scores in the most critical test (Test 9 – successive invisible displacement with two target boxes), suggests that this species is, in principle, able to mentally represent not only the existence but also the movements of unperceived objects (Mendes and Huber 2004). They were able to find food that was invisibly hidden behind two screens within the same trial, which corresponds to Piagetian Stage 6 of object permanence development (see de Blois et al. 1998). This is a surprising result, given that this level of object understanding and sensorimotor intelligence has previously been described only for great apes and humans. Even in humans, this ability is not present at birth, but develops through stages. Only after 18 months of age does a human infant have the capacity to solve invisible displacement tasks. Piaget (1937) believed that this remarkable cognitive development was possible only by using the uniquely human ability for symbolic representations of objects.

One lesson here is to be cautious in adopting a dichotomous view of the distribution of cognitive skills in primates (Old vs. New World monkeys or monkeys vs. apes). Rather, attention should be given to the other primates to gain a more accurate understanding of the distribution and evolution of cognitive skills in nonhuman primates. Admittedly, both the developmental paths and the nature of the representation might vary when we compare a wide range of species, with humans' understanding of objects growing exceptionally slowly early in development, but then reaching levels of high abstractness and flexibility. Finding hidden food and keeping track of mates, prey, or predators may be such a pervasive function for successful adaptation in nature that it may have been implemented by different neural systems during evolution. Recently, Gomez (2004) discussed the idea that primates have evolved a more complex representational system than other mammals for dealing with object location by integrating different subsystems of object representation.

10.4 Socially Influenced Foraging

One especially remarkable feature of the behavior of callitrichid monkeys is their cooperative breeding system, with breeding males and non-reproductive helpers assisting in infant caretaking. It has been suggested that their plasticity in vocal communication and their ability for social learning may be related to their complex helper system (Snowdon 2001). A significant feature of the cooperative breeding in marmosets, tamarins, and callimicos is the transfer of food from adults to infants (Feistner and Price 1991; see Fig. 10.1). Besides providing infants with additional nutrition or with food items that infants cannot obtain by themselves, food sharing is believed to facilitate the transmission of food preferences from adults to young animals (Brown et al. 2004). While most observed food transfers in callitrichids can be described as passive sharing or tolerated scrounging, active sharing initiated by the food possessor was reported for wild buffy-headed marmosets (*Callithrix flaviceps*, Ferrari 1987), golden lion tamarins (*Leontopithecus rosalia*, Brown and Mack

Fig. 10.1 Food transfer in common marmosets

1978), callimicos (*Callimico goeldii*) and pygmy marmosets (*Cebus pygmaea*, Feistner and Price 1991). Large insects such as orthopterans are the most frequently shared food items, but other animal matter and fruit are shared as well (Goldizen 1987; Ruiz-Miranda et al. 1999).

Food transfer to infants from parents and helpers influences both the preference for specific food categories and, more generally, a food's acceptance when encountered for the first time. Experiments in which family groups of common marmosets were provided with either novel or familiar food revealed that food transfers facilitated the acceptance of novel food items in infants and juveniles. Vitale and Queyras (1997) observed higher frequencies of food transfers from adults to juveniles when the offered food was novel for the juveniles. Brown and colleagues (Brown et al. 2005) reported that infant common marmosets exhibited more interest and begging with novel food items than with familiar ones, although food transfers were not more likely to occur with novel foods than with familiar ones.

In our laboratory, Voelkl et al. (2006) found that infants preferred to take the first of several presented food items from adults, rather then feeding on it independently only when it was novel (Fig. 10.1). Furthermore, the relative amount of food that was acquired socially was higher for novel than for familiar food. Such effects of food novelty on behavior are not expected if the only purpose of food transfers is provisioning the infants with an adequate diet. However, if another function of food transfers is to foster learning, then these transfers should preferably include food that is novel for the infants. These studies indicate that food transfers play an important role in the acquisition of novel food items into the diet of young marmosets. While adults do not show any signs of teaching (i.e., they do not share food novel to the infants more readily than familiar food), by begging more for novel than for familiar food, infants behave in a way that facilitates learning about unknown food sources.

In addition to food transfer, co-feeding of infants or juveniles with experienced adults can foster their acceptance of novel food items. Young primates often forage in close proximity to experienced adults, feeding from the same food sources or even taking scraps that the adults drop, creating ideal conditions for learning about novel food sources by social facilitation (King 1994). In an experimental study, Vitale and Queyras (1997) demonstrated that the mere visibility of feeding adults in an adjacent cage is sufficient to increase the juveniles' consumption of novel food.

Recently, Voelkl et al. (2006) found that in the absence of adults, infants either refused to probe unknown food items, or ate only small amounts, waiting longer and exploring more before feeding, than when food was familiar. Such differences in the treatment of novel and familiar food were much less pronounced when the adults were present during food presentation, suggesting that social facilitation helped to overcome neophobia in the infants. Yamamoto and Lopes (2004) reported the same effect for captive juvenile marmosets, while it was absent in subadults and adults.

A field study has confirmed that social influences on foraging are age-dependent, but it revealed that the ontogenetic development of foraging and the interplay between physical and cognitive maturation, independency, and social dynamics are even more complex than suggested by the laboratory studies. Observing free-living common marmosets, Schiel and Huber (2006) found that infants in their third and fourth month of age were more interested in the foraging activities of the subadult and adult group members than were all the other young individuals in the first year of life. Juveniles (5–10 months old), in contrast, were more independent in their foraging but showed stronger social influences on manipulating and processing food in those cases in which they had previously observed subadult or adult models.

These findings of subtle age-dependent differences in the effects of social foraging not only extend the assumption that young primates are constantly seeking information from adults (King 1994), but also suggest that the type of information acquired varies. After having detected a conspecific foraging for food in the vicinity, young animals may observe the conspecific's activity and become more explorative themselves. This motivational effect, called "social facilitation" (Zajonc 1965), does not lead to enduring improvement in foraging and has, therefore, been classified as social influence rather than social learning in the literature (Whiten and Ham 1992). For many theorists, social learning begins with stimulus enhancement, when the observer is attracted to the location or the object at which a skilled conspecific (who serves as a model) executes salient food manipulations (Thorpe 1956). As Schiel and Huber (2006) found, this may then lead to the observer following the model to the place where the model was feeding before, and may result in the observer exploring the same or a similar food item at the same time as the model. Clearly, the benefit for the observer here is that its foraging is adaptively canalized as its attention is drawn to some salient feature in the environment through observation of a conspecific. However, it appears that learning processes in the strict sense, that is, learning resulting in adaptive and persistent modifications in the behavior of the observer, often occur on an individual basis as a result of the consequences of manipulation of the environment by the observer through trial and error. Nevertheless, the preoccupation of researchers with the apparently more complex process of imitation (see below) means that stimulus enhancement is under-studied as a category of social (facilitation of) learning in its own right (Whiten et al. 2004).

10.5 Stimulus Enhancement

In our laboratory, adult common marmosets have proven to be attentive to the foraging behavior of conspecifics at a high degree as well as in a very selective manner. In an experimental study by Praschberger (2001), in which the monkeys were always confronted with two different stimuli, they preferred to observe feeding conspecifics over either food alone or non-feeding conspecifics. This suggests that the observer was really interested in the feeding activity of the other animal (the model), and not simply attracted by the food or stimulated by the presence of a conspecific. Furthermore, conspecifics extracting food from a novel apparatus received more attention than conspecifics that were feeding on freely available food, suggesting that the observer was motivated to collect new information. Seeing a skilled model manipulating a food box conferred a real advantage, because marmosets with this opportunity were more successful in operating the apparatus themselves than animals that only observed conspecifics that did not interact with the apparatus. The fact that the same results were observed regardless of whether the models were operating the apparatus successfully or not indicates that the observers were learning about the environment rather than about the specific opening techniques demonstrated by the model.

Stimulus enhancement effects may be even more specific, suggesting more fine-tuned cognitive processes. Caldwell et al. (1999) confronted captive common marmosets with either a trained marmoset demonstrator opening an apparatus to retrieve food (full demonstration) or a conspecific just taking food placed on top of the apparatus (partial demonstration). The full-demonstration observers contacted the particular parts of the apparatus that they had seen being touched by the demonstrator more often than the partial-demonstration observers, or animals that had no social demonstration. This result could be interpreted as a case of very "localized" stimulus enhancement (Caldwell et al. 1999), with an effect similar to what Huber et al. (2001) found in an experiment with keas (*Nestor notabilis*).

One further study with captive marmosets demonstrates that the observation of a skilled forager in the group leads to a more direct approach and efficient manipulation of a foraging substrate by the observer. Dungl and Huber (unpublished data) confronted marmosets with a box that had 14 movable devices (different kinds of lids, handles and levers) attached to it. The manipulation of one specific device opened access to the food inside the box. The marmosets, that observed experienced conspecific models, manipulated fewer devices before their first success and recovered food faster than the animals that had no social demonstration before being tested (Fig. 10.2).

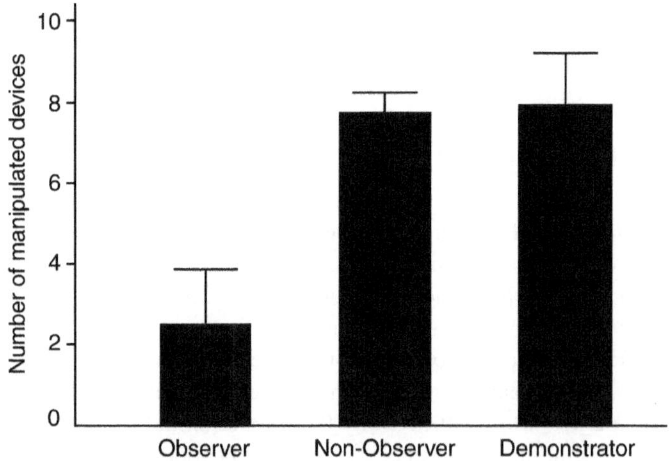

Fig. 10.2 Number of manipulations of devices before the first success of test subjects (observers) which observed skilled models, compared to those in subjects which had no social demonstration, and in the models during their first encounter with the boxes

This effect of stimulus enhancement is particularly strong when the observer has the opportunity to closely inspect the performance of a skilled conspecific, as is possible for callitrichids, who forage in tightly cohesive and highly tolerant groups (Schiel and Huber 2006). Caldwell and Whiten (2003) investigated the effect of social interaction on the social learning of an instrumental task (opening a small sliding door to get access to a food reward) in common marmosets. Those individuals that had the possibility to interact with a trained conspecific and to scrounge – defined by Barnard and Sibly (1981) as using in some ways the behavioural investment of another (producer) to obtain a limited recourse – learned significantly better than individuals that could only observe the trained model, or individuals with no social experience before testing.

10.6 Imitation

The findings of the studies described above provide multiple lines of evidence that social learning about the environment is a very powerful and beneficial form of learning in marmosets. As recently emphasized by Caldwell and Whiten (2004), if individuals are drawn to the functionally relevant parts of an object following observation of a demonstration, then this will allow them to solve foraging problems more quickly. A combination of enhancement effects and individual learning may suffice to solve a problem in most circumstances, particularly if the objects to be manipulated are simple. Only when specific and difficult manipulations are necessary will the advantage of learning by copying of a novel or otherwise improbable act or utterance, called imitation (Thorpe 1956), provide substantial benefits.

In the literature on social learning, imitation has received special attention, as its execution is believed to require special mental abilities (see Heyes and Ray 2000), and its function is regarded by some to be a prerequisite for culture (e.g., Whiten et al. 2003). Three experimental studies (Bugnyar and Huber 1997; Voelkl and Huber 2000; Caldwell and Whiten 2004) investigated the ability of common marmosets to imitate conspecific demonstrators in order to retrieve food enclosed in an "artificial fruit" (Whiten et al. 1996). All three studies used variants of the same experimental procedure (non-observer control): First, the subjects (observers) were allowed to observe a physically separated conspecific (demonstrator) opening a novel apparatus in order to retrieve food from the apparatus, and thereafter these subjects were given the opportunity to manipulate the apparatus on their own. The subjects' behavior was then compared with naïve animals that were confronted with the apparatus without prior observation of conspecifics (non-observers), and – in the Voelkl and Huber (2000) study – also with observers which saw another demonstrator opening the apparatus in a different way (two-action procedure).

In the first study of this kind, Bugnyar and Huber (1997) presented common marmosets a box with a pendulum door that could be either pushed or pulled to gain access to food inside the box. Observers were allowed to watch a conspecific demonstrator pull open the door. The observers showed less exploratory behavior than non-observers and, most importantly, two of them showed a strong tendency to use the demonstrated opening technique in the initial phase of the test. Only after some trials, in which they acquired their own experience of opening the pendulum door, did they begin to perform the simpler solution of pushing, which was preferred by the non-observers. The authors argued that pulling the door to get access to the food was not a simple act, but a compound action-pattern. The authors distinguished four independent elements plus one dependent element in the pulling performance of the demonstrator: (1) using the left hand, (2) taking the door from the right gap, (3) pulling, (4) holding the door wide open with one hand, and (5) taking the food. Two observer marmosets copied all of these actions in the appropriate order, which is very unlikely to be due to chance, considering the combined probability for spontaneous occurrence of these actions. Thus Bugnyar and Huber (1997) provided "suggestive, but not conclusive, evidence of imitation" (Heyes and Ray 2000, p. 232) in common marmosets.

In an attempt to provide data allowing a direct comparison between species, Caldwell and Whiten (2004) used a marmoset-sized version of an artificial fruit that has been designed for studies of imitation in children and chimpanzees (Whiten et al. 1996). One demonstrator ("full" demonstrator) was trained to open the apparatus by removing a handle, while the other demonstrator ("partial" demonstrator) simply ate the food from the lid of the apparatus. Unfortunately, none of the observers were successful in opening the apparatus – probably because of the technical sophistication of the opening mechanisms. However, the authors found clear response differences consistent with the different demonstration modes. Those animals that watched the "partial" demonstrator performing predominantly mouthing behavior, used their mouths more frequently, while those that watched the "full" demonstrator showing predominantly hand manipulation behavior used their hands more frequently. The authors described these findings as body part copying, but they pointed out that the behavior of the observers might have been dependent on several other social learning effects as well. For instance, it may be possible that reaching or grasping behaviors are in some way contagious (i.e., triggered by the same response) in marmosets, or the fact that the movement of the apparatus was clearly different for both observer groups could also account for the social learning seen.

Only a two-action method, which involves two demonstrators that differ in their body movements but create the same changes in the environment, controls for learning about the changes of state in the environment and, therefore, provides the most convincing evidence yet for imitative learning in animals. Voelkl and Huber (2000) applied this methodology, permitting two groups of marmosets to observe a demonstrator using one of two alternative techniques to remove the lids of baited film canisters and compared their initial test responses with one another and with a third group of marmosets that were never given the opportunity to observe a demonstrator. Furthermore, while one technique involved hand-opening behavior common to marmosets, the other technique consisted of a behavioral "peculiarity" (mouth-opening); that is, mouth opening was neither common in the animals under investigation nor necessary for lid removal. This requirement ensured that if the observers performed the technique, then they were, most probably, influenced by what they had witnessed.

In fact, both groups of observers preferred to open the canisters using the same method as their demonstrator. Since hand and mouth demonstrators brought about identical changes to the canister (opening, and exposing the food reward), the differential test behavior of the animals suggests that they indeed learned something about the demonstrator's behavior, rather than about certain properties of the canister. Furthermore, non-observers rarely opened the canister with the mouth, but they opened as many canisters as did members of both observer groups. An actual benefit to observer animals in terms of success rate could be found when the task was made more difficult by closing the lids of the canisters much more firmly. After this change, only the mouth-openers managed to open the canisters and retrieve the desired mealworms (Voelkl and Huber 2000). Therefore, even "slavish" copying (i.e., copying in the absence of insight) may have beneficial effects for observers (Huber 1998). In addition, as emphasized by Caldwell and Whiten (2004), social learning may provide particular practical benefits to individuals when it induces an individual to persist with unrewarded manipulations of an object, as individual learning (trial and error) is unlikely to be successful under such circumstances.

Our findings that common marmosets are able to replicate an action that they have observed being performed by another individual are supported by neurophysiological evidence. Imitation seems to require a matching system that allows animals to convert actions they observe others perform, into actions they execute themselves (Heyes and Ray 2000). This transformation of visual input into corresponding motor output has been detected at a neuronal level in the macaque brain. The so-called mirror neurons in the ventral premotor area F5 discharge both when the monkey performs an action and when it sees a similar action being performed by another conspecific (see Rizzolatti and Craighero 2004 for a recent review). With regard to the results found by Voelkl and Huber (2000), it is interesting that area F5 in the macaque brain contains mirror neurons for both hand and mouth actions (Ferrari et al. 2003).

At least two important issues remain unresolved at the moment. First, many theorists of imitation have claimed that an important requirement of imitation is the novelty of the learned action (see Huber 1998 for discussion). Simply matching the observed actions onto the internal motor repertoire would not be sufficient for imitation, but it may remain simply a case of "response facilitation." Secondly, some have argued that to produce true imitation, the individual needs to form a concept of the model's goals or intentions. Such an explicit understanding of imitation relies on attributing mental states to others, called "theory of mind" ability (Tomasello et al. 1993). However, no strong behavioral evidence for such abilities has been found in monkeys, to date (Tomasello and Call 1997). Even the recent finding that macaques recognize when they are being imitated does not necessarily imply the capacity to attribute imitative intentionality to the imitator (Paukner et al. 2005). Therefore, further research efforts are necessary to elucidate the physiological basis of imitation, its cognitive level, and its potential to extend the motor repertoire of monkeys.

10.7 Conclusion

Callitrichids are likely to (1) locate food by using some sort of cognitive map, (2) represent objects and their movements in an abstract manner, (3) benefit from social influences that aid in learning about new food by motivational and perceptual factors, and (4) learn new foraging techniques by observing others. Together, these findings provide evidence for quite a surprising level of understanding of the physical and social world in these monkeys and shake the long-held belief that callitrichids have low intelligence. These new insights suggest that the traditional monkey-to-ape shift in primate cognitive abilities should be reevaluated.

It has been claimed that monkeys are characterized by rapid learning, but only apes have a real understanding of the causal relationships involved in the physical and social world (Byrne 2000). However, current evidence suggests that at least two elements on which this opinion rests are being disproved. First, in contrast to repeated claims (e.g., Visalberghi and Fragaszy 2002), not only humans and apes, but common marmosets also are able to imitate (Voelkl and Huber 2000). Secondly, in addition to chimpanzees (McGrew 1992) and orangutans (van Schaik 2004), black-striped capuchin monkeys (*Cebus libidinosus*) manufacture and use tools routinely in the wild (Fragaszy et al. 2004a; Moura and Lee 2004). Furthermore, recent findings suggest that common marmosets and cotton-top tamarins, although never observed spontaneously using tools, are nevertheless equipped with a mechanism that picks out functionally relevant features of objects, and especially features that function to solve a means – end task (Spaulding and Hauser 2005). Spaulding and Hauser proposed that tamarins and marmosets are endowed with an innate mechanism for recognizing the functionally relevant features of tools, with marmosets appearing to require only minimal experience with a variety of both good and bad tools to determine which are effective for foraging tasks. Taken together, these new findings provide a new reference point for discussions regarding the evolution of tool use and material culture in primates (Fragaszy et al. 2004a).

Of course, it remains to be shown that common marmosets, with their potential to imitate, can acquire complex novel behavior by observation, and that capuchins or callitrichids can conceptualize the cause and effect relations between tool and task. Experiments addressing the first ability in common marmosets are currently running in our laboratory in Vienna, and research on the flexibility of tool behavior of capuchins is still in progress in Rome (E. Visalberghi, personal communication). Whether these lines of research will show that the apparent underlying cognitive superiority of great apes over monkeys is indeed a qualitative one remains to be determined. Testing whether monkeys have a human-like form of representation of instrumental behavior and mental states that allows them to undertake planning and to use others' behavior as a source of knowledge provides a challenge for the future.

References

Moura ACdA, Lee PC (2004) Capuchin stone tool use in Caatinga dry forest. Science 306:1909
Barnard CJ, Sibly RM (1981) Producers and scroungers: a general model and its application to captive flocks of house sparrows. Anim Behav 29:543–550
Brown K, Mack DS (1978) Food sharing among captive *Leontopithecus rosalia*. Folia Primatol 29:268–290
Brown GR, Almond REA, van Bergen Y (2004) Begging, stealing and offering: food transfer in nonhuman primates. Adv Stud Behav 34:265–295
Brown GR, Almond REA, Bates NJ (2005) Adult–infant food transfer in common marmosets: an experimental study. Am J Primatol 65:301–312
Bugnyar T, Huber L (1997) Push or pull: an experimental study on imitation in marmosets (*Callithrix jacchus*). Anim Behav 54:817–831
Byrne RW (1997) The technical intelligence hypothesis: an additional evolutionary stimulus to intelligence? In: Whiten A, Byrne RW (eds) Machiavellian intelligence II: extensions and evaluations. Cambridge University Press, Cambridge, pp 289–311
Byrne RW (2000) Evolution of primate cognition. Cogn Sci 24:543–570
Byrne RW, Whiten A (1988) Machiavellian intelligence. Social expertise and the evolution of intellect in monkeys, apes, and humans. Oxford University Press, New York
Caldwell C, Whiten A (2003) Scrounging facilitates social learning in common marmosets, *Callithrix jacchus*. Anim Behav 65:1085–1092
Caldwell CA, Whiten A (2004) Testing for social learning and imitation in common marmosets, *Callithrix jacchus*, using an artificial fruit. Anim Cogn 7:77–85
Caldwell CA, Whiten A, Morris KD (1999) Observational learning in marmoset monkey (*Callithrix jacchus*). AISB convention: symposium on imitation in animals and artifacts. The Society for the Study of Artificial Intelligence and the Simulation of Behaviour, Edinburgh, pp 27–31
Cheney DL, Seyfarth RM (1990) How monkeys see the world. University of Chicago Press, Chicago
Clutton-Brock T (ed) (1977) Primate ecology: studies of feeding and ranging behaviour in lemurs, monkeys and apes. Academic, London
de Blois ST, Novak MA, Bond M (1998) Object permanence in orangutans (*Pongo pygmaeus*) and squirrel monkeys (*Saimiri sciureus*). J Comp Psychol 112:137–152
Doré FY, Dumas C (1987) Psychology of animal cognition: Piagetian studies. Psychol Bull 102:219–233
Dumas C, Brunet C (1994) Permanence de l'objet chez le singe capucin (*Cebus apella*): Etude des déplacements invisibles. Rev Can Psychol Exp 48:341–357
Feistner ATC, Price EC (1991) Food offering in new world primates: two species added. Folia Primatol 57:165–168
Ferrari SF (1987) Food transfer in a wild marmoset group. Folia Primatol 48:203–206
Ferrari PF, Gallese V, Rizzolatti G, Fogassi L (2003) Mirror neurons responding to the observation of ingestive and communicative mouth actions in the monkey ventral premotor cortex. Eur J Neurosci 17:1703–14
Fleagle JG (1999) Primate ecology and evolution. Academic, New York
Fragaszy DM, Izar P, Visalberghi E, Ottoni EB, De Oliveira MG (2004a) Wild capuchin monkeys (*Cebus libidinosus*) use anvils and stone pounding tools. Am J Primatol 64:359–366
Fragaszy DM, Visalberghi E, Fedigan LM (2004b) The complete capuchin: the biology of the Genus *Cebus*. Cambridge University Press, Cambridge
Garber P (1989) Role of spatial memory in primate foraging patterns: *Saguinus mystax* and *Saguinus fuscicollis*. Am J Primatol 19:203–216

Garber P, Hannon P (1993) Modeling monkeys: a comparison of computer-generated and naturally occurring foraging patterns in two species of Neotropical primates. Int J Primatol 14:827–852

Goldizen AW (1987) Tamarins and marmosets: communal care of offspring. In: Smuts BB, Cheney DL, Seyfarth RM, Wrangham RW, Struhsaker TT (eds) Primate societies. The University of Chicago Press, Chicago/London, pp 34–43

Gomez JC (2004) Apes, monkeys, children, and the growth of mind. Harvard University Press, Cambridge, MA

Hauser MD, Williams T, Kralik JD, Moskovitz D (2001) What guides a search for food that has disappeared? Experiments on cotton-top tamarins (*Saguinus oedipus*). J Comp Psychol 115:140–151

Hershkovitz P (1977) Living New World monkeys (Platyrrhini), with an introduction to primates, vol 1. Chicago University Press, Chicago

Heyes CM, Ray ED (2000) What is the significance of imitation in animals? Adv Stud Behav 29:215–245

Huber L (1998) Movement imitation as faithful copying in the absence of insight. Behav Brain Sci 22:694

Huber L, Rechberger S, Taborsky M (2001) Social learning affects object exploration and manipulation in keas, Nestor notabilis. Anim Behav 62:945–954

Humphrey NK (1976) The social function of intellect. In: Bateson PPG, Hinde RA (eds) Growing points in ethology. Cambridge University Press, Cambridge, pp 303–317

Jolly A (1966) Lemur social behavior and primate intelligence. Science 153:501–506

King BJ (1994) The information continuum: evolution of social information transfer in monkeys, apes, and hominids. School of American Research Press, Santa Fe

Kummer H (1968) Social organization of hamadryas baboons. University of Chicago Press, Chicago

Martin R (1992) Patterns of primate evolution: classification and evolutionary relationship. In: Jones S, Martin R, Pilbeam D (eds) The Cambridge encyclopedia of human evolution. Cambridge University Press, Cambridge, pp 17–23

Mathieu M, Bouchard MA, Granger L, Herscovitch J (1976) Piagetian object-permanence in *Cebus capucinus*, *Lagothrica flavicauda* and *Pan troglodytes*. Anim Behav 24:585–588

McGrew WC (1992) Chimpanzee material culture. Implications for human evolution. Cambridge University Press, Cambridge

Mendes N, Huber L (2004) Object permanence in common marmosets (*Callithrix jacchus*). J Comp Psychol 118:103–112

Menzel EW, Juno C (1982) Marmosets (*Saguinus fuscicollis*): are learning sets learned? Science 217:750–752

Menzel EW, Menzel CR (1979) Cognitive, developmental and social aspects of responsiveness to novel objects in a family group of marmosets (*Saguinus fuscicollis*). Behaviour 70:250–279

Milton K (1981) Distribution patterns of tropical plant foods as an evolutionary stimulus to primate mental development. Am Anthropol 83:534–548

Parker ST (1990) Origins of comparative developmental evolutionary studies of primate mental abilities. In: Parker ST, Gibson KR (eds) "Language" and intelligence in monkeys and apes: comparative developmental perspectives. Cambridge University Press, Cambridge, pp 3–64

Parker ST, Gibson KR (1977) Object manipulation, tool use, and sensorimotor intelligence as feeding adaptations in cebus monkeys and great apes. J Hum Evol 6:623–641

Parker ST, McKinney ML (1999) Origins of intelligence. The evolution of cognitive development in monkeys, apes, and humans. The Johns Hopkins University Press, Baltimore/London

Passingham R (1982) The human primate. W. H. Freeman, San Francisco

Paukner A, Anderson JR, Borelli E, Visalberghi E, Ferrari PF (2005) Macaques (*Macaca nemestrina*) recognize when they are being imitated. Biol Lett 1:219–222

Piaget J (1937) La construction du réel chez l'enfant (The construction of the real in the child). Delachaux and Niestle, Oxford

Praschberger MC (2001) Aufmerksamkeitserhöhung und Stimmungsübertragung bei Weißbüschelaffen (Stimulus enhancement and social facilitation in common marmosets). University of Vienna, Vienna

Reader SM, Laland KN (2002) Social intelligence, innovation, and enhanced brain size in primates. Proc Natl Acad Sci U S A 99:4436–4441

Reader SM, MacDonald K (2003) Environmental variability and primate behavioural flexibility. In: Reader SM, Laland KN (eds) Animal innovation. Oxford University Press, Oxford, pp 83–116

Rizzolatti G, Craighero L (2004) The mirror-neuron system. Ann Rev Neurosci 27:169–92

Ruiz-Miranda CR, Kleiman DG, Dietz JM, Moraes E, Grativol AD, Baker AJ, Beck BB (1999) Food transfer in wild and reintroduced golden lion tamarins, *Leontopithecus rosalia*. Am J Primatol 48:305–320

Santos LR, Ericson B, Hauser MD (1999) Constraints on problem solving and inhibition: object retrieval in cotton-top tamarins. J Comp Psychol 113:186–193

Schiel N, Huber L (2006) Social influences on the development of foraging behavior in free-living common marmosets (*Callithrix jacchus*). Am J Primatol 68:1–11

Schino G, Spinozzi G, Berlinguer L (1990) Object concept and mental representation in *Cebus apella* and *Macaca fasciularis*. Primates 31:537–544

Seyfarth RM, Cheney DL, Marler P (1980) Monkey responses to three different alarm calls: evidence of predator classification and semantic communication. Science 210:801–803

Snowdon CT (2001) Social processes in communication and cognition in callitrichid monkeys: a review. Anim Cogn 4:247–257

Snyder DR, Birchette LM, Achenbach TM (1978) A comparison of developmentally progressive intellectual skills between *Hylobates lar*, *Cebus apella* and *Macaca mulatta*. In: Chivers DJ, Herbert J (eds) Recent advances in primatology, vol 3. Academic, New York, pp 945–948

Spaulding B, Hauser MD (2005) What experience is required for acquiring tool competence? Experiments with two callitrichids. Anim Behav 70:517–526

Spinozzi G (1989) Early sensoriomotor development in Cebus (*Cebus apella*). In: Antinucci F (ed) Cognitive structure and development in nonhuman primates. Lawrence Erlbaum Associates, Hilldsale, NJ, pp 55–66

Thorpe WH (1956) Learning and instinct in animals. Methuen, London

Tomasello M (2000) Primate cognition: introduction to the issue. Cogn Sci 24:351–361

Tomasello M, Call J (1997) Primate cognition. Oxford University Press, New York

Tomasello M, Kruger AC, Ratner HH (1993) Cultural learning. Behav Brain Sci 16:495–552

van Schaik CP (2004) Among orangutans: red apes and the rise of human culture. Belknap/ Harvard University Press, Boston

Vaughter RM, Smotherman W, Ordy JM (1972) Development of object permanence in the infant squirrel monkey. Dev Psychol 7:34–38

Visalberghi E, Fragaszy DM (2002) "Do monkeys ape?" Ten years after. In: Dautenhahn K, Nehaniv CL (eds) Imitation in animals and artifacts. MIT Press, Cambridge, MA, pp 471–499

Vitale A, Queyras A (1997) The response to novel foods in common marmoset (*Callithrix jacchus*): the effects of different social contexts. Ethology 103:395–403

Voelkl B, Huber L (2000) True imitation in marmosets. Anim Behav 60:195–202

Voelkl B, Schrauf C, Huber L (2006) Social contact influences the response towards novel food in infant marmosets. Anim Behav 72:365–372

Whiten A, Ham R (1992) On the nature and evolution of imitation in the animal kingdom: reappraisal of a century of research. In: Slater PJB, Rosenblatt JS, Beer C, Milkinski M (eds) Advances in the study of behavior, vol 21. Academic, New York, pp 239–283

Whiten A, Custance DM, Gomez J-C, Teixidor P, Bard KA (1996) Imitative learning of artificial fruit processing in children (*Homo sapiens*) and chimpanzees (*Pan troglodytes*). J Comp Psychol 110:3–14

Whiten A, Horner V, Marshall-Pescini S (2003) Cultural panthrology. Evol Anthropol 12:92–105

Whiten A, Horner V, Litchfield CA, Marshall-Pescini S (2004) How do apes ape? Learn Behav 32:36–52

Yamamoto ME, Lopes FdA (2004) Effect of removal from the family group on feeding behavior by captive *Callithrix jacchus*. Int J Primatol 25:489–500

Zajonc RB (1965) Social facilitation. Science 149:269–274

Section III
Ranging Behavior and Locomotion

Leila M. Porter, Susan M. Ford, Lesa C. Davis

Although data on ranging behavior have been available for *Callithrix jacchus* (the common marmoset) for a number of years, data on other marmosets and callimicos have not been published. Furthermore, although several studies of tamarin positional behavior have been conducted, equivalent data on marmoset and callimico species have been scarce. In this section, the authors present data on *Mico*, *Cebuella*, *Callithrix* and *Callimico* from genetic studies, laboratory experiments, and field observations. These chapters allow for a better understanding of how marmosets and callimicos have evolved different strategies for living in secondary and disturbed forest niches.

Gonçalves et al. (Chap. 11) use microsatellite markers to determine whether the silvery marmoset (*Mico argentatus*) has evolved mechanisms that allow it to maintain genetic diversity while colonizing patchy environments. By examining DNA samples of individuals from 10 populations, they conclude that it has not. Their study populations show very low levels of genetic diversity compared to other platyrrhines, and several populations sampled exhibit inbreeding depression. Thus, Gonçalves et al. demonstrate that understanding the effects of ranging and dispersal at the population-level can be important for developing long-term management strategies for marmosets' protection. In this case, they propose that it will be necessary to consider ways of increasing genetic heterozygosity in marmoset populations that have become isolated from one another.

While Gonçalves et al. consider the effects of ranging on the genetic diversity of a population, Veracini (Chap. 12) and Rehg (Chap. 13) examine ranging behavior through the observations of individual study groups. Veracini examines the ecological factors that shape the silvery marmoset's ranging behavior. Like marmosets of the genus *Callithrix*, Veracini finds that her study group of *Mico argentatus* primarily uses secondary growth, has a small home range (15.5 ha), and has important fruit and gum resources in the center of its range. This pattern of range use varies in several aspects from that of the callimico, as presented by Rehg. Rehg finds that callimicos also have the majority of their feeding sites in the interior of their home

Leila M. Porter
Department of Anthropology, Northern Illinois University, DeKalb, IL 60115, USA
e-mail: lmporter@niu.edu

range, but the home range is larger (59 ha) than that of *M. argentatus*. Furthermore, as callimicos generally eat fungi rather than exudates, callimicos (unlike *Mico*) prefer areas that contain tree falls, which create micro-habitats that often contain fungi. Both of these studies provide important data on the ranging behavior of these little known species, information that is essential for understanding what size and types of habitat and areas are needed to support these groups.

Garber et al. (Chap. 14) further explore the ways callimicos and marmosets use their environment, by examining how they are adapted to travel. Garber et al. examine how callimico, the common marmoset (*Callithrix jacchus*), and the pygmy marmoset (*Cebuella pygmaea*) use vertical supports for trunk-to-trunk leaping in an experimental, laboratory setting. They show that although pygmy marmosets can leap the farthest relative to their body size, the callimicos are able to jump most efficiently. The cost of leaping increases significantly as leaping distance increases for marmosets, but not for callimicos; thus, the diminutive pygmy marmoset avoids leaping long distances. Youlatos (Chap. 15) examines the positional behavior of *Cebuella pygmaea* in more detail, using data he collected from free-ranging animals in Ecuador. Youlatos finds that *Cebuella* uses different postures and modes of locomotion while engaged in different activities. While feeding on exudates, *Cebuella* relies primarily on claw-clinging, whereas while foraging for insects, it uses cantilever and quadrupedal stances. Chaps. 14 and 15 demonstrate that the marmosets and callimicos have evolved a range of postural and locomotor behavior that distinguishes them from one another (which are reflected in their anatomy – see Ford and Davis, Chap. 21).

The chapters in this section help fill in many details of the lesser-studied marmosets and callimicos; however, additional data on the ranging patterns of *Mico*, *Callibella*, and *Callithrix* are still needed, as well as from additional groups of *Cebuella* and *Callimico*. Furthermore, the study by Gonçalves et al. is an important reminder that although these monkeys are adapted to patchy environments, they may not be able to cope with the scale of habitat fragmentation and disturbance that is occurring in South America.

Chapter 11
Limited Dispersal and Genetic Structure of Silvery Marmosets (*Mico argentatus*) in the Fragmented Landscape of Central Amazonia

Evonnildo Gonçalves, Stephen F. Ferrari, Paulo E.G. Coutinho, Elytânia Menezes, Artur Silva, and Maria Paula C. Schneider

Abstract Three microsatellite loci were analyzed in ten remnant populations of silvery marmosets (*Mico argentatus*) in the fragmented landscape of central Amazonia, where surprisingly low levels of genetic diversity were observed (Ho=0.11–0.39). This pattern does not appear to be a result of habitat fragmentation, but rather of the ecological, behavioral, and reproductive characteristics of the species. Despite this, the results indicate that all the populations – including those resident in continuous forest – are subject to inbreeding and genetic drift. Overall, significant differentiation was found among populations (mean $R_{st}=0.22$, $P<0.001$), which indicates that they are no longer part of the same gene pool. However, no correlation was observed between genetic and geographic distances among sites, which suggests that populations of these marmosets are not panmictic, even in large areas of continuous forest. In this case, in addition to inbreeding in small, isolated populations, the original genetic structure, characterized by a Whalund effect, may account for the high frequency of deviation from Hardy–Weinberg equilibrium found in the populations analyzed. For eventual conservation management, a primary recommendation would be to mitigate the effects of genetic drift by promoting gene flow among populations.

Resumen Tres loci de microsatélites fueron analizados en diez poblaciones remanentes de la marmoseta plateada (*Mico argentatus*) en el paisaje fragmentado de la Amazonia central. Sorpresivamente se observaron niveles bajos de diversidad genética (Ho=0.11–0.39). Este patrón no parece ser el resultado de una fragmentación de hábitat, pero más bien de las características ecológicas, conductuales y reproductivas de la especie. A pesar de esto, los resultados indican que todas las poblaciones – incluyendo los residentes de bosques continuos – están sujetas a endogamia y deriva genética. De manera general, se encontraron diferencias genéticas significativas entre las poblaciones (R_{st} medio=0,22, $P<0,001$), indicando

E. Gonçalves (✉)
Department of Genetics, Universidade Federal do Pará, Caixa Postal 8607, 66.075-970, Belém, PA, Brazil
e-mail: ecostag@ufpa.br

S.M. Ford et al. (eds.), *The Smallest Anthropoids*, Developments in Primatology: Progress and Prospects, DOI 10.1007/978-1-4419-0293-1_11,
© Springer Science+Business Media, LLC 2009

que dichas poblaciones ya no son parte del mismo acervo genético. Sin embargo, no se observó una correlación entre las distancias genéticas y geográficas entre los sitios, lo cual sugiere que las poblaciones de estas marmosetas no son panmícticas, aún en áreas grandes de bosques continuos. En este caso, adicionalmente a la endogamia de las poblaciones pequeñas y aisladas, la estructura genética original está caracterizada por a efecto Whalund, el cual puede explicar la alta frecuencia de desviación del equilibrio Hardy–Weinberg encontrado en las poblaciones analizadas. Una recomendación primaria para el manejo para la conservación de la especie sería mitigar los efectos de la deriva genética promoviendo el flujo genético entre las poblaciones.

Resumo Da análise de três loci microssatélites em dez populações de sagüis prateados (*Mico argentatus*) remanescentes de uma paisagem fragmentada da Amazônia Central, foi observada uma diversidade genética surpreendentemente baixa (Ho=0,11 a 0,39). Este padrão não parece ser resultado da fragmentação de hábitat, mas de processos históricos relacionados principalmente às características ecológicas, comportamentais e reprodutivas da espécie. Por outro lado, os resultados indicam que todas as populações – incluindo aquela em floresta contínua – estão sujeitas ao endocruzamento e deriva genética. De maneira geral, foi observado um elevado e significante nível de diferenciação genética (R_{st} médio=0,22, P<0,001), indicando que estas populações não fazem mais parte do mesmo pool gênico. Contudo, não foi observada correlação entre as distancias genética e geográfica entre populações, sugerindo, que mesmo em uma grande área contínua os sagüis não constituem uma população panmítica. Neste caso, em adição ao endocruzamento nas populações pequenas e isoladas, a estrutura genética original, com caracterização de efeito Wahlund, pode explicar a elevada freqüência do desvio do equilíbrio de Hardy–Weinberg nas populações levantadas. Para o eventual manejo conservacionista, recomenda-se principalmente a mitigação dos efeitos de deriva genética através da promoção de fluxo gênico entre as populações.

11.1 Introduction

Despite the number of new species described in recent years (van Roosmalen et al. 2000), the ecology of the Amazonian marmosets, genus *Mico* (Rylands et al. 2000), remains relatively poorly known, especially in comparison with the Atlantic Forest species (*Callithrix*). To date, long-term field studies are available for only two species, the intermediate marmoset, *Mico intermedius* (Rylands 1986), and the silvery marmoset, *Mico argentatus* (Tavares and Ferrari 2002; Carvalho et al. 2005; Corrêa 2006; Veracini this volume, Chap. 12).

In addition to continuous lowland *terra firme* forest, *M. argentatus* is found in the Alter do Chão savanna formation on the right bank of the lower Tapajós river, where it inhabits relatively small, naturally-formed forest patches in a matrix of grassland and scrub (Miranda 1993). Further south, the landscape is dominated by two main features – a large area of continuous forest (the Tapajós National Forest, covering more than 600,000 hectares) and an even larger area of human colonization, initiated in the 1970s, characterized by extensive habitat fragmentation and the isolation of forest remnants of varying sizes. Continuous forest and both natural and anthropogenic habitat fragmentation can all be found within a 50 km radius (Fig. 11.1), providing an excellent opportunity for an analysis of the effects of habitat disturbance on primate populations (see Ferrari et al. 2003).

Habitat fragmentation interrupts normal demographic processes and gene flow within populations, with potentially deleterious consequences for the viability of populations over the long term (Lacy 1997). While habitat fragmentation has received a great deal of attention from primatologists and conservationists in recent years (e.g. Marsh 2003), there have been relatively few studies of its effects on genetic variation (Pope 1996, 1998; Dietz et al. 2000). An evaluation of specific effects is important because, while standard predictions include the loss of variability and inbreeding depression (Lacy 1997; Dietz et al. 2000), it is important to remember that phenomena such as outbreeding depression may also affect some species (e.g. Templeton 1986; Dudash and Fenster 2000).

The marmosets are especially interesting in this context, because their ecological specializations allow them to colonize disturbed and fragmented habitats successfully (Ferrari 1993). It would thus seem reasonable to assume that most marmoset

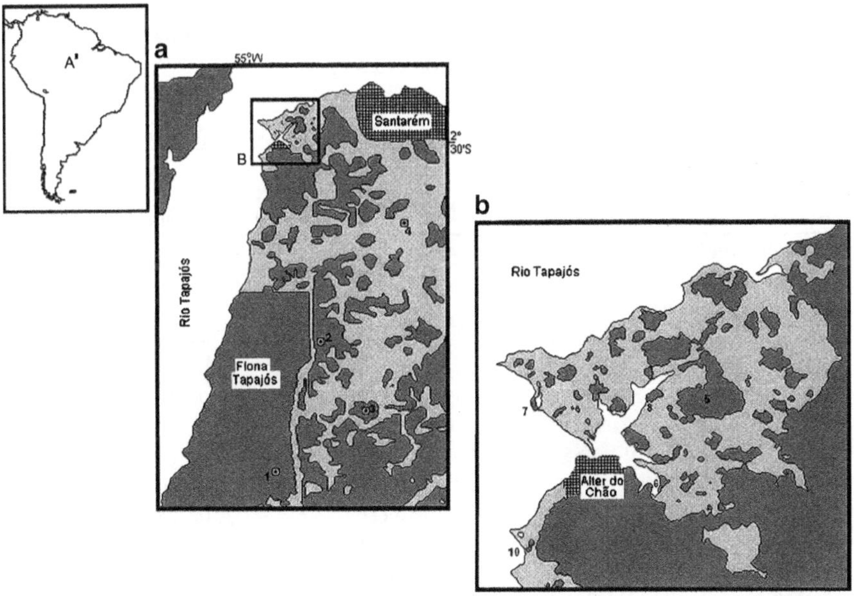

Fig. 11.1 Satellite-image based map of the study area, showing forested habitats (*terra firme* forest and savanna woodland) shaded in *dark gray*, and open habitats (savanna and anthropogenic clearings) in *light gray*. In the close-up of the Alter do Chão savanna (**b**), *dark* and *light shading* represent woodland and savanna habitats, respectively

species – especially those that inhabit savanna ecosystems – have adapted in some way to cope with the genetic effects of complex and variable demographic patterns. This study presents an analysis of the genetic variability of populations of *M. argentatus* from the lower Tapajós which inhabit continuous forest and natural and anthropogenic fragments of habitat of different sizes.

11.2 Methods

11.2.1 Study Area and Samples

Silvery marmosets were captured at a total of ten sites on the lower Tapajós (Table 11.1, Fig. 11.1), using manually-controlled traps baited with banana (see Gonçalves et al. 2003). For genetic analyses, samples of hair follicles were obtained using sterile tweezers and stored in plastic tubes (at least two samples per individual), with care being taken to avoid all contact with other substrates. Hair follicles were preferred over blood samples in order to eliminate the possible effects of chimerism (Dixson et al. 1992). Samples were maintained at room temperature before being transported to the laboratory for processing.

The ten sites correspond to continuous forest (Table 11.1: site 1); anthropogenic habitat fragments of different sizes (sites 2–4); and natural fragmentation within the savanna matrix (sites 5–10). Site 1 is representative of the original *terra firme* forest cover, and is used here as the control for inter-site comparisons. Sites are located between seventeen and fifty kilometers apart.

Table 11.1 Details of the *Mico argentatus* populations analyzed in the present study

Site [number on Fig. 11.1]	Area of forest (ha)	Number of specimens collected (groups captured)[a]	Geographic coordinates
Continuous forest			
Flona Tapajós [1]	>600,000	20 (3)	03°02 46″S, 54°57 18″W
Anthropogenic fragments			
Massafra [2]	4,500	7 (1)	02°50 06″S, 54°53 14″W
São Benedito [3]	900	22 (3)	02°57 05″S, 54°47 41″W
Tabocal [4]	30	34 (5)[b]	02°37 14″S, 54°43 51″W
Natural fragments in savanna			
Capiranga [5]	455	5 (1)	02°28 47″S, 54°55 58″W
Jardim [6]	4	4 (1)[b]	02°30 21″S, 54°55 48″W
Mureta [7]	11	6 (1)[b]	02°28 55″S, 54°58 24″W
Santa Luzia [8]	50	19 (2)[b]	02°28 55″S, 54°55 48″W
Sarú [9]	15	10 (2)[b]	02°29 42″S, 54°56 18″W
Tauá [10]	3	5 (1)[b]	02°32 35″S, 54°58 25″W
Total		132 (20)	

[a]Specimens are individual animals. Values in parentheses correspond to the number of groups from which at least one individual was captured

[b]All resident groups captured

11.2.2 Extraction of DNA and Microsatellite Analysis

Approximately 30 hair follicles were selected from each specimen for the extraction of DNA, following the method of Gagneux et al. (1997). Due to a lack of markers for *Mico*, PCRs were performed on the microsatellite loci published for the common marmoset, *Callithrix jacchus*, by Nievergelt et al. (1998), although most loci produced no fragments, even after testing with a touchdown PCR. This was probably due to the genetic distance between the two marmoset species. Efficient amplification was possible only for the loci Cj6, Cj11 and Cj14. The PCRs were performed in volumes of 20 μl containing 5–10 ng of genomic DNA, 50 mM KCl, 1.5 mM $MgCl_2$, 10 mM Tris–HCL, 20 μM of each dNTP, 0.5 μM of each primer and a unit of Taq DNA polymerase (Invitrogen). The reactions were carried out via the following protocol: 4 min at 94 °C for denaturation, followed by thirty-five 30 s denaturation cycles at 94 °C, annealing for 30 s at 53 °C for Cj6 and 55 °C for both Cj11 and Cj14, extension for 1 min at 72 °C, and a final cycle of 5 min at 72 °C to ensure complete extension of the PCR products. Analysis of the amplified fragments was carried out in an ALFexpress™ II automatic DNA processor using Allelinks version 1.0 software (Amersham Biosciences) after 6% denaturing polyacrylamide gel runs.

11.2.3 Data Analysis

Intrapopulational genetic diversity was quantified in terms of allele frequencies, the mean number of alleles per locus (*Na*), and observed (*Ho*) and expected (*He*) heterozygosity under Hardy–Weinberg equilibrium. These values were calculated using Pogene32 (Yeh et al. 1997). Given differences in samples sizes, allelic richness (*Rs*) was also estimated using FSTAT version 2.9.3.2 (Goudet 2001). The significance of differences in *Na*, *Rs* and *Ho* between pairs of populations was evaluated using Wilcoxon's signed-ranks test. Unbiased estimators of exact significance probabilities were generated to evaluate deviations from Hardy–Weinberg expectations, using a Markov chain (dememorization 5,000, 100 batches, 1,000 iterations per batch) performed in GENEPOP software (Raymond and Rousset 1995). Significance was calculated per locus, per population and over all loci and populations combined.

The significance of population differences was examined using the exact test of population differentiation proposed in GENEPOP, based on allele frequency variations (dememorization 5,000, 100 batches, 1,000 iterations per batch). Genetic differentiation among populations was also evaluated by calculating Rho, an unbiased estimator of Slatkin's R_{st}. The Rho statistic has been developed specifically for the analysis of microsatellite loci, because mutations at these loci appear to be characterized mainly by single-step deletions or the insertion of one repeat unit (Slatkin 1995). Estimates, based on a globally standardized dataset, were calculated for both individual loci and over all loci, across all populations and for pairwise population comparisons. These analyses were performed by the RST Calc package (Goodman 1997) with 1,000 iterations for both bootstrap and permutation tests.

The possible occurrence of recent population bottlenecks (severe reductions in effective population size) was evaluated using Wilcoxon's signed-ranks test (Piry et al. 1999). This test was conducted by BOTTLENECK (Cornuet and Luikart 1996; Luikart and Cornuet 1998), under three mutation models: the infinite allele model (IAM: Kimura and Crow 1964), the stepwise mutation model (SMM: Ohta and Kimura 1973), and the two-phase mutation model (TPM: Dirienzo et al. 1994). As recommended by Piry et al. (1999), the TPM was carried out with 95% strict single step mutations and 5% multistep mutations (with an exponential distribution of step numbers, and a mean of four steps).

11.3 Results

The number of alleles varied considerably among sites (Table 11.2) but, although sample size did influence *Na*, there was no clear overall pattern. Sites with small samples were at least as variable, if not more so, at some loci in comparison with some of those with larger samples. This was supported by *Rs* values, which are not influenced by sample size.

Perhaps the most surprising finding was the relatively inexpressive number of alleles recorded at Flona Tapajós, in continuous forest, despite the number of groups and individuals sampled (Table 11.1). This is probably related to the fact that the samples were collected at a single site within the continuous forest and are thus probably representative of a subset of its population, rather than its overall variability. Marmosets occur at relatively low overall densities in this forest (Ferrari et al. 2003), and appear to be distributed patchily, in local subpopulations. The populations found in the fragments – principally the anthropogenic ones – are presumably formed by the random isolation of such subpopulations.

Of the 26 alleles identified, only two are exclusive to a single site: allele J at the Cj6 locus of the Tabocal population (Table 11.3), and allele H of the CJ11 locus of the Capiranga population (Table 11.4). Predictably, they were found at the two most polymorphic loci, although the Tabocal allele was extremely rare. Some of the alleles are also exclusive to one of the two main study areas. In the case of the most variable locus (CJ6), for example, alleles A, H and I were recorded exclusively in savanna populations, whereas alleles E, K and L were restricted to forest populations. Similarly, allele A of the Cj11 locus was exclusive to savanna populations, as was allele A of the CJ14 locus (Table 11.5).

The identification of many exclusive alleles in the savanna populations is unexpected, given that fewer animals (49 vs. 83) and groups (8 vs. 12) were sampled, although perhaps the larger number of populations (6 vs. 4) is the more relevant factor. Either way, the fact that ten of the 26 alleles were exclusive to one or other study area may reflect the relatively long term isolation of the savanna populations from those that inhabit the continuous forest further south and east.

Table 11.2 Intrapopulational variability in silvery marmoset populations from central Amazonia

Site	Loci	*Na*	*Rs*	*Ho*	*He*	*HWE*
Flona Tapajós	Cj6	5.00	4.00	0.20	0.75	0.0000
	Cj11	3.00	2.23	0.00	0.40	0.0000
	Cj14	4.00	2.62	0.30	0.51	0.0072
	All loci	4.00	2.95	0.17	0.55	0.0000
Massafra	Cj6	3.00	2.57	0.29	0.56	0.1192
	Cj11	3.00	2.78	0.43	0.52	0.3312
	Cj14	3.00	2.93	0.43	0.62	0.0208
	All loci	3.00	2.76	0.38	0.57	0.0273
São Benedito	Cj6	5.00	3.83	0.00	0.74	0.0000
	Cj11	4.00	3.12	0.14	0.65	0.0000
	Cj14	4.00	3.10	0.18	0.64	0.0000
	All loci	4.33	3.35	0.11	0.68	0.0000
Tabocal	Cj6	9.00	5.08	0.47	0.85	0.0000
	Cj11	5.00	2.98	0.32	0.57	0.0000
	Cj14	4.00	3.28	0.18	0.71	0.0000
	All loci	6.00	3.74	0.32	0.71	0.0000
Capiranga	Cj6	2.00	2.00	0.00	0.48	0.0475
	Cj11	4.00	3.78	0.40	0.70	0.1890
	Cj14	3.00	2.80	0.40	0.58	0.3636
	All loci	3.00	2.86	0.27	0.57	0.0754
Jardim	Cj6	4.00	4.00	0.25	0.66	0.0271
	Cj11	3.00	3.00	0.50	0.41	1.0000
	Cj14	3.00	3.00	0.00	0.63	0.0275
	All loci	3.33	3.33	0.25	0.57	0.0254
Mureta	Cj6	3.00	2.00	0.17	0.54	0.1539
	Cj11	4.00	3.56	0.17	0.65	0.0033
	Cj14	3.00	2.91	0.00	0.61	0.0039
	All loci	3.33	2.82	0.11	0.60	0.0002
Santa Luzia	Cj6	6.00	4.47	0.63	0.80	0.0000
	Cj11	6.00	3.96	0.32	0.75	0.0000
	Cj14	3.00	2.82	0.21	0.63	0.0003
	All loci	5.00	3.75	0.39	0.73	0.0000
Sarú	Cj6	8.00	5.09	0.50	0.82	0.0100
	Cj11	4.00	3.51	0.10	0.67	0.0003
	Cj14	4.00	3.25	0.50	0.62	0.1680
	All loci	5.33	3.95	0.37	0.70	0.0001
Tauá	Cj6	4.00	3.80	0.40	0.72	0.1497
	Cj11	3.00	2.60	0.40	0.34	1.0000
	Cj14	1.00	1.00	0.00	0.00	–
	All loci	2.67	2.46	0.27	0.35	0.4340
All populations	Cj6	12.00	4.95	0.33	0.84	0.0000
	Cj11	8.00	3.86	0.24	0.71	0.0000
	Cj14	6.00	3.61	0.23	0.74	0.0000
	All loci	8.67	4.14	0.26	0.76	0.0000

Number of alleles (*Na*); allelic richness (*Rs*); observed heterozygosity (*Ho*); Nei's (1973) expected heterozygosity (*He*); and the *p* value for the Hardy–Weinberg equilibrium exact test (*HWE*)

Table 11.3 Allele frequencies of microsatellite locus Cj6 in silvery marmoset populations from central Amazonia

	Populations									
Allele	Sarú	Santa Luzia	Jardim	Mureta	Capiranga	Tauá	Tabocal	Flona Tapajós	São Benedito	Massafra
A[a]	0.0500	0.0789								
B	0.0500	0.2632		0.3333		0.3000	0.1471	0.2250	0.1364	
C	0.2000	0.2632		0.5833	0.6000	0.3000	0.1912	0.3750	0.2273	
D	0.3000	0.1316	0.2500	0.0833			0.0441		0.0455	
E[b]							0.1471		0.3636	0.5000
F	0.0500	0.1842			0.4000	0.3000	0.1912	0.1000	0.2273	0.4286
G	0.1500	0.0789	0.5000			0.1000	0.1618			0.0714
H[a]	0.1500		0.1250							
I[a]	0.0500		0.1250							
J[c]							0.0147			
K[b]							0.0588	0.1250		
L[b]							0.0441	0.1750		

[a]Allele exclusive to savanna populations
[b]Allele exclusive to forest populations
[c]Allele exclusive to the population

Table 11.4 Allele frequencies of microsatellite locus Cj11 in silvery marmoset populations from central Amazonia

	Populations									
Allele	Sarú	Santa Luzia	Jardim	Mureta	Capiranga	Tauá	Tabocal	Flona Tapajós	São Benedito	Massafra
A[a]		0.0526		0.5000						
B		0.2632	0.7500					0.0500		
C	0.5000	0.2895	0.1250		0.2000		0.0441	0.2000	0.1818	
D	0.2000	0.2895		0.2500	0.4000	0.8000	0.6029	0.7500	0.4773	0.2143
E		0.0789	0.1250	0.0833		0.1000	0.2353		0.2955	0.6429
F	0.1500	0.0263		0.1667	0.3000	0.1000	0.0882			
G	0.1500						0.0294		0.0455	0.1429
H[b]					0.1000					

[a]Allele exclusive to savanna populations
[b]Allele exclusive to the population

Table 11.5 Allele frequencies of microsatellite locus Cj14 in silvery marmoset populations from central Amazonia

	Population									
Allele	Sarú	Santa Luzia	Jardim	Mureta	Capiranga	Tauá	Tabocal	Flona Tapajós	São Benedito	Massafra
A[a]		0.0526		0.5000						
B		0.2632	0.7500					0.0500		
C	0.5000	0.2895	0.1250		0.2000		0.0441	0.2000	0.1818	
D	0.2000	0.2895		0.2500	0.4000	0.8000	0.6029	0.7500	0.4773	0.2143
E		0.0789	0.1250	0.0833		0.1000	0.2353		0.2955	0.6429
F	0.1500	0.0263		0.1667	0.3000	0.1000	0.0882			

[a]Allele exclusive to savanna populations

Mean observed heterogeneity (*Ho*) varied considerably among the different populations (Table 11.2), but once again, there was no clear overall trend with regard to fragment size or characteristics. One of the smallest fragments – Sarú – returned one of the highest values of *Ho*, whereas the value recorded at Flona Tapajós (continous forest) was the second smallest overall. With few exceptions, observed heterozygosity was much lower than expected and all three loci presented significant deviations from the Hardy–Weinberg equilibrium at a majority of sites.

Once again, this analysis presents unexpected results. All forest sites, but in particular Flona Tapajós, deviate significantly from the Hardy–Weinberg equilibrium, whereas two of the savanna populations (Capiranga and Tauá) are in equilibrium. It is possible that sample size is a factor here, given that the less deviant sites are those with smaller samples and fewer alleles. Alternatively, deviations may reflect a Wahlund effect, that is, a mixture of different gene pools. The fusion of genetically heterogeneous populations violates assumptions for Hardy–Weinberg equilibrium because resulting genotype frequencies are not the result of random mating. This effect would be more probable in the more recently-formed anthropogenic fragments.

Despite the large number of alleles that were exclusive to one or other study area (or type of fragment) and significant overall levels of genetic differentiation (mean $R_{st}=0.22$, $P<0.001$), there is no clear pattern of differentiation among the sites (Table 11.6). All the most significant ($P<0.001$) pairwise differences were found in comparisons between forest and savanna populations, but all comparisons between two savanna populations (Sarú and Capiranga) and the four forest populations were marked by a consistent lack of differentiation. Despite being less accentuated overall, significant differentiation was just as likely to be found between populations of the same group (forest or savanna) as between groups. There is also no clear pattern associated with geographic distance. The most isolated savanna population – Tauá – is more similar to São Benedito (a distant, anthropogenic forest fragment) than it is to most savanna populations, for example, and Tabocal, the most northerly forest locality, is no more similar to the nearest savanna populations than other forest populations.

No significant results were obtained in the bottleneck analysis (Wilcoxon's signed ranks test). This indicates that there have been no severe recent reductions in effective size for any of the marmoset populations studied here.

11.4 Discussion

11.4.1 Genetic Diversity

Overall, habitat fragmentation does not appear to have resulted in a major, universal loss of genetic diversity in the study populations. This may reflect the tolerance of habitat fragmentation in the species, although the reduced levels of heterozygosity are potentially problematic for the long-term survival of the populations.

Table 11.6 Genetic differentiation (R_{st}) for pairwise comparisons between ten silvery marmoset populations from central Amazonia. Table-wide significance was applied, using the sequential Bonferroni technique (Rice 1989), with $k=10$ and $\alpha=0.05$. All estimates are based on one thousand permutations

	Sarú [s]	Santa Luzia [s]	Jardim [s]	Mureta [s]	Capiranga [s]	Tauá [s]	Tabocal [a]	Flona Tapajós [c]	São Benedito [a]
Santa Luiza	0.1141*								
Jardim	0.3537*	0.3447**							
Mureta	0.3744**	0.1899*	0.2449						
Capiranga	0.1533	0.3228**	0.6101**	0.5047**					
Tauá	0.0277	0.1331	0.4963*	0.3127*	0.2246				
Tabocal	0.0311	0.3064***	0.4561**	0.4664***	0.0609	0.1874*			
Flona Tapajós	0.0764	0.1872**	0.3505	0.3327**	0.1018	0.2621**	0.1116**		
São Benedito	0.0282	0.2203***	0.5693***	0.4681***	0.0465	0.1346	0.0284	0.0911*	
Massafra [a]	0.0594	0.4259***	0.5010**	0.6555***	0.1760	0.3313*	0.0438	0.2050**	0.2010*

a anthropogenic fragment; *c* continuous forest; *s* savanna fragment
$^{*}P<0.05$
$^{**}P<0.01$
$^{***}P<0.001$

Few studies of the genetic structure of primate populations based on microsatellite markers are available, although the overall heterozygosity recorded here (*Ho* = 0.11–0.39 for three loci in ten populations) is relatively low in comparison with other platyrrhines. Grativol et al. (2001), for example, recorded mean *Ho* values of 0.34–0.65 based on the analysis of four microsatellite loci in four isolated populations of golden lion tamarins (*Leontopithecus rosalia*), whereas Bastos (2007) found values of 0.62–0.69 in three populations of Amazonian red-handed howlers (*Alouatta belzebul*), based on fifteen microsatellite loci. In three populations of the owl monkey *Aotus infulatus*, Melo (2004) recorded mean *Ho* values of 0.44–0.48 for five microsatellite loci. Differences among species most likely reflect ecological and demographic factors, in particular population size. It is nevertheless possible that the lower levels of heterozygosity recorded in the present study were related in part to the slightly smaller number of loci analyzed.

The failure of Flona Tapajós to provide a control model of expected genetic variability for the analysis of other samples almost certainly reflects relatively low levels of genetic diversity in the original, continuous population, as observed in *Leontopithecus* (Pope 1996). This probably reflects the patchy distribution of these marmosets in continuous forest, where they occupy specific habitat types (Tavares 1999; Veracini this volume, Chap. 12). In addition to reduced effective population size, the dizygotic twinning and bone marrow chimerism of the marmosets and other callitrichids may also contribute to lower genetic diversity in comparison with other species (Benirschke et al. 1962; Pope 1996).

Despite the limitations of the samples, in terms of the numbers of both loci and individuals, the sum of the evidence indicates clearly that the remnant populations of silvery marmosets in central Amazonia are no longer part of the same gene pool. This likely reflects regional differences in the genetic structure of the original population that have been reinforced after habitat fragmentation by factors such as inbreeding and genetic drift.

11.4.2 Deviations from Hardy–Weinberg Equilibrium

The relative scarcity of heterozygotes that provokes deviations from the Hardy–Weinberg equilibrium is most probably related to inbreeding in small and isolated populations. The data also indicate the influence of the Whalund effect.

Another possibility is the presence of null alleles, which are not amplified because of mutations at priming sites (Callen et al. 1993), resulting in increased homozygosity. Despite the fact that null alleles are relatively common in studies based on heterologous markers, the lack of specific markers for *M. argentatus* (markers for *C. jacchus* were used here: Nievergelt et al. 1998) is problematic for the confirmation of possible heterozygote deficiencies. However, amplification was rarely unsuccessful, and when it did fail, it normally did so for all three loci, which points to DNA of inadequate quality rather than the presence of null alleles. It thus seems reasonable to conclude that null alleles were either extremely rare or, more probably, absent from the loci and populations analyzed.

The sampling of more than one social group may have been a determinant for the excess of homozygotes in the populations from São Benedito, Tabocal, Flona, Sarú and Santa Luzia. This is because, while genotype frequencies may correspond to the Hardy–Weinberg equilibrium within a given breeding group, heterozygosity may be reduced at the population level (the Whalund effect). In addition to the Whalund effect, the organization of populations in relatively small, non-panmictic groups must also favor inbreeding, and both processes contribute to deviations from the Hardy–Weinberg equilibrium.

While inbreeding is normally avoided by dispersal from the natal group, a characteristic of all primate species (Strier 2000), this mechanism may be interrupted significantly by habitat fragmentation. In this context, one differentiating factor may be the greater connectivity of natural fragments in the savanna matrix, in comparison with anthropogenic fragments, which are located primarily in open pasture. Some between-fragment migrations were recorded during the study, although there are too few data to evaluate potential differences between natural and anthropogenic fragments.

11.4.3 Implications for Conservation

Populations of silvery marmosets in central Amazonia present one of the lowest levels of genetic diversity found in any primate species. This reduced diversity has been maintained following the process of habitat fragmentation, although some of the populations analyzed have become differentiated, possibly as a result of genetic drift. Observed levels of diversity indicate low viability of most populations over the long term, especially in very small, isolated populations, where the probability of inbreeding depression is high (Frankham et al. 2002).

However, like other marmosets, *M. argentatus* appears to be relatively tolerant of the ecological effects of habitat fragmentation (Tavares and Ferrari 2002; Carvalho et al. 2005; Corrêa 2006; Veracini this volume, Chap. 12), and it would thus seem reasonable to assume that they may also be relatively well-adapted to the potential effects of reduced genetic diversity. In this case, they may be similar to animals such as Speke's gazelle, *Gazella spekei* (Templeton and Read 1983), in which crossing closely-related individuals resulted in a reduction of lethal alleles in comparison with unrelated breeders. Given this, long-term metapopulation management of *M. argentatus* in the study area would not need to prioritize gene flow among all populations, but rather concentrate on strategies aimed at the reduction of genetic drift (increasing population size) and linking populations with similar genetic characteristics.

Acknowledgments This study was part of the Brazilian Environment Ministry's PROBIO initiative, financed by BIRD and CNPq, with additional support from FINEP/PRONEX, IBAMA/Santarém and the Kapok Foundation. The collection of specimens was authorized by IBAMA through special license 157/2001-DIFAS/DIREC. We are grateful to Adenilson Carvalho, Wellton Costa and Caroline Nievergelt for their help with data collection and analysis.

References

Bastos HB (2007) Estrutura genética de *Alouatta belzebul* (Primates, Platyrrhini) da área de influência da UHE-Tucuruí. Masters Thesis, Universidade Federal do Pará, Belém

Benirschke K, Anderson JM, Brownhill LE (1962) Marrow chimerism in marmosets. Science 138:513–515

Callen DF, Thompson AD, Shen Y, Phillips HA, Richards RI, Mulley JC, Sutherland GR (1993) Incidence and origin of "null" alleles in the (AC) n microsatellite markers. Am J Hum Genet 52:922–927

Carvalho AN, Corrêa HKM, Coutinho PEG, Ferrari SF (2005) Composição de grupos de *Mico argentatus* (Callitrichinae, Primates) em um fragmento florestal na Amazônia Central. Resumos do XI Congresso Brasileiro de Primatologia. Sociedade Brasileira de Primatologia, Porto Alegre, p 85

Cornuet JM, Luikart G (1996) Description and power analysis of two tests for detecting recent population bottlenecks from allele frequency data. Genetics 144:2001–2014

Corrêa HKM (2006) Ecologia de grupos de Mico argentatus em fragmentos florestais naturais na Amazônia central. Doctoral Dissertation, Goeldi Museum, Belém

Dietz JM, Baker AJ, Ballou JD (2000) Demographic evidence of inbreeding depression in wild golden lion tamarins. In: Young AG, Clarke GM (eds) Genetics, demography, and the viability of fragmented populations. Cambridge University Press, Cambridge, pp 203–211

DiRienzo A, Peterson AC, Garza JC, Valdes AM, Slatkin M, Freimer NB (1994) Mutational processes of simple-sequence repeat loci in human populations. Proc Natl Acad Sci USA 91:3166–3170

Dixson AF, Anzenberger G, Monteiro da Cruz MAO, Patel I, Jeffreys AJ (1992) DNA fingerprinting of free-ranging groups of common marmosets (*Callithrix jacchus jacchus*) in NE Brazil. In: Martin RD, Dixson AF, Wickings EJ (eds) Paternity in primates: genetic tests and theories. Basel, Karger, pp 192–202

Dudash MR, Fenster CB (2000) Inbreeding and outbreeding depression in fragmented populations. In: Young AG, Clarke GM (eds) Genetics, demography, and the viability of fragmented populations. Cambridge University Press, Cambridge, pp 35–53

Ferrari SF (1993) Ecological differentiation in the callitrichidae. In: Rylands AB (ed) Marmosets and tamarins. systematics, behaviour, and ecology. Oxford University Press, Oxford, pp 314–328

Ferrari SF, Iwanaga S, Ravetta AL, Freitas FC, Sousa BAR, Souza LL, Costa CG, Coutinho PEG (2003) Dynamics of primate communities along the Santarém-Cuiabá highway in southern central Brazilian Amazônia. In: Marsh LK (ed) Primates in fragments. Kluwer, New York, pp 123–144

Frankham R, Ballou JD, Briscoe DA (2002) Introduction to conservation genetics. Cambridge University Press, Cambridge

Gagneux P, Boesch C, Woodruff DS (1997) Microsatellite scoring errors associated with noninvasive genotyping based on nuclear DNA amplified from shed hair. Mol Ecol 6:861–868

Gonçalves EC, Ferrari SF, Silva A, Coutinho PEG, Menezes EV, Schneider MPC (2003) Effects of habitat fragmentation on the genetic variability of silvery marmosets, *Mico argentatus*. In: Marsh LK (ed) Primates in fragments. Kluwer, New York, pp 17–28

Goodman SJ (1997) Rst calc: a collection of computer programs for calculating estimates of genetic differentiation from microsatellite data and determining their significance. Mol Ecol 6:881–885

Goudet J (2001) FSTAT, a program to estimate and test gene diversities and fixation indexes. Version 2.9.3. http://www.unil.ch/izea/softwares/fstat.html

Grativol AD, Ballou JD, Fletcher R (2001) Microsatellite variation within and among recently isolated populations of golden lion tamarins. Conserv Genet 2:1–9

Kimura M, Crow JF (1964) The number of alleles that can be maintained in a finite population. Genetics 49:725–738

Lacy RC (1997) Importance of genetic variation to the variability of mammalian populations. J Mammal 78:320–335

Luikart G, Cornuet J-M (1998) Empirical evaluation of a test for identifying recently bottlenecked populations from allele frequency data. Conserv Biol 12:228–237

Marsh LK (2003) Primates in fragments. Kluwer, New York

Melo MAD (2004) Variabilidade e estrutura genética em três populações naturais de *Aotus infulatus* da Amazonia oriental. Masters Thesis, Universidade Federal do Pará, Belém

Miranda IS (1993) Estrutura do estrato arbóreo do cerrado amazônico de Alter do Chão, Pará, Brasil. Rev Bras Bot 16:143–150

Nievergelt CM, Mundy NI, Woodruff DS (1998) Microsatellite primers for genotyping common marmosets (*Callithrix jacchus*) and other callitrichids. Mol Ecol 5:1432–1434

Ohta T, Kimura M (1973) A model of mutation appropriate to estimate the number of electrophoretically detectable alleles in a finite population. Genet Res 22:201–204

Piry S, Luikart G, Cornuet J-M (1999) BOTTLENECK 1202: a computer program for detecting recent reductions in the effective population size using allele frequency data. J Hered 90:502–503

Pope TR (1996) Socioecology, population fragmentation, and patterns of genetic loss in endangered primates. In: Avise JC, Hamrick JL (eds) Conservation genetics: case histories from nature. Chase & Hall, New York, pp 119–159

Pope TR (1998) Genetic variation in remnant populations of the woolly spider monkey (*Brachyteles arachnoides*). Int J Primatol 19:95–109

Raymond M, Rousset F (1995) Genepop version 1.2: population genetics software for exact tests and ecumenism. J Hered 86:248–249

Rylands AB (1986) Ranging behaviour and habitat preference of a wild marmoset group, *Callithrix humeralifer* (Callitrichidae, Primates). J Zool Lond 210:489–514

Rylands AB, Schneider H, Langguth A, Mittermeier RA, Groves CP, Rodriguez-Luna E (2000) An assessment of the diversity of New World primates. Neotrop Primates 8:61–93

Slatkin M (1995) A measure of population subdivision based on microsatellite allele frequencies. Genetics 139:457–462

Strier KB (2000) Primate behavioral ecology. Allyn and Bacon, Boston

Tavares LI (1999) Estratégias de Forrageio de *Callithrix argentata* (Linnaeus 1771) na Estação Científica Ferreira Penna, Pará. Masters Thesis, Universidade Federal do Pará, Belém

Tavares LI, Ferrari SF (2002) Diet of the silvery marmoset (*Callithrix argentata*) at ECFPn: seasonal and longitudinal variation. In: Lisboa PLB (ed) Caxiuanã: Biodiversidade e Sustentabilidade. CNPq, Belém, pp 707–719

Templeton AR (1986) Coadaptation and outbreeding depression. In: Soulé ME (ed) Conservation biology: the science of scarcity and diversity. Sinauer, Sunderland, pp 105–116

Templeton AR, Read B (1983) The elimination of inbreeding depression in a captive herd of Speke's gazelle. In: Schonewald-Cox CM, MacBryde B, Thomas L (eds) Genetics and conservation. Benjamin Cummings Inc., Menlo Park, CA, pp 241–261

van Roosmalen MGM, van Roosmalen T, Mittermeier RA, Rylands AB (2000) Two new species of marmoset, genus *Callithrix* Erxleben, 1777 (Callitrichidae, Primates) from the Tapajós-Madeira interfluvium, saouth Central Amazonia, Brazil. Neotrop Primates 8:2–19

Veracini C (this volume) Habitat use and ranging behavior of the silvery marmoset (*Mico argentatus*) at Caxiuanã National Forest (eastern Brazilian Amazonia). In: Ford SM, Porter LM, Davis LC (eds) The smallest anthropoids: the marmoset/callimico radiation. Springer, New York, pp 221–240

Yeh FC, Yang RC, Boylet T (1997) POPGENE (version 1.32): Software microsoft window-based freeware for population genetic analysis. University of Alberta, Canada

Chapter 12
Habitat Use and Ranging Behavior of the Silvery Marmoset (*Mico argentatus*) at Caxiuanã National Forest (Eastern Brazilian Amazonia)

Cecilia Veracini

Abstract This work describes the ranging behavior and the habitat preferences of a wild group of silvery marmosets studied in the eastern Brazilian Amazonia for 11 months. The study group used secondary growth forests (*capoeiras*) for 78% of the observation time, flooded forest and *terra firme* forest for 9% of the time and edge areas for the rest of the time, without a significant seasonal variation. The *Mico argentatus* home range was 15.5 ha with an overlap of 24.2% with other groups. The area they used exclusively was situated in the center of the home range and included gum and fruit sources mainly. Variation in ranging behavior was observed among seasons. The study group ranged over a larger area in the early wet and late dry seasons. The estimated distance traveled by the group during the day varied from 630 m to 1710 m (average 1042 m, $n=83$) and the mean day range used was 2.73 ha. The group used 14 different sleeping sites; the more frequented were localized near their major food patches. The distribution of feeding trees had a major influence on the patterns of home range use. The more intensively used areas contained major exudate resources.

Resumen Este trabajo describe el comportamiento de desplazamiento y las preferencias de hábitat de un grupo silvestre de marmosetas plateadas (*Mico argentatus*) en la Amazonia Brasileña del este por un período de 11 meses. El grupo de estudio utilizó bosques de crecimiento secundario (*capoeiras*) el 78% del tiempo de observación, bosque inundado y bosque de tierra firme el 9% y áreas marginales por el resto del tiempo, sin variación estacional significativa. El ámbito hogareño de *M. argentatus* fue de 15.5 ha, con un traslape del 24.2% con otros grupos. Su área que este de uso exclusivo de este grupo estuvo situada en el centro del ámbito hogareño e incluyó las fuentes principales de exudados arbóreos y frutas. Variación en el comportamiento de desplazamiento fue observado entre las estaciones. El

C. Veracini (✉)
Laboratori di Antropologia, Dipartimento di Biologia Evoluzionistica "L. Pardi", Via del Proconsolo, 12. CAP, Firenze, 50122, Italy
e-mail: cpfveracini@yahoo.com

S.M. Ford et al. (eds.), *The Smallest Anthropoids*, Developments in Primatology: Progress and Prospects, DOI 10.1007/978-1-4419-0293-1_12,
© Springer Science+Business Media, LLC 2009

grupo estudiado recorrió un área más grande durante al principio de la estación húmeda y al final de la seca. La distancia de viaje estimada diaria para el grupo varió de 630 m a 1710 m (con un promedio de 1042 m, $n=83$) y la media del desplazamiento diario fue de 2.73 ha. El grupo utilizó 14 sitios dormideros diferentes; los más frecuentes fueron localizados cerca de las mayores zonas de comida. La distribución de los árboles alimenticios tuvo mayor influencia en los patrones de uso del ámbito hogareño. Las áreas utilizadas más intensivamente contenían los mayores recursos de exudados.

Resumo Este estudo descreve as áreas de uso e as preferências de habitat de um grupo de sagüis brancos estudados na Amazônia oriental do Brasil durante o período de 11 meses. O grupo estudado usou florestas secundárias (*capoeiras*) durante 78% do tempo de observação, florestas inundadas e de terra firme em 9% do tempo e áreas de bordas durante o resto do tempo, mas sem apresentar variação sazonal significativa. A área de uso do *Mico argentatus* foi de 15.5 ha. havendo uma sobreposição com a área de outros grupos em 24.2%. A área usada exclusivamente pelos animais estava situada no centro da área de uso e incluiu as fontes principais de goma e frutos. Foi observada variação no tamanho da área utilizada entre as estações. O grupo usou uma área maior no início da estação chuvosa e final da estação seca. A distância estimada viajada pelo grupo durante o dia variou entre 630 m e 1.710 m (média 1.042 m, $n=83$) e a área de uso diário foi de 2.73 ha. Foram registrados 14 locais de dormida diferentes; os mais freqüentes estavam localizados próximos à maior concentração de recursos alimentares. A distribuição das árvores de alimentação foi o principal fator que influenciou os padrões de utilização da área de uso. As áreas mais intensamente exploradas possuíam as principais fontes de exudatos.

12.1 Introduction

Marmosets are distributed over a large geographic area and are able to occupy a wide range of habitats (Hershkovitz 1977; Stevenson and Rylands 1988; Rylands et al. 1993; van Roosmalen et al. 2000). Despite their ecological flexibility, evidence from field studies show that marmosets have a marked preference for edge and secondary growth forests (Ferrari 1993). This preference seems related to several characteristics of these habitats: relative abundance of insect prey, dense vegetation for protection from predators, and the concentration of pioneer plant species, many of which are important fruit resources for marmosets (Terborgh 1983; Rylands 1986; Rylands and de Faria 1993).

Callitrichid-ranging behavior appears to be related to a wide array of factors including floristic composition and distribution of food resources (Rylands 1986; Peres 1994), foraging strategies (Garber 1984; Ferrari et al. 1996; Dietz et al. 1997), resource competition and intergroup resource defense (Garber et al. 1993;

Peres 2000), sleeping sites (Raboy and Dietz 2004), demography and reproductive strategies of the species (Digby and Barreto 1996), and human habitat disturbance (Alonso and Langguth 1989; Mendes Pontes and Monteiro da Cruz 1995). Home range size is quite variable in callitrichids, often showing differences in similar species and in populations of the same species (Rylands and de Faria 1993; Albernaz and Magnusson 1999; Miranda and de Faria 2001; Kierulff et al. 2002). Given the systematic use of plant exudates by marmosets and the behavioral adaptations linked to this behavior (Coimbra-Filho and Mittermeier 1976), many authors have suggested that the abundance and spatial distribution of gum trees may play an important part in determining the size and shape of marmoset groups' home range (Stevenson and Rylands 1988; Scalon et al. 1982; Maier et al. 1982). Rylands and de Faria (1993) argue that the relative proportions of exudates in the diet of different species of marmosets are related to marmoset home range size. On the other hand, Ferrari (1988) and Correa (1995) found that home range size of the southern marmosets *Callithrix flaviceps* (the buffy-headed marmoset) and *Callithrix aurita* (the white-eared marmoset) could be strongly influenced by invertebrate abundance.

Although the genus *Mico* (argentata-group marmosets) was traditionally considered less gummivorous than the genus *Callithrix* (jacchus-group) (Hershkovitz 1977; Stevenson and Rylands 1988; Rylands and de Faria 1993), some field studies show that *Mico argentatus* (the silvery marmoset) can have a very high percentage of exudates in the diet (Veracini 1997a; Tavares 1999). *M. argentatus* therefore is not expected to use as large a home range as other Amazonian, highly frugivorous marmoset species, such as *Mico intermedius* (the Aripuanã marmoset; Rylands 1986). Albernaz and Magnusson (1999), studying four groups of silvery marmosets living in a savanna area in Alter do Chão (central Brazilian Amazonia), found a high variability in the groups' home ranges (from 4 to 24 ha) and the groups' use of space was strongly influenced by the position of gum-producing trees.

Marmosets of the genus *Mico* are distributed in Amazonia, south of the Rio Amazonas and southeast of the Rio Madeira (de Vivo 1991; van Roosmalen et al. 2000; Rylands et al. Chapt. 2 this volume). Many aspects of the ecology and behavior of this genus are still poorly known. Until now only two species among the 14 taxa known to date have been subjects of long term ecological investigation: *M. intermedius* (Rylands 1982, 1986) and *M. argentatus*, (Veracini 1997b; Tavares 1999; Albernaz and Magnusson 1999).

The species *M. argentatus* is distributed in the lowland forests (under 200 m above sea level) of Pará state, between the mouth of the Rio Tocantins in the east and the Rio Tapajós and Cuparí in the west and extending south to the Rio Irirí as far as the lower Rio Curuá (Hershkovitz 1977; Ferrari and Lopes Ferrari 1990; Rylands et al. 1993, Chapt. 2 this volume). It is the only marmoset species occurring in eastern Brazilian Amazonia (east of Rio Xingú).

This work describes the habitat use and ranging behavior of a wild group of silvery marmosets studied through 11 months in a rainforest at the National Forest of Caxiuanã (Pa).

12.2 Materials and Methods

12.2.1 Study Area

This study was carried out in the "Estação Científica Ferreira Penna-Caxiuanã/CNPq," ECFP (1°42 30″S and 51°31 45) located within the 330,000 ha area of the National Forest of Caxiuanã, Pará, Brazil. The ECFP/Caxiuanã (33,000 ha) is a sparsely settled region composed of about 85% *terra firme* primary forest. The remaining territory is composed of *igapó* and *varzea* flooded forest, savannah (5%), and secondary growth forests. The forest of Caxiuanã is considered one of the richest biogeographic areas of eastern Brazilian Amazonia (Lisboa et al. 1997). The species of primates recorded at the ECFP are: *M. argentatus, Saguinus niger* (black tamarin), *Cebus apella* (brown capuchin), *Alouatta belzebul belzebul* (red-handed howler monkey), *Chiropotes utahicki* (Uta Hick's bearded saki), and *Aotus infulatus* (Kuhl's owl monkey).

In the study area (approximately 40 ha) there are four forest types (Lisboa et al. 1997; see Fig. 12.1): (a) The *terra firme* forest reaches 25–35 m in height with emergent trees of approx. 50 m. The canopy is dense and the lower strata and

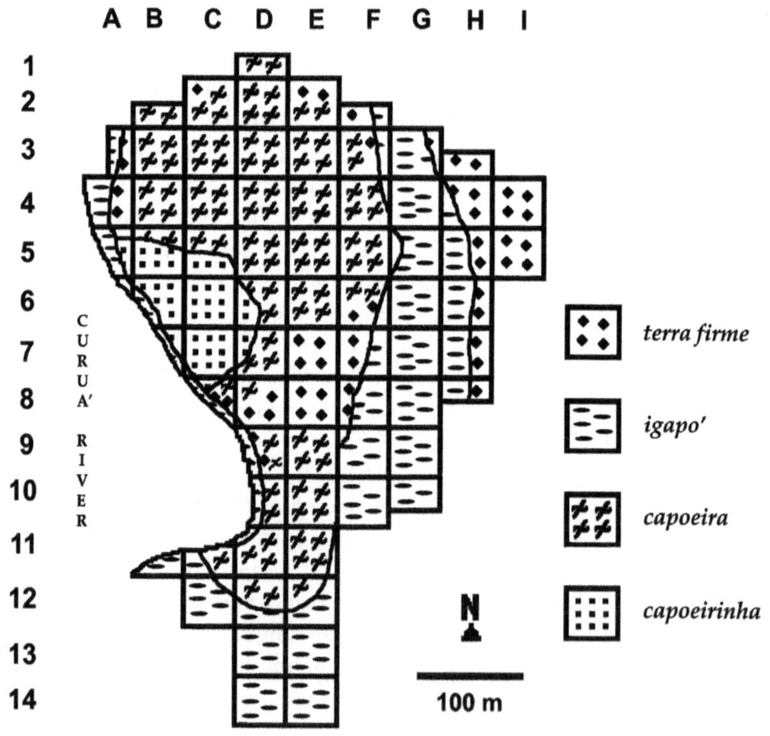

Fig. 12.1 Distribution of habitats in the home range of the study group

undergrowth are scattered. The *terra firme* is characterized by a great deal of diversity within the plant families Sapotaceae, Chrysobalanaceae, Lauraceae and Lecythidaceae. (b) The flooded forest of *igapó* has an average canopy of 25 m, and the predominant woody plant families are Fabaceae, Caesalpiniaceae, Mimosaceae, and Guttiferae. (c) The secondary growth forest (*capoeira*) was at different stages of re-growth during the study, from 15 up to 40 years. Its height was ca. 10–20 m depending on the ages of the regrowth. The *capoeira* mainly contained plants of the families Lecythidaceae, Melastomataceae and Lauracea with an abundance of the species *Myrcia fallax, Poecilanthe effuse,* and *Lacistema polistachyum.* The undergrowth was both very dense with patches of liana tangles and open with sparse vegetation with an abundance of the palm *Maximiliana maripa.* At the boundary with the *terra firme* (north side of the marmosets' home range), where the *capoeira* was older, Mimosaceae, Fabaceae, Guttiferae, and Lacistemaceae were predominant. (d) There was a secondary growth forest, about 3–4 years old (called *capoerinha*) of ca. 1.2 ha. This area had been recently cleared and at the time of the study reached about 5 m in height and was characterized by very dense bushy vegetation with an abundance of lianas. For a more complete description of the habitats, see Lisboa et al. (1997).

The state of Pará has a warm humid tropical climate with temperatures always over 18°C (Martorano et al. 1993). The region of Caxiuanã is classified as the climatic subtype Am3 (classification of Koppen) with an average annual rainfall from 2,000 to 2,500 mm and a relative humidity of 85%. The Caxiuanã region has a moisture deficit between the end of June and November. The wet season starts in January and lasts until June, with a great deal of rainfall during January and June (de Almeida et al. 1993). During 1996 the average annual rainfall was 2176.3 mm (Carvalho et al. 1997).

Data about plant phenology and fruit availability were obtained by quantitative and *ad libitum* observations (Veracini 1997b), and by botanical studies conducted in the ECFP (Ferreira et al. 1997). The plant phenology pattern of Caxiuanã forests appears similar to other Amazonian sites of the state of Pará (e.g., Pires-O'Brien 1993). Most trees flower either during the dry season or the transition from dry to wet season, and fruit mostly during the wet season. During the study, the period of major fruit availability of the forest of Caxiuanã was the rainy season (January to June) both in *terra firme* and *capoeira.* The *terra firme* forest showed a drastic fall of fruit resources starting in May and lasting for almost all the months of the dry season, with the exception of *Goupia glabra*, which fruited in August. The secondary growth forest showed high fruit availability for marmosets throughout most of the months of the wet season due to the concentration of many individuals of several pioneer plant species such as *Pouroma* sp., *Myrcia* sp., *Protium* sp., *Inga* sp., *Byrsonima aerugo*, and *Tapirira guianensis*. In the late wet and early dry seasons, a lower availability of fruit was recorded. In August an abundance of fruiting *G. glabra* was observed, and in September and October there was an abundant fructification of *Inga heterophilla* and *Inga laterifolia*. In the flooded forests, there was an abundance of flowers on *Symphonia globulifera* (Guttiferae) from the end of September until November.

12.2.2 *The Study Group*

The study group was initially composed of six individuals (May 1995): three adult females, one male, and two juveniles. During the study period, only the dominant female gave birth to twins, on three occasions at intervals of roughly 6 months. One adult female and a sub-adult disappeared, and one other adult female left the group for 3 months and then came back. The group composition at the end of the study was one adult male, two adult females, two dependent infants, two juveniles and two sub-adult females.

12.2.3 *Data Collection and Analysis*

During a preliminary period (6 months) of habituation of the study group, transects were opened in its home range. They formed squares of 50 m×50 m. Data on behavior, habitat use and ranging behavior were collected from January to November 1996. Quantitative data were collected using instantaneous scan sampling (Altmann 1974) in which a 1-min scan was collected at 5-min intervals following the procedure recommended by Ferrari and Rylands (1994) for the study of free-ranging callitrichids. For each scan the location of the group and the habitat they occupied were recorded. During a scan, the activity and the height at the moment of sighting of each visible animal were recorded. A minimum of 7–9 complete data days (from sleeping site to sleeping site) were collected each month. Data about ranging behavior were collected for a total of 83 complete observation days. All the sleeping sites, plant food species and sources (individual plants) were recorded, marked, and plotted on a map. All occurrences of intergroup encounters were documented.

Analysis: To determine daily path lengths, I mapped all the movements of the study group in each scan and calculated the sum of the straight-line distances traveled from one position to another during a full day of observation. Home range size was estimated by the sum of different squares (or a part of them) in which the group was observed for the total of the study period. The size of squares situated at the boundary of the home range, which were smaller than the others, were calculated following the relative size of the area that is 25%, 50%, 75% of a total square (Ferrari 1988; Passamani and Rylands 2000). To determine the day range, I calculated the total number of squares or the percentage of squares occupied by the study group in a full day of observation. The core area was defined as the area occupied for more than 40% of the observation days. The ranging behavior was estimated by the percentage of time (scans) that the group spent in each square. The analyses of habitat occupied by the group were based on the proportion of the number of scans. The data for each scan were combined to compare months, and months were combined to compare seasons: January–March (early wet), April–June (late wet), July–September (early dry), October–November (late dry). Nonparametric and parametric statistical tests (Siegel and Castellani 1988) were used, according to the sample size, to examine differences in the frequencies of records between seasons. Data were processed using the Statistica (5) software package.

12.3 Results

12.3.1 Habitat Use

The study group used very dense vegetation both in primary and in secondary habitats. The marmosets utilized the secondary growth forests of different ages (*capoeira* and *capoerinha*) for 78% of the observation time, the *terra firme* for 3%, the *igapó* for 6% and edge areas for the rest of time (Fig. 12.2), without significant seasonal variation (Kruskal–Wallis H [3, $n=11$]=2.485, $p=0.478$). The older secondary growth forest (*capoeira*) was intensely used all year and for all the activities of the group. The presence of fruit and exudate sources and the dense forested patches and liana tangles offered a very good habitat for marmosets. Edge areas and the 3-year secondary growth forest (*capoerinha*) were used mainly for foraging for animal prey. In addition, high trees (>15 m) of the *terra firme* forest densely festooned with creepers were frequently used for foraging for insects and resting. Although not statistically significant, marmosets showed a tendency to seasonally use the *terra firme* more during the early rainy season when most of the trees were fruiting. Marmosets showed a high preference (almost statistically significant $p=0.06$) for the *igapó* during the late wet and late dry season when they exploited the fruits and nectar of *Diosfaypos guianensis* and *Symphonia globlifera*, respectively. In that period the group spent many hours foraging for insects around these trees. Nevertheless, marmosets used the *igapò* for resting and social activities, sometimes arriving very near to the water.

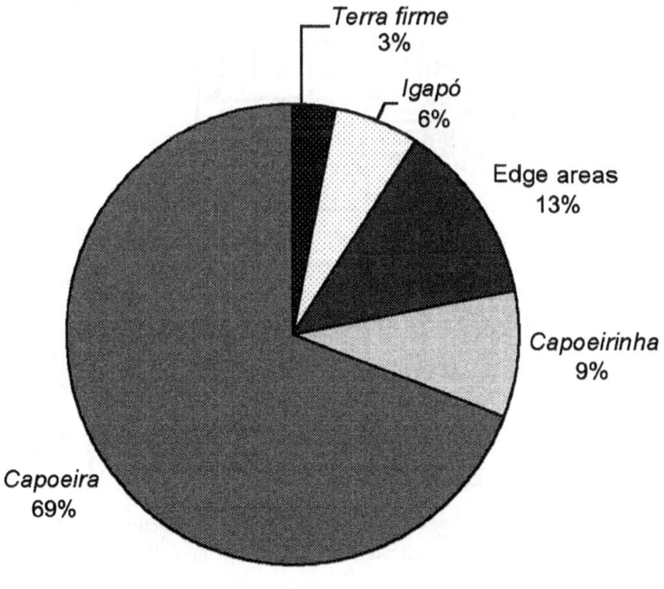

Fig. 12.2 Habitat use of the study group

12.3.2 *Ranging Behavior*

12.3.2.1 Ranging Behavior – General

During the study period the group utilized a total area of 15.5 ha (Fig. 12.3). The day ranges were much smaller than this, varying between 1 ha (a day in July) and 4.25 ha (a day in November) with an average of 2.73 ha per day (SD=0.76, n=83). On average the group traveled 1042 m (SD=223.03) each day (n=83) with a min of 630 m (a day in July) and max 1710 m (a day in March). Different parts of the home range were used with different intensity. Most of the boundary areas (5.76 ha) were frequented only for 0.57% of the total observation time. Sixteen squares, equivalent to 3.34 ha, were used for only 1 day equivalent to <0.1 of the total time records. These areas were mainly in the west-south periphery of the home range and included a large part of the flooded forests. In contrast, 10 squares equivalent to ca. 2.3 ha (14.83% of the total home range) were used for more than 40% of the

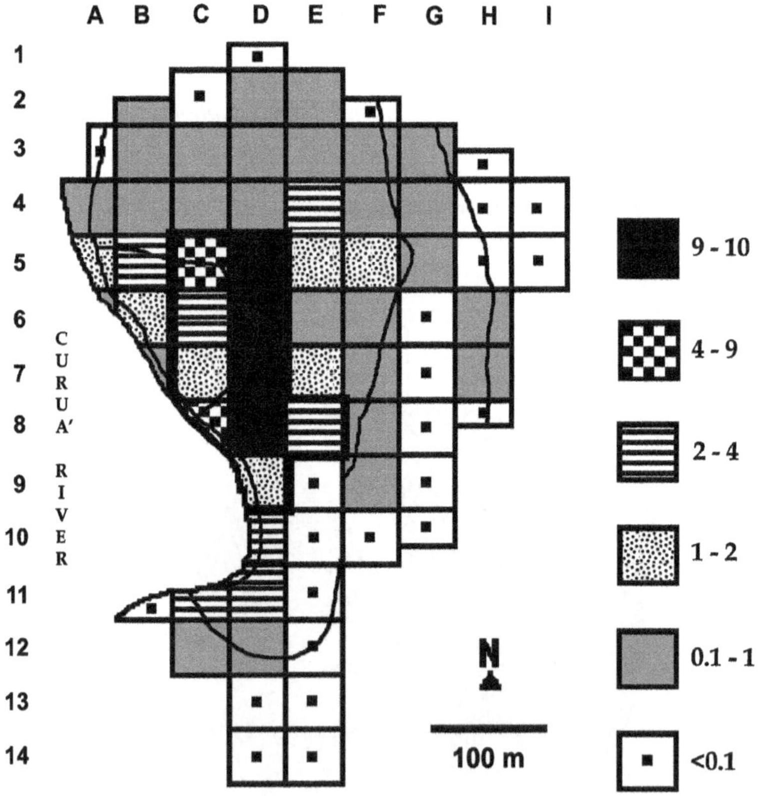

Fig. 12.3 Range size and percentages of occupancy records of the study group. The core area is indicated by the bold line. Numbers indicate percentage of scans – in relation to the total – spent in the square by the study group

observation days and for 66.7% of the total time records (core area, see Fig. 12.3 for percent of total time records or scans). This area includes 2 ha of the secondary growth forests with ca. 0.5 ha of the *capoerinha*, 0.25 ha of *terra firme*, and a very small area of *igapò*. Four squares (D5, D6, D7, D8) inside this area were visited for 90% of the observation days and for 47.6% of the total observation time. These four squares, which comprised the most heavily used core area, contained the exudate sources (one tree of *Parkia ulei* and many individuals of *T. guianensis*) and fruit sources (*Myrcia atramentifera*, *Pouroma* sp., *G. glabra*, *Protium* sp.) commonly used by the study group (Fig. 12.4). The preferential use of the core area could also be linked to the dense vegetation, which characterized the boundary among the different habitats, and to the presence of three emergent trees, important sleeping sites of the study group. The remaining parts of the home range (7.44 ha) were used with different intensity throughout the year.

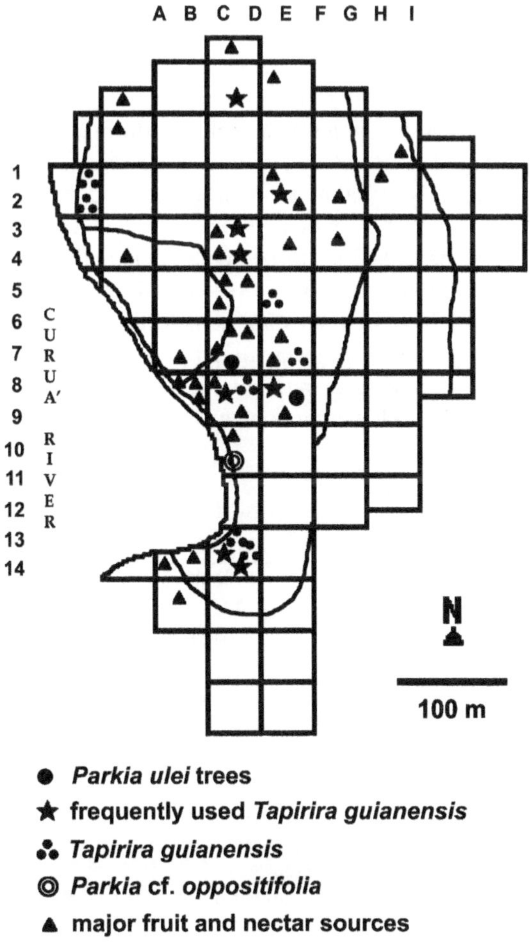

Fig. 12.4 Common feeding trees of the study group. The nectar and fruit sources indicated in the map are in accordance with the monthly top ranked taxa in the *M. argentatus* diet. All the *T. guinanesis* trees used by the marmosets are indicated in the map: "*T. guianensis* frequently used" means used in all the seasons throughout the year

The home ranges of two or three other groups overlapped with the study group. Aggressive encounters were particularly frequent with the group living in an area north of the home range. The study group defended its territory vigorously and had exclusive use of a large part of its home range. The percentage of overlap with the other marmoset groups was ca. 3.75 ha, equivalent to 24.2% of the total area. In contrast, the overlap with a group of sympatric tamarins (*S. niger)* was almost total except for the 3-year secondary growth forest (*capoerinha*) where the tamarins were never observed. Encounters and interactions between the two species were very rare. I recorded only 18 interactions between the two species in the study period. Most interactions were agonistic, consisting of aggressive acts (threat, attack, chase). The aggressive acts were initiated by members of both species in the same proportion. The most aggressive acts of *M. argentatus* against *S. niger* were in proximity or above their most important gum deposits, while *S. niger* attacked and threatened *M. argentatus* nine times during foraging and near *T. guianensis* trees.

12.3.2.2 Ranging Behavior by Seasons

Interseasonal differences in range use were observed (Table 12.1). The group ranged over a larger area each day during the early wet and late dry seasons than during the other two seasons. Both day range area and daily path length were statistically different between seasons and they were correlated ($r=0.51, p=0.001$). In the early wet and late dry season, the study group used larger home range areas (11.42 ha and 12 ha, respectively) than in the late wet and early dry season (9.87 ha and 10.06 ha, respectively; see Fig. 12.6). Figure 12.5 shows this variation month-by-month. It is interesting to note that the total area covered in a month varied independent of the average daily values (see, e.g., the January to March period). This indicates that individual days were spent more in different areas during the periods with lower day ranges but higher monthly home range areas.

Table 12.1 Range use by season of the study group

Seasons[a]	Mean day path length (m)	Mean day range (ha)[b]
Early wet	1136.88	2.99
Late wet	930.52	2.36
Early dry	1004.36	2.51
Late dry	1124.14	3.23
Total mean	1042.00	2.73
n	83	83
SD	223.03	0.76
[c]F (3, 79)	4.79	6.59
p	0.004	0

[a]Jan–Mar: early wet season; Apr–Jun: late wet season; July–Sep: early dry season; Oct–Nov: late dry season

[b]Calculated by the number of full or partial 50 m × 50 m squares entered

[c]Anova for testing differences among seasons

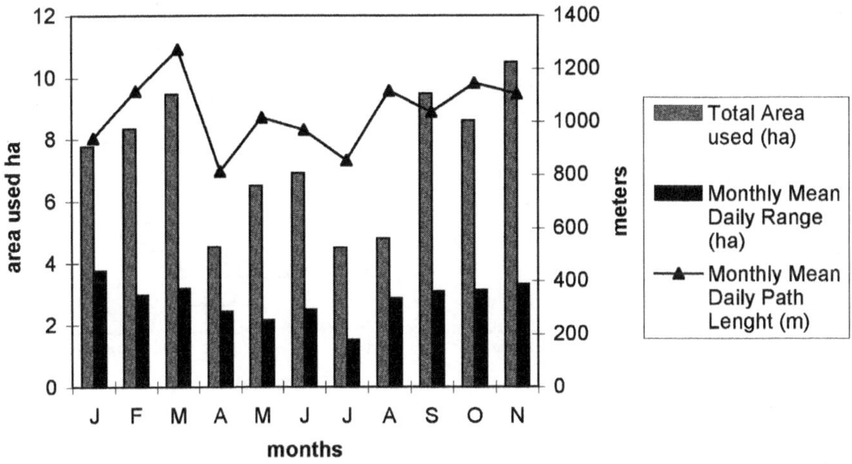

Fig. 12.5 The monthly home range size and mean monthly day range for the study group, calculated by the number of full or partial squares entered. Jan–Mar: early wet season; Apr–Jun: late wet season; July–Sep: early dry season; Oct–Nov: late dry season

Although marmosets focused their activity in the center of their home range, they varied across seasons in the percentage of square occupation (Fig. 12.6). Six squares showed significantly different values between seasons (Table 12.2). The pattern of range use was likely linked to fruit and nectar seasonal availability. In the early wet season, during fruit abundance, marmosets exploited more widely dispersed fruit sources. The number of fruit taxa utilized monthly was significantly higher in that period than during the rest of the year (Veracini 1997b). Species diversity of plants in the diet was the highest in February and March. The significantly greater use of E4 in the early wet season was correlated with the high exploitation of the fruits of *T. guianensis* in March. In the late dry season, as already mentioned, marmosets fed intensely on nectar and shifted their range to the *igapó* (Fig. 12.6d). In accordance, the highest values for squares G3, G4 and C12 were observed during the period of nectar abundance. Squares D5, D7, D8 which contained the main exudate sources were visited very frequently throughout the year, sometimes several times per day. Square D8 showed a statistically significantly lower frequency in the early wet and late dry seasons. A greater use of all this area (D5-D7-D8) was observed in the late wet season and in the early dry season in accordance with the lack of fruit resources. In that period, marmosets showed a frenetic activity of tree gouging that corresponded to an increase in the consumption of exudates (Veracini 1997b). In July marmosets fed almost exclusively on exudates, in particular on *P. ulei*; in this month I recorded their smallest average monthly day range (1.52 ha) and square D7 was used to the maximum for any month. The utilization of square D7 (where the *P. ulei* tree is located) was negatively correlated with the average monthly day length. The higher use of D9 in the late wet season is likely correlated with the exploitation of seed pod exudates of *P.* cf. *oppositifolia* in May.

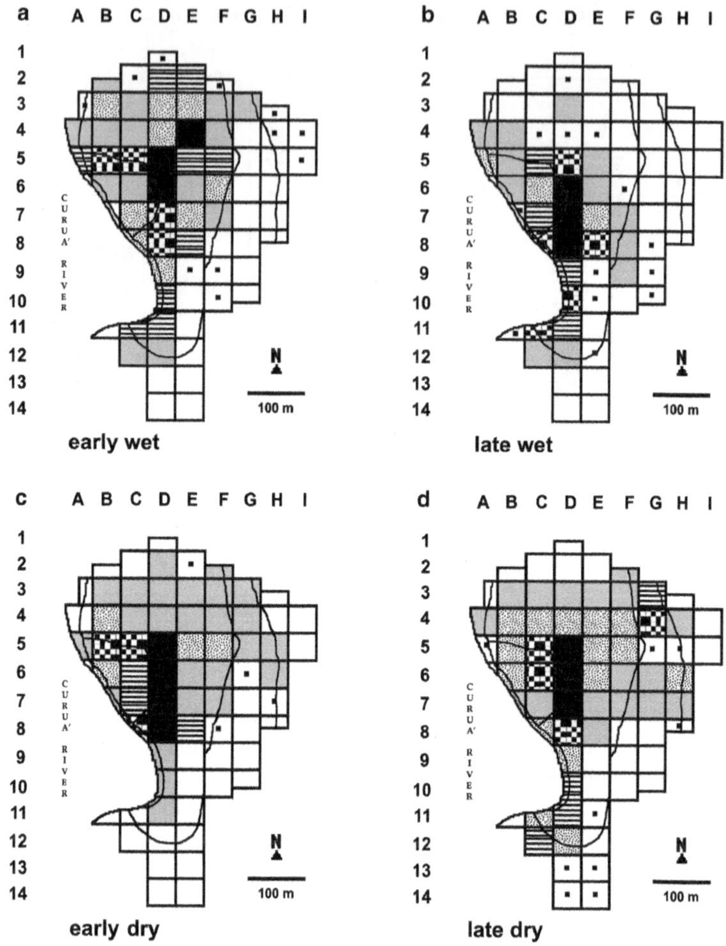

Fig. 12.6 Ranging behavior by seasons according to the percentage of occupancy records (see Fig. 12.3). Jan–Mar: early wet; Apr–Jun: late wet; July–Sep: early dry; Oct–Nov: late dry

Table 12.2 Squares showing different seasonal[a] occupation

Squares entered	Early wet	Late wet	Early dry	Late dry	F (gl1,2) 3.7[b]	*p*
D9	1.13	3.59	0.78	1.42	8.098	0.011
D8	4.68	17.08	16.40	6.68	5.777	0.026
C12	0.24	0.24	0.00	2.87	14.322	0.002
C2	0.11	0.00	0.00	0.00	167.474	0.000
E4[c]	10.92	0.38	1.02	1.51	4.029	0.069
E3	1.96	0.07	0.14	0.10	5.737	0.034
G3	0.34	0.00	0.34	2.94	10.757	0.005
G4	0.00	0.00	0.58	4.21	11.146	0.005

[a]Jan–Mar: early wet season; Apr–Jun: late wet season; July–Sep: early dry season; Oct–Nov: late dry season

[b]Anova for testing differences among seasons ($n = 11$)

[c]Not statistically different but indicative of a trend

12.3.3 Sleeping Sites

In the study period the group utilized 14 different sleeping sites (Fig. 12.7). Those used most frequently were distributed inside or in proximity to the core area. The sites in D8 and D9 was 46.7% of the total observation. The site in D8, a tall tree of ca. 40 m, was the most frequented with 22% of the records. Three sleeping sites outside the core area were used only once (see Fig. 12.7). On many occasions marmosets chose the sleep trees nearest to the last feeding site of the day. My data show that 61 days in 82 they used a sleeping site situated at distance of 100 m or less from the last feeding site of the day. This pattern was more evident in the late wet and dry seasons when marmosets often chose sleeping sites near the main gum trees. In that

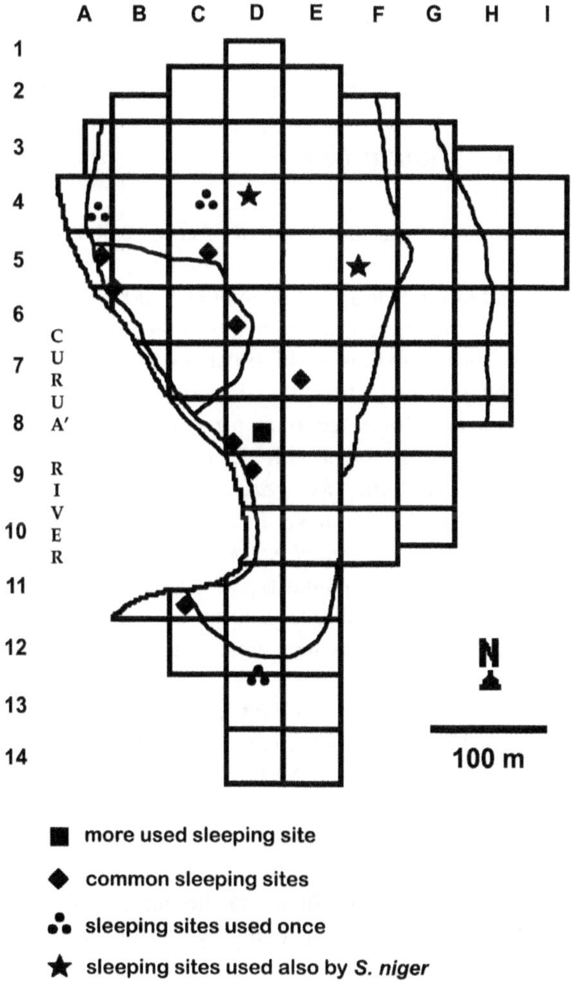

Fig. 12.7 Sleeping site locations

period they usually fed on gum before sleeping and after awakening. The preferred sites were often used consecutively, sometimes for more than three nights. Two sleeping trees outside the core area were also used by *S. niger.* The sleeping sites were mostly (11 in 14) located 13 m above the ground. The majority of the sleep trees were isolated or emergent trees densely covered in tangles of lianas or other climbing plants. On many occasions palm trees and a hole in a tree at 17 m were used; nevertheless marmosets used some unusual sites such as a hanging fallen tree trunk at ca. 16 m and a dead palm of 7 m covered by lianas located in the *capoerinha.*

The preference of a particular sleeping site seems influenced by the following factors: (a) the pattern of trees (presence of lianas and height); (b) the proximity of the feeding sites in a determinate day or period of time (particularly at the end of the day); (c) their position inside the home range (particularly if it is in the core area); and (d) the ranging behavior of the day (sometimes an unusual or atypical site was chose).

12.4 Discussion

12.4.1 Habitat Use

Data about habitat utilization of *Mico* revealed that they are able to occupy a wide range of different habitats, although dense vegetation or disturbed forests are used more intensely. van Roosmalen et al. (2000) observed *Mico manicorensis* in dense primary *terra firme* rainforest and in secondary forest surrounding plantations and fields. They noted a very high density of *M. manicorensis* in *seringal*, areas managed by humans where the density of rubber trees is very high. Stalling and Mittermeier (1983) recorded *Mico melanurus* inhabiting tall forest (canopy at 20–30) and scrub forest with a canopy of 5–10 m in the *chaco* of Paraguay. Branch (1983) observed *Mico humeraliferus* on the Rio Tapajos most frequently in secondary growth forest, and never in flooded areas of *varzea.* Rylands (1986) observed a preference of *M. intermedius* for the disturbed forest although the study area had a variety of different habitat types. Albernaz and Magnusson (1999) found a wide range of different habitats used by *M. argentatus* and noted that the presence of continuous forest appeared more important than food availability in determining home range size.

M. argentatus at the ECFP/Caxiuanã showed a clear preference for the secondary growth forest. This habitat supported a high density of *T. guianensis* trees and many individuals of the Myrtaceae, Mimosaceae, and Cecropiaceae families. Its dense bushy vegetation offered great availability of insect prey and protective cover from predators. Predator pressure seemed very high at Caxiuanã, where a great variety of potential predators (snakes and raptors) have been observed in large numbers (Cardoso da Silva and Pimentel Neto 1997; pers. obs.). A high rate of attempts at predation, especially by raptors, and consequent alarm calls and freezing behavior (up to 50 min) was recorded (Veracini 1997b). The study group showed a relatively high use of flooded areas, in contrast to groups of the same species studied by Albernaz and Magnusson (1999) and to other previous studies of this genus. The group's less intense use of *terra firme* forest was probably due to the sparse undergrowth of this habitat and to widely dispersed fruiting trees. Nevertheless, the study group frequently used *terra firme* for foraging for insects, resting, and sleeping. These results are in accordance with Ferrari and Lopes (1996), who found the density of *M. argentatus* at the ECFP/Caxiuanã very low in comparison with that of the sympatric tamarin *S. niger*, and observed marmoset groups occur only in regions that contain disturbed areas. In most of these areas, a high concentration of *T. guianensis* trees with holes produced by marmosets was found although marmosets also used a high percentage of exudate sources of mature trees of the primary *terra firme* forest such as *P. ulei* and *Carapa guianensis* exudates and seed pod exudates of *Parkia* cf. *oppositifolia* (Veracini 1997a). Ferrari et al. (2003) also reported a relative abundance of the silvery marmoset in man-made forest fragments and their patchy distribution in continuous forests in the south-central regions of Amazonia.

12.4.2 Ranging Behavior

The current study group's home range size falls within the values found by Albernaz and Magnusson (1999) for the same species (4–24 ha), and it is smaller than that found by Rylands (1986) for *M. intermedius* (23.5–18 ha) and for the southern marmosets *C. aurita* and *C. flaviceps* (Ferrari et al. 1996), the latter two considered relatively frugivorous marmoset species among the jacchus-group (Rylands and de Faria 1993). The mean monthly day range (2.73 ha) and the mean day path length (1042 m) were smaller than in *M. intermedius* (5–8.9 ha and 1459 m, respectively) and in *C. flaviceps* (5.4 ha and 1222.5 m). On the whole, the ranging pattern observed in *M. argentatus* in Caxiuanã is actually more similar to values recorded for more gummivorous marmosets such as *Callithrix kuhlii*, *C. geoffroyi* and *C. penicillata* (Rylands 1989; Miranda and de Faria 2001; Passamani and Rylands 2000). To date, the extent to which different food source distributions may affect variation in the home range size in marmosets is still not completely understood. As Ferrari et al. (1996) pointed out, local habitat characteristics could be more important than species-specific adaptations.

With regard to seasonal variation, the study group used an area significantly larger in the early wet and late dry seasons. Decreased fruit availability and the associated increased use of exudates in the late wet and early dry seasons likely caused the reduction of the home range size. *M. argentatus* in Alter do Chão (Albernaz and Magnusson 1999) and *M. intermedius* (Rylands 1986) also showed a smaller home range during periods of fruit scarcity. In the late dry season the study group appeared to expand its home range searching for nectar in the flooded areas. Nectar of *S. globulifera* is a key resource for Amazonian callitrichids in the dry season (Peres 1993; Garber 1993; Veracini 2000). Thus, mean daily path lengths were significantly different between seasons, unlike in studies of other species of marmosets (Rylands 1986; Stevenson and Rylands 1988; Ferrari 1988; Correa 1995).

The study group's home range overlapped 24.2% with its neighbors. As expected, intergroup conflicts occurred principally along boundary zones, especially on the north side where some feeding trees were located. In several callitrichid species there is evidence that neighboring groups compete for access to feeding trees located in shared areas of their range (Stevenson and Rylands 1988). Garber et al. (1993) noted that intergroup encounters may facilitate the transfer of migrants between groups. During this study a female of a neighboring group attempted to enter the study group during an intergroup encounter, and two subadults disappeared on one of these occasions.

I lacked sufficient data to test the hypothesis that the presence of *S. niger* influences the ranging behavior of the silvery marmoset. The question of possible competition for resources between these two callitrichids is still open (Ferrari and Lopes Ferrari 1990; Ferrari 1993). Nevertheless, an ecological niche separation seems to better describe their behavior and interactions at the ECFP/Caxiuanã (Veracini 1998).

12.4.3 Sleeping Sites

Sleeping habits reflecting antipredator adaptations have been suggested for arboreal New World monkeys (Ferrari 1988; Heymann 1995). Isolated trees, a high number of sleep trees, and a low frequency of re-use appear to be the main features for reducing the risk of predators. The sleeping sites used in this study did not fit this pattern at least in two aspects. The study group used a relatively low number of sleeping trees when compared with *M. intermedius* (Rylands 1986) that had quite a comparable home range size. *M. argentatus* also had the habit of sleeping in the same sites for consecutive nights even though many sites were available. Actually the study group seemed to select sleeping sites that were close to the last feeding sites of the day. This behavior was also observed by Stevenson and Rylands (1988) in *C. jacchus* and by Day and Elwood (1999) in *Saguinus midas*. These authors argued that the vicinity of food sources was the most important factor in the selection of a sleeping site. There may be a strong energy conservation factor, associated with these marmosets' small body size, driving the selection of sleeping sites near important feeding trees.

12.5 Conclusion

In conclusion, despite the fact that *M. argentatus* showed a marked preference for the disturbed areas, this study confirmed the tendency of the genus to adapt to a large variety of habitats. It appears that the distribution of fruit, nectar, and exudate resources within the study group's home range had a major influence on the marmosets' movement patterns. The more intensively used areas were directly related to the presence of major exudate resources that constituted a great percentage of the diet throughout the year. The home range dimensions, the daily routes, and the concentration of the daily activity around the exudate sources found in *M. argentatus* at the ECFP/Caxiuanã, are characteristic of marmosets that are more exudativorous than frugivorous (Stevenson and Rylands 1988). The importance of gum trees for *M. argentatus* was also pointed out by the study of Albernaz and Magnusson (1999) in Alter do Chão (Pa), as predicted by Rylands and de Faria (1993) for Amazonian marmoset populations living in open and dryer habitats. Although the region of Caxiuanã is very rich in terms of plant diversity and fruit resources, it is possible that the long dry season of eastern Brazilian Amazonia forces marmosets to make a greater use of exudates than in south-central Amazonia where *M. intermedius* is found. To date we do not have data on other Amazonian marmoset species for further comparisons. Considering the extent of their geographical range and the different habitats they may occupy, it is likely that we shall see a great deal of variation in adaptations and ecological strategies in Amazonian marmoset populations.

Acknowledgements I thank the editors for inviting me to contribute to this book and for their important comments and help. This study was supported by the Italian MURST (Ministero per l'Università e per lo Sviluppo Tecnologico e Scientifico) and by a grant of the Capes/CNPq, Brazil. I also thank the Estação Cientifica Ferreira Penna of Caxiuanã, Dr. P. Lisboa and Prof. Bento Mascarenhas. I thank Prof. Stephen F. Ferrari for inviting me to study callitrichids in the ECFP and for hosting me at the Universidade Federal do Pará. I thank also Drs S. Almeida and L. Silva, and Mr. Nelson Arajuo Rosa for the botanical help and identification of botanical samples.

References

Albernaz AL, Magnusson WE (1999) Home-range size of the bare-ear marmoset (*Callithrix argentata*) at Alter do Chão, Central Amazonia, Brazil. Int J Primatol 20(5):667–677

Alonso C, Langguth A (1989) Ecologia e Comportamento de *Callithrix jacchus* (Primates: Callitrichidae) numa ilha de Floresta Atlantica. Rev Nordestina Biol 6(2):107–137

Altmann J (1974) Observational study of behaviour: sampling methods. Behaviour 49:227–267

Branch L (1983) Seasonal and habitat differences in the abundance of primates in the Amazon (Tapajos) National Park, Brazil. Primates 24:424–431

Carvalho de Moraes J, José de P R da Costa, da Rocha EJP, da Silva IMO (1997) Estudos hidrometeorológicos na bacia do rio Caxiuanã. In: Lisboa PL (ed) Caxiuanã. MPEG/CNPq, Bol. MPEG, Belém, pp 85–95

Coimbra-Filho AF, Mittermeier RA (1976) Exudate-eating and tree-gouging in marmosets. Nature 262:630

Correa K (1995) Ecologia e comportamento alimentar de um grupo de Saguis-da-serra-escuros (*Callithrix aurita*, E'. Geoffroy) no parque Estadual da Serra do Mar, Nucleo Cunha SP (Brasil). Master's Thesis Univ Federal Minas Gerais, Belo Horizo

Cardoso da Silva JM, Pimentel Neto DC (1997) As Aves. In: Lisboa PL (ed) Caxiuanã. MPEG/CNPq, Bol. MPEG, Belém, pp: 403–416

Day RT, Elwood RW (1999) Sleeping site selection by the golden-handed tamarin *Saguinus midas midas*: the role of predation risk, proximity to feeding sites and territorial defense. Ethology 105:1035–1051

de Almeida SS, Lisboa PLB, Silva ASL (1993) Diversidade Floristica de uma Comunidade arborea na Estação Científica Ferreira Penna, em Caxiuanã (Pará). Bol Mus Para Emilio Goeldi serie Bot 9(1):93–188

de Vivo M (1991) Taxonomia de *Callithrix* Erxleben, 1977 (Callitrichidae). Fundação Biodiversitas, Belo Horizonte

Dietz JM, Peres CA, Pinder L (1997) Foraging ecology and use of space in wild golden lion tamarins (*Leontopithecus rosalia*). Am J Primatol 41:289–305

Digby LJ, Barreto CE (1996) Activity and ranging patterns in common marmosets (*Callithrix jacchus*). In: Norconk M, Rosenberger AL, Garber PA (eds) Adaptive radiations of neotropical primates. Plenum, New York, pp 173–185

Ferrari SF (1988) The behaviour and ecology of the buffy-headed marmoset, *Callithrix flaviceps* (O. Thomas, 1903). Unpublished PhD Thesis, University College, London.

Ferrari SF (1993) Ecological differentiation in the Callitrichidae. In: Rylands AB (ed) Marmosets and tamarins. systematics, behaviour, and ecology. Science, Oxford, pp 314–328

Ferrari SF, Lopes Ferrari MA (1990) A survey of Primates in central Pará. Mus Bol Pará Emilio Goeldi, Ser Zool 6(2):169–179

Ferrari SF, Rylands AB (1994) Activity budget and differential visibility in field studies of three marmosets (*Callithrix* spp.). Folia Primatol 63(2):78–83

Ferrari SF, Correa K, Coutinho PG (1996) Ecology of the "Southern" Marmosets (*Callithrix aurita* and *Callithrix flaviceps*): how different, how similar? In: Norconk M, Rosenberger AL, Garber PA (eds) Adaptive radiations of neotropical primates. Plenum, New York, pp 157–171

Ferrari SF, Iwanaga S, Ravetta AL, Freitas FC, Sousa R, Souza LL, Costa CG, Coutinho PEG (2003) Dynamics of primate communities along the Santarém-Cuiaba highway in south-central Brazilian Amazonia. In: Marsh LK (ed) Primates in fragments. Ecology and conservation. Kluwer Academic/Plenum Publisher, New York, pp 123–157

Ferreira L, Almeida S, Rosário C (1997) As areas de inundação. In: Lisboa PL (ed) Caxiuanã. MPEG/CNPq, Bol. MPEG, Belém, pp 195–211

Ferrari SF, MA Lopes (1996) Primate populations in eastern Amazonia. In MA. Norconk, AL Rosenberger, and PA Garber, eds., Adaptive Radiations of the Neotropical Primates, Plenum Press New York, pp 53–67

Garber PA (1984) Use of habitat and positional behavior in a neotropical primate, *Saguinus oedipus*. In: Rodman P, Cant J (eds) Adaptations for foraging in non human primates. Columbia University Press, New York, pp 112–133

Garber PA (1993) Feeding ecology and behaviour of the genus *Saguinus*. In: Rylands AB (ed) Marmosets and tamarins. Systematics, behaviour, and ecology. Science, Oxford, pp 273–295

Garber PA, Pruetz DJ, Isaacson J (1993) Patterns of range use, range defense and intergroupspacing in moustached tamarin monkeys. Primates 34(1):11–25

Hershkovitz P (1977) Living New World Monkeys, part1 (Platyrrhini), with an introduction to Primates. Chicago University Press, Chicago

Heymann EW (1995) Sleeping habitus of tamarins, *Saguinus mystax* and *Saguinus fuscicollis* (Mammalia, Primates, Callitrichidae), in north-eastern Perù. J Zool 237:211–237

Kierulff MCM, Raboy BE, da Oliveira PP, Miller K, Passos FC, Prado F (2002) Behavioral ecology of lion tamarins. In: Kleina DG, Rylands AB (eds) Lion tamarins biology and conservation. Smithsonian Institution Press, Washington, pp 157–187

Lisboa PLB, da Silva ASL, de Almeida SS (1997) Florística e estrutura dos ambientes. In: Lisboa PL (ed) Caxiuanã. MPEG/CNPq, Bol. MPEG, Belém, pp 163-193

Maier W, Alonso C, Langguth A (1982) Field observations on *Callithrix jacchus jacchus* L. Z Saugetierk 47:334–346

Martorano LG, Pereira LC, Nachet D (1993) Tipologia climatica do estado do Pará adaptação do Metodo de Koppen. Bol Geogr Teor 23:45–4

Mendes Pontes AR, Monteiro da Cruz MAO (1995) Home range, intergroup transfer, and reproductive status of common marmosets *Callithrix jacchus* in a forest fragment in North-Eastern Brazil. Primates 36(3):335–347

Miranda GHB, de Faria DS (2001) Ecological aspects of Black-pincelled marmoset (*Callithrix penicillata*) in the *cerradão* and dense *cerrado* of the Brazilian Central Plateau. Braz J Biol 61(3):397–404

Passamani M, Rylands AB (2000) Home range of a Geoffroy's marmoset group, *Callithrix geoffroyi* (Primates, Callitrichidae) in South-Eastern Brazil. Rev Brasil Biol 60(2):275–281

Peres CA (1993) Diet and feeding ecology of saddle-back (*Saguinus fuscicollis*) and moustached (*Saguinus mystax*) tamarins in an Amazonian *terra firme* forest. J Zool Lond 230:567–592

Peres CA (1994) Primate responses to phenological changes in an Amazonian *terra firme* forest. Biotropica 26(1):98–112

Peres CA (2000) Territorial defence and the ecology of group movement in small-bodied neotropical primates. In: Boinsky S, Garber PA (eds) On the move: How and why animals travel in groups. University of Chicago Press, Chicago, pp 100–123

Pires-O'Brien MJ (1993) Phenology of tropical trees from Jari, lower Amazon. I. Phenology of eight forest communities. Bol Museo Paraense Emilio Goeldi ser Bot 9(1):67–91

Raboy BE, Dietz JM (2004) Diet, foraging, and use of space in wild golden-headed lion tamarins. Am J Primatol 63:1–15

Rylands AB (1982) The behaviour and ecology of three species of marmosets and tamarins (Callitrichidae, Primates) in Brazil. Unpublished PhD Thesis, University of Cambridge, Cambridge.

Rylands AB (1986) Ranging behaviour and habitat preference of a wild marmoset group, *Callithrix humeralifer* (Callitrichidae, Primates). J Zool Lond 210:489–514

Rylands AB (1989) Sympatric Brazilian callitrichids: the black tufted-ear marmoset, *Callithrix kuhli*, and the golden-headed lion tamarin, *Leontophitecus chrysomelas*. J Hum Evol 18:679–695

Rylands AB, de Faria DS (1993) Habitats, feeding ecology, and home range size in the genus *Callithrix*. In: Rylands AB (ed) Marmosets and tamarins. Systematics, behaviour, and ecology. Science, Oxford, pp 262–272

Rylands AB, Coimbra-Filho AF, Mittermeier RA (1993) Systematics, geographic distribution, and some notes on the conservation status of the Callitrichidae. In: Rylands AB (ed) Marmosets and tamarins. Systematics, behaviour, and ecology. Science, Oxford, pp 11–77

Rylands AB, Mittermeier RA, Coimbra-Filho AF (this volume) The systematics and distribution of the marmosets (*Callithrix, Callibella, Cebuella,* and *Mico*) and callimico (*Callimico*) (Callitrichidae, Primates). In: Ford SM, Porter LM, Davis LC (eds) The smallest anthropoids: The marmoset/callimico radiation. Springer Press, New York, pp 25–61

Scalon CE, Chalmers NR, Monteiro da Cruz MAO (1982) Home range use and the exploitation of gum in the Marmoset *Callithrix jacchus jacchus*. Int J Primatol 10(2):123–136

Siegel CE, Castellani NJ (1988) In: Caracciolo E. (ed) Statistica non parametrica. Mc-Graw-Hill Libri Italia Srl, Milano.

Stalling JR, Mittermeier RA (1983) The black-tailed marmoset (*Callithrix argentata melanura*) recorded from Paraguay. Am J Primatol 4:159–163

Stevenson MF, Rylands AB (1988) The marmoset, genus *Callithrix*. In: Rylands AB, Coimbra-Filho AF, da Fonseca GAB (eds) Ecology and behavior of neotropical primates, vol 2. World Wildlife Fund, Washington DC, pp 131–222

Tavares J (1999) Estratégias de Forrageio de um grupo silvestre de Sagui-Branco (*Callithrix argentata*, Linnaeus 1771) na Estação Científica Ferreira Penna–MPEG/CNPq, Melgaço. Univ Federal do Pará, Belém, Pará. Master's Thesis

Terborgh J (1983) Five New World Primates: a study in comparative ecology. University Press, Princeton

van Roosmalen MGM, van Roosmalen T, Mittermeier RA, Rylands AB (2000) Two new species of marmosets, genus *Callithrix* Erxleben, 1777 (Callitrichidae, Primates), from the Tapajos/Madeira interfluvium, South Central Amazonia, Brazil. Neotrop Primates 8(1):2–18

Veracini C (1997a) O comportamento alimentar de *Callithrix argentata* (Linnaeus 1771) (Primata, Callitrichinae). In: Lisboa PL (ed) Caxiuanã. MPEG/CNPq, Bol. MPEG, Belém, pp 437–446
Veracini C (1997b) Ecologia e comportamento di *Callithrix argentata* (Linnaeus, 1771) e le sue interazioni con *Saguinus midas niger* (E' Geoffroy, 1803). Unpublished PhD Thesis, University of Florence, Florence
Veracini C (1998) Ecology of, and Interactions between *Callithrix argentata* and *Saguinus midas niger* at Caxiuanã Forest, Eastern Amazonia. Vol. of Abstracts, XVIIth Congress of the International Primatological Society, Antananarivo, Madagascar, August 10–14
Veracini C (2000) Dados preliminares sobre a ecologia de *Saguinus niger* (É. Geoffroy, 1803) na Estação Científica Ferreira Penna-Caxiuanã (Pará). Neotrop Primates 8(3):108–113

Chapter 13
Ranging Patterns of *Callimico goeldii* (callimico) in a Mixed Species Group

Jennifer A. Rehg

Abstract Ranging patterns are behavioral adaptations that relate to demographic and ecological variables, such as population density and resource distribution. Intraspecific variation in ranging may reflect responses to variable environmental conditions. A study of *Callimico goeldii,* the callimico or Goeldi's monkey, in association with *Saguinus fuscicollis*, the saddle-backed tamarin, and *Saguinus labiatus*, the red-bellied tamarin, was conducted at a *terra firme* forest in northwestern Brazil periodically between April 1999 and August 2003. Data were collected primarily on one relatively stable mixed species group. The joint range size of this mixed species group was 48 ha over 10 months, and after 13 months, it increased to 56 ha for the tamarins and to 59 ha for the callimicos; there was approximately 26% range overlap with neighboring groups of tamarins, and 3% overlap with neighboring callimico groups. The locations of plant food resources, resting sites, and treefalls were found to be significantly related to range use by *Callimico*. There was no evidence that *Callimico* spent more time along the periphery of the range than would be expected by chance, and range use was patchily distributed throughout the range. Five encounters were observed among the study group of callimico and neighboring callimicos. No direct intergroup encounters were observed among the tamarins, and they did not participate in the callimico encounters. All encounters were within 100 m of a plant food resource visited on that day, although there was no additional evidence that the encounters related directly to defending access to these feeding sites. Based on these data, the callimicos did not appear to demonstrate resource or range defense. Previous research on other groups shows different ranging and association patterns in *Callimico*; additional investigation of intersite variation in resources or demography may help explain these differences.

J.A. Rehg (✉)
Department of Anthropology, Southern Illinois University Edwardsville, Box 1451, Edwardsville, IL, 62026, USA
e-mail: jrehg@siue.edu

S.M. Ford et al. (eds.), *The Smallest Anthropoids*, Developments in Primatology: Progress and Prospects, DOI 10.1007/978-1-4419-0293-1_13,

© Springer Science+Business Media, LLC 2009

Resumen Los patrones de desplazamiento son adaptaciones del comportamiento relacionados a variables demográficas y ecológicas, tales como la densidad de la población y la distribución de recursos. La variación intraespecífica en el desplazamiento puede reflejar respuestas a condiciones ambientales variables. Un estudio de *Callimico goeldii*, el calimico o mono de Goeldi, en asociación con *Saguinus fuscicollis* y *S. labiatus*, se llevó a cabo en un bosque de *terra firme* al noroeste de Brasil entre abril de 1999 y agosto de 2003. Los datos fueron recolectados principalmente de un grupo de especies mixtas relativamente estable. La tamaño del ámbito conjunto de este grupo de especies mixtas fue de 48 ha durante los primeros 10 meses y después de los 13 meses incrementaron a 56 ha para *Saguinus* y a 59 ha para *Callimico*; hubo aproximadamente 26% de traslape de ámbito con los grupos vecino de *Saguinus* y 3% de traslape con los grupos vecinos de *Callimico*. La ubicación de plantas utilizadas como recurso alimenticio, de los sitios de descanso y de las caídas de árboles estuvo significativamente relacionada al ámbito utilizado por *Callimico*. No hubo evidencia de que *Callimico* pasara más tiempo a lo largo de la periferia de su ámbito de la que hubiese sido esperada al azar y el uso del ámbito estuvo distribuido en forma de parches a través de su área. Cinco encuentros fueron observados entre el grupo de calimicos estudiado y los grupos de calimicos vecinos. No se observaron encuentros directos intergrupales entre los grupos de tamarinos y estos tampoco participaron en los encuentros entre calimicos. Todos los encuentros ocurrieron dentro de los 100 m de distancia de una planta de recursos alimenticios visitada en ese día, aunque no hubo evidencia adicional que apoyara que los encuentros se relacionaran directamente a defender el acceso a esos sitios de alimentación. Basados en estos datos, los calimicos no parecen defender sus recursos o su ámbito. Investigaciones previas de otros grupos muestran distintos patrones de desplazamiento y de asociación en *Callimico*; investigaciones adicionales de variación de recursos o demografía entre sitios podría explicar dichas diferencias.

Resumo Padrões de uso de área são adaptações comportamentais relacionadas com variáveis demográficas tais como densidade populacional e distribuição de recursos. Variações intra-específicas no padrão de uso da área podem refletir resposta às condições ambientais variáveis. Um estudo com *Callimico goeldii* (sagüi-de-goeldi) em associação com *Saguinus fuscicollis* (sagüi-de-cara-suja) e *Saguinus labiatus* (sagüi-de-bigode) foi realizado em diferentes períodos em floresta de terra firme no noroeste do Brasil entre abril de 1999 e agosto de 2003. Os dados foram coletados primariamente em um grupo de espécies mistas relativamente estável. A área conjunta de uso do grupo foi de 48 ha ao longo de dez meses e após 13 meses cresceu para 54 ha para os tamarinos e 59 para os calimicos. Foi verificada uma sobreposição de aproximadamente 26% da área com grupos vizinhos de tamarinos e 3% de sobreposição com grupos vizinhos de *Callimico*. A localização das árvores de alimentação e de dormida e queda de árvores foram significativamente correlacionada com a área de uso de *Callimico*. Não foi encontrada evidência de que *Callimico* passava mais tempo nas bordas do território do que o esperado ao acaso e que o uso da área tenha sido distribuído em ilhas dentro da área de uso total. Foram registrados cinco encontros entre o grupo de callimico estudado e grupos de calimicos vizinhos. Nenhum encontro intergrupo foi registrado entre os tamarinos; além do mais não tiveram participação nos encontros dos grupos de callimicos. Todos os encontros foram num raio de 100 m de uma árvore de alimentação visitada no dia, embora não houvesse nenhuma evidência adicional de que os encontros estivessem relacionados à defesa no acesso a esses sítios de alimentação. Baseado nesses dados, callimico parece defender território ou recursos. Pesquisas anteriores com outros grupos mostram diferentes padrões de utilização de áreas de uso e de associação em callimico; pesquisas adicionais sobre as variações entre ou inter-sítios com relação aos recursos ou à demografia poderão ajudar a explicar essas diferenças.

13.1 Introduction

Ranging patterns, including home range or territory size, intergroup range overlap, and spatiotemporal variation in range use, vary greatly within and among primate species. Such variation may relate to many factors, including population density, group size and composition, intergroup relatedness, resource availability, and habitat quality (Dawson 1979; Mitani and Rodman 1979; Rylands 1986; Chapman 1988; Ferrari and Lopes Ferrari 1989; Cheney 1987; Lazaro-Perea 2001). Documenting variation in ranging patterns, and understanding the relationship between ranging and ecological, social, and demographic variables, are important in characterizing species' adaptations. Such information also has practical applications for conservation, as key information on the quality and quantity of spatial requirements is needed in designing conservation strategies targeted at populations or species. The effects of habitat loss, fragmentation, and other anthropogenic disturbances on primate habitats and resources may impact ranging patterns as a consequence of changes in the ecological, social, and demographic environment; thus, research on ranging patterns provides information on the impact of such disturbances.

The Callitrichidae is a diverse clade of primates, especially in its ranging patterns. Home ranges of less than 5 ha have been recorded for some groups of *Callithrix* and less than a hectare for *Cebuella*, while ranges of 30–40 ha or larger appear common for species of *Saguinus* and *Leontopithecus* (Hubrecht 1985; Soini 1982; Rylands 1989; Scanlon et al. 1989; Mendes-Pontes and Monteiro da Cruz 1995; Terborgh 1983; Norconk 1986; Terborgh and Stern 1987; Garber 1988; Rylands 1989; Peres 1992, 2000; Dietz et al. 1997; Albernaz and Magnusson 1999; Lazaro-Perea 2001). Intraspecific variation is also notable. Within *Saguinus*, reported range size varies from as little as 7 ha to more than 140 ha (Peres 1992, 2000; Garber et al. 1993), and in a single population of *Mico argentatus* (silvery marmosets), reported ranges varied from about 5 to 24 ha (Albernaz and Magnusson 1999).

In addition, the proportion of range overlap and the number of neighboring groups sharing these areas are variable, as are the frequency and duration of intergroup encounters and types of behavior (tolerant or agonistic) directed toward conspecifics of other groups (Izawa 1978; Soini 1982; Terborgh 1983; Lopes and Ferrari 1994; Hubrecht

1985; Norconk 1986; Rylands 1986; 1989; Garber 1988; Peres 1992; de la Torre et al. 1995; Mendes-Pontes and Monteiro da Cruz 1995; Ferrari et al. 1996). Also variable are spatiotemporal patterns of ranging. A few studies have reported the most intense use of the home range along its periphery (Norconk 1986; Peres 1989, 1992); other studies indicate concentrated use of "core areas" or interior parts of the range (Soini 1982; Terborgh 1983; Crandlemire-Sacco 1986). Ranging patterns may also change throughout the year or across years (Dawson 1979; Soini 1982; Terborgh 1983; de la Torre et al. 1995; Soini 1987; Crandlemire-Sacco 1986; Digby and Barreto 1993).

Observed variation among callitrichids may be to some extent phylogenetically based, as part of the "niche" of a particular species or genus, but is also likely to be highly responsive to local conditions (ecological, social, and demographic). For example, heterogeneous spatial and temporal ranging patterns have been shown to be affected by the distribution of food resources (Soini 1982, 1987; Crandlemire-Sacco 1986; Dietz et al. 1997; Albernaz and Magnusson 1999). Some studies report territoriality and the monitoring of range boundaries as equally important influences on range use (Terborgh 1983; Norconk 1986; Soini 1987; Peres 1989, 1992); changes in the demographic or social environment can, thus, affect ranging.

Callimico goeldii (callimico) is the only currently recognized species of its genus, and differs from other callitrichids in anatomical, reproductive, behavioral, and ecological features (Hill 1959; Heltne et al. 1981; Porter 2000; Garber and Leigh 2001; Dettling 2002; Rehg 2003; Porter and Garber 2007). It is geographically distributed from the Pando Department of northern Bolivia, into northwestern Brazil, southeastern Peru, and southern Colombia (Hershkovitz 1977; Ferrari et al. 1999). Despite this broad geographic distribution, it exists at low population densities at most sites (Izawa 1979; Pook and Pook 1981; Christen 1999), and is considered "Near Threatened" by the IUCN (2004). It forms mixed species groups with *Saguinus* in areas of sympatry, and has been mainly studied in associations with *Saguinus labiatus* (red-bellied-tamarin) and *Saguinus fuscicollis* (saddle-backed tamarin) (Buchanan-Smith 1991; Hanson 2000; Porter 2000; Barry 2002; Azevedo-Lopes and Rehg 2003; Rehg 2006a). A study of the ranging patterns of callimico may help to clarify the factors that influence its patchy occurrence, and to develop an integrated view of its adaptations. Information on callimico ranging patterns also has applications for interpreting the polyspecific associations among *Saguinus* and *Callimico*, as postulated benefits of callitrichid mixed species groups include joint range and resource defense (Norconk 1986; Garber 1988; Peres 1992).

In this study, I investigated the ranging patterns of *C. goeldii* (callimico) in mixed species groups with two sympatric tamarins, *S. labiatus*, the red-bellied tamarin, and *S. fuscicollis*, the saddle-backed tamarin, at a field site in northwestern Amazonian Brazil. The analyses included spatiotemporal variation in use of the range, the effects of feeding and resting sites on range use, and the frequency and context of intergroup encounters and how behaviors possibly related to territoriality or resource defense might affect ranging. I addressed two main questions: (1) What were the ranging patterns, in terms of home range size and intergroup range overlap, of callimico compared with those of sympatric, associated tamarins? (2) How was range use by callimico related to the location of key resources and possible territorial behaviors, such as patrolling range boundaries and intergroup encounters?

13.2 Methods

13.2.1 Study Site and Study Group

Data were collected periodically between April 1999 and August 2003 at the Fazenda Experimental Catuaba (FEC), an 820 ha reserve managed by the Universidade Federal do Acre (UFAC). The FEC (S10°04, W067°36) is located about 25 km from Rio Branco, the capital city of the state of Acre in northwestern Brazil (Fig. 13.1). This region of Brazil receives an average of approximately 2,000 mm of rain annually, forty percent of which falls during the wet season from January to March (IMAC 1991). There is also a marked dry season from June to August (IMAC 1991). The site is *terra firme* forest of various stages of succession, including primary and secondary

Fig. 13.1 Location of the field site, Fazenda Experimental Catuaba, approximately 25 km east of Rio Branco, the capital city of Acre, Brazil

forest, and bamboo (Rehg 2003). Primates at the FEC include *C. goeldii*, *S. labiatus*, *S. fuscicollis*, *Saimiri sciureus*, *Cebus albifrons*, *C. apella*, *Callicebus cupreus*, *Pithecia irrorata*, and *Aotus nigriceps*. At the FEC, the estimated density of *Callimico* is 1.2 groups/km^2; the estimated density of *Saguinus* (each species) is 2 groups/km^2 (Rehg 2003). A more detailed description of the study site is given elsewhere (Rehg 2003).

Data collection focused on one main study group of *Callimico* in a mixed species group with *S. labiatus* and *S. fuscicollis* (collectively known as Stream Group) from September 1999 to March 2000, December 2000, and June through August 2002 (hereafter referred to as the main study period). The callimicos associated with the same mixed tamarin group throughout the study and were not known to associate with other groups; all three species were associated an average of 61% of observation time (Rehg 2006a). The group size for the three constituent species of Stream Group changed throughout the course of the study because of the birth of infants as well as disappearances (from death or emigration). Group sizes fluctuated between 7 and 9 individuals for *C. goeldii*, 6 and 7 individuals for *S. fuscicollis*, and 4 and 11 individuals for *S. labiatus* (Rehg 2003). Additional data were collected on other groups of callitrichids opportunistically, and included isolated sightings as well as follows of several hours.

13.2.2 Data Collection

Follows of Stream Group ranged from a few minutes to over 10 consecutive hours (mean = 4.3 h, SD = 2.7 h, n = 97). Nearly 420 h of data were collected on the main study group (n = 4,581 samples). During follows, an instantaneous scan sampling method was used to collect behavioral data (Altmann 1974; Lehner 1979). Scans began every 5 min and included a duration of 2 min of observation. During scans, data were collected on association status and the behavior of sampled individuals. Other data, including observations of social interactions and observations of intergroup encounters, were collected *ad libitum*.

Among the data collected during scans was the location of the group in the forest based on a mapped trail system within a 330 ha region of the study site. These trails were mapped using a Brunton transit compass and a tape measure, and synthetic, colored flags were attached to trees with trail identification markers at mapped points (typically intervals of between 4 and 20 m, depending on visibility). The analyses and results that follow focus on these locational data.

13.2.3 Map Construction

All the mapped data obtained from the field were entered into Autosketch (vers. 6.0), an autocad-style computer program, to generate a map of the site. The home range of Stream Group callimicos and tamarins was reconstructed using the minimal convex polygon method. This same method was used to measure partial ranges of other callitrichid groups. Range overlap among neighboring groups was calculated by measuring the area formed by the intersecting range polygons of groups.

13.2.4 Data Analyses

To evaluate spatiotemporal variation in ranging, a 50×50 m grid was superimposed on the map of the home range, and the use of each 2,500 m^2 quadrat was determined by the number of scan samples (to represent time) that *Callimico* was located in the quadrat. Four categories were designated to quantify intensity of use: lowest (1% or less total time spent in that quadrat); moderate (1–2% total time spent in that quadrat); highest (greater than 2% total time spent in that quadrat); and visited, but use unquantified (this category represents use of areas without sufficient mapped trails to quantify use).

To evaluate the effect of resource distribution on ranging patterns, an analysis of covariance was performed with quadrat use as the dependent variable; independent variables included the number of callimico plant feeding sites, the number of callimico fungus feeding sites, the presence/absence of day-time resting sites, and the presence/absence of treefalls in each quadrat. These variables were chosen for analysis because of their relatively permanent spatial locations and their hypothesized importance to *Callimico*. Plant and fungus feeding sites were stationary food resources with discrete locations (unlike animal prey); fruit and fungus feeding together composed the majority of the callimicos' diet throughout the year (Rehg 2003). Particular day-time resting sites were observed to be repeatedly visited. Although previous analyses indicated that different habitat types throughout the home range were used by *Callimico* in proportion to their availability (Rehg 2006b), treefalls appeared to be focal points for certain activities, specifically fungus feeding and resting. Data were analyzed without regard to association status because of small sample sizes for *Callimico* alone and in certain association combinations; therefore, the possible effects of associations on ranging are not identified here.

To evaluate whether *Callimico* exhibited evidence of territorial behavior, I examined ranging with regard to possible evidence for patrolling of range boundaries and regions of range overlap, using a method similar to that of Garber et al. (1993). The range of Stream Group callimicos (and tamarins) was subdivided into three regions: (1) periphery – the area located <50 m from the range boundary; (2) interior – the area located >50 m from the boundary; and (3) areas of intergroup range overlap – parts of the range that are known to have been visited or used by neighboring groups of *Callimico* and *Saguinus* (Fig. 13.2). The proportion of the range area composing each region was used as the criterion of expected use (Table 13.1), and the proportion of scan samples in which *Callimico* was observed in these regions was calculated as observed use. This criterion of expected use assumes that if there are no preferences for certain regions, the proportion of time spent in each region will be proportional to its area.

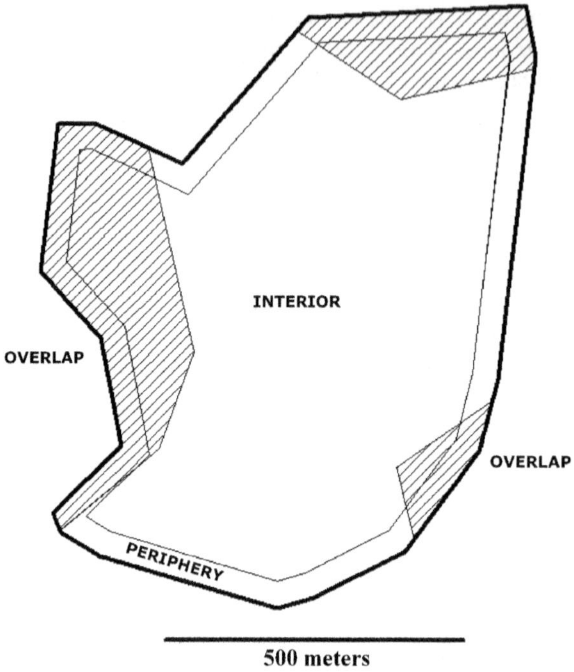

Fig. 13.2 Stream Group's range was subdivided into three "regions": (1) periphery: <50 m from range boundary; (2) interior: >50 m from boundary; (3) areas of overlap

Table 13.1 Proportion of observations of Stream Group *Callimico* in different regions of its home range

Zones	*Callimico goeldii*	Proportion of range area
Interior	0.85	0.65
Periphery	0.07	0.11
Overlap	0.08	0.24

Bootstrapping, a resampling method, was used to test the null hypothesis that *the callimicos* used their range without regard to location relative to the range boundary (i.e., region). Resampling methods allow the probability of results from the data to be evaluated without making assumptions about the distribution of the population, and when random sampling has been violated (Crowley 1992; Edgington 1980; Potvin and Roff 1993), as would be the case with successive 5-min scan samples. In this study, results using the resampling method should be interpreted as being more conservative than those from traditional statistics; these methods are less likely to result in incorrect rejection of null hypotheses, given the failure of the data to meet assumptions for standard statistical tests. The bootstrapping program used in this study was written and executed in SAS (Statistical Analysis System, Carey, North Carolina), and the criterion for statistical significance is equivalent to an alpha level of 0.05 using standard statistical analyses. A more detailed explanation of this method is provided in Rehg (2003, 2006a).

I also examined the frequency and context of intergroup encounters involving Stream Group callimicos. I calculated the expected rates of intergroup encounters (c.f. Waser 1987), with estimates of population density, group spread, and daily path length for groups at the study site from Rehg (2003). This method (Waser 1987), modeled on Brownian motion, assumes that movement by groups is random within the study area. This assumption is incorrect, in that we know animals' movements are affected by many factors, such as resource availability, but this method allows a baseline minimum of interactions to be estimated. This measure has been used in previous studies of callitrichids to provide a null hypothesis of the encounter rate that should occur by chance (Peres 1989, 1992; Garber et al. 1993).

13.3 Results

13.3.1 Home Range Size and Overlap

The range of Stream Group (tamarins and callimicos) through the 2000 study period was 42 ha, increasing in the 2002 period to 48 ha, and in the 2003 study period the range size for the tamarins increased to 56 ha and for *Callimico* to 59 ha (Fig. 13.3). As the estimated range area increased, the boundaries of the range generally were

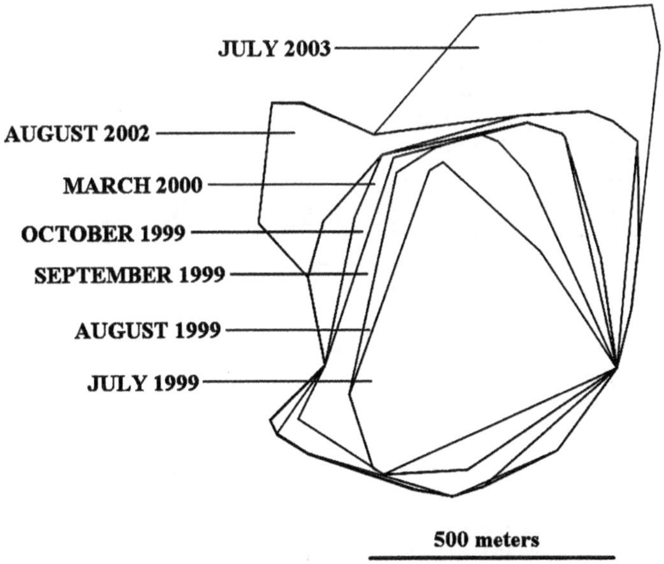

Fig. 13.3 Changes in the home range of Stream Group *Callimico* over the course of the study

extended all around its exterior, except where limited on the western edge by an area of treeless pasture. Observed increases in the home range size over this time probably do not represent a real expansion in the extent of the range, but likely reflect increased information on range use with more observation hours.

The home ranges of Stream Group *Callimico* and *Saguinus* were identical in size and location, except the additional 3 ha used by *Callimico*, which was visited on a single afternoon by the callimicos after separating from the tamarins. The shared range of Stream Group callimicos and tamarins overlapped the ranges of at least three tamarin mixed groups, totaling 14.7 ha (26% of the tamarins' range area), and at least one other group of *Callimico*, amounting to 1.7 ha (3% of the callimicos' range area). Partial home ranges of neighboring groups of tamarins and callimicos varied from 11 to 37 ha, and are based on multiple sightings and short, opportunistic follows; the complete extent of these ranges is not known.

13.3.2 Ranging and Resources

Nearly all the parts of the Stream Group callimicos' home range were used during the observation periods (Fig. 13.4), and quadrats of more intense use were scattered

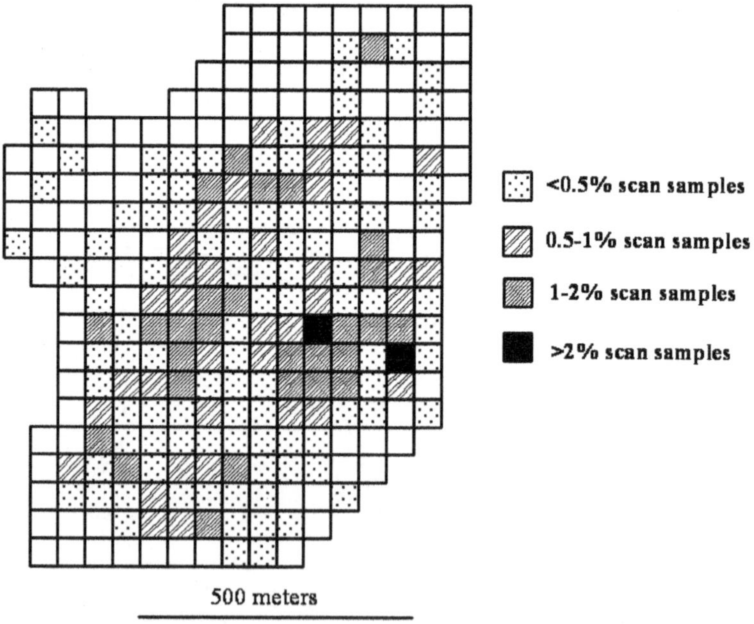

Fig. 13.4 Four categories were designated to quantify intensity of use: lowest (1% or less 1% of the total samples observed in the quadrat); moderate (>1%–2% of the total samples observed in the quadrat); highest (>2% total samples observed in the quadrat); and visited, but use unquantified due to lack of sufficient mapped trails at the time of use

throughout the range. Based on a visual inspection of this pattern, range use was not concentrated at the periphery; rather, intensely used quadrats were located throughout the more interior areas.

Stream Group callimicos were observed feeding at 71 different individual plant resources (e.g., fruit trees, lianas), and 16 different fungus sites. Twenty-one different day-time resting areas were observed to be used. Both feeding and resting sites were located in all parts of the range. Based on the ANCOVA, the number of plant feeding sites, the number of fungus feeding sites, and the presence/absence of a resting site used by *Callimico*, as well as the presence/absence of a treefall in a quadrat, all positively influenced the total observed time in quadrats by *Callimico*. As feeding and resting were substantial time-consuming activities of *Callimico* (13% and 33%: Rehg 2003), and as the criterion for a resting site was partly time-based, these variables are not independent of the time-based measure of quadrat use.

13.3.3 Ranging and Monitoring Boundaries

Callimico spent less time in overlap and peripheral regions, and more time in the interior than expected. Analyses of bootstrapped data sets comparing the expected versus the observed use of regions rejected the null hypothesis that Stream Group callimicos used their range randomly with respect to regions.

13.3.4 Intergroup Encounters

During the main study period, five intergroup encounters were observed between Stream Group callimicos and at least one other callimico group (mean duration 39 min, range 16–75 min). Based on the proximity of the encounters (all occurred within 500 m of each other), it is possible that all were with a single neighboring group. Three encounters occurred during the wet season (October and November 1999), and two occurred during the dry season of 2002 (June and July 2002). No encounters were observed in the dry season of 2003. All encounters occurred between 7:00 and 11:00 h. The encounters were characterized by agonistic vocalizations, chasing, and brief aggressive contact. At times during encounters, individuals of both groups engaged in other activities (resting, foraging, autogrooming) while still in proximity.

As the animals were not marked, it is not possible to know if all the individuals in the group participated in the encounters. However, at the time of the first three encounters there was a dependent infant callimico in Stream Group, and at no time during the encounter was the infant, or an adult carrying the infant, seen participating in encounters. All encounters were located within 100 m of a fruit feeding site visited on the day of the encounter. However, most feeding sites used that day or within 5 days before and after an encounter were beyond 100 m of the encounter sites.

It is not known if the neighboring callimico group ever visited the feeding sites closest in proximity to the intergroup encounters.

In all the cases, Stream Group tamarins were with *Callimico* prior to the encounters and remained nearby, but did not participate. During the second encounter, the tamarins were resting about 30 m away. No tamarin "face-to-face" encounters were observed, although in two encounters a neighboring group of *S. labiatus* and *S. fuscicollis* accompanied the neighboring callimicos; Stream Group tamarins and the neighboring tamarins must have been aware of each other.

Based on Waser's (1987) method, encounters among groups of callimicos should occur approximately once every 600–1,200 h. These rates are very low, thus observed rates were greater than expected, assuming completely random movement. Encounters among groups of tamarins are expected to occur once every 400–800 h; while no face-to-face tamarin encounters were observed, the two occasions on which Stream Group tamarins and the neighboring mixed tamarin group were in proximity could perhaps be considered encounters in which tolerance was exhibited, since no interaction occurred.

13.4 Discussion

13.4.1 Home Range Size and Overlap

The range size for Stream Group callimicos at the FEC is smaller than the 118 ha and 150 ha areas used by groups studied in Bolivia (Porter 2000; Porter and Garber 2007), but equivalent to an estimated 50–60 ha range suggested in a briefer study at another Bolivian site (Pook and Pook 1981). The area used by Stream Group *S. labiatus* and *S. fuscicollis* is similar to that used by mixed groups of these species at other sites (Buchanan-Smith 1990; Porter 2000). Stream Group *Callimico* and *Saguinus* were in association most of the time (Rehg 2006a), so it is not surprising that their home ranges were essentially identical. Although the estimated home range size for *Callimico* has only been reported for a few groups at equally few sites, the degree of variation for *Callimico* (about 50–150 ha), while notable, is less than that observed intraspecifically for certain *Saguinus* (Peres 1992; Garber et al. 1993; Norconk 1986), and similar to the proportion of intraspecific variation for certain marmosets (Hubrecht 1985; Albernaz and Magnusson 1999; Lazaro-Perea 2001). Additional research is necessary to understand the specific factors influencing the intersite range size variation in *Callimico*, although the existence of seasonally important habitat (bamboo and stream edges) and availability of critical dry season fungus resources may be key variables (Porter 2000). While Vasarhelyi (2000) has suggested that the monotypic *Callimico* may represent more than one distinct population (subspecies or species), the Bolivian sites in the Pando are within approximately 300 km of the FEC in Acre, and sites in the two countries are separated by a single river, the Rio Acre; it is possible that Acrean and Bolivian groups are of a single metapopulation.

Range overlap among groups of *Callimico* has not previously been explicitly reported. Researchers have suggested that low population densities may be associated with callimico groups being separated by up to several kilometers (Izawa 1979; Pook and Pook 1981); however, Porter et al. (2001) reported several intergroup encounters involving a main study group of callimicos, implying some overlap, or at least contiguous ranges. The few hectares of home range that groups of callimico at the FEC were known to share represent a very small proportion of the range (less than 3%). As reported, overlap in tamarins and marmosets is highly variable (none to more than 80%), range overlap among Stream Group tamarins and neighboring tamarin groups is well within this range.

13.4.2 Spatiotemporal Variation in Ranging

Time spent in different quadrats showed that Stream Group *Callimico* used its range in a spatially uneven manner, and use was significantly affected by the location of key resources. Based on an ANCOVA, plant and fungus feeding sites were significantly related to the use of different quadrats throughout the study. The location of daytime resting sites was also positively related to the use of quadrats, which is not surprising as resting made up nearly a third of the callimicos' activity budget, and a number of resting areas were repeatedly visited. These variables are expected to covary with occupation of quadrats because these activities require time, and occupy a substantial proportion of the activity budget. Evaluating the effect of the distribution of potential food resources throughout the group's home range (i.e., fungus-producing locations or fruit trees of utilized species), and not just specific feeding sites, will require additional sampling efforts to identify and map these resources throughout the site.

The presence of one or more tree falls was also significantly related to the time spent in a quadrat. As all tree falls that were opportunistically encountered throughout the home range ($n=48$) were recorded, this variable does not suffer from the same bias as the feeding and resting sites. Although there was no systematic attempt to locate all tree falls, I moved through most parts of the range sufficiently to encounter most of the tree falls in this area. As tree falls were preferred resting spots and the primary source of dietary fungus for *Callimico* (Rehg 2003), tree falls, fungus feeding sites, and resting sites are presumably related variables from a callimico perspective.

Range use by tamarins and marmosets is typically reported as unevenly distributed spatially, based on resources in a way similar to that observed for callimico. (Soini 1982; Terborgh 1983; Hubrecht 1985; Crandlemire-Sacco 1986; Rylands 1986; Garber 1988; Garber et al. 1993). For example, Crandlemire-Sacco (1986) reported that the ranging of a main study group appeared to follow the distribution of groves of *Salacia*, and resting sites were frequently chosen based on proximity to these trees.

Other analyses indicated that *Callimico* was observed in habitat types in proportion to habitat availability (Rehg 2006b), suggesting that habitat preferences generally did not influence ranging patterns. The heterogeneity of habitats in Stream Group's home range may help explain this finding. For example, of 215 quadrats within the home range of Stream Group through 2002 (for which data on habitat availability were collected), 81% included more than one type of habitat. Thus, even remaining in a single quadrat, or a 2,500 m^2 area, the monkeys would frequently be able to encounter different types of habitats.

Use of the home range based on proximity to the range boundaries by "regions" indicated a preference for the range interior by *Callimico*. While areas of most intense use were patchily distributed (not contiguous), they tended to be located in the interior of the home range (Rehg 2003). An analysis of the location of the feeding sites demonstrated that more sites were found in the interior of the home range; as a result, more information is needed on the location of potential food resources to determine whether the callimicos preferred to remain in the interior because it was interior, or were simply using their range based on resource availability.

Nevertheless, this pattern suggests that the callimicos (and tamarins) were not patrolling the range boundaries. As the three species of this polyspecific group remained together for the majority of the activity periods in which they were observed, there was no evidence of interspecific differences in ranging with regard to range defense (Rehg 2003). This does not necessarily indicate a lack of range defense – it may be that this ranging behavior was sufficient to allow monitoring for intrusions by other groups without spending more time along the boundaries (c.f. Terborgh 1983). Alternatively, the observed ranging pattern could indicate avoidance of range periphery and neighboring groups, or that range defense generally was not a concern for the group. It has been suggested that tamarin groups might benefit from remaining in a frequently used core area where the location of resources is well-known and foraging presumably more efficient, instead of searching for resources in less-familiar, peripheral areas (Terborgh 1983). More data on the synchronous ranging patterns of multiple callitrichid groups are needed to distinguish between alternative hypotheses.

Only five encounters among Stream Group callimicos and one (or possibly two) neighboring groups of callimicos were witnessed, and no interactive confrontations between groups of *S. labiatus* or *S. fuscicollis* were observed, despite proximity on two occasions to a neighboring tamarin mixed group. Compared to field studies of other tamarins in which intergroup encounters were frequent (Norconk 1986; Garber 1988; Peres 1992), the paucity of observed encounters in *Saguinus* is atypical (although see Lopes and Ferrari 1994).

The infrequency of intergroup encounters involving Stream Group suggests that neighboring groups do not often encounter each other in the areas of range overlap or at the range boundaries at the FEC. Mathematical estimates of the expected rates of intergroup encounters involving Stream Group suggest that encounters should be relatively infrequent, assuming the groups use their ranges without respect to the location of neighbors. In this sense, the lack of observed intergroup encounters among the tamarins and the few among the callimicos would support the interpretation that Stream Group was not directing range use to encounter neighboring groups at the FEC.

Although in all cases, encounters were located within 100 m of a plant food resource visited on the day of the encounter, the encounters were not necessarily directly related to defending food resources. The majority of plant feeding sites used on the days of the encounters and within 5 days before or after each encounter were greater than 100 m from encounter locations. As noted previously, the location of the plant feeding sites was a primary factor affecting range use by Stream Group; thus, it is not surprising that groups may encounter one another near the sites of food resources. The low rate of encounters, and the lack of participation by the tamarins in encounters with the callimicos, suggest that a primary benefit of the nearly daily mixed species groups was not joint defense of specific feeding sites.

At the FEC, as reported in other studies (Porter et al. 2001), agonism among callimico groups occurred during the encounters. Given the fact that the individuals were not identifiable, the breeding status of the individuals in the group was unknown, and no sexual interactions were observed during the intergroup encounters, it cannot be determined if the encounters related in part to mate defense, although that is certainly possible. It is also possible that the encounters may allow individuals to monitor the structure of other groups to evaluate dispersal and breeding opportunities, or serve to reinforce group solidarity, as has been suggested previously for callitrichids (Hubrecht 1985; Garber et al. 1993; Lazaro-Perea et al. 1999; Lazaro-Perea 2001). A lack of data on the group compositions and intergroup relationships of Stream Group and the neighboring groups does not allow these hypotheses to be addressed at this time. Future research on the callimico population at the FEC and resource availability will allow the question of the importance of demographic, social, and ecological factors to be better understood.

Acknowledgements I wish to thank the Conselho Nacional de Desenvolvimento Científico e Tecnológico (CNPq) and the Universidade Federal do Acre (UFAC) for granting me permission to conduct the research and to work at the Fazenda Experimental Catuaba. My thanks are due to all the researchers, administrators, and technicians of the Universidade Federal do Acre and the Herbário do Parque Zoobotânico, and in particular, Marco Antônio de Oliveira, Armando Muniz Calouro, and Maria Aparecida de Oliveira Azevedo Lopes for their help. Thanks also to Steve Leigh, Paul Garber, Stan Ambrose, and Marina Cords. This research was funded in part by the Graduate College of the University of Illinois.

References

Albernaz AL, Magnusson WE (1999) Home-range size of the bare-ear marmoset (*Callithrix argentata*) at Alter do Chão, Central Amazonia, Brazil. Int J Primatol 20(5):665–677

Altmann J (1974) Observational study of behavior: sampling methods. Behaviour 49:227–267

Azevedo-Lopes MAO, Rehg JA (2003) Observations of *Callimico goeldii* with *Saguinus imperator* in the Serra do Divisor National Park, Acre, Brazil. Neotrop Primates 11(3): 181–183

Barry B (2002) A pilot study to assess the benefits for saddleback tamarin (*Saguinus fuscicollis*) when in polyspecific association with Goeldi's monkey (*Callimico goeldii*) and emperor tamarin (*Saguinus imperator*) in the Manu Region, south-eastern Peru. Masters Thesis, Oxford Brookes University, England

Buchanan-Smith H (1990) Polyspecific association of two tamarin species, *Saguinus labiatus* and *Saguinus fuscicollis*, in Bolivia. Am J Primatol 22:205–214
Buchanan-Smith H (1991) Field observation of Goeldi's monkey, *Callimico goeldii*, in Northern Bolivia. Folia Primatol 57:102–105
Chapman CA (1988) Patterns of foraging and range use by three species of neotropical primates. Primates 29:177–194
Cheney D (1987) Interactions and relationships between groups. In: Smuts BB, Cheney DL, Seyfarth FM, Wrangham RW, Struhsaker TT (eds) Primate Societies. University of Chicago Press, Chicago, pp 267–281
Christen A (1999) Survey of Goeldi's monkeys (*Callimico goeldii*) in northern Bolivia. Folia Primatol 70:107–111
Crandlemire-Sacco J (1986) The ecology of the saddle-backed tamarin, *Saguinus fuscicollis*, of southeastern Peru. PhD Thesis. University of Pittsburgh, Pittsburgh
Crowley PH (1992) Resampling methods for computation-intensive data analysis in ecology and evolution. Annu Rev Ecol Syst 23:405–447
Dawson GA (1979) The use of time and space by the Panamanian tamarin, *Saguinus oedipus*. Folia Primatol 31:253–284
de la Torre S, Campos F, De Vries T (1995) Home range and birth seasonality of *Saguinus nigricollis graellsi* in Ecuadorian Amazonia. Am J Primatol 37:39–56
Dettling AC (2002) Reproduction and development in Goeldi's monkey (*Callimico goeldii*). Evol Anthropol Suppl 1:207–210
Dietz JM, Peres CA, Pinder L (1997) Foraging ecology and use of space in wild golden lion tamarins (*Leontopithecus rosalia*). Am J Primatol 41:289–305
Digby LJ, Barreto CE (1993) Social organization in a wild population of *Callithrix jacchus*. Folia Primatol 61:123–134
Edgington ES (1980) Randomization tests. Marcel Dekker, New York
Ferrari SF, Lopes Ferrari MA (1989) A re-evaluation of the social organisation of the Callitrichidae, with reference to the ecological differences between genera. Folia Primatol 52:132–147
Ferrari SF, Corrêa HKM, Coutinho PEG (1996) Ecology of the "southern" marmosets (*Callithrix aurita* and *Callithrix flaviceps*): how different, how similar? In: Norconk MA, Rosenberger AL, Garber PA (eds) Adaptive Radiation of Neotropical Primates. Plenum, New York, pp 157–171
Ferrari SF, Iwanaga S, Ramos EM, Messias MR, Ramos PCS, Cruz Neto EH (1999) Expansion of the known distribution of Goeldi's monkey (*Callimico goeldii*) in south-western Brazilian Amazonia. Folia Primatol 70:112–116
Garber PA (1988) Diet, foraging patterns, and resource defense in a mixed species troop of *Saguinus mystax* and *Saguinus fuscicollis* in Amazonian Peru. Behaviour 105:18–33
Garber PA, Leigh SR (2001) Patterns of positional behavior in mixed-species troops of *Callimico goeldii*, *Saguinus labiatus*, and *Saguinus fuscicollis* in northwestern Brazil. Am J Primatol 54:17–31
Garber PA, Pruetz JD, Isaacson J (1993) Patterns of range use, range defense, and intergroup spacing in moustached tamarin monkeys (*Saguinus mystax*). Primates 34(1):11–25
Hanson AM (2000) Habitat use in relation to diet, with particular emphasis on mycophagy, by *Callimico goeldii* in Pando, Bolivia. Unpublished Masters Thesis. State University of New York, Stony Brook
Heltne PG, Wojcik JF, Pook AG (1981) Goeldi's monkey, genus *Callimico*. In: Coimbro-Filho AF, Mittermeier RA (eds) Ecology and behavior of neotropical primates. Academia Brasileira de Ciencias, Rio de Janeiro
Hershkovitz P (1977) Living New World monkeys (Platyrrhini), vol. 1. The University of Chicago Press, Chicago
Hill WCO (1959) The anatomy of *Callimico goeldii* (Thomas). Trans Am Phil Soc 49(5):1–116
Hubrecht RC (1985) Home-range size and use and territorial behavior in the common marmoset, *Callithrix jacchus jacchus*, at the Tapacura Field Station, Recife, Brazil. Int J Primatol 6(5):533–550
IMAC (1991) Núcleo de Cartografia (Rio Branco, AC). Rio Branco, AC: Atlas Geográfico Ambiental do Acre

IUCN (2004) 2004 IUCN Red List of threatened species. <http://www.iucnredlist.org>. Downloaded on 10 May 2005

Izawa K (1978) A field study of the ecology and behavior of the black-mantle tamarin (*Saguinus nigricollis*). Primates 19:241–274

Izawa K (1979) Studies on peculiar distribution pattern of *Callimico*. Kyoto University Overseas Research Reports: 1–19

Lazaro-Perea C (2001) Intergroup interactions in wild common marmosets, *Callithrix jacchus*: territorial defence and assessment of neighbors. Anim Behav 62:11–21

Lazaro-Perea C, Snowdon C, de Fátima Arruda M (1999) Scent marking in wild groups of common marmosets (*Callithrix jacchus*). Behav Ecol Sociobiol 46:313–324

Lehner PN (1979) Handbook of ethological methods. Garland STPM, New York

Lopes MA, Ferrari SF (1994) Foraging behavior of a tamarin group (*Saguinus fuscicollis weddelli*) and interactions with marmosets (*Callithrix emiliae*). Int J Primatol 15(3):373–387

Mendes-Pontes AR, Monteiro da Cruz MAO (1995) Home range, intergroup transfers, and reproductive status of common marmosets *Callithrix jacchus* in a forest fragment in north-eastern Brazil. Primates 36(3):335–347

Mitani JC, Rodman PS (1979) Territoriality: the relation of ranging pattern and home range size to defendability with an analysis of territoriality among primate species. Behav Ecol Sociobiol 5:241–251

Norconk MA. (1986) Interactions between primate species in a Neotropical forest: mixed-species troops of *Saguinus mystax* and *Saguinus fuscicollis* (Callitrichidae). Unpublished PhD Thesis. University of California, Los Angeles

Peres CA (1989) Costs and benefits of territorial defense in wild golden lion tamarins, *Leontopithecus rosalia*. Behav Ecol Sociobiol 25:227–233

Peres CA (1992) Consequences of joint territoriality in a mixed-species group of tamarin monkeys. Behaviour 123:220–246

Peres CA (2000) Territorial defense and the ecology of group movements in small-bodied neotropical primates. In: Boinski S, Garber PA (eds) On the move. University of Chicago Press, Chicago, pp 100–123

Pook AG, Pook G (1981) A field study of the socio-ecology of the Goeldi's monkey (*Callimico goeldii*) in Northern Bolivia. Folia Primatol 35:288–312

Porter LM (2000) The behavior and ecology of the Goeldi's monkey (*Callimico goeldii*) in northern Bolivia. PhD Thesis. State University of New York, Stony Brook

Porter LM, Garber PA (2007) Niche expansion of a cryptic primate, *Callimico goeldii*, while in mixed species troops. Am J Primatol 69:1340–1353

Porter LM, Hanson AM, Becerra EN (2001) Group demographics and dispersal in a wild group of Goeldi's monkeys (*Callimico goeldii*). Folia Primatol 72(2):108–110

Potvin C, Roff DA (1993) Distribution-free and robust statistical methods: viable alternatives to parametric statistics? Ecology 74(6):1617–1628

Rehg, JA (2003) Polyspecific associations of *Callimico goeldii*, *Saguinus labiatus*, and *Saguinus fuscicollis* in Acre, Brazil. PhD Thesis. University of Illinois, Urbana-Champaign

Rehg JA (2006a) Seasonal variation in polyspecific associations among *Callimico goeldii*, *Saguinus labiatus*, and *S. fuscicollis* in Acre, Brazil. Int J Primatol 27(5):1399–1428

Rehg JA (2006b) Hábitats utilizados por três espécies de primatas, *Callimico goeldii*, *Saguinus labiatus* e *Saguinus fuscicollis*, na Fazenda Experimental Catuaba, Acre, Brasil. In: Drummond P (ed) Fauna do Acre. Série Estudos e Pesquisas, 2. EDUFAC, Rio Branco, pp 147–172

Rylands AB (1986) Ranging behaviour and habitat preference of a wild marmoset group, *Callithrix humeralifer* (Callitrichidae, Primates). J Zool 210:489–514

Rylands AB (1989) Sympatric Brazilian callitrichids: the black tufted-ear marmoset, *Callithrix kuhli*, and the golden-headed lion tamarin, *Leontopithecus chrysomelas*. J Hum Evol 18:679–695

Scanlon CE, Chalmers NR, Monteiro da Cruz MAO (1989) Home range use and the exploitation of gum in the marmoset *Callithrix jacchus jacchus*. Int J Primatol 10(2):123–136

Soini P (1982) Ecology and population dynamics of the pygmy marmoset, *Cebuella pygmaea*. Folia Primatol 39:1–21

Soini P (1987) Ecology of the saddle-back tamarin *Saguinus fuscicollis illigeri* on the Rio Pacaya, Northeastern Peru. Folia Primatol 49:11–32
Terborgh J (1983) Five New World primates: a study in comparative ecology. Princeton University Press, Princeton
Terborgh J, Stern M (1987) The surreptitious life of the saddle-backed tamarin. Am Sci 75:260–268
Vasarhelyi K (2000) Is *Callimico* monotypic? A reassessment in the light of new data. Dodo 36:20–29
Waser PM (1987) Interactions among primate species. In: Smuts BB, Cheney DL, Seyfarth FM, Wrangham RW, Struhsaker TT (eds) Primate societies. University of Chicago Press, Chicago, pp 210–226

Chapter 14
A Comparative Study of the Kinematics of Trunk-to-Trunk Leaping in *Callimico goeldii*, *Callithrix jacchus*, and *Cebuella pygmaea*

Paul A. Garber, Ana Sallenave, Gregory E. Blomquist, and Gustl Anzenberger

Abstract Callitrichids are characterized by a highly variable positional repertoire, which includes prehensile behaviors on small, flexible supports, and claw-clinging, scansorial travel, and trunk-to-trunk leaping that commonly occurs on large noncompliant supports. In this study, we examine the kinematics of trunk-to-trunk leaping in callimico (*C. goeldii*), the common marmoset (*Callithrix jacchus*), and the pygmy marmoset (*Cebuella pygmaea*). In the wild, each species is reported to exploit vertical trunks during locomotor and feeding behavior. These species differ, however, in body mass and limb proportions. Compared to common marmosets and pygmy marmosets, callimicos are heavier and have relatively longer hindlimbs.

Video sequences of leaping to and from noncompliant vertical poles were collected on captive family groups of callimicos and marmosets housed at the Primate Facility of the Anthropological Institute, University of Zuerich, Switzerland. Data from 438 leaps were analyzed using equations provided by Warren and Crompton (1998a).

During all trunk-to-trunk leaps, *Callimico*, *Callithrix*, and *Cebuella* were characterized by forelimb-first landing. Results indicate that *C. jacchus* failed to leap a distance of 2 m between vertical poles, whereas *Callimico* leaped this distance frequently. The maximum distance leaped by *Cebuella* was 1.4 m. Two patterns of trunk-to-trunk leaping were identified. In both common marmosets and pygmy marmosets, leaps were characterized by a relatively low angled body orientation at takeoff, low takeoff velocity, an in-air trajectory characterized by significant vertical height loss, and relatively high compressive forces during landing. In contrast, callimico's long powerful hindlimbs acted to generate higher velocities at takeoff, an in-air trajectory resulting in a height gain during all but the longest leaps, and a reduction in travel velocity when arriving at the landing substrate. The mechanical cost of transport (*C*) was relatively constant across different leaping distances for callimico, but increased significantly with increased leaping distance in the other two marmoset species. Overall, it appears that these three marmoset taxa are

P.A. Garber (✉)
Department of Anthropology, University of Illinois, 109 Davenport Hall, 607 South Mathews Ave, Urbana, IL, 61801, USA
e-mail: p-garber@uiuc.edu

S.M. Ford et al. (eds.), *The Smallest Anthropoids*, Developments in Primatology: Progress and Prospects, DOI 10.1007/978-1-4419-0293-1_14,

© Springer Science+Business Media, LLC 2009

characterized by different patterns of positional behavior and exploit vertical trunks in different ways.

Resumen Los calitrícidos están caracterizados por un repertorio de posturas altamente variable que incluye tanto conductas prensiles en pequeños soportes flexibles, así como el agarre y ascenso con las garras y el salto de tronco a tronco que comúnmente ocurre sobre soportes grandes y difíciles. En este estudio, examinamos las cinemáticas del salto de tronco en tronco en los calimicos (*Callimico goeldi*), las marmosetas comunes (*Callithrix jacchus*) y las marmosetas pigmeas (*Cebuella pygmaea*). En estado silvestre, se reporta que cada especie utiliza troncos verticales durante las conductas de locomoción y alimentación. Estas especies difieren, sin embargo, en cuanto a la masa corporal y en la proporción de miembros. Comparados con marmosetas comunes y las marmosetas pigmeas, los calimicos son más pesados y tienen miembros traseros relativamente más largos.

Secuencias de video de salto hacia y desde postes verticales difíciles fueron colectadas de grupos familiares cautivos de calimicos y marmosetas resguardados en las instalaciones para primates del Instituto Antropológico de la Universidad de Zurich, Suiza. Los datos de 438 saltos fueron analizados utilizando las ecuaciones proporcionadas por Warren y Crompton (1998a).

Durante todos los saltos de tronco a tronco, *Callimico*, *Callithrix* y *Cebuella* estuvieron caracterizados por aterrizajes con los miembros anteriores. Los resultados indican que *C. jacchus* falló al saltar a una distancia de 2 metros entre los postes verticales, mientras que *C. goeldii* saltó frecuentemente dicha distancia. La distancia máxima saltada por *C. pygmea* fue de 1.4 m. Dos patrones de salto de tronco a tronco fueron identificados. En ambas marmosetas, los saltos fueron caracterizados por una orientación del cuerpo en un ángulo relativamente bajo al despegue, una velocidad baja de despegue y una trayectoria en el aire caracterizada por la pérdida significativa de altura vertical y fuerzas compresivas relativamente altas durante el aterrizaje. En contraste, los poderosos miembros largos traseros de los calimicos actuaron para generar velocidades más altas al despegue, una trayectoria en el aire resultando en una ganancia de altura durante todos los saltos, excepto los más largos, y una reducción en la velocidad de viaje al llegar al sustrato aterrizaje. El costo mecánico de transporte (C) fue relativamente constante a través de las diferentes distancias de saltos para los calimicos, pero aumentó significativamente con el incremento de la distancia de salto para las dos especies de marmosetas. En general, parece que estos tres taxa de marmosetas están caracterizados por diferentes patrones de comportamiento postural y explotan los troncos verticales de maneras distintas.

Resumo Os callitriquíneos são caracterizados por um repertório posicional altamente variável que inclui tanto comportamentos preênseis em suportes flexíveis de pequena dimensão como a locomoção com auxílio das garras em troncos verticais e os saltos tronco-a-tronco que comumente ocorrem em suportes verticais inflexíveis de grande dimensão. Neste estudo, examinamos a cinemática dos saltos tronco-a-tronco em micos-de-Goeldi (*C. goeldii*), micos-estrela (*Callithrix acchus*) e sagüis-leãozinho (*Cebuella pygmaea*). Na natureza, estas espécies exploram troncos verticais durante a locomoção e a alimentação. No entanto, elas apresentam diferenças quanto ao tamanho corporal e as proporções dos membros. Comparados aos micos-estrela e sagüis-leãozinho, os micos-de-Goeldi são mais pesados e têm membros anteriores relativamente maiores.

Seqüências de vídeo de saltos para e a partir de postes verticais inflexíveis foram coletadas em grupos familiares de micos-de-Goeldi e sagüis mantidos em cativeiro no Setor de Primatas do Instituto Antropológico da Universidade de Zurique, Suíça. Dados de 438 saltos foram analisados usando equações fornecidas por Warren e Crompton (1998a).

Durante todos os pulos tronco-a-tronco, *Callimico*, *Callithrix* e *Cebuella* caracterizam-se por aterrisarem primeiro os membros anteriores. Os resultados indicaram que *C. jacchus* não consegue saltar uma distância de 2 metros entre postes verticais, enquanto *Callimico* salta freqüentemente esta distância. A distância máxima saltada por *Cebuella* foi de 1,4 m. Dois padrões de salto tronco-a-tronco foram identificados. Em micos-estrela e sagüis-leãozinho, os saltos foram caracterizados por uma orientação corporal com ângulo relativamente baixo na decolagem, baixa velocidade de decolagem, uma trajetória no ar caracterizada por uma perda de altura vertical significativa e forças compressivas relativamente altas na aterrisagem. Em contraste, os membros posteriores longos e poderosos dos micos-de-Goeldi geraram altas velocidades de decolagem, uma trajetória no ar resultando em ganho de altura durante todos os saltos, exceto os mais longos, e uma redução na velocidade de deslocamento à medida que o indivíduo se aproxima do substrato de aterrissagem. O custo mecânico do transporte (C) nos micos-de-Goeldi foi relativamente constante entre saltos a diferentes distâncias, mas aumentou significativamente com o aumento da distância de salto nas outras duas espécies. Em resumo, parece que estes três táxons são caracterizados por diferentes padrões de comportamento posicional e exploram troncos verticais de maneiras diferentes.

14.1 Introduction

Tamarins, marmosets, lion tamarins, pygmy marmosets, and callimicos are distinct from other New World primates in exhibiting small body mass and the possession of claw-like nails (tegulae) on the ends of all digits excluding the hallux, which bears a flattened nail. Claw-like nails enable callitrichids to exploit resources found on tree trunks such as bark refuging insects, plant exudates, fungi, and small vertebrates, as well as to cling to large vertical supports while scanning the ground for potential prey (Garber 1992; Porter and Garber 2004). In addition, callitrichids are reported to frequently use small diameter supports during travel (climbing and leaping), and when foraging for plant and invertebrate food items located in the periphery of the tree crown (Garber 1994; Garber 1991; Garber and Leigh 2001). In this regard, callitrichids are characterized by an usually varied positional repertoire that includes both prehensile and "claw" dominated locomotor and postural behaviors.

However, even among closely related species "marked differences in patterns of locomotion, substrate preferences, vertical ranging, and locomotor anatomy" are evident (Garber and Leigh 2001, p 27), with several species having undergone independent evolutionary decreases or increases in body mass, changes in limb, hand, and trunk proportions, as well as gut adaptations (Coimbra-Filho and Mittermeier 1978; Garber 1992; Garber et al. 1996; Garber and Leigh 1997; Hamrick 1998; Bicca-Marques 1999; Porter and Garber 2004). For example in the case of tree gouging and exudate feeding in marmosets, it has been proposed "that intense specialization on plant gums evolved locally [independently] in some forms of *Callithrix* and *Cebuella* (Garber et al. 1996, p 90)."

In this study, we compare the kinematics of trunk-to-trunk leaping in *Callimico goeldii*, *Callithrix jacchus*, and *Cebuella pygmaea*. Recent evidence based on nuclear and mitochondrial DNA consistently places *Callimico goeldii* as a sister taxon to the marmoset clade (Horovitz and Meyer 1997; Pastorini et al. 1998; Canavez et al. 1999). *C. goeldii* is reported to be the largest of the three study species (adult body mass 350–535 g: Encarnación and Heymann 1998; Garber and Leigh 2001). It is characterized by highly elongated hindlimbs reduced head-body length, moderately sized forelimbs, and the lowest intermembral index among all callitrichid species (Table 14.1; Garber and Leigh 2001; Davis, 2002). In contrast, *Cebuella* is the smallest species. Adult body mass in *Cebuella* is 100-130 g (Soini 1988). Pygmy marmosets possess relatively short hindlimbs and the highest intermembral index of any marmoset species (Jungers 1985; Davis 2002). Common marmosets are intermediate in body mass and limb proportions (Table 14.1).

In the wild, these three species also differ in the manner in which vertical supports are exploited as foraging and traveling substrates (Tables 14.2a and 14.2b, Soini 1988;

Table 14.1 Mass, body length, and limb proportions

Species[a]	Body mass (gm)	Head body length (gm)	Femur length (mm)	IMI[b]	BRI	HFI
Cebuella pygmaea	115.1	137.9	37.6	82.3	89.9	87.8
Callithrix jacchus	320.2	220.0	56.6	74.5	86.4	80.7
Callimico goeldii	492.0	224.0	73.2	70.6	89.7	74.4

[a] Data for *Cebuella* are from Soini (1988) and Davis (2002). Data for *Callithrix jacchus* are from Araújo et al. (2000), Hershkovitz (1977), and Davis (2002). Data for *Callimico* are from Ford and Davis (1992), Hershkovitz (1977), and Davis (2002)

[b] *IMI* intermembral index, *BRI* brachial index, *HFI* humerofemoral index

Table 14.2a Positional behavior during travel in *Callimico* and *Cebuella*[a]

Species	QWalk/ Qbound	Hop	Climb/ clamber[b]	T-T leap	Leap	Scansorial/ cling	Other	*n*
Cebuella pygmaea	32.3	0.0	5.7	12.9	22.2	23.7	2.9	418
Callimico goeldii	17.0	17.0	5.7	23.0	18.8	15.9	2.3	351

[a] Data for *Cebuella* are from Youlatos (1999) and data for *Callimico* are from Garber and Leigh 2001

[b] Climb/Clamber include vertical ascent/descent involving hand grasp (Youlatos 1999). *T-T leap* trunk-to-trunk leaping. Hop is an exaggerated form of hindlimb dominated bounding that is present in Callimico but is not reported in other marmosets. Leap including all other forms of leaping other than T-T leap.

Table 14.2b Patterns of substrate utilization (support diameter) during travel in *Callimico* and *Cebuella*[a]

Species	Small	Medium	Large	*n*
Cebuella pygmaea	52.1	29.2	17.0	418
Callimico goeldii	17.1	46.7	35.9	525

Data for *Cebuella* are from Youlatos (1999) and data for *Callimico* are from Garber and Leigh 2001

[a] Small: ~2 cm in diameter; Medium: >2 and ~10 cm in diameter, Large: >10 cm in diameter

Stevenson and Rylands 1988; Garber 1992; Porter 2000; Porter and Garber 2004; Youlatos 2005, Chapt. 15 this volume).

In order to better understand how species differences in body mass, anatomy, and feeding ecology affect callitrichid positional behavior, we compared the kinematics of trunk-to-trunk leaping in *Callimico*, *Callithrix*, and *Cebuella* and address the following questions:

1. Are there differences in body orientation, travel speed, leaping performance, and the distance leaped among the three study species?
2. How do factors such as body mass and limb proportions influence the kinematics of trunk-to-trunk leaping in callitrichids?
3. What do species differences in vertical support use tell us about the evolution of trunk foraging adaptations in New World primates?

14.2 Methods

14.2.1 Animals, Research Facility and Research Design

Video data were collected on trunk-to-trunk leaping in family groups of six callimicos, nine common marmosets, and five pygmy marmosets housed at the Primate Facility of the Anthropological Institute, University of Zuerich, Switzerland. All recordings were done in the same outdoor facility measuring 60.3 m^3 (6.5 m × 3.5 m × 2.65 m) that contained natural vegetation and dead tree branches, which served as locomotor pathways. Only one group at a time was present in the outdoor facility during data collection.

A set of two wooden fixed, noncompliant vertical supports were introduced into the enclosure. These supports had diameters measuring 2.5, 6, and 15 cm, and are referred to as small, medium, and large, respectively. These size classes were selected because they are similar to support class sizes used by wild callitrichids during trunk-to-trunk leaping (Youlatos 1999, Chapt. 15 this volume; Porter 2000; Garber and Leigh 2001). Centimeter grids attached to each pole enabled us to accurately record take off and landing heights, changes in the relative position of body segments (hands, head, feet), and to calculate changes in height-gain or height-loss during each leap. Landing and takeoff support diameter and the horizontal distance between vertical trunks were varied systematically. The minimum distance between vertical poles was 0.7 m and the maximum distance was 2 m. The smallest diameter poles flexed minimally when used as landing or takeoff platforms. Individuals of each species that were video-taped leaping, were weighed after the experiments were terminated. Mean adult body mass was 468.7±48.9 g for *Callimico* ($n=4$), 379.4±42.1 g for *Callithrix* ($n=9$), and 166.6±5.7 g for *Cebuella* ($n=3$).

Two digital video cameras (a SONY DCR – THV 900E PAL, 3CCD Progressive Scan, 48X digital Zoom and a SONY DCR-TRV320-, NTSC, 450X digital zoom) were positioned inside the enclosure to document leaping behavior. Camera one was at a fixed distance of 3 m and positioned perpendicular to the direction of the leap. This camera captured both the takeoff and landing phases of each locomotor sequence. The second camera also was positioned perpendicular to the direction of the leap, but at a distance of 1.3 m from one of the vertical trunks. This was done in order to obtain larger close-up images of landings and takeoffs. Standard NTSC film speed (30 frames per second) was used (see Garber et al., 2005) for a more complete description of methods). We only included film sequences in which full leg extension and body angle at takeoff were clearly visible. Mealworms, raisins, fruit, and gum arabic were used to encourage the marmosets to jump between the vertical supports.

14.2.2 Kinematic Analysis

Leaping sequences were captured and edited using either Adobe Premiere 6.0 or iMovie 3.03. Seven points and one angle were measured directly from captured video sequences using QuickImage, a modification of the public domain NIH Image program (developed at the U.S. National Institutes of Health and available on the Internet at http://www.rsb.info.nih.gov/nih-image/) for the Apple Macintosh (QuickImage is available at http://www.usm.maine.edu/~walker/software.html). The points were: head takeoff and landing height, hand takeoff and landing height, right foot takeoff height, left foot takeoff height, and both feet landing height. The angle of body takeoff was measured on the frame closest in time to the full extension of the hindlimbs. This angle was measured by following an imaginary line running from the animal's head down the body midline to a vertex at the animal's feet and up the vertical support. The complement of this angle was used in the calculations below. The height gained or lost in flight was calculated as the difference of head landing and takeoff vertical positions. A minor correction factor was included to account for slight angling of the recorded video. Additionally, frames were counted from takeoff to landing to estimate horizontal velocity (see Garber et al. 2005).

Vertical height change of the specimen's head, h; takeoff angle from horizontal, α (alpha); horizontal leaping distance, s; and body mass, M_b; were used in the calculation of kinetic energy, E_{KE}; potential energy, E_{PE}; mechanical energy (E_{total}); kinetic cost of transport (C_{KE}); mechanical cost of transport, C; and takeoff (v_0) and landing velocity (v_f) for each leap through kinematic equations (see below) similar to those given in Warren and Crompton (1998a). Additional variables are noted as follows: t is calculated flight duration, β (beta) the landing angle, v_x is horizontal velocity remaining constant throughout the leap, and v_y the vertical velocity calculated separately at takeoff and landing.

Initial conditions of the leap are expressed in terms of v_0,

$$v_x = v\phi \cos\alpha$$
$$v_y = v\phi \sin\alpha - gt$$

with $t = 0$ at takeoff, and $g = 9.80665\ \mathrm{m/s^2}$ in all calculations

$$t = \frac{s}{v_x} = \frac{s}{v_\phi \cos\alpha}$$

$$h = v_y t - \frac{1}{2} g t^2 = v_\phi \sin\alpha \left(\frac{s}{v_\phi \cos\alpha} \right) - \frac{1}{2} g \left(\frac{s}{v_\phi \cos\alpha} \right)^2$$

v_0 is then solved for by substitution and simplification.

$$h = s\left(\frac{\sin\alpha}{\cos\alpha} \right) - \frac{gs^2}{2v_\phi^2 \cos^2\alpha}$$

$$h = s\tan\alpha - \frac{gs^2}{2v_\phi^2}\left(1 + \tan^2\alpha\right)$$

$$\left(s\tan\alpha\right) - h = \frac{gs^2}{2v_\phi^2}\left(1 + \tan^2\alpha\right)$$

$$v_\phi^2\left[\left(s\tan\alpha\right) - h\right] = \frac{gs^2}{2}\left(1 + \tan^2\alpha\right)$$

$$v_\phi^2 = \frac{gs^2\left(1 + \tan^2\alpha\right)}{2\left(s\tan\alpha - h\right)}$$
$$v_\phi = \frac{gs^2\left(1 + \tan^2\alpha\right)}{2\left(s\tan\alpha - h\right)}$$

Calculations for β (beta) and v_f are as follows.

$$v_x = v_\phi \cos\alpha$$

$$v_y = v_\phi \sin\alpha - gt = v_\phi \sin\alpha - \frac{gs}{v_\phi \cos\alpha}$$

$$v_f = \sqrt{v_x^2 + v_y^2}$$

$$\beta = \cos^{-1}\frac{v_x}{v_y}$$

The kinetic energy required for each jump was calculated using the following equation and substitution. Given difficulties in determining the identity of each individual leaping, the mean body mass of the leaping animal for each species was used in all calculations.

$$E_{\mathrm{KE}} = \frac{M_{\mathrm{b}}}{2} v_{\phi}^{2}$$

$$E_{\mathrm{KE}} = \frac{M_{\mathrm{b}}}{2} \left(\frac{gs^2 \left(1 + \tan^2 \alpha\right)}{2\left(s \tan \alpha - h\right)} \right)$$

This simplifies to

$$E_{\mathrm{KE}} = \frac{M_{\mathrm{b}} gs^2 \left(1 + \tan^2 \alpha\right)}{4\left(s \tan \alpha - h\right)}$$

Vertical height change and body mass were used to find potential energy change of each leap.

$$E_{\mathrm{PE}} = M_{\mathrm{b}} gh$$

Total mechanical energy of the leap was the sum of E_{KE} and loss in E_{PE}.

Finally, the cost of transport, a mass and distance standardized measure of mechanical energy, was calculated.

$$C = \frac{E_{\mathrm{KE}} - E_{\mathrm{PE}}}{M_{\mathrm{b}} s}$$

Statistical analyses were conducted using Statistica 5.5 (Statsoft Inc. 2000). Because species differed substantially in size, comparisons of leaping kinematics among species should be based primarily on mass-standardized values. In our study C_{ke} and C satisfy this requirement, as the kinetic energy (C_{ke}) or combined kinetic and potential energy (C) of moving 1 kg of mass 1 m horizontally. Due to small sample sizes for particular leaps, nonnormality and heteroscedasticity, nonparametric tests (Kruskal–Wallis test, Spearman rank correlation, Mann–Whitney U test) were run to compare sample distributions. Statistical significance was set at $p<0.001$.

14.3 Results

Unlike trunk-to-trunk leaping in many prosimians, in which the hindlimbs typically contact the landing support in advance of the forelimbs, *Callimico*, *Cebuella*, and *Callithrix* were found to land hand-first during all trunk-to-trunk leaps (Fig. 14.1).

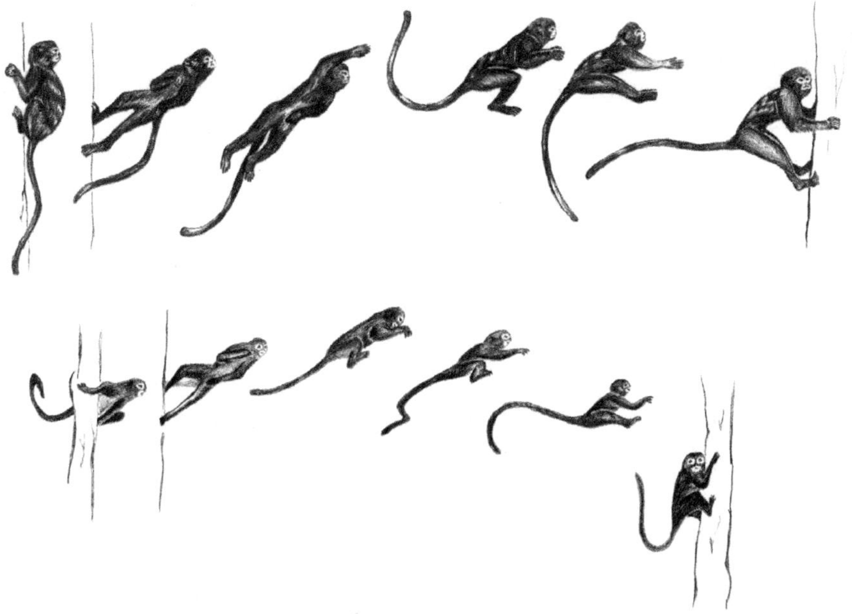

Fig. 14.1 An illustrated comparison of trunk-to-trunk leaping in *Callimico* (*top*) when crossing a distance of 2 m and *Cebuella* (*below*) when crossing a distance of 1 m highlighting differences in leaping trajectories and height gain and loss. The body size differences between *Callimico* and *Cebuella* are not drawn to scale. See text for additional details

Callimico leapt the longest distance (2 m), and could do so repeatedly over the course of a few seconds. In contrast, *Callithrix* would not attempt to leap 2 m, and was cautious and hesitant on leaps of 1.7 m. Finally, the longest leaps in *Cebuella* covered a distance of 1.4 m. We could not coax pygmy marmosets to leap distances of 1.5, 1.7, or 2.0 m. The maximum distance leaped was 7.7 times head-body length in *C. jacchus*, 8.9 times head-body length in *Callimico*, and 10.1 times head-body length in *Cebuella*.

14.3.1 Quantitative Analysis of Trunk-to-Trunk Leaping

Kinematic information based on a total of 438 trunk-to-trunk leaps (122 leaps in *Callimico goeldii*, 180 in *Callithrix jacchus* and 136 in *Cebuella pygmaea*) of between 0.7 and 2 m from vertical supports of small, medium, and large diameter was analyzed. Results indicate that the three study species differed considerably in the trajectory and manner of leaping (Fig. 14.2). Regardless of the distance traversed, both *Callithrix* and *Cebuella*, experienced a significant loss in travel height (Table 14.3). Height lost consistently increased with distance leaped in these two

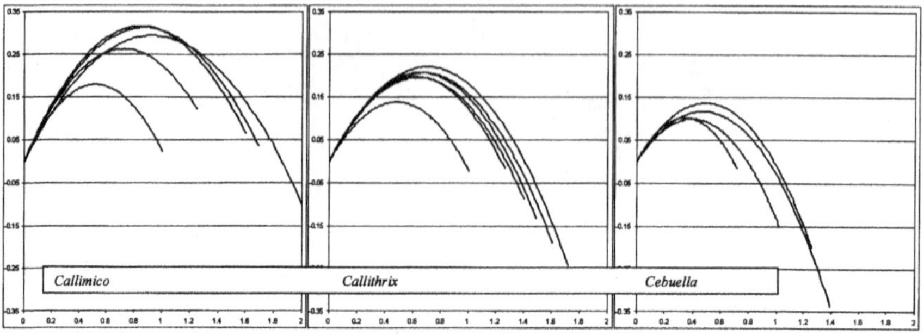

Fig. 14.2 Each panel in this figure shows the leap trajectory produced by the recorded median takeoff velocity ($v_{\varnothing}$) and angle (α) for a single study species (*Callimico goeldii, Callithrix jacchus, Cebuella pygmaea*). Axes are equivalent among panels and record horizontal distances traveled on the *x*-axis and vertical height changes on the *y*-axis. Both are recorded in meters. *Callimico* typically adopts high takeoff angles and velocities except at a distance of 2 m. However, *Callithrix* and *Cebuella* tend to lose height particularly during leaps that approach the maximum they would attempt

genera (*Callithrix*: $R=-0.54$, $p<0.0001$; *Cebuella*: $R=-0.83$, $p<0.0001$). In contrast, *Callimico* was found either to gain height or to maintain height when leaping distances of 1.0–1.7 m, and to lose height only during the longest leaps (2 m). As a result, there was evidence of a weaker overall relationship between height loss and travel distance in Goeldi's monkey ($R=-0.22$, $p=0.004$).

Takeoff and landing angles also were examined for each leap. Takeoff angles reflect the orientation of the body midline relative to the horizontal plane of the leap. These angles are low when the body is nearly horizontal at takeoff. They are high when the body midline is close to the vertical takeoff support. In general, *Callimico* adopted higher take off angles (median: 35°) than did *Callithrix* (median: 32°) and *Cebuella* (median: 28°) (Table 14.3), however, these differences were not significant in leaps of shorter distance such as those of 1.0 m and 1.25 m. Leaping with the body oriented more vertically (higher takeoff angle) requires that more of the available work is put into lifting the animal's mass against gravity and therefore takeoff angle and velocity are expected to be inversely correlated. However, no evidence was found for a significant decrease in takeoff velocity at higher angles for any of the species (*Callimico*: $R=0.02$, $p=0.78$, *Callithrix*: $R=0.08$, $p=0.23$, *Cebuella*: $R=-0.29$, $p=0.0004$). Takeoff angles, however, were positively correlated with changes in vertical height during the leap, with higher angles at takeoff associated with a net gain in vertical height and lower angles associated with a net decrease in vertical height ($R=0.59$, $p<0.0001$). Although in the wild, changes in height during leaping may represent an important locomotor strategy associated with patterns of vertical ranging, habitat preference, and foraging strategies, in the present study height gain or height loss during leaping served no advantage in this regard. Instead, patterns of height gain and height loss were likely to represent species-specific biomechanical constraints associated with generating sufficient takeoff and landing forces required to leap particular distances.

Landing angles (Table 14.3) were also found to differ between species, with *Cebuella* (median 40.4°) and *Callithrix* (median: 39.9°) experiencing the highest landing angles and *Callimico* the lowest landing angles (median: 33.6°). Higher or steeper landing angles corresponded to a loss in height during leaping and lower landing angles corresponded to height increases ($R=-0.79$, $p<0.0001$). Similarly, velocity at landing increased as a function of height loss ($R=-0.57$, $p<0.0001$) due to increased acceleration aided by gravity. This effect was more evident for *Callithrix jacchus* and *Cebuella pygmaea*, as these species adopted lower takeoff angles and took advantage of the downward displacement aided by gravity to arrive at the landing substrate. This occurred, however, at the cost of increased landing velocity and presumably greater compressive forces on the forelimbs during impact. In contrast, *Callimico*, was found to adopt a parabolic trajectory closer to the ballistic optimum of 45° (the most energy efficient posture for leaping; Crompton et al. 1993) and to gain or not lose height during all but the longest leaps (2 m). At comparable leaping distances, *Callimico* arrived at the landing substrate at lower landing velocities than did *Callithrix* and *Cebuella* (Table 14.3).

Finally, a comparison of the kinematics of leaping at different distances (Table 14.3) indicated that both takeoff and landing velocities were strongly influenced by distance leaped (v_0: $R=0.76$, $p<0.0001$; v_f: $R=0.88$, $p<0.0001$). In *Callimico*, takeoff velocity ranged from a median of 3.3 m/s in the shortest leaps to 4.49 m/s in the longest leaps. In *Callithrix*, these values ranged from 3.32 to 3.91 m/s, and in *Cebuella* from 2.78 to 3.40 m/s. As indicated in Table 14.3, when leaping the same distance (i.e., 1.25 m) *Callimico* generated a greater takeoff velocity than did *Callithrix* or *Cebuella*. Similarly, when leaping a horizontal distance of 1.7 m (*Cebuella* did not jump this distance), *Callimico* generated a significantly greater median takeoff velocity than did *Callithrix* (4.25 vs. 3.91 m/s, $p<0.0001$). The ability of *Callimico* to generate higher takeoff velocities compared to both common marmosets and pygmy marmosets at similar leaping distances is likely to reflect the mechanical advantage of hindlimb elongation in this species.

Landing velocity also was significantly correlated with vertical distance traveled ($R=-0.57$, $p<0.0001$), being higher when the animal lost height. A comparison of landing velocity between *Callimico*, *Callithrix*, and *Cebuella* when leaping a distance of 1.25 m indicate significant species differences. Landing velocity was highest in *Cebuella*, which also was characterized by the greatest losses in vertical height. A similar pattern was found when comparing changes in vertical height and landing velocity in *Callithrix* and *Callimico* during leaps of 1.7 m ($U=55$, $p<0.0001$). At this distance, common marmosets experienced a vertical height loss of 22.7 cm and arrived at the landing platform at a speed of 4.46 m/s, whereas *Callimico* experienced a small increase in vertical height (2.7 cm) and contacted the landing substrate at a speed of 4.2 m/s.

Increases in takeoff velocities with distance leaped resulted in increases in total mechanical energy ($R=0.82$, $p<0.0001$) and as a consequence an increase in the cost of transport. The relationship between the cost of transport (C) (a relative measure of the energy required to move 1 kg of body mass a horizontal distance of 1 m) and distance leaped differed among species. This was due primarily to the fact

Table 14.3 Analyses of kinematic data calculated during trunk-to-trunk leaping (distances leaped are 0.7–2 m)[a]

Cebuella pygmaea

	Overall median	0.7 m median	1 m median	1.25 m median	1.4 m median	Kruskal–Wallis	
						H	p
n	136	41	37	32	26		
α (deg)	28.12	30.77	26.89	29.28	26.57	17.65	0.0005
β (deg)	40.48	31.35	38.55	41.31	44.66	44.44	<0. 0001
v_0 (m/s)	3.15	2.78	3.10	3.36	3.40	92.57	<0. 0001
H (m)	–0.154	–0.010	–0.145	–0.216	–0.337	94.68	<0. 0001
v_f (m/s)	3.62	2.85	3.53	3.94	4.30	121.20	<0. 0001
E_{ke} (J)	0.83	0.64	0.80	0.94	0.96	92.57	<0. 0001
E_{pe} (J)	–0.251	–0.016	–0.238	–0.353	–0.551	94.68	<0. 0001
E_{total} (J)	1.09	0.68	1.04	1.29	1.54	121.20	<0. 0001
C_{ke} (m/s^2)	4.68	5.51	4.82	4.50	4.12	67.10	<0. 0001
C (m/s^2)	2.25	0.81	2.25	2.63	3.34	87.98	<0. 0001

Callithrix jacchus

	Overall median	1 m median	1.25 m median	1.4 m median	1.5 m median	1.6 m median	1.7 m median	Kruskal–Wallis	
								H	p
n	180	35	42	32	26	24	21		
α (deg)	31.80	29.81	32.61	31.63	32.50	31.59	32.18	7.45	0.1891
β (deg)	39.93	31.72	33.11	38.06	44.18	45.49	48.38	52.07	<0.0001
v_0 (m/s)	3.72	3.32	3.66	3.74	3.76	3.85	3.91	72.99	<0.0001
H (m)	–0.097	–0.024	–0.012	–0.057	–0.181	–0.214	–0.227	56.74	<0.0001
v_f (m/s)	3.93	3.32	3.69	3.95	4.13	4.32	4.46	142.15	<0.0001
E_{ke} (J)	2.62	2.09	2.54	2.66	2.68	2.81	2.89	72.99	<0.0001
E_{pe} (J)	–0.361	–0.088	–0.045	–0.212	–0.671	–0.795	–0.845	56.74	<0.0001
E_{total} (J)	2.93	2.23	2.58	2.97	3.23	3.54	3.78	142.18	<0.0001
C_{ke} (m/s^2)	4.94	5.51	5.36	5.00	4.72	4.63	4.49	48.23	<0.0001
C (m/s^2)	3.34	2.22	2.61	3.15	3.78	4.09	4.20	84.41	<0.0001

Callimico goeldii							Kruskal–Wallis	
	Overall median	1 m median	1.25 m median	1.6 m median	1.7 m median	2 m median	*H*	*p*
n	122	26	38	12	29	17		
α (deg)	35.37	34.65	35.86	36.72	35.60	32.28	8.69	0.0693
β (deg)	33.68	32.45	28.13	34.35	34.97	36.85	10.07	0.0392
v_0 (m/s)	4.06	3.30	3.87	4.16	4.25	4.49	68.81	<.0001
H (m)	0.041	−0.007	0.112	0.049	0.027	−0.175	17.44	0.0016
v_f (m/s)	3.91	3.38	3.62	4.03	4.20	4.79	99.03	<0.0001
E_{ke} (J)	3.86	2.55	3.51	4.06	4.23	4.72	68.81	<0.0001
E_{pe} (J)	0.186	−0.034	0.515	0.225	0.125	−0.804	17.44	0.0016
E_{total} (J)	3.58	2.67	3.07	3.81	4.14	5.37	99.03	<0.0001
C_{ke} (m/s^2)	5.42	5.44	5.99	5.42	5.31	5.03	14.93	0.0048
C (m/s^2)	5.33	5.70	5.25	5.08	5.19	5.73	13.94	0.0075

[a] α *(deg)* takeoff angle; β *(deg)* landing angle; v_0 *(m/s)* takeoff velocity; v_f *(m/s)* landing velocity; *H(m)* the loss or gain in height during the leap; E_{KE} *(J)* the kinetic energy of leaping; E_{PE} the potential energy change of leap; E_{total} total mechanical energy of leap; C_{KE} *(m/s^2)* kinetic cost of transport; *C(m/s^2)* the mechanical cost of transport. See methods for additional details

that pygmy marmosets and common marmosets lost height during leaping – a loss in potential energy. *C* increased with distance leaped in *Callithrix* and *Cebuella*, while remaining generally constant across leaping distances in *Callimico* (Fig. 14.1, Table 14.3).

14.4 Discussion

In this study we examined the kinematics of trunk-to-trunk leaping in three closely related genera of callitrichid primates. Pygmy marmosets, common marmosets, and callimicos appear to represent a monophyletic marmoset clade (Horovitz and Meyer 1997; Pastorini et al. 1998; Singer et al. 2003) whose members are reported to commonly exploit vertical supports as foraging, feeding, and locomotor substrates (Soini 1988; Stevenson and Rylands 1988; Porter 2000; Garber and Leigh 2001; Garber et al. 2005; Youlatos 2005; Table 14.2a). However, these species differ markedly in the manner in which vertical supports are used. In this regard it is important to distinguish among three patterns of postural and locomotor behavior that traditionally have been lumped into a single positional category termed vertical clinging and leaping (Kinzey et al. 1975). Vertical clinging represents a postural behavior generally associated in callitrichids with exudate feeding and insect and small vertebrate foraging on large diameter vertical substrates (Garber 1992). Many marmoset species are reported to spend much of their day clinging to tree trunks while gouging holes into the bark to stimulate the flow of exudates (Lacher et al. 1984; Coimbra-Filho and Mittermeier 1978; Ramirez et al. 1978; Stevenson and Rylands 1988; Youlatos 2005). Scansorial locomotion is a form of "claw-dominated" vertical ascent and occasionally vertical descent on large trunks. This involves a lateral sequence gait in which the forelimbs act in concert to pull the body forward and then the hindlimbs act in concert to provide a propulsive thrust. Both vertical clinging and scansorial locomotion are facilitated in marmosets and tamarins by directly embedding their claw-like nails into the bark. Trunk-to-trunk leaping represents a different form of positional behavior associated with strong hindlimb propulsion required to propel an individual across gaps in the forest understory. In callitrichids, these leaps principally are reported to begin and end on noncompliant supports (see Youlatos 2005 for a discussion of "terminal/vertical" leaping in *Cebuella*) and are associated with forelimb-first landing (Garber 1991; Garber and Leigh 2001). Other taxa of platyrrhines (i.e., *Pithecia*: Walker 1993) and prosimians (i.e., *Tarsius, Lepilemur*, *Propithecus*, *Avahi*, and *Indri*, see Warren and Crompton 1998b) that are reported to engage in trunk-to-trunk leaping do not possess claw-like nails and principally land hindlimb-first.

There is evidence of considerable variation in the manner and degree to which different callitrichid species exploit large vertical and/or sharply inclined substrates. In the case of *Cebuella*, Kinzey et al. (1975) report that 77% of feeding and 57% of resting occurred while clinging to vertical or sharply inclined trunks and lianas. Data from Youlatos (1999, Chapt. 15 this volume) indicate that in pygmy marmosets, trunk-to-trunk leaping accounted for 13% of positional behavior during travel whereas scansorial locomotion and vertical clinging accounted for 24% (Table 14.2a). Overall, however, quadrupedal walk and quadrupedal bound (32.3%) were the most common forms of travel (Table 14.2a). In Youlatos' (1999) study, the majority of *Cebuella* travel occurred on small supports (52.1%: Youlatos 1999; Table 14.2b). In a more recent study of positional behavior of *Cebuella* inhabiting a dense liana forest, however, Youlatos (2005, p 228) states that "pygmy marmosets use claw climbing, claw clinging, and vertical supports more than any other callitrichid."

In contrast, *Callimico* is reported to restrict much of their travel to medium and large diameter supports (82.6%) and to more frequently engage in trunk- to-trunk leaping (23%) and less frequently engage in scansorial locomotion and vertical clinging postures than *Cebuella* (Porter 2000, Garber and Leigh 2001, Tables 14.2a and 14.2b). Presently, there exist no published quantitative data on substrate preference and positional behavior in wild common marmosets. Based on qualitative accounts, *C. jacchus* is reported to frequently adopt vertical clinging postures and scansorial behavior when exploiting trunk resources (e.g., plant exudates). However, travel in common marmosets is described as principally quadrupedal with most leaping occurring on small diameter supports in the tree canopy (Leslie Digby pers. comm., Maria Adelia Oliveira Monteiro da Cruz pers. comm.). Trunk-to-trunk leaping appears to be a much less common mode of travel in *Callithrix jacchus* than in either *Cebuella* or *Callimico*.

The results of the present study suggest that trunk-to-trunk leaping in callitrichids reflects a compromise between the mechanical demands associated with acceleration at takeoff and compressive forces encountered during landing. Two major patterns of trunk-to-trunk leaping were identified and are described below (Fig. 14.1).

Hindlimb propulsion is a critical component of all leaping. Long hindlimbs extend the duration of time over which the body can be accelerated during takeoff (Crompton et al. 1993, Demes et al. 1995, Terranova 1996). Relative to body size, pygmy marmosets have extremely reduced hindlimbs. Data from Davis (2002) indicate that relative femur length and relative tibia length are shorter in pygmy marmosets than in all other callitrichids except *Callithrix penicillata*. Similarly relative to head-body length, *Callithrix jacchus* is characterized by a short femur whereas *Callimico* has a relatively short body and a long femur (see Table 14.1 for references).

Relatively short hindlimbs in common marmosets and pygmy marmosets resulted in trunk-to-trunk leaps characterized by a lower angle and lower velocity at takeoff, significant vertical height loss, a considerable loss of potential energy, and relatively high compressive forces during landing compared to callimicos. In *Cebuella*, this pattern is evident even during leaps of 1 m. In *Callithrix*, the pattern is most evident during leaps greater than or equal to 1.4 m in horizontal distance. In contrast, long powerful hindlimbs in *Callimico* act to generate high takeoff velocity resulting in minimal height loss and a relatively low landing velocity. It is only during leaps of 2 m (and presumably greater) that *Callimico* adopts a leaping trajectory similar to *Callithrix* and *Cebuella* (decrease in takeoff angle, increase in landing angle, decrease in vertical height, increase in landing velocity, decrease in the mechanical cost of transport) resulting in gravity-aided high compressive forces during landing.

Our study also highlights the fact that the smaller-bodied species such as *Cebuella* and *Callithrix jacchus* were characterized by considerably greater additional costs of transport during even short increases in the distance leaped (Table 14.3). In *Cebuella*, calculations of the mechanical cost of transport (a standardized measure of the cost of moving a unit body weight a unit distance) increased from 0.81 when leaping 0.7 m, to 2.25 when leaping 1.0 m, to 3.34 when leaping 1.4 m. A similar pattern occurs in *C. jacchus* (*C* increases from 2.22 when leaping 1.0 to 4.20 when leaping 1.7 m; Table 14.3). In contrast, the cost of leaping in *Callimico* varied little during both short and longer distance leaps (Table 14.3).

It also is interesting to note that relative to other callitrichids (except *Callimico*), pygmy marmosets and saddleback tamarins (*Saguinus fuscicollis*) are the only species for which quantitative data indicate frequent trunk-to-trunk leaping (Garber 1991; Youlatos 1999; Chapt. 15 this volume, Garber and Leigh 2001) in the forest understory. Quantitative data on distance leaped in these taxa, however, are unavailable. *S. fuscicollis* is characterized by relatively long forelimbs. Assuming *S. fuscicollis* adopts a pattern of leaping or a leaping trajectory similar to *Cebuella*, elongated forelimbs may aid in dissipating the high compressive forces encountered during forelimb-first landing on noncompliant supports (Garber and Leigh 2001). In the case of *Cebuella*, it remains less clear the degree to which the aspects of forelimb morphology are best understood as an adaptation to forelimb-first trunk-to-trunk leaping, vertical clinging, or scansorial travel (Davis 2002; Ford and Davis Chapt. 21 this volume).

Cebuella appears to be more closely related to the Amazonian marmosets of the genus *Mico* than to the northern and eastern Atlantic dry forest marmosets of the genus *Callithrix* (Rylands et al. 2000; Cortés-Ortiz, Chapt. 1 this volume). Thus, despite the fact that both common marmosets and pygmy marmosets exhibit a suite of traits associated with intensive exploitation of plant gums (procumbent lower incisors, elongated and complex hindgut, year-round trunk gouging), differences in limb proportions, body mass, and patterns of positional behavior suggest each may have converged on this adaptive pattern independently. Moreover, it appears that trunk-to-trunk leaping is a more common locomotor pattern in *Cebuella* than in *Callithrix*.

In *Callimico*, elongated hindlimbs may function to generate sufficient force at takeoff during trunk-to-trunk leaping to facilitate a straight or upward locomotor trajectory at most distances. This, in combination with aerodynamic movements of the tail and body may slow the speed of landing in order to reduce the forces on the forelimbs during impact (Garber et al. 2005). At most leaping distances evaluated in this study, *Callimico*'s landing velocity was lower than that of *Cebuella* and *Callithrix*. It was only when leaping 2 m, that *Callimico* adopted a leaping trajectory more similar to the pygmy and common marmosets. This may reflect a limit or threshold on mechanically efficient leaping for *Callimico*. In the wild, Garber and Leigh (2001) report that only 13% of trunk-to-trunk leaps in *Callimico* were equal to or greater than 2 m in horizontal distance. In a more recent study in which the actual distances between take-off and landing platforms were measured in the field, over 40% of callmico's trunk-to-trunk leaps were greater than 2 m (Garber and Porter, unpub. data). Leaps of 2 m on noncompliant vertical supports in the present study, resulted in considerably greater height loss and landing velocity, and presumably higher compressive forces impacting the forelimbs during landing than leaps of shorter distances.

In conclusion, although pygmy marmosets, common marmosets, and callimicos are reported to commonly exploit large-diameter vertical supports as feeding and locomotor substrates, there is evidence of important species differences in positional behavior and anatomy. These patterns appear to reflect species differences in ecology that are associated with independent evolutionary changes in trunk foraging adaptations. Additional studies of the biomechanics of forelimb-first landing, vertical clinging, and scansorial locomotion in callitrichines are needed to better understand the adaptive radiation of this primate subfamily.

Acknowledgments Funding for this project was provided by the University of Illinois and the A.H. Schultz Foundation (Zuerich). Assistance in stimulating the pygmy marmosets to leap was provided by Jenni Garber and Sara Garber. PAG wishes to thank both Jenni and Sara for working with him on this research project and for their love and continued support. He also wishes to thank Chrissie McKenney for her love and inspiration. GA acknowledges the outstanding efforts of Heinz Galli in animal care and in maintaining the primate colony, and the efforts of Belinda Benz and Claudia Rudolf von Rohr in animal habituation and training. AS wishes to thank Emilio White for collaborating in the video-analysis. We also wish to thank Susan Ford, Lesa Davis, and Leila Porter for inviting us to contribute to this important volume.

References

Araújo A, Arruda MF, Alencar AI, Albuquerque F, Nascimento MC, Yamamoto ME (2000) Body weight of wild and captive common marmosets (*Callithrix jacchus*). Int J Primatol 21:317–324

Bicca-Marques JC (1999) Hand specialization, sympatry, and mixed-species associations in callitrichines. J Hum Evol 36:349–378

Canavez FC, Moreira MAM, Simon F, Parham P, Seuánez HN (1999) Phylogenetic relationships of the callitrichinae (Platyrrhini, Primates) based on beta2-microglobulin DNA sequences. Am J Primatol 48:225–236

Coimbra-Filho AF, Mittermeier RA (1978) Tree-gouging, exudates eating, and the "short-tusked" condition in *Callithrix* and *Cebuella*. In: Kleiman DG (ed) The biology and conservation of the Callitrichidae. Smithsonian Institution Press, Washington DC, pp 105–115

Cortés-Ortiz L (this volume) Molecular phylogenetics of the Callitrichidae with an emphasis on the marmosets and *Callimico*. In: Ford SM, Porter LM, Davis LC (eds) The smallest anthropoids: The marmoset/callimico radiation. Springer Press, New York, pp 3–24

Crompton RH, Sellers WI, Gunther MM (1993) Energetic efficiency and ecology as selective factors in saltatory adaptation of prosimian primates. Proc R Soc London (B) 254:41–45

Davis LC (2002) Functional morphology of the forelimb and long bones in the Callitrichidae (Platyrrhini: Primates). PhD Thesis, Southern Illinois University at Carbondale, IL

Demes B, Jungers WL, Gross TS, Fleagle JG (1995) Kinetics of leaping primates: influence of substrate orientation and compliance. Am J Phys Anthropol 96:419–430

Encarnación F, Heymann EW (1998) Body mass of wild *Callimico goeldii*. Folia Primatol 69:368–371

Ford SM, Davis LC (1992) Systematics and body size: implications for feeding adaptations in New World monkeys. Am J Phys Anthropol 88:415–468

Ford SM, Davis LC (this volume) Marmoset postcrania and the skeleton of the dwarf marmoset, *Callibella humilis*. In: Ford SM, Porter LM, Davis LC (eds) The smallest anthropoids: The marmoset/callimico radiation. Springer Press, New York, pp 411–448

Garber PA (1991) A comparative study of positional behavior in three species of tamarin monkeys. Primates 32:219–230

Garber PA (1992) Vertical clinging, small body size, and the evolution of feeding adaptations in the Callitrichinae. Am J Phys Anthropol 88:469–482

Garber PA (1994) Phylogenetic approach to the study of tamarin and marmoset social systems. Am J Primatol 34:99–219

Garber PA, Leigh SR (1997) Ontogenetic variation in small-bodied New World primates: Implications for patterns of reproduction and infant care. Folia Primatol 68:1–22

Garber PA, Leigh SR (2001) Patterns of positional behavior in mixed-species troops of *Callimico goeldii*, *Saguinus labiatus*, and *Saguinus fuscicollis* in northwestern Brazil. Am J Primatol 54:17–31

Garber PA, Rosenberger AL, Norconk MA (1996) Marmoset misconceptions. In: Norconk MA, Rosenberger AL, Garber PA (eds) Adaptive radiations of neotropical primates. Plenum, New York, pp 87–95

Garber PA, Blomquist G, Anzenberger G (2005) Kinematic analysis of trunk-to-trunk leaping in Goeldi's monkey (*Callimico goeldii*). Int J Primatol 26:217–234

Hamrick MW (1998) Functional and adaptive significance of primate pads and claws: evidence from New World anthropoids. Am J Phys Anthropol 106:113–128

Hershkovitz P (1977) Living New World Monkeys (Platyrrhini), with an Introduction to the Primates, Vol 1. Chicago University Press, Chicago

Horovitz I, Meyer A (1997) Evolutionary trends in the ecology of New World monkeys inferred from a combined phylogenetic analysis of nuclear, mitochondrial, and morphological data. In: Givnish TJ, Sytsma KJ (eds) Molecular evolution and adaptive radiation. Cambridge University Press, Cambridge, pp 189–224

Jungers WL (1985) Body size and scaling of limb proportions in primates. In: Jungers WL (ed) Size and scaling in primate biology. Plenum, New York, pp 345–381

Kinzey WG, Rosenberger AL, Ramirez M (1975) Vertical clinging and leaping in a neotropical primate. Nature 255:327–328

Lacher T, Bouchardet G, Alves C, Magalhaes-Castro B (1984) Parasitism of trees by marmosets in a central Brazilian gallery forest. Biotropica 16:128–129

Pastorini J, Forstner MRJ, Martin RD, Melnick DJ (1998) A reexamination of the phylogenetic position of *Callimico* (Primates): incorporation of new mitochondrial DNA sequence data. J Mol Evol 47:32–41

Porter LM (2000) The behavior and ecology of the Goeldi's monkey (*Callimico goeldii*) in northern Bolivia. PhD Thesis, State University of New York at Stony Brook, NY

Porter LM, Garber PA (2004) Goeldi's monkeys: a primate paradox? Evol Anthropol 13:104–115

Ramirez MF, Freese CH, Revilla CJ (1978) Feeding ecology of the pygmy marmoset, *Cebuella pygmaea*, in northeastern Peru. In: Kleiman DG (ed) The biology and conservation of the Callitrichidae. Smithsonian Institution Press, Washington DC, pp 91–104

Rylands AB, Schneider H, Langguth A, Mittermeier RA, Groves CP, Rodriguez-Luna E (2000) An assessment of the diversity of New World primates. Neotrop Primates 8:61–93

Rylands AB, Mittermeier RA, Coimbra-Filho AF (this volume) The systematics and distribution of the marmosets (*Callithrix, Callibella, Cebuella,* and *Mico*) and Callimico (*Callimico*) (Callitrichidae, Primates). In: Ford SM, Porter LM, Davis LC (eds) The smallest anthropoids: The marmoset/callimico radiation. Springer Press, New York, pp 25–61

Singer SS, Schmitz J, Schwiegk C, Zischler H (2003) Molecular cladistic markers in New World monkey phylogeny (Platyrrhini, Primates). Mol Phylogenet Evol 26:490–501

Soini P (1988) The pygmy marmosets, genus *Cebuella*. In: Mittermeier RA, Rylands AB, Coimbra-Filho AF, da Fonseca GA (eds) Ecology and behavior of Neotropical primates, vol 2. World Wildlife Fund, Washington DC, pp 79–129

StatSoft Inc (2000) STATISTICA for Windows [Computer program manual]. Tulsa, OK

Stevenson MF, Rylands AB (1988) The marmosets, genus *Callithrix*. In: Mittermeier RA, Rylands AB, Coimbra-Filho AF, da Fonseca GA (eds) Ecology and behavior of Neotropical primates, vol 2. World Wildlife Fund, Washington DC, pp 131–222

Terranova CJ (1996) Variation in the leaping of lemurs. Am J Primatol 40:145–165

Walker SE (1993) Positional adaptations and ecology of the Pitheciini. PhD Thesis, City University of New York

Warren RD, Crompton RH (1998a) Diet, body size, and energy costs of locomotion in saltatory primates. Folia Primatol 69:86–100

Warren RD, Crompton RH (1998b) Locomotor ecology of *Lepilemur edwardsi* and *Avahi occidentalis*. Am J Phys Anthropol 104:471–486

Youlatos D (1999) Positional behavior of *Cebuella pygmaea* in Yasuní National Park, Ecuador. Primates 40:543–550

Youlatos D (2005) Locomotion, postures, and habitat use by pygmy marmosets (*Cebuella pygmaea*). Am J Phys Anthropol 40:228

Youlatos D (this volume) Locomotion, postures, and habitat use by pygmy marmosets (*Cebuella pygmaea*). In: Ford SM, Porter LM, Davis LC (eds) The smallest anthropoids: The marmoset/callimico radiation. Springer Press, New York, pp 279–297

Chapter 15
Locomotion, Postures, and Habitat Use by Pygmy Marmosets (*Cebuella pygmaea*)

Dionisios Youlatos

Abstract Pygmy marmosets (*Cebuella pygmaea*) are the smallest representatives of the Callitrichidae and are characterized by morphological correlates to extensive exudativory and to frequent claw climbing, clinging, quadrupedalism, and leaping on vertical supports. This morpho-behavioral complex is unique and crucial for understanding the evolutionary history of the family. I studied the positional behavior and habitat use of a group of pygmy marmosets in a terra-firme rainforest in Amazonian Ecuador. During the study period, the pygmy marmosets confined their movements to the understory of a dense liana forest, where they largely utilized lianas and tree trunks, showing a strong preference for the latter. In addition, they extensively used and preferred supports >10 cm in diameter, and almost two thirds of these supports were vertical. Feeding bouts (finding, manipulating, and eating plant foods) on exudates and foraging (finding, manipulating, and eating arthropod prey) occurred at around 5 m above ground. The dominant feeding posture was claw clinging on vertical trunks. In contrast, the dominant foraging postures were quadru-tripedal stand and cantilever, with extensive use of small and medium sized lianas. Traveling occurred lower to the ground than did foraging or feeding, and locomotion was dominated by claw climbing and terminal leaping, and low proportions of quadrupedal activities and vertical leaps. Claw climbing was mainly used for movements within trees while the other forms of locomotion were used for crossing between trees. Landing support use was similar for both short and long terminal leaps. However, both short and long vertical leaps initiated from and ended on larger supports more frequently than terminal leaps. In addition, long vertical leaps were initiated on larger supports than were shorter vertical leaps. Field observations show that pygmy marmosets use claw climbing, claw clinging, and vertical supports more than any other callitrichid, but they use less leaping and vertical leaping than *Callimico*. In general, these positional data are not entirely

D. Youlatos (✉)
Department of Zoology, Aristotle University of Thessaloniki, School of Biology, 54124, Thessaloniki, Greece
e-mail: dyoul@bio.auth.gr

S.M. Ford et al. (eds.), *The Smallest Anthropoids*, Developments in Primatology: Progress and Prospects, DOI 10.1007/978-1-4419-0293-1_15,
© Springer Science+Business Media, LLC 2009

coupled with morphological predictions for pygmy marmosets, but likely reflect adaptations that evolved recently within the evolutionary radiation of callitrichids.

Resumen Las marmosetas pigmeas (*Cebuella pygmea*) son los representantes más pequeños de los Callitrichidae y están caracterizados por sus correlaciones morfológicas de alimentación extensiva de exudados y de frecuentes ascensos y agarres con garras, cuadrupedalismo y saltos sobre soportes verticales. Este complejo morfológico-conductual es único y crucial para comprender la historia evolutiva de la familia. Yo estudié el comportamiento postural y el uso del hábitat de un grupo de marmosetas pigmeas en la selva de terra-firme en la Amazonia del Ecuador. Durante el periodo de estudio, las marmosetas pigmeas limitaron sus movimientos en el sotobosque de un denso bosque de lianas, en donde utilizaron principalmente lianas y troncos de árboles, mostrando una fuerte preferencia por estos últimos. Adicionalmente, utilizaron y prefirieron soportes >10 cm de diámetro, siendo verticales casi dos tercios de estos soportes. Los episodios de alimentación de exudados y de forrajeo (encontrando, manipulando y comiendo presas de artrópodos) ocurrieron a una altura aproximada de 5 m del suelo. La postura dominante de alimentación fue el agarre con garras de troncos verticales. En contraste, la postura dominante de forrajeo fue el postura cuadru-trípeda y voladiza, con uso extensivo de lianas de tamaño pequeño y mediano. El desplazamiento ocurrió más cerca al suelo que el forrajeo o la alimentación y la locomoción estuvo dominada por el asenso con garras y saltos terminales, así como por proporciones bajas de actividades cuadrúpedas y saltos verticales. El asenso con garras fue usado principalmente en el movimiento dentro del mismo árbol, mientras que las otras formas de locomoción se usaron para cruces entre los árboles. El uso de soportes de aterrizaje fue similar para saltos terminales tanto cortos como largos. Sin embargo, los saltos verticales cortos y largos, iniciaron y terminaron en soportes más grandes con mayor frecuencia que los saltos terminales. Adicionalmente, los saltos verticales largos también se iniciaron en soportes más grandes que los saltos verticales cortos. Observaciones de campo muestran que las marmosetas pigmeas utilizan el ascenso con garras, el agarre con garras y los apoyos verticales más que cualquier otro calitrícido, pero utilizan menos saltos y saltos verticales que *Callimico*. En general, estos datos posturales no están enteramente acoplados con las predicciones morfológicas para las marmosetas pigmeas, pero posiblemente reflejan adaptaciones que evolucionaron recientemente dentro de la radiación evolutiva de los calitrícidos.

Resumo Sagüis-leãozinho (*Cebuella pygmaea*) são os menores representantes dos Callitrichidae e são caracterizados por terem uma morfologica correlata à sua extensiva exudativoria e ao frequente uso das garras para subir, agarrar, locomoção quadrupedal e salto em suportes verticais. Este complexo morfo-comportamental é único e crucial para entendermos a história evolutiva da família. Eu estudei o comportamento postural e uso do habitat de um grupo de sagüis-leãozinho na floresta de terra firme da Amazônia Equatoriana. Durante o periodo de estudo, os sagüis leãozinho confinaram seus movimentos na parte mais baixa da floresta de liana densa, onde eles usaram amplamente as lianas e os troncos de árvores, mostrando uma forte preferência pelos últimos. Mais ainda, eles usaram extensivamente e preferencialmente suportes >10 cm de diâmetro, e quase dois terços destes suportes eram verticais. A alocação de tempo à alimentação (procura, manipulação e predação de artópodos) ocorreu cerca de 5 m acim a do solo. A postura de alimentação dominante foi qaudru-tripedal em pé e "cantilever," com extensivo uso de lianas de pequeno e médio porte. A movimentação ocorreu mais abaixo e próxima ao solo do que o forrageamento ou alimentação e a locomoção foi dominada por subidas com garras e saltos terminais, e baixa proporção de atividades quadrupedais e saltos verticais. Subindo com garras foi principalmente utilizado em movimentos em cada árvore individualmente, enquanto as outras formas de movimentação foram usadas para atravessar entre árvores. O uso de suporte de aterragem foi similar tanto para saltos longos como curtos. Entretanto, tanto saltos curtos como longos na vertical iniciaram e terminaram em suportes maiores do que os saltos terminais. Adicionalmente, saltos verticais mais longos foram iniciados em suportes mais largos mais frequentemente do que saltos terminais. Observações de campo mostram que os sagüis-leãozinho usam as garras para subir e agarrar nos troncos, e suportes verticais mais do que qualquer outro callitriquídeo, mas usam menos saltos horizontais e verticais do que *Callimico*. Em geral, estes dados de postura corporal não são inteiramente congruentes com as previsões morfologicas para sagüis-leãozinho, mas provavelmente refletem adaptações que evoluiram recentemente dentro da radiação evolutiva dos calitriquídeos.

15.1 Introduction

Callitrichids are the most diminutive of all living anthropoids, ranging from the smallest of the group, the pygmy marmosets, *Cebuella pygmaea*, at 150 g, to the largest of the group, the golden lion tamarins, *Leontopithecus* spp., at 700 g (Garber 1992). All these New World anthropoids are small bodied, have claw-like nails or tegulae, have lost or reduced the hypocone on maxillary molars, have lost or reduced the mandibular and maxillary third molars, and frequently have twins (Garber 1992). These morphological traits are partly shared by callimicos (*Callimico goeldii*) that are now considered to be part of the same family, with close ties to marmosets (Neusser et al. 2001; Porter and Garber 2004; Cortés-Ortiz Chap. 1 this volume). This suite of characters was once considered to be primitive for anthropoids (Hershkovitz 1977), but recent investigations have suggested that the aforementioned traits, along with a high ratio of fetal weight to maternal body weight, represent derived traits and are the result of phyletic dwarfism and the constraints of small body size on diet, foraging patterns, and reproductive success (Rosenberger 1977; Ford 1980, 1986).

Callitrichids have species-specific differences in dental morphology, body size, and diet, but all are distinguishable from other anthropoids by their ability to claw climb, cling, and leap from and to large vertical supports, behaviors that enable them to exploit food sources such as tree exudates and bark dwelling insects. Clawed positional modes are used less frequently in tamarins (*Saguinus* spp., *Leontopithecus* spp.) than in marmosets (*Cebuella pygmaea, Callithrix* spp., *Mico* spp.), and callimicos (*Callimico goeldii*). The positional behavior of tamarins is mainly characterized by the quadrupedal walk, run and bound, and the leap between terminal branches. Vertical claw climbing and clinging activities on large supports are usually less common in tamarins than in marmosets, and are frequently associated with opportunistic and occasional exudate feeding and insect capturing, especially in the smaller species (Fleagle and Mittermeier 1980; Garber 1980, 1991; Garber and Sussman 1984; Yoneda 1984; Rosenberger and Stafford 1994; Garber and Pruetz 1995; Youlatos 1999a; Garber and Leigh 2001; Youlatos and Gasc 2001; Arms et al. 2002; Porter 2004; Cunha et al. 2006; Jackson and Ford 2006). The positional behavior of marmosets is still poorly described quantitatively, but given their more specialized plant exudate feeding behavior, it has been assumed that they incorporate higher proportions of vertical clawed locomotor and postural activities and, possibly, vertical leaping than the tamarins (Sussman and Kinzey 1984; Garber 1992; but see Garber et al. Chap. 14 this volume). The only quantitative positional data for marmosets are of the smallest representative of the family, the pygmy marmoset (*Cebuella pygmaea*), which exhibits frequent vertical claw climbing and claw clinging on large supports during feeding, as well as terminal and vertical leaping (Kinzey et al. 1975; Soini 1988; Youlatos 1999b; Jackson and Ford 2006). Similarly, callimicos also appear to exhibit high frequencies of vertical activities, especially vertical leaping and claw climbing (Garber and Leigh 2001; Porter 2004), locomotor styles that are functionally associated with specific postcranial traits (Davis 2002).

Despite the tendencies of both pygmy marmosets and callimicos to engage in vertical activities, they share few postcranial traits (Davis 2002, Ford and Davis Chap. 21 this volume). Pygmy marmosets appear to be unique among callitrichids, mainly due to their very small size, high rates of gummivory, liana forest and lower forest strata utilization, and occupation of their own eco-morphological space (Rosenberger 1992; Davis 2002; Youlatos 2004). This may be related to the extreme evolutionary size reduction of pygmy marmosets (Ford and Davis 1992) and the fact that in some callitrichid lineages behavioral and morphological adaptations associated with the use of large vertical supports may have evolved independently (Kinzey et al. 1975; Youlatos 1999b; Garber et al. Chap. 14 this volume). In this context, it is necessary to consider the relationship among morphological traits, postural and locomotor behavior, and habitat utilization, in order to understand the evolution of the callitrichids. Quantitative data on the positional behavior of pygmy marmosets have been previously reported (Youlatos 1999b). This chapter provides a more detailed analysis of the locomotor and postural behavior of a group of pygmy marmosets in southeastern Ecuador, along with additional, previously unpublished, information on habitat use and preference.

15.2 Study Site, Subjects, and Methods

The study was conducted in a 350 ha site of undisturbed *terra firme* hilly forest in the Yasuni National Park in southeastern Ecuador. The site (00°42′01″ S, 76°28′05″ W) is located at kilometer marker 47 on the Pompeya Sur-Iro road of Maxus Ecuador Inc. The data presented here were collected between August and October 1995. Data were collected on one group of pygmy marmosets (one adult male, one adult female, one juvenile male, one infant female). The animals were habituated 3–4 days prior to data collection, and remained in a patch of liana forest of 1 ha for the entire period of the study.

Data were collected using focal animal methods wherein an adult individual was followed for 15 min, shifting afterwards to the next available subject. If the focal animal was lost from sight for 5 min, the next available adult was followed. Three sets of data were collected: (a) 5-min general behavior and habitat use data, (b) 20-s locomotor and postural behavior data, and (c) ad libitum data on feeding and foraging The data collected using the 5-min instantaneous sampling method concerned (i) general behavior, (ii) height of the animal, (iii) support type, (iv) support size, and (v) support inclination. The twenty-second instantaneous sampling concerned data on (i) general behavior, (ii) locomotor or postural mode, (iii) associated context, (iv) support type, (v) support size and (vi) support inclination. Lastly, the ad libitum feeding and foraging parameters included (i) food type, (ii) plant food source type, (iii) height of the animal (in m), (iv) DBH of food source (in cm), and (v) bout duration (in sec). Sampled categories of behavior, positional modes, associated context, support type, size, and inclination, food type, and plant food source type are presented in Table 15.1.

In order to assess support preference by pygmy marmosets, I estimated support availability in the study patch. Dense liana forests are characterized by a closely packed network of randomly arranged supports with a relatively regular density at lower heights (Fig. 15.1). For these reasons, I used five cuboid sampling volumes, 20 m long, 1 m wide, and 2 m high (space equivalents of sampling quadrats), within the liana forest patch exploited by the animals during the study period (Warren 1997). The area of the sampling cuboids was delimited by flagging tape on the ground, and the height was also marked by flagging tape around the major lianas and tree trunks included in the area. The sampled area represented almost 10% of the home range of the studied group. Along transects, I measured the number of available supports considering a single measured unit of support, every time one of the three concerned parameters (type, size, inclination) changed. At the end, a total of 1377 supports were measured, and availability was expressed as percents of total supports calculated for each parameter. Jacobs' D index was used to estimate preference for different support categories (Jacobs 1974):

$$D = u–a/u + a–2ua$$

where u is the proportion of use of support, and a is the proportion of availability of support. The value of D ranges from –1 for avoidance to +1 for preference, whereas values close to 0 indicate neutrality of choice.

Table 15.1 Definition of sampled categories of behavior, associated context, locomotor and postural modes, support type, support size, and support inclination, food type, and plant food source type

Behavior	
Travel	Moving between feeding and/or resting sites
Feed	Active search, acquisition, manipulation, and eating of plant foods
Forage	Active search, acquisition, manipulation, and eating of arthropods
Rest	Resting, pausing, as well as social interactions, and other non-active behaviors
Associated context	
Across	Moving across or between trees or lianas
Within	Moving within a single tree or liana
Locomotion/Postures	
Quadrupedal Walk	Symmetrical slow progression along single horizontal and moderately inclined supports
Quadrupedal Bound	Asymmetrical fast progression along single horizontal and inclined supports
Vertical Grasp	Quadrupedal upward and downward progression along single vertical supports involving hand grasp
Clamber	Quadrupedal progression to various directions across multiple supports
Bridge	Gap crossing mode involving active or passive compliance of initial and landing supports
Claw Climb	Quadrupedal progression using the tegulae along large vertical supports
Terminal Leap	Gap crossing mode involving an airborne phase with the body held horizontally or inclined
Vertical Leap	Leaping from and to vertical supports with body held mainly inclined or subvertical
Sit	A seated posture with strongly flexed hindlimbs
Quadru-tripedal Stand	Standing on four or three lightly flexed limbs
Bipedal Stand	Standing on two flexed hind limbs
Cantilever	The feet anchor the lower body to a near-vertical support while the trunk is extended and the forelimbs seize arthropods
Claw Cling Up	Vertical or subvertical clinging using the tegulae with the head upwards
Claw Cling Down	Vertical or subvertical clinging using the tegulae with the head downwards
Support type	*Trunk; Branch; Liana; Palm Frond*
Support size	<2 cm; 2–5 cm; 5–10 cm; >10 cm
Support inclination	
Vertical	Up and down vertical within 22.5° of true vertical
Horizontal	Within 22.5° of true horizontal
Oblique	Between the vertical and horizontal classes
Food type	*Exudates; Arthropods*
Plant food source type	*Lianas; Trees*

In total, I collected 181 records of general behavior, 418 records of travel locomotion, 202 records of feeding locomotion, 224 records of feeding postures, 89 records of foraging locomotion, and 61 records of foraging postures.

Fig. 15.1 Liana forest in Yasuni National Park; a typical *terra firme* habitat of pygmy marmosets in Amazonian Ecuador

In addition, I collected a total of 73 records of feeding/foraging bouts on 10 different trees and lianas. I used ANOVA with Tukey HSD post-hoc test to estimate the significance of the difference of mean heights between different behavioral contexts and between different food types, as well as the bout duration between different food types (Zar 1996). Differences between the frequencies of behaviors were calculated using log-likelihood ratio *G*-tests enhanced through Monte Carlo procedures (Zar 1996). In all cases, *p* values of 0.05 and less were considered significant.

15.3 Results

15.3.1 Habitat Utilization

During the study period, the pygmy marmosets confined their activities to a small area within a dense liana forest (see Fig. 15.1). Table 15.2 provides the percentages of availability of different support characteristics. In the area exploited by the pygmy marmosets, lianas were the dominant supports (see Table 15.2). In terms of support size, supports <2 cm and 2–5 cm composed more than two thirds of the estimated available supports (Table 15.2). Lastly, oblique supports represented almost half of all the inclination categories, with vertical supports corresponding to 37.4% of the available supports. (Table 15.2).

Considering all the active behaviors, the pygmy marmosets mainly used lianas (54.9%) and vertical trunks (43.6%), and appeared to strongly prefer the

Table 15.2 Proportions of available supports within the liana forest exploited by pygmy marmosets in Yasuni National Park, Ecuador

Support types	(%)	Support sizes	(%)	Support inclinations	(%)
Liana	80.6	<2 cm	47.5	Vertical	37.4
Branch	12.2	2–5 cm	30.9	Oblique	49.7
Trunk	3.9	5–10 cm	10.9	Horizontal	12.9
Palm frond	3.3	>10 cm	10.7		

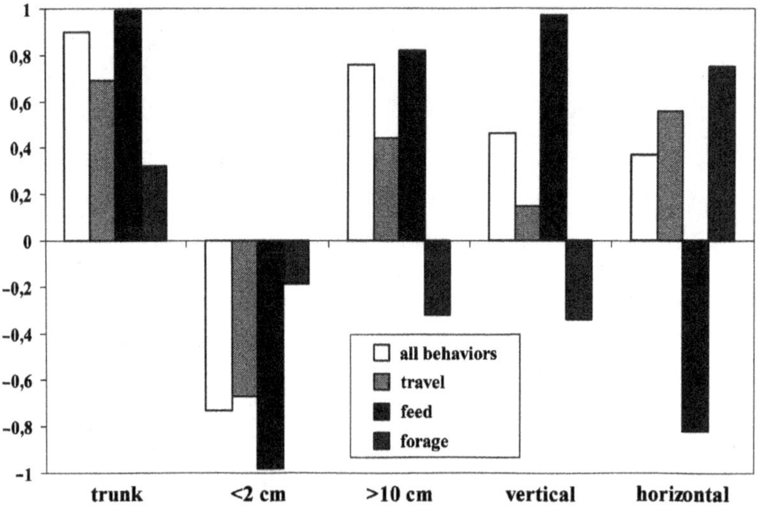

Fig. 15.2 Jacobs's *D* values of preference for selected categories of support type (trunk), support size (<2 cm, >10 cm), and support inclination (vertical, horizontal) during all behaviors, travel, feed, and forage (>0 preference, <0 avoidance, 0=neutral)

latter ($D=0.9$; see Fig. 15.2). As far as support size is concerned, the pygmy marmosets extensively used and showed a net preference for supports >10 cm (36.8% of the supports) (see Fig. 15.2). In contrast, supports <2 cm were used relatively infrequently (12.3%) and appeared to be avoided by the study animals (see Fig. 15.2). Vertical supports were used frequently (62.0%), but the animals showed only a slight preference for them ($D=0.46$). In contrast, horizontal supports were only moderately used (24.3%) according to availability ($D=0.37$).

Feeding was the main activity of pygmy marmosets (26.5%) during the study period. Traveling and foraging accounted for 10.5% and 6.6% of the activity budget, respectively. The rest of the active period was devoted to resting (including play and grooming). When all the behavioral activities are grouped together it is clear that animals spend the majority of their time in the lower strata of the forest with an overall mean height of 6.2 ± 2.5 m above ground. Strata use varied across behaviors; however, the mean feeding height was 5.25 ± 2.28 m, and was significantly higher than the mean traveling height (3.26 ± 1.33 m; Tukey HSD:

$p=0.009$). In contrast, foraging occurred at a mean height of 4.7 ± 1.6 m above the ground and was not significantly different from either traveling or feeding. Only resting appeared to occur significantly higher than all the other behaviors (7.3 ± 2.1 m; Tukey HSD: rest vs. feed $p<0.001$; rest vs. forage $p<0.001$; rest vs. travel $p<0.001$).

15.3.2 Feeding and Foraging

During the period of data collection, observations of feeding were exclusively on exudates. Exudate feeding bouts ($n=61$) lasted 112.2 ± 106.8 s, and were significantly shorter than insect foraging bouts ($n=12$, mean = 173.6 ± 88.6 s; $F=1.75$ $p=0.049$). During exudate feeding, the study animals exploited seven different lianas with a mean DBH of 8.7 ± 8.9 cm, which was significantly smaller than that of the tree trunks exploited for exudates ($n=3$, mean DBH = 30.05 ± 15.5; $F=2.12$, $p=0.023$).

Both feeding locomotion and postures occurred principally on trunks (see Table 15.3), and the pygmy marmosets showed a very strong preference for this type of support ($D=0.99$, see Fig. 15.2). Supports >10 cm were very frequently used (see Table 15.3) and were clearly preferred (see Fig. 15.2). In contrast, the pygmy marmosets infrequently used supports <2 cm (see Table 15.3) and appeared to avoid them (see Fig. 15.2). In terms of support inclination, vertical supports almost monopolized both feeding locomotion and postures (see Table 15.3), and were strongly preferred ($D=0.97$, see Fig. 15.2). In contrast, horizontal supports were seldom used (Table 15.3) and were avoided (see Fig. 15.2).

During feeding, claw climb was by far the dominant locomotor mode (93.6%) with vertical leaping ranking second (see Table 15.4). In a similar manner, feeding postures were largely dominated by vertical claw clinging (Table 15.5), that occurred primarily on supports >10 cm (see Fig. 15.3).

Table 15.3 Percents of use of different support categories in different behavioral locomotor and postural contexts

		Feed (%)		Forage (%)	
	Travel (%)	Locomotion	Postures	Locomotion	Postures
Trunk	18.4	84.2	85.7	7.9	6.6
Lianas	77.8	15.8	14.3	92.1	93.4
Branches	2.1	0.0	0.0	0.0	0.0
Palm fronds	1.7	0.0	0.0	0.0	0.0
<2 cm	15.3	1.5	0.0	37.1	39.3
2–5 cm	36.8	5.4	6.7	37.1	47.5
5–10 cm	29.2	23.8	21.0	21.3	8.2
>10 cm	18.7	69.4	72.3	4.5	5.0
Vertical	44.7	97.5	97.4	25.8	18.0
Oblique	20.8	1.0	1.3	23.6	31.2
Horizontal	34.5	1.5	1.3	50.6	50.8

Table 15.4 Percents of locomotor modes of pygmy marmosets during travel, feeding, foraging, and all behaviors combined (travel vs. feed: $G=315.8$, $p<0.001$; travel vs. forage: $G=28.02$, $p<0.001$; feed vs. forage: $G=205.1$, $p<0.001$)

	Feed	Forage	Travel	All behaviors
Quadrupedal walk	0.5	34.8	17.5	14.8
Quadrupedal bound	0.5	13.5	17.7	12.3
Vertical grasp	0.5	3.4	3.1	2.4
Clamber	0.0	12.3	2.6	3.1
Bridge	0.0	1.1	0.2	0.3
Claw climb	93.6	16.8	23.7	42.5
Terminal leap	1.5	14.6	22.2	15.4
Vertical leap	3.5	3.4	12.9	9.1
Samples	202	89	418	706

Table 15.5 Percents of postural modes of pygmy marmosets during feeding, foraging, and all behaviors combined (Feed vs. forage: $G=226$, $p<0.001$)

	Feed (%)	Forage (%)	All behaviors (%)
Sit	1.3	16.4	4.6
Quadrupedal/tripedal stand	0.0	39.3	8.4
Bipedal stand	0.9	0.0	0.7
Cantilever	0.0	34.4	7.4
Claw cling up	81.7	6.6	65.6
Claw cling down	16.1	3.3	13.3
Samples	224	61	285

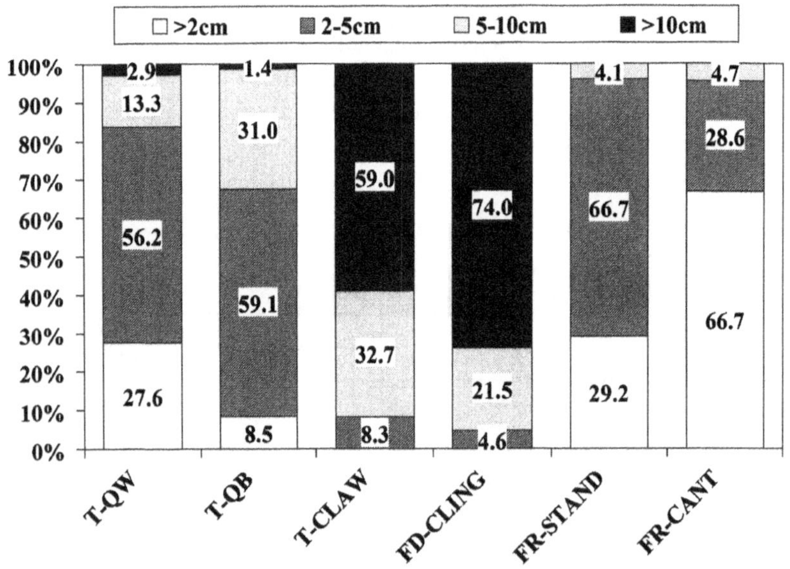

Fig. 15.3 Percents of use of different support size categories during selected locomotor and postural modes (T-QW: quadrupedal walk in travel; T-QB: quadrupedal bound in travel; T-CLAW: claw climb in travel; FD-CLING: claw cling in feed; FR-STAND: quadru/tripedal stand in forage; FR-CANT: cantilever in forage)

Foraging bouts concerned active search and capturing of arthropods on dense liana entanglements. Lianas were preferred for all locomotion and postures during foraging (Table 15.3; $D=0.55$). During foraging activities, the pygmy marmosets extensively used supports under 5 cm in diameter. These accounted for almost two thirds of support use (see Table 15.3). Horizontal supports represented half of the utilized supports (Table 15.3) and were preferred by the study animals (see Fig. 15.2).

During foraging, the quadrupedal walk was the dominant locomotor mode, with claw climbing ranking next (see Table 15.4). In addition, the quadrupedal bound, vertical leaping and clamber also were used frequently during foraging (see Table 15.4). The quadrupedal stand and cantilever were the dominant foraging postures (see Table 15.5). Quadrupedal postures occurred primarily on supports of 2–5 cm, whereas almost two thirds of the supports used in cantilever were <2 cm (see Fig. 15.3).

15.3.3 Travel Locomotion

During travel locomotion, lianas were the dominant supports (77.8%). However, vertical trunks, which ranked second (see Table 15.3), appeared to be particularly preferred (see Fig. 15.2). Supports with diameters of between 2–5 cm and 5–10 cm were also frequently used (Table 15.3). In contrast, small (<2 cm) and very large (>10 cm) supports were used infrequently (see Table 15.3), and the former appeared to be avoided by pygmy marmosets ($D=-0.67$, see Fig. 15.2). Vertical and horizontal supports were commonly used (Table 15.3), but the latter appeared to be slightly preferred (see Fig. 15.2).

Claw climbing and terminal leaping were the dominant traveling modes (see Table 15.4). The quadrupedal walk, quadrupedal bound and vertical leaping were also relatively common (see Table 15.4). When the pygmy marmosets traveled within a single tree, they mainly employed clawed locomotion (see Fig. 15.4), whereas the other modes were used frequently for crossing between trees (see Fig. 15.4). The claw climb principally occurred on relatively large supports, of which more than half were >10 cm (see Fig. 15.3). On the other hand, the quadrupedal walk occurred primarily on supports less than 5 cm (see Fig. 15.3). In contrast, support use during the quadrupedal bound differed significantly from the quadrupedal walk (walk vs. bound: $G=17.2$, $p=0.001$), as the supports during bounding were commonly of a diameter between 2 and 10 cm (see Fig. 15.3).

In terms of terminal leaps, 54.1% of recorded leaps covered distances <0.5 m. When all terminal leaps were compared, the size of both initial and landing supports were statistically similar, and all occurred on supports of <5 cm. A similar profile in initial and terminal support size use was derived, when short (horizontal distance <0.5 m) and long (horizontal distance >0.5 m) leaps were compared (Table 15.6). In terms of support inclination, short leaps were initiated more from oblique supports than horizontal supports, whereas longer leaps were initiated more from horizontal supports than oblique supports (see Fig. 15.5). These differences

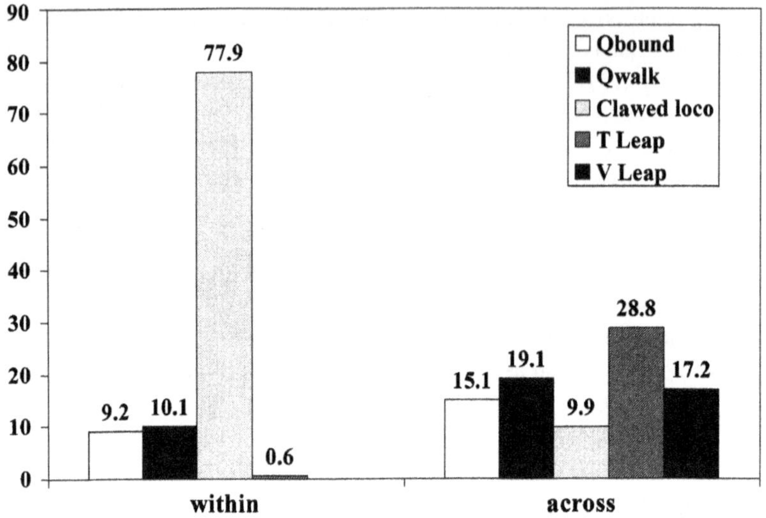

Fig. 15.4 Percents of major locomotor modes used within and between trees

Table 15.6 Percents of use of initial and landing support size categories in short terminal leaps (<0.5 m) and long terminal leaps (>0.5 m)

	Leaps <0.5 m		Leaps >0.5 m	
	Initial (%)	Landing (%)	Initial (%)	Landing (%)
<2 cm	25.4	23.7	22.0	28.0
2–5 cm	40.7	44.1	40.0	34.0
5–10 cm	20.3	18.6	14.0	16.0
>10 cm	13.6	13.6	24.0	22.0
Samples	59	59	50	50

were statistically significant (G=8.61, p=0.014). Landing frequently occurred on horizontal and oblique supports for both distance categories, showing a significantly lower use of oblique supports in short leaps (G=6.1, p=0.046; see Fig. 15.5).

In terms of vertical leaps, 51.6% of the recorded leaps were <0.5 m in distance, and initial and landing supports were between 2 and 10 cm. As indicated above, terminal leaps occurred significantly more frequently on smaller supports than during vertical leaps (vertical leaps vs. terminal leaps: initial supports: G=33.1, p<0.001; landing supports: G=48.04, p<0.001). When horizontal leaps of different distances were compared (see Table 15.7), it is clear that short vertical leaps are initiated from smaller supports (5–10 cm) than longer vertical leaps (initial supports: short vertical vs. long vertical: G=16.2, p=0.013). However, both short and long vertical leaps frequently terminated on supports 2–10 cm, with no significant differences between the two (see Table 15.7).

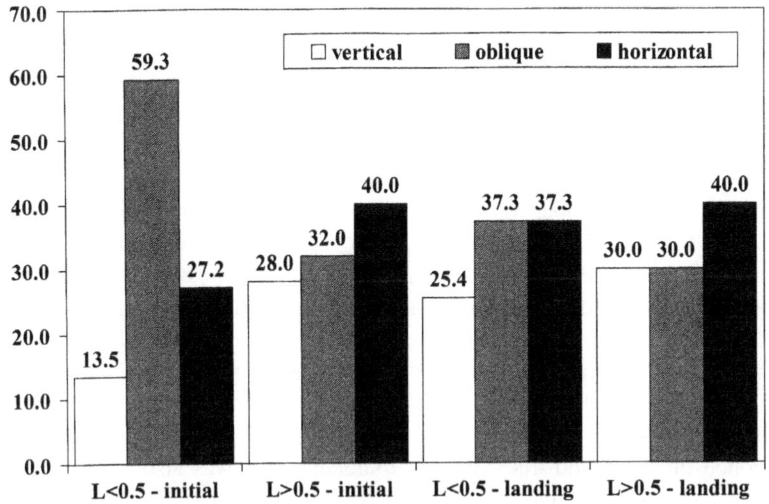

Fig. 15.5 Percents of use of initial and landing support inclination categories in short terminal leaps (<0.5 m) and long terminal leaps (>0.5 m)

Table 15.7 Percents of use of initial and landing support size categories in short vertical leaps (<0.5 m) and long vertical leaps (>0.5 m)

	Leaps <0.5 m		Leaps >0.5 m	
	Initial (%)	Landing (%)	Initial (%)	Landing (%)
<2 cm	3.0	0.0	0.0	0.0
2–5 cm	30.3	45.4	22.6	22.6
5–10 cm	51.5	30.3	29.0	58.1
>10 cm	15.1	24.3	48.3	19.3
Samples	33	33	31	31

15.4 Discussion

This study on habitat utilization and positional behavior of pygmy marmosets has revealed important insights into how these small callitrichids exploit their environment. In the *terra firme* liana forest of Yasuni National Park in Ecuadorian Amazonia, pygmy marmosets confined their activities to the lower strata of the understory. Traveling occurred very close to the ground, while feeding and foraging took place only slightly higher. These results are in accordance with previous observations on the activity patterns of pygmy marmosets in flooded forests (Soini 1988; Townsend 2001). Overall, pygmy marmosets showed consistent use and strong preference for vertical supports of mainly large size while both traveling and feeding. This is also the case for groups of pygmy marmosets in other forests (Kinzey et al. 1975; Soini 1988; Townsend 2001). Significant use of large vertical supports has also been reported for other callitrichids, but the pygmy marmosets in this study showed more frequent use of vertical supports than all other species, including callimico (Garber 1980; Garber and Pruetz 1995; Youlatos 1999a; Youlatos and Gasc 2001; Porter 2004; Cunha et al. 2006; Garber et al. Chap. 14 this volume). In terms of support size use, pygmy marmosets mainly used supports 2–10 cm in diameter. Direct comparisons of support use with other studies are not feasible given the differences in the definition of size categories. Large callitrichids such as callimicos and red-handed tamarins (*Saguinus midas*) do not appear to exhibit a preference for such large diameter supports (Youlatos and Gasc 2001; Porter 2004). On the other hand, the large moustached tamarins (*Saguinus mystax*), which engage in high frequencies of quadrupedal activities, as well as the common marmoset (*Callithrix jacchus*), which engage in both quadrupedal and leaping activities, appear to show a comparable profile of large-sized support use (Garber and Pruetz 1995; Cunha et al. 2006).

During the study period, pygmy marmosets fed mainly on exudates while positioned on vines and trees of large diameter. This was also the case for other groups of pygmy marmosets in flooded forests (Kinzey et al. 1975; Soini 1988; Townsend 2001). In contrast, foraging involved longer bouts for the active search of arthropods within dense liana entanglements, a pattern also observed in other studies (Kinzey et al. 1975; Soini 1988). These striking differences in microhabitat use during different behavioral contexts resulted in significant differences in the respective postural and locomotor behavior.

During exudate feeding, locomotion was primarily performed by claw climbing upon vertical trunks. Vertical leaping was rather uncommon. Consumption of exudates was executed through claw clinging on the vertical trunks of the feeding sources. Frequent claw climbing and claw clinging is also encountered in other small bodied exudate feeding primates (Charles-Dominique 1977; Crompton 1984; Garber, 1992; Off and Gebo 2005) and is usually associated with specific postcranial morphofunctional adaptations. Claw clinging postures on large vertical supports allow the establishment of a relatively firm lever system that is required for gouging activities (Cartmill 1974, 1985; Kinzey et al. 1975; Garber 1992; Hamrick 1998). Additionally, vertical claw clinging allows small primates to scan for arthropods that either dwell upon the bark of trees and vines or on the ground litter (Charles-Dominique 1977; Crompton 1984; Garber 1992). Finally, they may also serve as launching or landing platforms for traveling, especially for those animals that may infrequently forage on vertical trunks for either arthropods or exudates such as callimicos (Porter 2004) or squirrels (Youlatos 1999c).

In pygmy marmosets, arthropod foraging on small-sized liana entanglements apparently requires slow and fast quadrupedal activities. The walk, bound, and clamber appear to enable pygmy marmosets to engage in cautious and rapid movements within this dense and unstable microhabitat. Subsequently, mobile prey was accessed and ultimately captured by active cantilever. The same active postural mode is also used in high frequencies by many other small-bodied primates that habitually forage on small gregarious prey (Charles-Dominique 1977; Crompton 1984). On the other hand, prey manipulation and ingestion is accomplished by adopting quadrupedal and/or tripedal standing, a foraging posture that provides relative stability while handling prey, especially when occurring on relatively larger supports as is the case for pygmy marmosets.

Traveling between feeding and foraging sites was performed by the use of relatively shared proportions of claw climbing, leaping, and quadrupedal fast and slow progression, whereas vertical leaping was less common. Similar profiles have been previously reported for pygmy marmosets in flooded forests (Kinzey et al. 1975; Soini 1988). These modes compose the principal positional profiles for almost all callitrichids that have been studied so far (Fleagle and Mittermeier 1980; Garber 1980, 1991; Garber and Sussman 1984; Garber and Pruetz 1995; Youlatos 1999a; Garber and Leigh 2001; Youlatos and G; asc 2001; Arms et al. 2002; Porter 2004; Cunha et al. 2006), but pygmy marmosets appear to claw climb and claw cling more than tamarins and callimicos. However, the lack of detailed quantitative data on the positional behavior of other marmosets (*Callithrix, Mico, Callibella*) renders comparisons across callitrichids impossible. The very specialized exudativory of other marmosets implies that frequent vertical clawed postural and locomotor activities must be quite common (Garber 1992), although this remains to be tested in the field.

The presence of claws is advantageous for all arboreal mammals that need to exploit large vertical supports for traveling and feeding (Cartmill 1974; Hamrick 1998). Nevertheless, the vertical support niche that pygmy marmosets appear to prefer imposes mechanical constraints that are relative to body size: smaller mammals would face greater problems than larger ones in grasping vertical supports regularly encountered in the forest understory (Cartmill 1974, 1985). Thus, although claws enable arboreal mammals of all sizes to negotiate large vertical supports, smaller ones tend to engage in higher frequencies of positional activities that involve claws, such as the claw climb and claw cling in all directions (see Youlatos 1999c).

In pygmy marmosets, the relatively long forelimbs, the short radii, and the specialized shoulder joint anatomy, reflect their tendency to move frequently through vertical claw climbing and clinging (Davis 2002; Ford and Davis Chap. 21 this volume). The lateral placement of the forelimbs should facilitate both firm immobility and powerful movements upon large vertical supports (Cartmill 1985; Thorington and Thorington 1989). However, no similar trend has been observed among tamarins which frequently claw climb and cling vertically (Garber 1991; Garber and Leigh 2001). On the other hand, pygmy marmosets appear to lack certain functional specializations of the scapula that characterize other vertically clinging callitrichids (Davis 2002; Ford and Davis Chap. 21 this volume). It is very likely that the particularly reduced size of pygmy marmosets has imposed different mechanical constraints on the postcranium. In addition, the fact that pygmy marmosets also move frequently using the quadrupedal walk, bound, and clamber in both travel and foraging may account for the differences between pygmy marmoset morphology and that of the other callitrichids.

Contrary to previous assumptions (Kinzey et al. 1975), pygmy marmosets are not particularly frequent leapers compared to the other callitrichids. Leaping between terminal supports was the main gap crossing traveling mode for pygmy marmosets, and all terminal leaps (both short and long) appeared to be both initiated and terminated on supports <5 cm. This pattern of initial and landing support use, in what is usually termed as acrobatic leaps, is also encountered in the other callitrichids that use high proportions of leaping behavior (Garber 1992; Garber and Pruetz 1995; Garber and Leigh 2001; Cunha et al. 2006; Garber et al. Chap. 14 this volume). In effect, most tamarins employ similar (or greater) frequencies of terminal leaping as the pygmy marmosets do, but only a few species engage in similar frequencies of vertical leaping (Garber 1991; Youlatos 1999a; Garber and Leigh 2001; Youlatos and Gasc 2001; Porter 2004). In contrast, callimicos are unique among all the callitrichids, in their common use of a very specialized bounding hop, as well as the terminal and vertical leaps (Rosenberger and Stafford 1994; Garber and Leigh 2001). In the latter, they appear to be particularly proficient (see Garber et al. 2005). These differences in leaping behavior may explain the particular postcranial morphology of pygmy marmosets which lacks considerable adaptations related to vertical leaping compared to the highly specialized hind limb morphology of callimico (Davis 2002).

In contrast to terminal branch leaping, vertical leaps were initiated from larger supports. In addition, longer vertical leaps initiated on larger supports than shorter vertical leaps, and all primarily terminated on supports of a diameter under 10 cm. Vertical leap initiation from large diameter supports appears to be a means to reduce energy loss since support compliance decreases as support diameter increases (Crompton et al. 1993). Leapers tend to prefer large initial supports, particularly small mammals whose take-off forces are higher than those of larger mammals (Demes et al. 1995). This energy avoidance strategy may also explain why pygmy marmosets initiate longer vertical leaps from significantly larger supports than shorter vertical leaps. Overall, it can be hypothesized that smaller leaping callitrichids would tend to prefer larger initial supports (at least relative to their body sizes) in higher frequencies than larger leaping primates. Unfortunately, detailed data for similar comparisons are unavailable, but this pattern was not supported by observations in sympatric tamarins and callimicos (Garber and Leigh 2001).

Demes et al. (1995) also suggest that, as landing forces are reduced by support compliance, arboreal leapers tend to select landing supports that are usually smaller in diameter. This pattern was partly observed in the vertical leaps of pygmy marmosets. That would imply that pygmy marmosets most likely try to reduce these forces by opting for the more suitable supports in the understory, trying to reduce physical exertion as an energy saving strategy.

The analysis of the positional behavior of pygmy marmosets has shown that claw climbing, claw clinging, and vertical leaping may not be specifically related to each other and may be adaptations for different behavioral contexts. In callitrichids, extensive use of claw climbing and clinging on large vertical supports appears to be related to a specific foraging strategy of exploiting vertical resources, such as bark dwelling insects and exudates (Garber 1992). This is further supported by the relatively low percents of clawed activities in callimicos that engage infrequently in exudate feeding (but see Porter et al., submitted) or foraging on bark-dwelling arthropods (Garber and Leigh 2001; Porter and Garber 2004). In the case of pygmy marmosets, the extreme evolutionary reduction of body size is very likely to be associated with the remarkably high frequency of these vertical clawed activities (see also Ford 1986). On the other hand, in other species, claws may be used primarily for traveling within the understory where vertical supports are abundant and travel is facilitated by vertical clawed activities. For example, squirrels that use the lowest forest layers frequently engage in vertical leaping (Youlatos 1999c), and callimicos which exploit certain food resources found on tree trunks, such as bamboo fungus, and use vertical supports to locate insects, small vertebrates, and fungi on or near the ground, often engage in vertical travel in the understory (Porter and Garber 2004).

The results of the present report do not lend support to clear relationships between morphology and positional behavior among callitrichids. Further and more detailed analyses of postcranial morphology, coupled with experimental approaches of vertical leaping kinematics and biomechanics (see Garber et al. Chap. 14 this volume) and detailed analyses of the positional behavior of marmosets will establish more robust, morpho-functional associations. In addition, more detailed data on the manipulative behavior of these primates during foraging will also help clarify the relationship between morphology and behavior (see Bicca-Marques 1999). Moreover, the newly discovered dwarf marmoset (*Callibella humilis*), which seems to be closely related to both the Amazonian marmosets (*Mico*) and pygmy marmosets, appears to be of intermediate morphology between the two genera (Van Roosmalen and van Roosmalen 2003, Ford and Davis Chap. 21 this volume) and should help clarify how adaptations in morphology and behavior have evolved among the callitrichids.

Acknowledgments I am particularly indebted to Drs. SM Ford, LC Davis, and LM Porter who invited me to contribute to this volume. Field research was made feasible by the financial support of NSF (SBR 9222526), attributed to Dr. JGH Cant, University of Puerto Rico, to whom my warm thanks are addressed. In addition, I am indebted to Dr. PS Rodman, University of California at Davis, for his assistance in locating the study site and logistics, Dr. L Albuja V, Escuela Politécnica Nacional in Quito, for a wide variety of essential assistance, Mr. I Cornejo P, for his help in transportation, Dr. L Arcos Teren, Pontificia Universidad Católica del Ecuador, for arranging lodging at the research station, and the Ecuadorian Institute of Forestry and Natural Areas (INEFAN) for permission to conduct study in the Yasuni National Park. Many thanks are due to Dr. WE Pozo for his help during data collection, and to Drs. A DiFiore and L Dew for their assistance in the field.

References

Arms A, Voges D, Fischer M, Preuschoft H (2002) Arboreal locomotion in small New World monkeys. Z Morphol Anthropol 83:243–263

Bicca-Marques JC (1999) Hand specialization, sympatry, and mixed-species associations in callitrichines. J Hum Evol 36:349–378

Cartmill M (1974) Pads and claws in arboreal locomotion. In: Jenkins FA Jr (ed) Primate locomotion. Academic Press, New York, pp 45–83

Cartmill M (1985) Climbing. In: Hildebrand M, Bramble DM, Liem KF, Wake KD (eds) Functional vertebrate morphology. Belknap Press, Cambridge, pp 73–88

Charles-Dominique P (1977) Ecology and behaviour of nocturnal primates. Columbia University Press, New York

Cortés-Ortiz L (this volume) Molecular phylogenetics of the Callitrichidae with an emphasis on the marmosets and *Callimico*. In: Ford SM, Porter LM, Davis LC (eds) The smallest anthropoids: The marmoset/callimico radiation. Springer Press, New York, pp 3–24

Crompton RH (1984) Habitat structure, foraging, and locomotion in two species of *Galago*. In: Rodman PS, Cant JGH (eds) Adaptations for foraging in non-human primates. Columbia University Press, New York, pp 73–111

Crompton RH, Sellers WI, Gunther MM (1993) Energetic efficiency and ecology as selective factors in the saltatory adaptation of prosimian primates. Proc R Soc Lond B 254:41–45

Cunha AA, Vieira MV, Grelle CEV (2006) Preliminary observation on habitat, support use and diet in two non native primates in an urban Atlantic forest fragment: the capuchin monkey (*Cebus* sp.) and the common marmoset (*Callithrix jacchus*) in the tijuca forest, Rio de Janeiro. Urban Ecosyst 9:351–359

Davis LC (2002) Functiona morphology of the forelimb and long bones in the Callitrichidae (Platyrrhini, Primates). PhD Dissertation, Southern Illinois University, Carbondale

Demes B, Jungers WL, Gross TS, Fleagle JG (1995) Kinetics of leaping primates: Influence of substrate orientation and compliance. Am J Phys Anthrop 96:416–429

Fleagle JG, Mittermeier RA (1980) Locomotor behavior, body size, and comparative ecology of seven Surinam monkeys. Am J Phys Anthrop 52:301–314

Ford SM (1980) Callithricids as phyletic dwarfs, and the place of the Callithricidae in the Platyrrhini. Primates 21:31–43

Ford SM (1986) Comments on the evolution of claw-like nails in Callithricids (marmosets/tamarins). Am J Phys Anthrop 70:25–26

Ford SM, Davis LC (1992) Systematics and body size: implications for feeding adaptations in New World monkeys. Am J Phys Anthropol 88:415–468

Ford SM, Davis LC (this volume) Marmoset postcrania and the skeleton of the dwarf marmoset, *Callibella humilis*. In: Ford SM, Porter LM, Davis LC (eds) The smallest anthropoids: The marmoset/callimico radiation. Springer Press, New York, pp 411–448

Garber PA (1980) Locomotor behavior and feeding ecology of the Panamanian tamarin (*Saguinus oedipus geoffroyi*, Callithricidae, Primates). Int J Primatol 1:185–201

Garber PA (1991) A comparative study of positional behavior in three species of tamarin monkeys. Primates 32:219–230

Garber PA (1992) Vertical clinging, small body size, and the evolution of feeding adaptations in the Callitrichinae. Am J Phys Anthrop 88:469–482

Garber PA, Leigh SR (2001) Patterns of positional behavior in mixed-species troops of *Callimico goeldi*, *Saguinus labiatus*, and *Saguinus fuscicollis* in northwestern Brazil. Am J Primatol 54:17–31

Garber PA, Pruetz JD (1995) Positional behavior in moustached tamarin monkeys: Effects of habitat on locomotor variability and locomotor stability. J Hum Evol 28:411–426

Garber PA, Sussman RW (1984) Ecological distinctions in sympatric species of *Saguinus* and *Sciurus*. Am J Phys Anthrop 65:135–146

Garber PA, Blomquist G, Anzenberger G (2005) Kinematic analysis of trunk-to-trunk leaping in *Callimico goeldii*. Int J Primatol 26:223–240

Garber PA, Sallanave A, Blomquist G, Anzenberger G (this volume) A comparative study of the kinematics of trunk-to-trunk leaping in *Callimico goeldii*, *Callithrix jacchus*, and *Cebuella pygmaea*. In: Ford SM, Porter LM, Davis LC (eds) The smallest anthropoids: The marmoset/callimico radiation. Springer Press, New York, pp 259–277

Hamrick MW (1998) Functional and adaptive significance of primate pads and claws: Evidence from New World anthropoids. Am J Phys Anthrop 106:113–127

Hershkovitz P (1977) Living New World monkeys (Platyrrhini), vol 1. University of Chicago Press, Chicago

Jackson CP, Ford SM (2006) Contexts of positional behavior in captive pygmy marmosets (*Cebuella pygmaea*). Am J Phys Anthrop S42:109

Jacobs J (1974) Quantitative measurement of food selection. Oecologia 14:413–417

Kinzey WG, Rosenberger AL, Ramirez M (1975) Vertical clinging and leaping in a neotropical anthropoid. Nature 255:327–328

Neusser M, Stanyon R, Bigoni F, Wienberg J, Muller S (2001) Molecular cytotaxonomy of New World monkeys (Platyrrhini): comparative analysis of five species by multi-color chromosome painting gives evidence for a classification of *Callimico goeldii* within the family Callitrichidae. Cytogenet Cell Genet 94:206–215

Off EC, Gebo DL (2005) Galago locomotion in Kibale National Park, Uganda. Am J Primatol 66:189–195

Porter L (2004) Forest use and activity patterns of *Callimico goeldi* in comparison to two sympatric tamarins, *Saguinus fuscicollis* and *Saguinus labiatus*. Am J Phys Anthrop 124:139–153

Porter LM, Garber PA (2004) Goeldi's monkeys: A primate paradox? Evol Anthropol 13:104–115

Rosenberger AL (1977) *Xenothrix* and ceboid phylogeny. J Hum Evol 6:461–481

Rosenberger AL (1992) Evolution of feeding niches in New World monkeys. Am J Phys Anthropol 88:525–562

Rosenberger AL, Stafford BJ (1994) Locomotion in captive *Leontopithecus* and *Callimico*: a multimedia study. Am J Phys Anthrop 94:379–394

Soini P (1988) The pygmy marmoset, genus *Cebuella*. In: Mittermeier RA, Rylands AB, Coimbra-Filho A, Fonseca GAB (eds) Ecology and behavior of neotropical primates. WWF, Washington, DC, pp 79–130

Sussman RW, Kinzey WG (1984) The ecological role of the Callithricidae: A review. Am J Phys Anthrop 64:419–449

Thorington RW Jr, Thorington EM (1989) Postcranial proportions of *Microsciurus* and *Sciurillus*, the American pygmy tree squirrels. Adv Neotrop Mammal 1989:125–136

Townsend WR (2001) *Callithrix pygmaea*. Mammal Spec 665:1–6

Van Roosmalen MGM, van Roosmalen T (2003) The description of a new marmoset genus, *Callibella* (Callitrichinae, Primates), including its molecular phylogenetic status. Neotrop Prim 11:1–10

Warren RD (1997) Habitat use and support preference of two free-ranging saltatory lemurs (*Lepilemur edwardsi* and *Avahi occidentalis*). J Zool Lond 241:325–341

Yoneda M (1984) Comparative studies on vertical separation, foraging behavior and traveling mode of saddle-backed tamarins (*Saguinus fuscicollis*) and red-chested moustached tamarins (*Saguinus labiatus*) in northern Bolivia. Primates 25:414–422

Youlatos D (1999a) Comparative locomotion of six sympatric primates in Ecuador. Ann Sci Nat Zool Biol Anim 20:161–168

Youlatos D (1999b) Positional behavior of *Cebuella pygmaea* in Yasuni National Park, Ecuador. Primates 40:543–550

Youlatos D (1999c) Locomotor and postural behavior of *Sciurus igniventris* and *Microsciurus flaviventer* (Rodentia, Sciuridae) in eastern Ecuador. Mammalia 63:405–416

Youlatos D (2004) A multivariate analysis of organismal and habitat parameters in two neotropical primate communities. Am J Phys Anthrop 123:181–194

Youlatos D, Gasc J-P (2001) Comparative positional behaviour of five primates. In: Bongers F, Charles-Dominique P, Forget M, Théry M (eds) Nouragues: Dynamics and plant-animal interactions in a neotropical rain forest. Kluwer Academic Publishers, Dordrecht, pp 103–114

Zar JH (1996) Biostatistical analysis. Prentice-Hall, London

Section IV
Anatomy

Susan M. Ford, Leila M. Porter, Lesa C. Davis

Many aspects of the gross anatomy of the common marmoset, *Callithrix jacchus*, have been reported, as it has been widely used in laboratory studies for many decades. W.C.O. Hill published a detailed gross anatomical study of the callimico (*Callimico goeldii)* in 1959, based on limited study. But little is known of the anatomy of most other marmosets, and only a few comparative studies have been attempted. The contributions in this section include both functional and descriptive studies and comparisons, looking at reproductive features, cranial shape, structure, and function, and postcranial morphology.

The marmoset and tamarin pattern of twinning with a simplex uterus and the anastomosis of fetal blood flow is unique among mammals, and many of the specific details of uterine morphology and fetal development that may accompany this distinctive reproductive pattern remain a mystery. Rutherford and Tardiff (Chap. 16) compare placental weights from twin and triplet fetuses of the common marmoset (*Callithrix jacchus*), demonstrating that placental size reflects the conflicts between maternal and fetal resource needs. Triplets appear to pursue different strategies for resource capture, strategies that may ultimately prove detrimental to post-uterine survival.

Two chapters compare the cranial morphology of callimicos and marmosets, asking slightly different questions and using different analytical techniques. Marroig and Cheverud (Chap. 17) explore the allometric trajectories, finding that many of the cranial differences can be explained by the size differences across this clade marked by increasing miniaturization, and they suggest prenatal size-diminution may have been the critical heterochronic process. Aguiar and Lacher (Chap. 18) provide a thorough review of what is known (and not known) of the newest marmoset, the dwarf *Callibella humilis,* and its discovery. Both studies confirm that *Callimico* exhibits a distinctive cranial shape, and pygmy marmoset (*Cebuella*) crania are more like those of *Callithrix* than *Mico*, possibly related to continuum in degrees of tree-gouging behavior. The dwarf marmoset, *Callibella,* is still known only by a very few specimens and is reported to be from a tiny geographic range.

Susan M. Ford
Department of Anthropology, Southern Illinois University, Carbondale, IL 62901, USA
e-mail: sford@siu.edu

Although most like the pygmy marmoset, its cranium is clearly distinct, particularly in its mandibular morphology.

Tree gouging in marmosets and exudativory present special challenges to the cranial structure of these diminutive forms, which are met in part by the lack of lingual enamel on their lower incisors and by very large gapes. Here, Taylor et al. (Chap. 19) and Vinyard et al. (Chap. 20) continue their ongoing studies of the structural correlates of tree-gouging behavior. Taylor et al. find that muscle fiber length and architecture of the chewing muscles (masseter and temporalis) facilitate stretch and large gape but do not support increased force or load resistance. Vinyard et al. document that, indeed, marmosets are not exerting extreme forces during gouging, whether in the lab or in the field, and their bony architecture supports wide gape but not force. It appears that it is not the effort or depth of the gouge that brings forth exudate flow, but something related to the wide gape, long excursion, and ability to apply light but continuous force efficiently throughout the gouging activity that allows marmosets to be successful exudate harvesters.

In the final chapter of this section, Ford and Davis (Chap. 21) compare the appendicular skeletons of marmosets and callimicos. Skeletal features document even greater differentiation between the genera than do cranial features, with *Callimico* and each of the marmosets being quite distinct. A scenario of marmoset evolution is offered, with *Callithrix* least specialized postcranially, and the Amazonian taxa progressively more unique, but in different ways. The appendicular skeleton of the pygmy marmoset, *Cebuella*, is highly derived for long periods of clinging, upside down and under branch locomotion, but limited and only short-distance leaps. Although only a single skeleton of *Callibella* is available for study, it is distinctive in a number of aspects.

These studies together document the unique nature of marmoset adaptations to both diet and posture, the strong support anatomically for four marmoset genera including separate status for the poorly known but distinctive dwarf marmoset (*Callibella humilis)*, and the marked structural differences between *Callimico* and the marmosets. The discovery that placental structure shows adaptations to twinning suggests new avenues for research as we try to discover if callimicos had twins in their past. And the unique cranial and, particularly, postcranial adaptations of the strikingly patterned dwarf marmoset (see Fig. 1 in the Preface), apparent even from the few individuals available for study, make it clear that this monkey needs further study and conservation of its small known range.

Reference

Hill WCO (1959) The anatomy of *Callimico goeldii* (Thomas). Trans Am Phil Soc 49:1–116

Chapter 16
Mother's Little Helper? The Placenta and Its Role in Intrauterine Maternal Investment in the Common Marmoset (*Callithrix jacchus*)

Julienne N. Rutherford and Suzette Tardif

Abstract Litter size variation within the Callitrichidae is the result of complex interactions among genetic and environmental factors, and occurs across many facets of the reproductive cycle, from ovulation number to intrauterine litter size reduction to neonate lactation competition. Selection appears to have acted upon the callitrichine ability to make "decisions" relative to maternal nutritional state and litter size in a way that has yielded a highly sensitive and plastic system. Natural variation in marmoset litter size, birth weights and placental weights, and maternal condition create an opportunity to test hypotheses related to intrauterine growth retardation and maternal investment. We present evidence suggesting that differences in fetal/placental weight ratios in marmosets represent distinct strategies of intrauterine resource solicitation by members of litters of different sizes as a result of prenatal parent-offspring conflict. Individual triplets are associated with a smaller share of the placenta by weight than are twins, suggesting a mechanism by which triplets increase placental efficiency in the face of finite maternal resources and uterine space constraints. Twin and triplet fetuses appear to pursue different intrauterine strategies for maximizing allocation of the maternal resources via the placenta. Since complete triplet litters are almost never successfully reared to weaning, maternal limitations of energy intake and investment in offspring from conception to weaning appear to be in conflict with the triplet strategy of optimizing the intrauterine environment through placental development and function.

Resumen El tamaño de la variación entre los Callitrichidae es el resultado de una compleja interacción entre factores genéticos y de medio ambiente, y parece ocurrir a lo largo de muchas facetas del ciclo reproductivo, desde el número de ovulación

J.N. Rutherford (✉)
Department of Oral Biology, College of Dentistry, University of Illinois at Chicago, 801 S Paulina, Chicago, IL 60612,
e-mail: ruther4d@uic.edu

S.M. Ford et al. (eds.), *The Smallest Anthropoids*, Developments in Primatology: Progress and Prospects, DOI 10.1007/978-1-4419-0293-1_16,

© Springer Science+Business Media, LLC 2009

a la reducción del tamaño de la fuente intrauterina a la competencia lactante del recien nacido. La selección parece haber actuado sobre la habilidad de los calitricidos de tomar "decisiones" relacionadas con el estado nutricional maternal y el tamaño de las camada en una forma que ha producido un sistema altamente sensitivo y plástico. La variación del tamaño de las camadas marmoset, sus pesos al nacer y pesos de las placentas, junto a la condición materna crearon la oportunidad de poner a prueba la hipótesis relacionada con el retraso del crecimiento intrauterino y la inversión materna. Este estudio presenta evidencias que sugieren que las diferencias en los índices de peso fetal y de placenta entre los marmosets podrían representar distintas estrategias en los recursos de incitación intrauterina por miembros de camadas de diferentes tamaños; todo ello como resultado de conflicto prenatal entre padres y descendencia. Los individuos trillizos están asociados a mayor compartimiento de la placenta que los gemelos, sugiriendo un mecanismo en el cual los trillizos solicitan crecimiento placental compensatorio en cara a recursos maternales finitos. Los fetos gemelos y trillizos parecen seguir diferentes estrategias intrauterinas para maximizar para la repartición de la placenta. Desde que las camadas de trillizos son casi nunca exitosas en la crianza durante el destete, las limitaciones maternas de energia admitida y la inversión en las crias desde la concepcion hasta el destete parecen estar en conflicto con las estrategias de los trillizos de optimizar el medio ambiente intrauterino a través del crecimiento de la placenta.

Resumo A variação entre tamanho da prole na família Callitrichidae é o resultado de situações complexas entre fatores genéticos e ambientais e parece ocorrer por meio de muitas facetas do ciclo reprodutivo, desde o número de ovulações até a redução do tamanho da prole intra-útero e a competição dos neonatos pela lactação. A seleção parece ter atuado sobre a habilidade dos calitriquídeos de tomar "decisões" relativas ao estado nutricional materno e tamanho da prole, que resultou na produção de um sistema altamente sensível e plástico. A variação natural no tamanho da prole em sagüi, do peso ao nascimento e do peso placentário e das condições maternas cria uma oportunidade para testar hipóteses relacionadas ao retardo no crescimento intra-uterino e ao investimento materno. Nós apresentamos evidências sugerindo que as diferenças na razão peso placentário/peso fetal em sagüi podem representar diferentes estratégias de solicitação de recursos intra-uterinos pelos membros das proles de diferentes tamanhos, resultante de um conflito pais-prole durante o período pré-natal. Tendo como base o peso, indivíduos de uma prole de triplos compartilham a placenta pelo peso mais do que os gêmeos sugerindo um mecanismo pelo qual os triplos requerem um crescimento compensatório da placenta em função dos recursos finitos da mãe. Fetos duplos e triplos parecem utilizar diferentes estratégias no útero da mãe para maximizar a alocação da placenta. Uma vez que as proles de triplos quase nunca são criadas com sucesso até o desmame, as limitações da mãe na ingestão de energia e investimento na prole, da concepção ao desmame, parece gerar um conflito com as estratégias de triplos de otimizar o ambiente intra-uterino por meio do crescimento placentário.

16.1 Introduction

When Trivers (1974) introduced the genetic explanation for parent–offspring conflict theory, he set in motion the dismantling of romantic notions of pregnancy as a time of unparalleled harmony and synchrony between the mother and fetus. There are real conflicts between the mother and fetus in terms of resource allocation throughout gestation. Since a fetus is more related to itself than to its mother or to nonmonozygous siblings, in the context of inclusive fitness it can be predicted to pursue a strategy of exploitation of maternal resources that may be at odds with the mother's interest in long term reproductive investment (Haig 1993, 1996; Long 2005). If fetal demands outstrip maternal resources, fetal growth can be compromised due to maternal strategies of prenatal investment. Long term effects of this prenatal conflict may include the programming of the fetus for susceptibilities to adult onset of diseases such as diabetes, obesity, and cardiovascular disease (Leon et al. 1996; Barker 1998; Barker et al. 2002; Kuzawa and Adair 2003), and possibly even serious mental illness (Wahlbeck et al. 2001; Mittendorfer-Rutz et al. 2004).

The function and structure of the organ supporting pregnancy, the placenta, and its role as one of the unusual features of gestation in the Callitrichidae are often given only a cursory mention in most primate biology or anatomy texts (e.g., Ankel-Simons 2000). However, the study of the unique structure and function of the callitrichid placenta has the potential to yield tantalizing evidence regarding phylogeny, maternal investment, and the development of powerful models of fetal programming and parent–offspring conflict.

In this chapter, we provide an overview of marmoset placental structure and function, then describe relations among placental measures, litter size, fetal growth, and maternal condition. These findings are discussed in relation to theories regarding control of maternal investment. Mothers may need to make a series of physiological "decisions" that will have a negative impact on fetal growth. This negotiation of maternal investment in the current pregnancy can lead to disruption of normal placental development and consequential and/or concomitant disruption of fetal development.

Evidence suggests that the marmosets and tamarins have similar features of growth and development of the fetus and placenta, albeit with variations in event timing, but placental and fetal growth is most extensively studied in *Callithrix jacchus* (common marmosets) (Chambers and Hearn 1985; Merker et al. 1988; Rutherford and Tardif 2008, 2009). The Southwest National Primate Research Center (SNPRC) houses the largest known collection of common marmoset naturally-delivered term placentas and serves as the basis for our studies. The description and theoretical interpretation of placentation events and features derived from this and other studies of *Callithrix* should be viewed as relevant for tamarins as well. Because *Callimico goeldii* (the callimico) has singleton births and because not enough is known about the timing and nature of developmental events in this species, the description of fetal and placental weights and their relations as well as discussions about intrauterine resource allocation should be taken to apply primarily to twin-bearing marmosets and tamarins, despite the close evolutionary relationship between *Callimico* and *Callithrix* (see review by Cortés-Ortiz, Chap. 1 this volume).

16.2 The Marmoset Placenta

16.2.1 The Marmoset Placenta in Comparative Context

The marmoset placenta, like that of most platyrrhines and many catarrhines regardless of litter size, is comprised of two separate placental discs (Fig. 16.1), whereas baboons, apes, and humans all have a single discoid placenta (Mossman 1987), except in the case of some multiple births (e.g., diamnionic/dizygotic twins with individual placentas). The presence of the relatively common bidiscoid placenta along with a simplex uterus and a single pair of pectorally located nipples is important evidence in the construction of mammalian phylogeny and the description of callitrichid twinning as a derived rather than a primitive condition (Ford 1980; Leutenegger 1980). The two discs of the callitrichine placenta fuse early in development and are connected by extensive vascular anastomoses, blood vessels that radiate from the umbilicus of each fetus out over the surface of the discs and traversing the placenta membranes, joining the discs to each fetus (Fig. 16.1; Wislocki 1932, 1939; Benirschke et al. 1962). Anastomotic placentas in human multiple pregnancy is a serious developmental abnormality that often underlies clinical presentations of twin-to-twin transfusion syndrome, a frequently fatal condition in

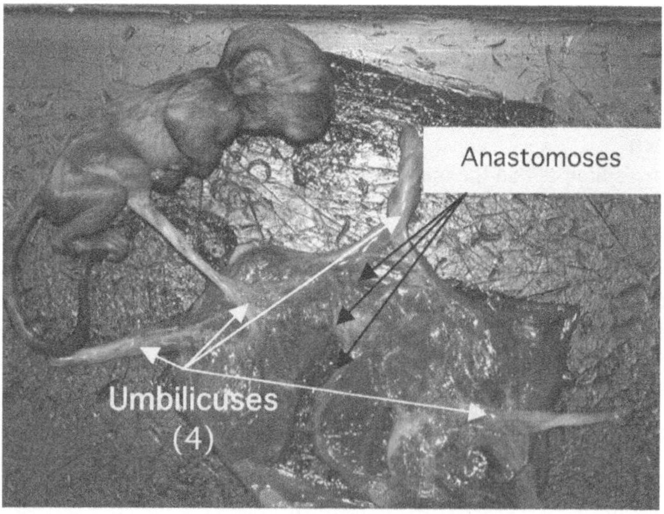

Fig. 16.1 Photo illustrating multiple (4) umbilical attachments to highly anastomosed bidiscoid placenta (photo by D. Layne Colòn)

which one twin receives an inordinate blood supply shunted from a "donor" twin frequently leading to the intrauterine demise of one or both twins (Benirschke 1995). This condition is not reported to occur in marmoset littermates. The extensive nature of the callitrichine placental anastomoses creates a more dispersed gradient for blood flow between the two discs, thus obviating the risk of a single vascular interconnection causing a unidirectional flow from one twin to the other. Haig (1999) suggests that the evolution of extensive placental anastomoses in marmosets may have occurred in response to the potentially lethal consequences of transfusion syndrome, and can thus be considered yet another component of the litter-bearing complex.

The result of this shared placental circulation early in embryonic development is a "prenatal exchange of circulating hematopoietic tissue" (Benirschke et al. 1962, p 513). Marmosets and tamarins are thus hematopoietic chimeras, meaning that each individual is a composite of cells from multiple individuals, (Benirschke et al. 1962; Benirschke and Brownhill 1962; Gengozian et al. 1964), a phenomenon that is recognized as one of the unique aspects of marmoset biology (Haig 1999). Chimeric cells are found in the blood, bone marrow, and spleen in the form of the nucleated blood cells (e.g., lymphocytes). Chimerism has been demonstrated rarely in human twins (Benirschke and Brownhill 1962) and is not known to be a feature of normal fetal twin development in any species other than marmosets and tamarins. In cattle, when heterosexual twin pairs occur, shared placental circulation results in female exposure to male fetal androgenic steroids, leading to the masculinization of genitalia and gonads and ultimately causing sterility, i.e., "freemartinism" (Ono 1969). This does not occur in the marmoset, suggesting there is insensitivity to fetal androgens on the part of the female fetus. The absence of freemartinism in the marmoset has been interpreted as evidence that dizygotic twinning is a specialized trait in this primate group (Haig 1999).

Gengozian et al. (1980) rejected the idea that chimerism extended beyond blood cells to include the germ line. However, Ross and colleagues (Ross et al. 2007) have recently identified chimeric cells in a broad range of somatic tissues of *Callithrix kuhlii*, including the germ cells. If this observation bears out, the implication is that relatedness and inclusive fitness in this species is a far more complex phenomenon than has ever been considered among mammals. This would obviously have a significant impact on our understanding of marmoset reproductive biology and the evolution of twinning and cooperative infant care; this body of work is fascinating in its potential and further inquiry in the interest of substantiating these interesting results and their attendant theoretical ramifications is underway.

Some researchers have referred to individual placental discs as primary or secondary on the basis of weight (e.g., Merker et al. 1988). However, we discourage this usage as these appellations imply functional distinctions that are not likely to exist with any regularity. Underscoring the unity of function demonstrated by the two placental discs (beyond their highly anastomotic nature) is the random way in which multiple marmoset fetuses are attached to the discs. An early observation by Wislocki (1932) suggested that in a twin marmoset litter, the smaller of the two fetuses would be attached to the smaller of the two discs. However, the work of Wislocki (1939)

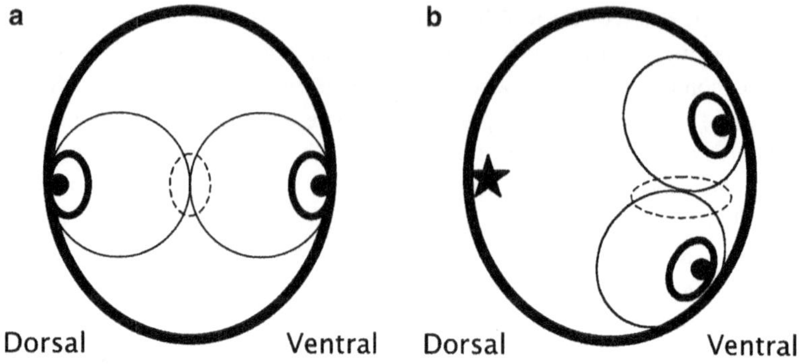

Fig. 16.2 Blastocyst implantation in the marmoset. Dashed circle illustrates region of eventual chorionic fusion and rupture, thus creating common exocoelomic cavity. (**a**) Blastocysts implant on opposing walls of the uterus. Each implantation site forms a separate placental disc and the developing embryos attach to their own disc. (**b**) The blastocysts implant side by side on the ventral wall and attach to a single disk. The star indicates the site of secondary disc formation at the abembryonic pole

revealed that patterns of placental attachment relate to blastocyst implantation, such that blastocysts that implant adjacent to each other on the same portion of the uterine wall (e.g., both on the ventral wall) share umbilical attachment to a single disc. In contrast, blastocysts implanting at points opposite each other (both the ventral and dorsal walls) will develop umbilical attachments separately (Fig. 16.2). A recent analysis of common marmoset placentas has shown that there is no significant difference in weight between the two discs (Rutherford 2007), confirming the finding of an earlier work by Chambers and Hearn (1985). Finally, of five triplet placentas for which individual disc weights and umbilical cord insertion number have been recorded at SNPRC, only one case had the greater number of cord insertions in the larger of the two discs (Rutherford, unpublished observations). These observations combined with demonstrations of microstructural identity between discs (Enders and Lopata 1999; Wynn et al. 1975) lead us to argue that the two placental discs are in effect a single functionally integrated organ supporting marmoset pregnancy, and analyses of function of an individual disc can be applied to interpretations of the function of the whole.

16.2.2 Implantation and Early Placental Growth and Development

Like other anthropoid primate placentas (including those of humans), the marmoset placenta is hemochorial, meaning that a single chorionic layer separates the fetal and maternal circulations (Luckett 1974; see Fig. 16.3). At term, the hemochorial placental barrier consists primarily of one layer of trophoblast cells and the closely apposed endothelium of the fetal capillaries. These two cell layers share a fused

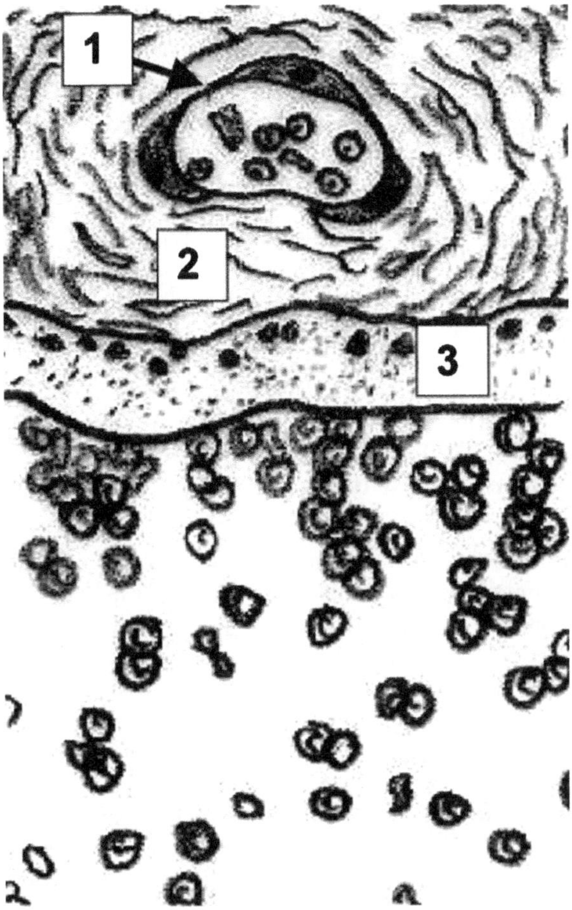

Fig. 16.3 The hemochorial placenta, separating the fetal and maternal circulations by three tissue layers: (1) Fetal capillary endothelium. (2) Villous mesoderm. (3) Trophoblast

basement membrane. Fetal tissues are directly nourished by the gas and nutrient content of maternal blood. This is the most intimate of placental arrangements (Mossman 1987) and may be related to the increased demand for gases and nutrients later in gestation due to primate brain growth. However, other mammals not noted for a primate-like fetal brain/body ratio also have hemochorial placentas (e.g., guinea pigs, rabbits), and some cetaceans, with brain/body ratios equal to or in excess of that of primates, have an epitheliochorial placenta (Haig 1993). This form introduces another barrier between maternal and fetal circulations, suggesting some effects of phylogenetic or perhaps even locomotor and mechanical constraints on placental development.

The callitrichid placenta differs from other anthropoid placentas by taking on a trabecular arrangement of the tissue at the maternal–fetal interface. Whereas the tissue in contact with the maternal blood supply is ramified into small fingerlike-projections (i.e., villi) in the placentas of the old world monkeys, apes, and humans, the callitrichid placenta is finely layered in thin interconnected sheets called trabeculae, with only the regions closest to the maternal surface appearing as free villous structures. Although the cellular components of the chorion are the same in all anthropoid primates, this trabecular arrangement is a distinguishing feature of the callitrichid placenta. Benirschke (2002) describes the placenta of squirrel monkeys as having a trabecular/villous arrangement and Mossman (1987) describes the howler monkey placenta as trabecular (but see Benirschke 2002), suggesting the possibility that this structural feature of the placenta was shared by a common ancestor of the New World primates, but most extensively expressed by the marmosets and tamarins.

16.3 Litter Size Variation in Marmosets and Its Relation to Maternal Investment

Twinning in the marmosets and tamarins is one of a suite of derived traits including loss of the third molar, reduced complexity of the upper molars, clawlike nails, and allometric scaling of the eye, and are widely regarded as related to phyletic dwarfism, or a reduction in body size over evolutionary time (Ford 1980; Martin 1992). A variety of mechanisms for the evolution of twinning in the Callitrichidae have been proposed. These can broadly be divided into predation, ecological, and obstetric explanations, the tenets of which are not necessarily exclusive from one another. Eisenberg (1981) has argued that twinning is a response to increased predation pressures arising from decreased body size. Twinning would be favored if high infant mortality were common, and would co-evolve with behavioral strategies related to predation avoidance, such as increased intragroup cooperation (Caine 1993). The cryptic antipredation adaptations in these small-bodied primates suggest that cooperation, particularly with reference to infant care, is necessary; individuals carrying offspring must usually hide and remain still to avoid predation and such behavior would have a deleterious effect on foraging efficiency if infant care were restricted to only one individual (Tardif and Jaquish 1994). However, predation pressures and cooperation may be more relevant to the evolution of cooperative care of infants than to twinning per se. For example, *Callimico* exhibits a high degree of social cooperation and communal infant care that is decoupled from twinning (see Porter and Garber, Chap. 4 this volume).

Leutenegger (1973, 1980) proposed the obstetric hypothesis, suggesting that twinning in the Callitrichidae evolved directly as a response to phyletic dwarfism. The ratio of fetal weight to maternal weight is high for small-bodied primates; Leutenegger (1973) suggested that this large fetal size makes delivery difficult and thus total fetal mass is divided across multiple fetuses to allow the mother to carry to term the maximum mass possible while limiting head circumference. Whereas stabilizing selection on the relation between fetal head circumference and the dimensions of the vaginal canal would constrain fetal size, particularly in small-bodied species that produce infants with relatively large brains, it is not clear how this relates to the number of fetuses produced per reproductive event (Goldizen 1990; Martin 1992). Small body size does not exhibit a default relation to twinning within all members of the Callitrichidae. *Callimico goeldii* exhibits adult weights similar to those found in the twin-bearing *Callithrix* (marmosets), *Saguinus* (tamarins), and *Leontopithecus* (lion tamarins), but produces a single infant. Birth weights for these single neonates are on average absolutely larger than, but relative to adult body weight and do not differ from those for callitrichine twins (Garber and Leigh 1997). Among other primates, the ratio of fetal weight to maternal weight is highest in two species of tarsiers (*Tarsier bancanus* and *T. syrichta*), higher than that seen in any of the callitrichid primates (Martin 1992). Martin (1992) points out that these primates would be expected to have difficulty during parturition and should produce smaller twins rather than a single large infant but this is not the case. Further, when marmosets do give birth to singletons, although birth weights tend to be higher (Chambers and Hearn 1985; Jaquish et al. 1995), they do not appear to be related to abnormal gestations and deliveries caused by high fetal weight and large head circumference. Such complications include dystocia, a condition in which delivery of the fetal shoulder is obstructed by the mother's pubic symphysis. This situation is frequently exacerbated by abnormally large fetal size, which in humans is a frequent complication of gestational diabetes (Clausen et al. 2005). In marmosets, however, only a single case of dystocia has been reported in the literature (Lunn 1980) and is not known to be a sequelum of singleton deliveries. Indeed, dystocia leading to fetal and maternal death has been observed in a triplet pregnancy in which both maternal and fetal weights were abnormally high (Rutherford and Tardif, unpublished observations). In sum, marmosets happen to be small and happen to twin, but there is poor evidence indicating that the former causes the latter. Goldizen (1990) suggests there must be benefits of twinning other than constraining fetal size, because a more parsimonious solution would be to evolve a smaller single fetus.

It seems likely that twinning is related to a change in resource availability that supported an increase in maximum intrinsic rate of population growth (r_{max}) in a way that is related to small body size, possibly as small-bodied marmosets and tamarins began to exploit an insectivorous/gummivorous ecological niche (Martin 1992). These resources are available in niches such as low shrubs and terminal branches that would be unable to support the body weight of larger primates (Eisenberg 1981). Further, exudates are a relatively stable resource in space and time. In a recent cladistic analysis of the Callitrichidae, Ah-King and Tullberg (2000) found that the appearance of twinning is linked to exudate feeding but not to reduction in body size. They point out that the callimico, which is similar in size to the other callitrichids but does not twin, also does not exploit plant gums as a food resource (Ah-King and Tullberg 2000). Exploitation of a temporally stable feeding niche unavailable to larger-bodied primates may have favored an increase in population growth. If the environment in which marmosets and tamarins evolved favored an increase in reproductive rate, a genetic or epigenetic trait for variable litter size could rapidly spread through time and be shaped by selection into a highly specialized system supported by changes in behavioral traits related to infant care.

Although marmosets and tamarins regularly produce twins, triplets are common in captivity, and quadruplets and even quintuplets have been observed (Rothe et al. 1992; Tardif et al. 2003; Rutherford and Tardif, unpublished observations), making it possible to investigate questions relating to litter size variation. Tardif and Jaquish (1997) found that ovulation number was related to body size, such that larger females had higher ovulation numbers, and that within females, individuals weighed more when ovulating 3–4 ova than when ovulating only 1–2. This finding suggests that potential reproductive output is sensitive to immediate energy availability. Sensitivity to environmental conditions in the form of current maternal energy supply may be one of the key determinants of reproductive output in marmosets. This interpretation is further supported by the low repeatability of both ovulation number ($r=0.081$, Tardif and Jaquish 1997) and litter size ($r=0.128$, Jaquish et al. 1991) within individual females, suggesting low heritability for these traits.

Litter size reduction throughout gestation is common (Jaquish et al. 1996; Windle et al. 1999). Jaquish et al. (1996) found that all singleton litters in a sample of pregnancies followed by serial ultrasonography started out as twin litters, and two cases in which triplet litters ultimately produced twins. Windle et al. (1999) have used both serial ultrasonography and hysterotomy to demonstrate litter size reduction during both early embryonic development and the later fetal period. At term, some littermates may be stillborn whereas other infants in the same litter are born healthy and survive to weaning. Postnatal mortality in one or more littermates with the survival of the rest is common as well, with triplets experiencing higher mortality rates than twins (Jaquish et al. 1991). There appear to be many points along the path from ovulation to weaning during which litter size in marmosets can be negotiated. This variability of reproductive output has likely been an important point of selection throughout the evolutionary history of the marmosets and tamarins (Tardif and Jaquish 1997). Callimicos, producing singletons but possibly having an earlier age of first reproduction than marmosets and tamarins (Martin 1992), may have exploited different pathways to respond to selective forces to optimize reproductive output.

16.4 The Model: Intrauterine Growth Retardation and Elevated Conflict Over Resource Allocation

16.4.1 Maternal Effects on Fetal, Postnatal, and Placental Growth

In marmosets, aspects of fetal and postnatal growth are related to maternal condition (e.g., weight and age) in complicated ways. Tardif and Jaquish (1994) found that larger females have higher ovulation numbers, and larger mothers have larger litters (Tardif and Jaquish 1994, 1997). Whereas birth weight is unrelated to maternal nonpregnant adult weight, it is related to maternal age, with older mothers having heavier offspring (Tardif and Bales 2004). Birth weight is also related to litter size, such that it decreases as litter size increases (Chambers and Hearn 1985). Since larger mothers have larger litters, fetal growth (as indicated by weight at birth) is a function, at least to some degree, of maternal weight.

Tardif and Bales (2004) also found that older mothers seemed able to support greater fetal long bone growth as reflected by knee-heel length, but only if they were also in the medium-to-high weight category. Low birth weight infants exhibited low initial postnatal weight gain (g/day). Being born to a low adult weight mother flattened but did not eliminate this effect, such that the difference in g/day between low and high birth weight infants was much greater in offspring of high weight mothers. High birth weight offspring of high weight mothers put on more weight per day than high birth weight offspring of low weight mothers, even when birth weights were the same. The significance of these relations is currently unclear, but underscores the complicated, multigenerational effects of birth weight on reproductive outcome and fetal development, and ultimately, the evolution of life histories (e.g., Kuzawa 2005).

Comparative and clinical studies suggest a role for maternal age and weight in placental development as well. Steven-Simons et al. (1995) found that placental weight in humans is significantly positively related to both prepregnant weight and to pregnant weight gain. Women starting pregnancy at higher weight produce heavier placentas. Taricco et al. (2003) reported that placental weights are increased in pregnancies complicated by gestational diabetes, a condition often associated with maternal obesity. Rahima and Bruce (1987) found that older rats have significantly heavier placentas compared to younger rats, even though fetal weights do not differ significantly.

16.4.2 The Fetal/Placental Weight Ratio and Intrauterine Conflict

Coall and Chisholm (2003), using parent–offspring conflict theory (Trivers 1974) as their foundation, predicted that if a mother attempts to restrict allocation of resources to her fetus, the fetus will respond by resisting that restriction. An increase in the number of fetuses may mimic this kind of restriction, particularly if not met by a concomitant increase in energy intake on the part of the mother. In the face of this potential shortfall, the triplet fetus may be capable of launching a resistant counter-strategy. Haig (1993) suggests that a fetus could respond to restriction of maternal resources by "increasing its absolute allocation to placental growth" (p 500). In effect, placental overgrowth should be solicited if an increase in the overall amount of placental tissue confers some in utero survival benefit to the fetus. This overgrowth would reduce the ratio of fetal to placental weight, (i.e., in this condition, one gram of placenta would produce fewer grams of fetus than in the "normal" state), providing a glimpse into the relative quality of the intrauterine environment. This ratio is known to be reduced in cases of intrauterine growth retardation in humans as a result of maternal hypoxia (Ali 1997), anemia (Howe 1994; Wheeler 1994), and famine (Lumey 1998), as well as in experimentally induced maternal nutritional restriction (rats: Langley-Evans et al. 1996; Doherty et al. 2003; sheep: Robinson et al. 1994). Maternal undernutrition can result from restriction of energy intake, but this restriction of energy available to fetal development could also lie at the interface of increased fetal demand as a function of litter size and relative per fetus resource allocation by the mother. In marmosets, mothers do not increase energy intake during gestation (Nievergelt and Martin 1999), so an additional fetus could represent restriction of resources available for fetal development, thus creating an environment of elevated conflict over maternal investment. Further, differences in maternal condition could relate to the quality of the intrauterine environment, mitigating or elevating conflict.

16.5 The Study: Assessing the Effects of Litter Size and Maternal Condition on Fetal and Placental Relations in the Marmoset

To test predictions of an intrauterine conflict model in marmosets, we first explored the hypothesis that triplet marmosets, experiencing potential restriction to maternal resources and at greater risk of mortality both pre- and postnatally, will be associated with a relatively larger share of the placenta than will twins (Rutherford and Tardif 2008). Those findings are summarized here. Second, we present our investigation of how maternal age, nonpregnant weight, pregnant weight gain, and maternal birth weight ("maternal condition") are related to each other, and to placental weight, neonate weight, and fetal/placental weight ratios. We expect that the greatest conflict between mother and offspring over resource allocation would occur in triplet pregnancies carried by small females because this would seem to be a particularly energetically costly condition; therefore we predict that these pregnancies will be characterized by the lowest F/P ratios. No doubt, variance in intrauterine sibling competition for resources plays an important role in shaping this environment, but those interactions are not addressed here.

16.5.1 Methods

To determine the nature of litter size variation in the ability of the placenta to support fetal growth, we totaled birth weights within litters, for total litter weight, and divided this by the total placental weight. This is the fetal/placental weight (F/P) ratio. It should be stated that since these were weights taken at birth, the term "neonatal weight" is more accurate, but in the interest of nomenclature conventions within the literature, we will call it "fetal weight." The ratio of fetal weight to placental weight is commonly termed placental efficiency since this relationship describes the amount of fetal mass supported by a unit of placental mass (e.g., Wilson et al. 1999;

Wilson and Ford 2001; Mesa et al. 2003; Dwyer et al. 2005). Only pregnancies for which we had both placental and fetal weights were included in analyses of relations to the F/P ratio, yielding a sample of 28 pregnancies from 19 dams. A sample of 29 pregnancies from a total of 19 dams was included in the analyses of maternal birth weight and age. Nineteen pregnancies from a total of 13 dams were included in analyses of nonpregnant weight. Fifteen pregnancies from a total of 9 dams were used for analyses of pregnant weight gain. Because repeatability of ovulation number and litter size is low (Tardif and Jaquish 1997), each pregnancy is treated as an independent event.

To differentiate the effects of nonpregnant weight on placental and fetal growth regardless of litter size, we divided pregnancies into two categories: those produced by heavier-than-average females (>=409.26 g; $n=8$) and lighter-than-average females (<409.26 g, $n=12$). To test the prediction that pregnancies of lighter mothers carrying triplets will be marked by significantly different ratios of fetal to placental ratios because of costs due to greater conflict over resources, we divided triplet pregnancies into two groups on the basis of the mean for nonpregnant weight: heavier (>=429.64 g) and lighter (<429.64).

Data were analyzed to determine to what extent and in which direction twin and triplet pregnancies, as well as those of mothers of different weights, differed. All data were normally distributed and were analyzed using independent samples *t*-tests and Pearson's correlations. ANCOVA was used to control the effects of placental weight when analyzing litter size patterns of difference in the fetal/placental weight ratio. Statistical analyses were performed using SPSS software, version 13.0.

16.5.2 Results

16.5.2.1 Litter Size and the Fetal/Placental Weight Ratio

Rutherford and Tardif (2008) conducted an analysis of differences in marmoset placental structure and function according to litter size and some of those results are summarized here. The means for individual fetal, total litter, placental weights, and the fetal/placental weight ratio are given in Table 16.1. Differences in placental and litter weights according to litter size are shown in Table 16.2. Individual triplets were smaller than twins (28.09 vs. 31.53 g, $F=0.794$, $df=26$, $p=0.044$), but as a whole, triplet litters weighed significantly more (84.34 vs. 63.36 g, $F=0.4.128$, $df=26$, $p<0.001$; Fig. 16.4). Despite striking differences in fetal and litter weights between the two litters, there was no significant difference in placental weight between twins and triplets ($F=0.016$, $df=25$, $p=0.283$; Fig. 16.5). Placental weight and total litter weight were strongly correlated ($r=0.518$, $p=0.007$, not shown), but although triplet litters were significantly heavier than their twin counterparts, their placentas were not. As a consequence, the fetal/placental weight ratio was strongly correlated with placental weight in a negative direction ($r=-0.700$, $p<0.001$). The fetal/placental weight ratio was unrelated to differences in total litter weight ($r=0.254$, $p=0.221$).

Table 16.1 Descriptive statistics for individual study components

	n	Range	Minimum	Maximum	Mean	SD	Variance
Maternal variables							
Maternal age	29	51	25	76	49.10	13.080	171.096
Maternal adult weight	19	170.00	337.00	507.00	409.263	52.976	2806.427
Maternal birth weight	29	13.00	25.00	38.00	30.345	3.446	11.877
Weight gained during pregnancy	15	123.00	45.00	168.00	92.467	31.904	1017.838
Placental weight							
Placental weight	27	12.92	5.73	18.65	9.513	3.056	9.339
Fetal weights							
Total litter weight	28	52.00	53.00	105.00	76.00	15.305	234.234
Average fetal weight	28	21.50	18.00	39.50	29.439	4.450	19.805
Fetal: placental weight ratio							
(placental efficiency)	25	7.14:1	4.77:1	11.91:1	8.304:1	1.966	3.865

Table 16.2 Independent samples *t*-tests comparing placenta, fetal, and total litter weights between litter size categories

	Twin litters		Triplet litters	
	n	Weight*	*n*	Weight*
Average fetal weight (g)	11	31.52 ± 3.64	17	28.09 ± 4.49**
Total litter weight (g)	11	63.36 ± 7.05	17	84.34 ± 13.44***
Placental weight (g)	10	8.68 ± 2.91	17	10.01 ± 3.12

*Weights are shown as the mean ± SD

**$p<0.05$

***$p<0.01$

Because placental weight strongly correlates with placental efficiency, an ANCOVA was performed to assess how much impact litter size has on differences in this variable once the effects of placental weight are accounted for. The grouping variable was litter size and placental weight was the covariate. Raw and estimated marginal means are presented in Table 16.3. The raw measures are labeled "Unadjusted Means." The estimated marginal means are means that have been adjusted for the covariate and are labeled "Adjusted Means." The triplet fetal/placental weight ratio was significantly higher than that for twins (adjusted marginal means: 9.04 g fetus/g placenta vs. 7.2 g fetus/g placenta), meaning that per gram, the triplet placenta supports more fetal growth than does the twin placenta, i.e., is relatively smaller than the twin placenta with respect to support of total fetal growth. This result is the opposite of that predicted by the overgrowth hypothesis. Triplet litters, although significantly heavier than twin litters, were associated with a decrease in placental mass relative to fetal mass.

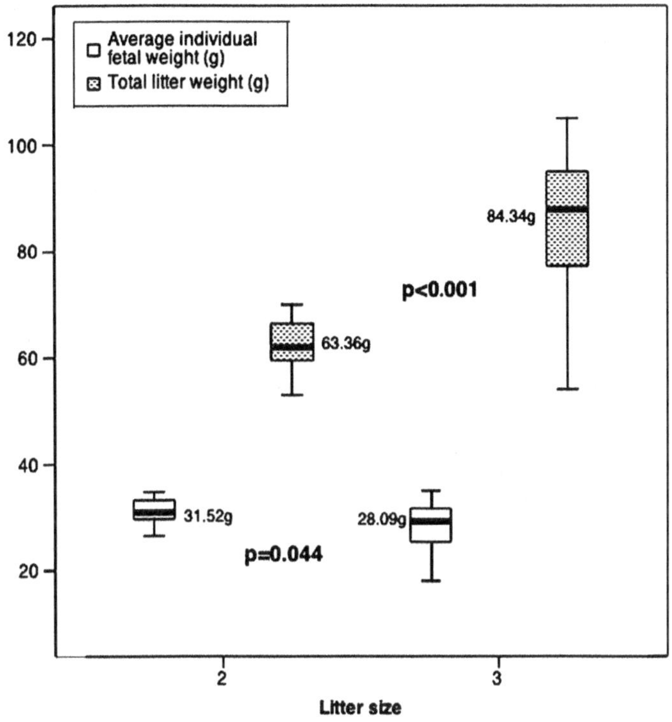

Fig. 16.4 Average fetal and total litter weights by litter size

16.5.2.2 Relations between Maternal Condition and Litter and Fetal Characteristics

In the following sections, we present the results of our analyses of maternal condition and its relation to litter and placental characteristics. The average age for the mothers in this sample was 49.10 months, and mean adult weight was 409.26 g with an average weight gain during pregnancy of 92.47 g (Table 16.1).

Correlations among all maternal condition variables and between maternal condition and litter and placental weights are shown in Table 16.4. Weight gain during pregnancy was significantly correlated with litter weight ($r=0.838$, $p<0.001$). Maternal adult weight was significantly correlated both with total litter weight ($r=0.483$, $p=0.042$) and with weight gain during pregnancy ($r=0.522$, $p=0.046$), so that larger females produced heavier litters and gained more weight during pregnancy.

Whereas there were strong relations between a) fetal and placental weights and b) maternal variables and fetal weights, there was no relation between placental weight and any maternal variable (Table 16.4). Neither age nor any aspect of maternal weight throughout her life history had any obvious bearing on placental weight.

In terms of differences between twins and triplets, neither maternal age nor birth weight differentiated the two litter categories (Table 16.5). However, maternal

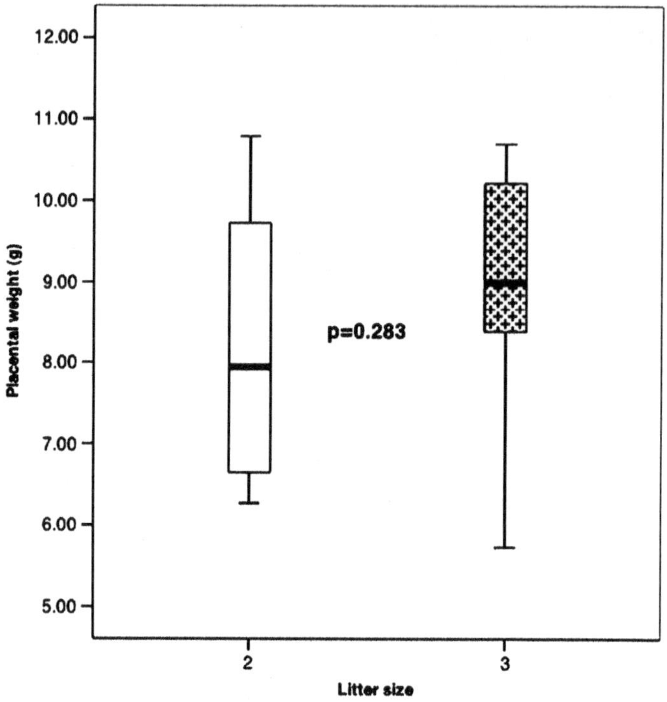

Fig. 16.5 Placental weight by litter size

Table 16.3 Analysis of covariance (ANCOVA) for Fetal/Placental weight ratio

	Dependent variable = Fetal/Placental weight ratio (g fetus/g placenta)		
MODEL	Independent variable = Litter size		
GROUP	Unadjusted mean (SD)	N	
Twin	7.765 (1.860)	10	
Triplet	8.663 (2.014)	15	
ANCOVA Results	Significance	Partial Eta Squared	
Placental weight (covariate)	0.001*	0.392	
		95% Confidence interval	
GROUP	Adjusted mean (SE)	Upper	Lower
Twin	7.203 (0.371)	6.434	7.972
Triplet	9.037 (0.300)	8.415	9.660

*Model significant at $p<0.01$

adult weight was significantly greater in mothers of triplets (Fig. 16.6), as was weight gain during pregnancy. Lighter-than-average mothers (<409.26 g) gave birth to both twin ($n=5$) and triplet ($n=6$) litters. In contrast, seven of the eight heavier-than-average mothers (~409.26 g) gave birth only to triplet litters.

Table 16.4 Correlations among maternal variables and placental, fetal, and total litter weights

		Placental weight	Total litter weight	Average fetal weight	Maternal age	Maternal adult weight	Maternal birth weight	Weight gained during pregnancy
Maternalage	Pearson Corr.	0.209	0.028	0.445 (*)		–0.009	–0.381 (*)	–0.023
	Sig. (2-tailed)	0.295	0.887	0.018		0.970	0.041	0.934
	N	27	28	28		19	29	15
Maternal adult weight	Pearson Corr.	0.276	0.483 (*)	0.206	–0.009		0.480 (*)	0.522 (*)
	Sig. (2-tailed)	0.268	0.042	0.412	0.970		0.038	0.046
	N	18	18	18	19		19	15
Maternal birth weight	Pearson Corr.	–0.317	–0.146	–0.621 (**)	–0.381 (*)	0.480 (*)		0.329
	Sig. (2-tailed)	0.107	0.459	0.000	0.041	0.038		0.231
	N	27	28	28	29	19		15
Weight gained during pregnancy	Pearson Corr.	0.231	0.838 (**)	0.335	–0.023	0.522 (*)	0.329	
	Sig. (2-tailed)	0.427	0.000	0.223	0.934	0.046	0.231	
	N	14	15	15	15	15	15	

*Correlation is significant at $p<0.05$ level

**Correlation is significant at $p<0.01$ level

Table 16.5 Independent *t*-tests comparing maternal age and weights between litter size categories

	Twin litters		Triplet litters	
	n	Weight*	*n*	Weight*
Maternal age (months)	11	53.36±14.05	18	46.50±12.12
Maternal adult (nonpregnant) weight (g)	6	376.67±26.62	13	424.31±56.02**
Maternal weight gained during pregnancy (g)	6	65.00±16.15	9	110.78±25.99***
Maternal birth weight (g)	11	29.00±3.13	18	31.17±3.45

*Values are shown as the mean±SD
** $p<0.05$
*** $p<0.01$

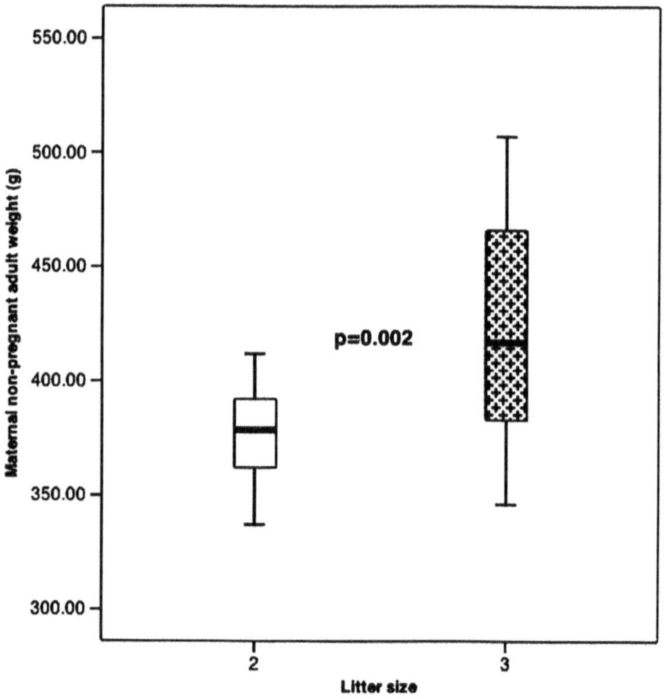

Fig. 16.6 Maternal nonpregnant adult weight by litter size

In addition to producing primarily triplet litters, heavier mothers fall into two categories in terms of placental weight, those producing larger (>13 g, $n=2$) or smaller (<9 g, $n=3$) placentas (Fig. 16.7). Due to the small size of this sample, significance of this relation could not be assessed. However, it is interesting to note that large mothers with smaller placentas gave birth to neonates whose average weights fall within a range very close to the overall sample mean weight of 29.4 g.

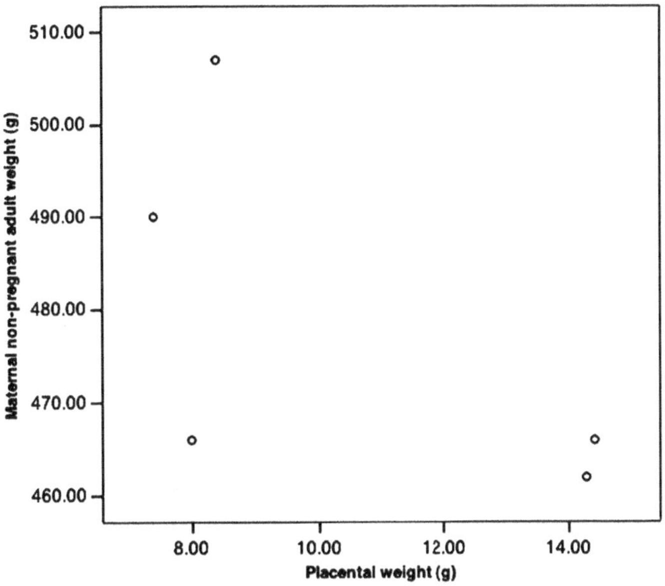

Fig. 16.7 Relation between maternal weight and placental weight in heavier-than-average (~424 g) mothers of triplets

The average birth weights of those infants born to large mothers producing large placentas ranged from average (31.3 g) to very low (22.7 g).

16.5.2.3 Maternal Birth Weight

Maternal birth weight in this sample was significantly and positively related to nonpregnant adult weight ($r=0.480$, $p=0.038$). It was also significantly correlated with birth weight of the next generation, but this association was negative ($r=-0.621$, $p<0.001$) (Table 16.4; Fig. 16.8). Maternal birth weight was not related to total litter weight (Table 16.4) or significantly different between twin and triplet litters (Table 16.5). Maternal birth weight was related to placental weight but not significantly ($r=-0.317$, $p=0.107$), and there was no relation to total litter weight ($r=-0.146$, $p=0.459$).

16.5.2.4 Maternal Age

Maternal age was not related to placental weight, although it was significantly correlated with average neonate weight ($r=0.445$, $p=0.018$), such that older mothers have larger neonates (Table 16.4). Neither total litter weight nor litter size was related to age.

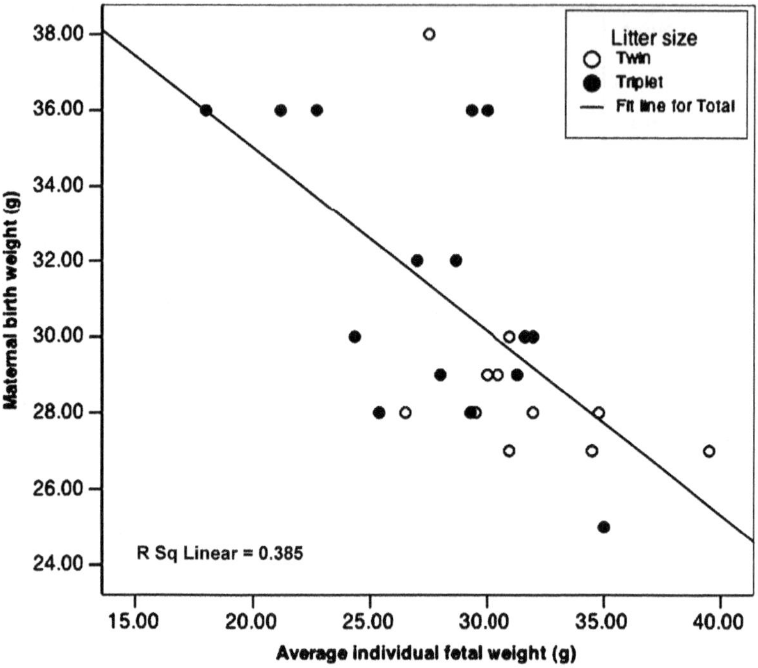

Fig. 16.8 Relation between maternal birth weight and offspring birth weight

Table 16.6 Correlations between the fetal/placental weight ratio and maternal variables

		Maternal age	Maternal adult weight	Maternal birth weight	Weight gained during pregnancy
Fetal/placental weight ratio (g fetus/g placenta)	Pearson correlation	–0.180	0.170	0.256	0.702 (*)
	Sig. (2-tailed)	0.389	0.515	0.217	0.005
	N	25	17	25	14

*Correlation is significant $p<0.01$ level

16.5.2.5 Maternal Variables and the Fetal/Placental Weight Ratio

As shown in Table 16.6, the fetal/placental weight ratio was significantly correlated with maternal weight gain during pregnancy ($r=0.702$, $p=0.005$), but unrelated to maternal adult weight ($r=0.170$, $p=0.515$). The fetal/placental ratio increased as maternal weight gain increased (Fig. 16.9). We found no differences in fetal/placental weight ratio between heavier and lighter mothers having triplet litters (Table 16.7).

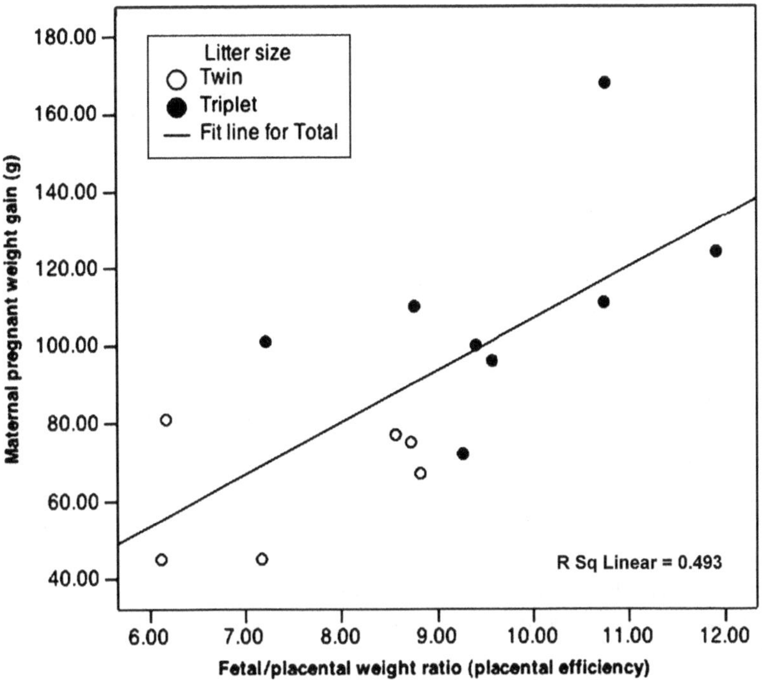

Fig. 16.9 Relation between maternal weight gain and the fetal/placental weight ratio

Table 16.7 Independent samples *t*-test comparing the fetal/placental weight ratio from pregnancies produced by lighter- and heavier-than average mothers of triplets

	Lighter mothers(<424 g)		Heavier mothers(±424 g)	
	n	Ratio[a]	*n*	Ratio[a]
Fetal/placental weight ratio	7	9.135:1 ± 1.060	4	8.485:1 ± 3.390[b]

[a]Ratios are shown as the mean ± SD
[b]ns

16.6 Discussion

Several intriguing patterns relevant to models of maternal investment and intrauterine resource allocation emerge, creating a foundation for further model building and testing. The average nonpregnant weight for the mothers in this sample was 409.63 g, which is higher than that found in studies by Tardif and colleagues of the same colony (357.7 g, Tardif et al. 2001; 376.5 g, Tardif et al. 2002). The sample used in the current study is restricted to those females for whom pregnancy data included both fetal and placental weights. Placental collection is a relatively recent addition to colony protocol, and thus the maternal weights in the current study may be more representative of recent secular trends in adult weight. However, the pattern of relations among maternal weight variables and between maternal weight and fetal variables is consistent with previous reports. Total litter weight is associated with nonpregnant weight gain and with litter size. In other words, heavier mothers have larger litters of heavier infants. Also, older mothers have larger infants. These findings are similar to those of Tardif and colleagues (Tardif and Jaquish 1994, 1997; Tardif and Bales 2004).

Maternal birth weight was negatively related to neonate weight, and positively related, though nonsignificantly ($p=0.101$) to litter size. This may be the result of a cascade of interrelated variables. The direction of these relations predicts the following: a female's birth weight is positively related to her nonpregnant adult weight, and maternal nonpregnant weight is positively associated with litter size (e.g., small mothers have smaller litters), and litter size is negatively related to neonate size (e.g., small litters comprise relatively larger neonates, even when total litter weight is lower).

This is only one way in which a female born small could give birth, on average, to larger neonates. Tardif and Bales (2004) demonstrate that litter size is an important and sometimes confounding determinant of adult weight and so it is likely that the addition of maternal litter size and total litter weight into the analyses of reproductive outputs and outcomes will provide further nuance, and provoke more questions. In humans, heritability of birth weight is relatively high (i.e., 20–45% of variation in birth weight explained by genetic factors; Vlietinck et al. 1989; Clausson et al. 2000). In marmosets, litter size variation complicates the genetic components of this relation, suggesting a critical role for environmental effects in fetal development. These environmental effects are multi-level and multigenerational, encompassing the mother's own fetal experience, resource availability prior to ovulation and during pregnancy, intrauterine competition between siblings, and the differential hormonal milieu associated with litter size variation.

Placental growth in marmosets seems not to be controlled by any variable relating to maternal adult weight. Whereas Steven-Simons et al. (1995) found that placental weight in humans was related to both nonpregnant weight and pregnant weight gain, we found no relation with either nonpregnant weight or weight gain. It appears that weight gain during marmoset pregnancy is a function entirely of litter size and total litter weight, not placental growth.

We report that heavier marmoset mothers produce triplets much more frequently than twins. Tardif and Jaquish (1997) demonstrated that ovulation number is a function of body weight, such that large females produce the greatest number of ova per ovulatory event. This in turn is related to an increase in litter size, and the findings of this study, albeit involving a much smaller sample size, offer further support to an already well-supported phenomenon. Interestingly, the distribution of placental size in this group is strongly dichotomous. Heavier marmoset mothers produce either smaller or larger placental weights. The neonates associated with the larger placentas in the group (>9 g) were quite variable in birth weight. This variation may be the result of earlier litter-size reduction, a very common phenomenon in marmosets (Jaquish et al. 1996; Windle et al. 1999). In particular, the triplet litter that combined a very low average birth weight with a high nonpregnant weight and a heavy placenta may be indicative of a quadruplet litter that was reduced to triplets later in gestation. The range in birth weights may reflect the timing of this reduction, such that the heavier infants were a result of early litter size reduction and the lightest infants a result of litter reduction taking place sometime later.

In contrast to the findings by Rahima and Bruce (1987) in rats, we did not find any relation between maternal age and placental weight. They report that older mothers produce larger placentas, even when fetal weight is controlled for. This means that older rat mothers would have pregnancies with lower F/P ratios than younger mothers, and this suggests that maternal age, at least in rats, is associated with greater placental dysfunction as a result of increased conflict over resource allocation. This could be related to the very different life history of the rat, with a much faster developmental path to reproduction and senescence than the marmoset. Lurie et al. (1999) found a positive correlation between fetal/placental weight ratio and age in a study of 431 human females, suggesting the opposite effect. Young mothers (17–19 years) had the highest fetal/placental weight ratios and the lowest birth weight babies, indicating that this may be the greatest period of conflict over resource allocation and is likely related to the prolonged juvenile growth period that is the hallmark of human development. Maternal age in marmosets is not related to the fetal/placental weight ratio, an indication that age has no bearing on placental growth. It could also suggest that the range of ages of the females in this study was not large enough to capture differences related to elevated maternal age.

Our hypothesis that the fetal/placental weight ratio would be different in the pregnancies of lighter females carrying triplet litters was not supported. There was no difference within either twin or triplet litters between the fetal/placental weight ratio in pregnancies of lighter and heavier mothers. This may be related to the lack of a significant relation between placental weight and maternal weight. It could also signal that the burden of acquiring and maintaining access to placental tissue, and thus maternal resources, is on the fetus, and that the difference between gestating twins and triplets, at least in terms of investing in placental growth, is not particularly costly for the mother (Rutherford & Tardif 2008). Maternal condition does not appear to mitigate, nor directly elevate, allocation of placental tissue.

In marmosets, higher litter weight triplets are associated with a significantly smaller proportion of the total placental weight relative to fetal weight than are twins. Put another way, per gram, the triplet placenta supports more fetal growth than the twin placenta, suggesting some increase in placental efficiency. Therefore, if the triplet marmoset pregnancy represents a stressed, possibly nutrient-restricted state from the perspective of the individual fetus, the placenta is not responding by overgrowth as has been hypothesized by Haig (1993) and Coall and Chisholm (2003). Studies in mice have shown increases in active amino acid transport by placentas that have been restricted in growth due to the deletion of the gene coding for insulin-like growth hormone, a hormone produced by both the fetus and placenta that regulates placental growth (Constancia et al. 2002). In similar ways, triplets may in fact be engaging in strategies at the metabolic and cellular level that increase access to maternal resources, rather than strategies that increase overall placental growth.

Whereas triplet birth weight appears to be most limited late in gestation (Chambers and Hearn 1985), placental growth maxima may be determined earlier (Fig. 16.10).

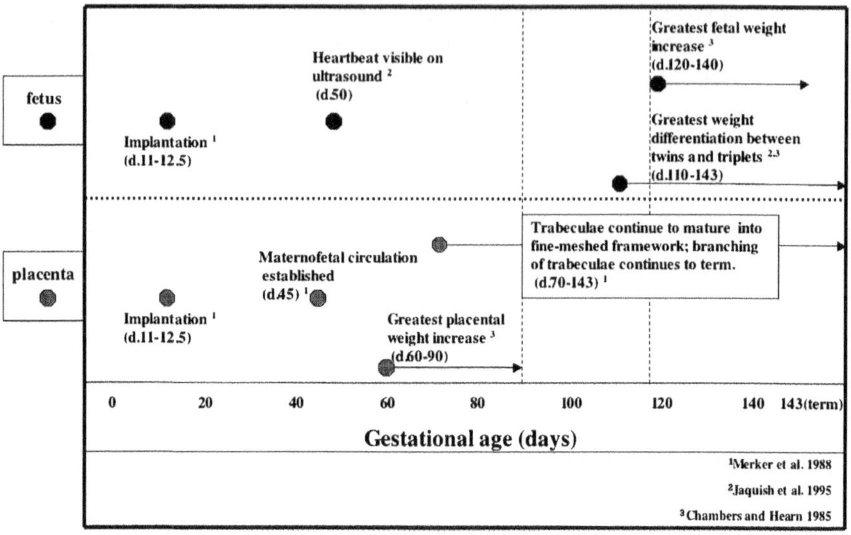

Fig. 16.10 Timeline of fetal and placental milestones during marmoset development

The maximum increase in placental weight is completed by day 90 (Chambers and Hearn 1985), a month earlier than the period of greatest increase in fetal growth, occurring between days 120–140 (Jaquish et al. 1995). Triplet fetuses experience growth restriction compared to twin trajectories relatively late in gestation, with the maximum difference between twin and triplets occurring around day 130 of a 143-day gestation (Chambers and Hearn 1985). Whereas overall placental size cannot be altered at this late point in development, the maturation of the placental trabeculae continues until term (Merker et al. 1988). Therefore, the stressed triplet pregnancy may be able to compensate for increasingly contested resources by expanding the tissue at the microscopic level of the maternal–fetal interface, or otherwise altering its function, thereby pursuing developmental pathways that increase efficiency at the cellular level, an outcome not captured by the fetal/placental weight ratio. In support of this hypothesis, we have recently shown that marmoset triplet placentas have an expanded microscopic surface area available for nutrient transport and endocrine signaling (Rutherford and Tardif 2009).

Compensatory growth of the placenta, like that described by Lumey (1998) in humans as being the result of famine experienced during early gestation and predicted by Haig (1993) to be evidence for conflict over resources, may not be an appropriate model for comparison to the triplet marmoset pregnancy. In the compensatory model, restriction of maternal energy intake early in gestation has an immediate and persistent effect on fetal and placental development because the pregnancy begins in a state of energetic burden to the mother. In contrast, a marmoset mother may not perceive the addition of a fetus early in gestation as an energetic burden. However, as fetal mass accrues and metabolic needs become more demanding, the fetuses in the triplet litter may experience a per capita restriction of maternal resources, compared to their twin counterparts. Restriction of resources available to the fetus on the part of the mother may be passive; Nievergelt and Martin (1999) report that marmoset females do not increase energy intake during pregnancy even when carrying triplet litters. This surprising finding suggests that from the mother's perspective, gestating triplets is no more costly than gestating twins, or at least that her investment in the form of energy intake (not taking into account starting reserves in the form of body weight) is finite. However, the experience from the perspective of the fetus is much different. Postnatal survivorship for individual triplet neonates is much lower than that for twins, suggesting that *in utero* competition for resources is elevated. Fetal pathways (such as activation of placental hormonal signaling systems) that increase access or at least maintain normative planes of nutritional and metabolic support via the placenta to maintain pregnancy to term are separate from maternal patterns of pre- and postnatal energetic and behavioral investment. In this way, the relatively smaller, potentially more efficient triplet placenta represents an imperfect solution to the problem of offspring survival because it is at odds with the mother's strategy of balancing resources she may prefer to invest in an optimal litter number of two, the lactation burden of her current offspring, and fat reserves that relate to ovulation number for future reproductive events. Understanding the range of adaptive pathways available to the placenta in the context of litter size variation will enhance our growing appreciation that the callitrichids as a group have evolved to exhibit great flexibility in reproductive output.

16.7 Conflicting Demands, Competing Strategies

Marmoset females have developed a variety of postnatal strategies for tailoring investment in their offspring according to litter size (Tardif et al. 2003). Marmoset mothers are limited to caring for two infants due to the number of nipples (Schultz 1948) and energetic constraints of lactation (Nievergelt and Martin 1999; Tardif et al. 2001). Mothers do not appear to increase their total behavioral investment in triplet litters versus twin litters in carrying or nursing bouts, but rather have finite time and energetic resources to devote to their offspring for which the offspring must actively compete (Tardif et al. 2002). Triplets do initiate infant-carrying bouts at a greater rate than twins do (Tardif et al. 2002), but these bouts are shorter in duration, and so are lactation bouts (Tardif et al. 2001), and are met with a higher degree of maternal harassment (Tardif et al. 2001). Fite et al. (2005a, b) have suggested that mothers will decrease caregiving efforts in the event of conception during intense lactation, further jeopardizing survival for triplets. An aggressive strategy to court mother's time is no guarantee of garnering more resources, but because individual fitness costs for triplets are high, it is a strategy worth pursuing.

This strategy for acquiring allocation of maternal resources is one that begins early in gestation and differences in the relation between birth weight and placental weight in triplet and twin litters suggest triplets may employ strategies to maximize access to maternal resources in a way that is unavailable to them postnatally. We have shown that being a triplet is associated with relative placental undergrowth, but this leaves open the door on mechanisms that increase efficiency at the metabolic level. We have also demonstrated that maternal nonpregnant weight and birth weight interact with litter size in complicated ways to impact fetal growth and outcomes. Placental growth in marmosets appears to be under the control of the fetal soma, rather than maternal factors, and is potentially both responsive to and responsible for the growth of the individual fetus. These exciting findings are consistent with clinical models of intrauterine growth retardation and evolutionary predictions of parent–offspring conflict theory, and suggest that marmosets and the other callitrichids present unique opportunities to investigate the ways in which these processes interact in the intrauterine environment.

When there are finite resources to be devoted to fetal development, additional fetuses will strain availability. There are limits to maternal investment, and these limits are detrimental to individual marmoset triplets, whose most powerful ally in the tug-of-war over maternal resources is the placenta.

Acknowledgments The authors would like to thank Donna Layne Colòn, L.A.T. for her ongoing excellent management of the SNPRC marmoset colony and her skillful execution of data collection and participation in research design. Our development of this research has benefited from conversations and correspondence with Christopher Kuzawa (Northwestern University) and David Haig (Harvard University). Comments from the editors of this volume, two anonymous reviewers, and Dr. Haig have greatly improved the final manuscript. Funding for SNPRC colony management and research has been provided by NIH grants R01-RR02022 and P51-RR1396 (SDT), and grants to JNR from the American Society of Primatologists, the Center for the Integrative Study of Animal Behavior (Indiana University) and the Indiana University Graduate School Grant-in-Aid of Doctoral Research program have funded the research on which this chapter is based.

References

Ah-King M and Tullberg BS (2000) Phylogenetic analysis of twinning in Callitrichinae. Am J Primatol 51: 135–41

Ali KMZ (1997) The association between spontaneous preterm birth and placental histology at high and low altitude areas of southern Saudi Arabia. Saudi Med J 18(4):349–352

Ankel-Simons F (2000) Primate anatomy, 3rd edn. Academic Press, San Diego

Barker D (1998) In utero programming of chronic disease. Clin Sci 95:115–128

Barker D, Eriksson JG, Forsen T, Osmond C (2002) Fetal origins of adult disease: Strength of effects and biological basis. Int J Epidemiol 31:1235–1239

Benirschke K (1995) The biology of the twinning process: how placentation influences outcome. Semin Perinatol 19(5):342–350

Benirschke K (2002) Comparative placentation website. Download date: April 7, 2006. http://medicine.ucsd.edu/cpa/indxfs.html

Benirschke K, Brownhill LM (1962) Further observations on marrow chimerism in marmosets. Cytogenetics 1:245–247

Benirschke K, Anderson JM, Brownhill LE (1962) Marrow chimerism in marmosets. Science 138:513–515

Caine N (1993) Flexibility and co-operation as unifying themes in *Saguinus* social organization and behaviour: The role of predation pressures. In: Rylands AB (ed) Marmosets and Tamarins: systematics, behaviour, and ecology. Oxford Science Publications, Oxford, pp 200–219

Chambers PL, Hearn JP (1985) Embryonic, foetal and placental development in the common marmoset monkey (*Callithrix jacchus*). J Zool 207:545–561

Clausen T, Burksi TK, Oyen N, Godang K, Bollerslev J, Henriksen T (2005) Maternal anthropometric and metabolic factors in the first half of pregnancy and risk of neonatal macrosomia in term pregnancies: a prospective study. Eur J Endocrinol 153(6):887–894

Clausson B, Lichtenstien P, Cnattingius S (2000) Genetic influence on birth weight and gestational length determined by studies in offspring of twins. Br J Obstet Gynaecol 107(3): 375–381

Coall DA, Chisholm JS (2003) Evolutionary perspectives on pregnancy: maternal age at menarche and infant birth weight. Soc Sci Med 57:1771–1781

Constancia M, Hemberger M, Hughes J, Dean W, Ferguson-Smith A, Fundele R, Stewart F, Kelsey G, Fowden A, Sibley C, Reik W (2002) Placental-specific IGF-II is a major modulator of placental and fetal growth. Nature 417: 945–8

Cortés-Ortiz L (this volume) Molecular phylogenetics of the Callitrichidae with an emphasis on the marmosets and *Callimico*. In: Ford SM, Porter LM, Davis LC (eds) The smallest anthropoids: The marmoset/callimico radiation. Springer Press, New York, pp 3–24

Dwyer CM, Calvert SK, Farish M, Donbavand J, Pickup HE (2005) Breed, litter and parity effects on placental weight and placentome number, and consequences for the neonatal behaviour of the lamb. Theriogenology 63: 1092–1110

Eisenberg J (1981) The Mammalian radiations: An analysis of trends in evolution, adaptation, and behavior. Chicago, University of Chicago Press

Enders AC, Lopata A (1999) Implantation in the marmoset monkey: Expansion of the early implantation site. Anat Rec 256:279–299

Fite JE, French JA, Patera KJ, Hopkins EC, Rukstalis M, Ross CN (2005a) Elevated urinary testosterone excretion and decreased maternal caregiving effort in marmosets when conception occurs during the period of infant dependence. Horm Behav 47:39–48

Fite JE, Patera KJ, French JA, Rukstalis M, Hopkins EC, Ross CN (2005b) Opportunistic mothers: Female marmosets (*Callithrix kuhlii*) reduce their investment in offspring when they have to, and when they can. J Hum Evol 49:122–142

Ford SM (1980) Marmosets and tamarins as phyletic dwarfs, and the place of the Callitrichidae in Platyrrhini. Primates 21(1):31–43

Garber PA, Leigh SR (1997) Ontogenetic variation in small-bodied New World primates: implications for patterns of reproduction and infant care. Folia Primatol 68:1–22

Gengozian N, Batson JS, Eide P (1964) Hematologic and cytogenetic evidence for hematopoietic chimerism in the marmoset *Tamarinus nigricollis*. Cytogenetics 10:384–393

Gengozian N, Brewen JG, Preston RJ, Batson JS (1980) Presumptive evidence for the absence of functional germ cell chimerism in the marmoset. J Med Primatol 9(1–2):9–27

Goldizen AW (1990) A comparative perspective on evolution of tamarin and marmoset social systems. Int J Primatol 11: 63–83

Haig D (1993) Genetic conflicts in human pregnancy. Q Rev Biol 68:495–532

Haig D (1996) Altercation of Generations: genetic conflicts of pregnancy. Am J Reprod Immunol 35:226–232

Haig D (1999) What is a marmoset? Am J Primatol 49: 285–96

Howe DT (1994) Maternal factors, fetal size, and placental ratio at 18 weeks: Their relationship to final size. In: Ward RHT, Smith SK, Donnai D (eds) Early fetal growth and development. Royal College of Obstetricians and Gynaecologists Press, London, pp 345–354

Jaquish CE, Gage TB, Tardif SD (1991) Reproductive factors affecting survivorship in captive Callitrichidae. Am J Phys Anthropol 84:291–305

Jaquish CE, Toal RL, Tardif SD, Carson RL (1995) Use of ultrasound to monitor prenatal growth and development in the common marmoset (*Callithrix jacchus*). Am J Primatol 36:259–275

Jaquish CE, Tardif SD, Toal RL, Carson RL (1996) Patterns of prenatal survival in the common marmoset (*Callithrix jacchus*). J Med Primatol 25:57–63

Kuzawa C (2005) Fetal origins of developmental plasticity: Are fetal cues reliable predictors of future nutritional environments? Am J Phys Anthropol 17:5–21

Kuzawa C, Adair LS (2003) Lipid profiles in adolescent Filipinos: Relation to birth weight and maternal energy status during pregnancy. Am J Clin Nutr 77:960–966

Langley-Evans SC, Phillips GJ, Benediktsson R, Gardner DS, Edwards CRW, Jackson AA, Seckl JR (1996) Protein intake in pregnancy, placental glucocorticoid metabolism and the programming of hypertension in the rat. Placenta 17:169–172

Leon DA, Koupilova I, Lithell HO, Berglund L, Mohsen R, Vagero D, Lithell U-B, McKeigue PM (1996) Failure to realize growth potential in utero and adult obesity in relation to blood pressure in 50 year old Swedish men. Br J Med 312:401–416

Leutenegger W (1973) Maternal-fetal weight relations in Primates. Folia Primatol 20:280–293

Leutenegger W (1980) Monogamy in Callitrichids: A consequence of phyletic dwarfism? Int J Primatol 1(1):95–98

Long TA (2005) The influence of mating system on the intensity of parent-offspring conflict in primates. J Evol Biol 18:509–515

Luckett WP (1974) Comparative development and evolution of the placenta in primates. In: Luckett WP (ed) Contributions to primatology, vol. 3: Reproductive biology of the primates. S Karger, Basel, pp 142–234

Lumey LH (1998) Compensatory placental growth after restricted maternal nutrition. Placenta 19:105–111

Lunn SF (1980) A case of placenta praevia in a common marmoset (*Callithrix jacchus*). Vet Rec 106:414

Lurie S, Feinstein M, Mamet Y (1999) Human fetal-placental weight ratio in normal singleton near-term pregnancies. Gynecol Obstet Invest 48(3):155–157

Martin RD (1992) Goeldi and the dwarfs: The evolutionary biology of the small New World monkeys. J Hum Evol 22:367–393

Mesa H, Safranski TJ, Johnson RK, and Lamberson WR (2003) Correlated response in placental efficiency in swine selected for an index of components of litter size. J Anim Sci 81: 74–79

Merker H-J, Bremer D, Csato W, Heger W, Gossrau R (1988) Development of the marmoset placenta. In: Neubert D, Merker H-J, Hendrickx A (eds) Non-human primates – developmental biology and toxicology. Ueberreuter Wissenschaft, Berlin, pp 245–272

Mittendorfer-Rutz E, Rassmussen F, Wasserman D (2004) Restricted fetal growth and adverse maternal psychosocial and socioeconomic conditions as risk factors for suicidal behaviour of offspring: a cohort study. Lancet 364(9440):1102–1104

Mossman HW (1987) Vertebrate fetal membranes. MacMillan, Houndmills

Nievergelt C, Martin RD (1999) Energy intake during reproduction in captive common marmosets (*Callithrix jacchus*). Physiol Behav 65(4/5):849–854

Ono S (1967) The problem of the bovine freemartin. J Reprod Fert 7(Suppl):53–61

Porter L, Garber PA (this volume) Social behavior of callimicos: Mating strategies and infant care. In: Ford SM, Porter LM, Davis LC (eds) The smallest anthropoids: The marmoset/callimico radiation. Springer Press, New York, pp 87–101

Rahima A, Bruce NW (1987) Fetal and placental growth in young, primiparous and old, multiparous rats. Exp Gerontol 22:257–261

Robinson JS, Owens JA, DeBarro T, Lok F, Chidzanja S (1994) Maternal nutrition and fetal growth. In: Ward RHT, Smith SK, Donnai D (eds) Early fetal growth and development. Royal College of Obstetricians and Gynaecologists Press, London, pp 317–328

Ross CN, French JA, Orti G (2007) Germ-line chimerism and paternal care in marmosets (*Callithrix kuhlii*). Proc Natl Acad Sci USA 104(15):6278–6282

Rothe H, Darms K, Koenig A (1992) Sex ratio and mortality in a laboratory colony of the common marmoset (*Callithrix jacchus*). Lab Animal 26(2):88–99

Rutherford JN (2007) Litter size effects on placental structure and function in common marmoset monkeys (*Callithrix jacchus*): implications for intrauterine resource allocation strategies. Unpublished doctoral Dissertation, Indiana Universit

Schultz AH (1948) The number of young at birth and the number of nipples in primates. Am J Phys Anthropol 6(1):1–23

Steven-Simons C, Metlay L, McAnarney E (1995) Maternal prepregnant weight and pregnant weight gain: relation to placental microstructure and morphometric oxygen diffusion capacity. Am J Perinatol 12(6):407–412

Tardif SD, Bales K (2004) Relations among birth condition, maternal condition, and postnatal growth in captive common marmoset monkeys (*Callithrix jacchus*). Am J Primatol 62:83–94

Tardif SD, Jaquish CE (1994) The common marmoset as a model for nutritional impacts upon reproduction. Ann NY Acad Sci 709:214–215

Tardif SD, Jaquish CE (1997) Number of ovulations in the marmoset monkey (*Callithrix jacchus*): Relation to body weight, age and repeatability. Am J Primatol 42:323–329

Tardif SD, Power M, Oftedal O, Power R, Layne DG (2001) Lactation, maternal behavior and infant growth in common marmoset monkeys (*Callithrix jacchus*): Effects of maternal size and litter size. Behav Ecol Sociobiol 51:17–25

Tardif SD, Layne DG, Smucny DA (2002) Can marmoset mothers count to three? Ethology 108:825–836

Tardif SD, Smucny DA, Abbott DH, Mansfield K, Schultz-Darken N, Yamomato ME (2003) Reproduction in captive common marmosets (*Callithrix jacchus*). Comp Med 53(4):364–368

Taricco E, Radaelli MS, Nobile de Santis MS, Cetin I (2003) Foetal and placental weights in relation to maternal characteristics in gestational diabetes. Placenta 24:343–347

Trivers RL (1974) Parent-offspring conflict. Am Zool 14:249–264

Vlietinck R, Derom R, Neale MC, Maes H, van Loon H, Derom C, Thiery M (1989) Genetic and environmental variation in the birth weight of twins. Behav Genet 19(1):151–161

Wahlbeck K, Forsen T, Osmond C, Barker DJ, Eriksson JG (2001) Association of schizophrenia with low maternal body mass index, small size at birth, and thinness during childhood. Arch Gen Psychiatry 58(1):48–52

Wheeler T (1994) Influences of the maternal environment on placental growth and function. In: Ward RHT, Smith SK, Donnai D (eds) Early fetal growth and development. Royal College of Obstetricians and Gynaecologists Press, London, pp 257–265

Wilson ME, Biensen NJ, Ford SP (1999) Insight into the control of litter size in pigs, using placental efficiency as a selection tool. J Anim Sci 77:1654–1658

Wilson ME and Ford SP (2001) Comparative aspects of placental efficiency. Reproduction Suppl 58: 223–32

Windle CP, Baker HF, Ridley RM, Oerke A-K, Martin RD (1999) Unrearable litters and prenatal reduction of litter size in the common marmoset (*Callithrix jacchus*). J Med Primatol 28:73–83

Wislocki GB (1932) Placentation in the marmoset (*Oedipomidas geoffroyi*), with remarks on twinning in monkeys. Anat Rec 52(4):381–399

Wislocki GB (1939) Observations on twinning in marmosets. Am J Anat 64:445–483

Wynn RM, Richards SC, Harris J (1975) Electron microscopy of the placenta and related structures of the marmoset. Am J Obstet Gynecol 122(1):60–69

Chapter 17
Size and Shape in Callimico and Marmoset Skulls: Allometry and Heterochrony in the Morphological Evolution of Small Anthropoids

Gabriel Marroig and James M. Cheverud

Abstract *Callithrix, Mico, Cebuella, Callibella,* and *Callimico* (the marmosets and callimicos) are part of a large radiation of species and genera, the Callitrichidae, within the New World Monkeys (NWM). The diversification of this phylogenetically well supported clade comprises five genera and around 60 species. The evolutionary history of the whole clade is to a large extent, a history of a secondary reduction in body size. Here we analyze the size and shape variation in skulls of marmosets and callimicos in order to understand the evolution of these small anthropoids and their potential life-history correlates. Our results indicate that allometric patterns are conserved in the groups analyzed and that a large fraction of their differentiation is size-related. After correcting for size-scale differences, callimicos are still very different from all marmosets, but interestingly, pygmy marmosets are more similar to marmosets of the eastern Brazil group (*Callithrix*). Marmosets, callimicos, and pygmy marmosets deviate significantly downward in birth weight from the expected value given their gestation lengths in comparison with the rest of the NWM. These results suggest that the evolution of marmosets/callimicos as a small-body clade, reaching its extreme in the pygmy marmosets, occurred due to a heterochronic process where prenatal growth rates were reduced. This contrasts with other pygmy mammals where miniaturization is primarily postnatal.

Resumen *Callitrhix*, *Cebuella* y *Callimico* son parte de una amplia radiación de especies y géneros (los calitricidos) de los monos del Nuevo Mundo. La diversificación de dicho clado bien apoyado filogenéticamente, comprende 5 géneros y alrededor de 60 especies. La historia evolutiva de todo el clado es en su gran extensión la historia de una reducción secundaria en el tamaño corporal. En el presente estudio analizamos la variación de el tamaño y forma en esqueletos de marmosets y callimicos con el fin de entender la evolución de dichos pequeños antropóides y su potencial correlación con su historia natural. Nuestros resultados indican que los

G. Marroig (✉)
Departamento de Genética e Biologia Evolutiva, Instituto de Biociências, Universidade de São Paulo, CP 11.461, CEP 05422-970, São Paulo, Brazil
e-mail: Gmarroig@ib.usp.br

S.M. Ford et al. (eds.), *The Smallest Anthropoids*, Developments in Primatology: Progress and Prospects, DOI 10.1007/978-1-4419-0293-1_17,

© Springer Science+Business Media, LLC 2009

patrónes alométricos son conservados en los grupos analizados y que una larga fracción de su diferenciación está relacionada con el tamaño. Luego de corregir las diferencias de escala de tamaño, los callimicos son muy diferentes de todos los marmosets, pero interesantemente, los marmosets pigmeos son muy similares a los marmosets del grupo Jacchus. Los marmosets y los marmosets pigmeos, cuando son comparados con el resto de los monos del Nuevo Mundo, presentan pesos de nacimiento que desvían significativamente para abajo en relación con el valor esperado dado su tiempo de gestación. Dichos resultados conjuntamente sugieren que la evolución de los marmosets como un clado de tamaño corporal pequeño, alcanzan el extremo en los marmosets pigmeos, y ocurren debido al proceso heterocrónico en donde el índice de crecimiento prenatal fue reducido. Ello contrasta con otros mamíferos pigmeos en donde la miniaturización es primariamente postnatal.

Resumo *Callithrix, Mico, Cebuella, Callibella,* e *Callimico* (sagüis, sagüis-leãozinho e micos-preto) fazem parte de uma ampla irradiação de espécies e gêneros (os calitriquíneos) dentro dos macacos do Novo Mundo. A diversificação deste clado, que é bem suportado por análises filogenéticas, engloba 5 gêneros e cerca de 60 espécies. A história evolutiva do clado como um todo é em grande parte uma história de redução secundária no tamanho corporal. Uma análise da variação de tamanho e forma nos crânios dos sagüis e micos é apresentada com o objetivo de entender a evolução destes pequenos antropóides e sua correlação potencial com sua história-natural. Nossos resultados indicam que os padrões alométricos são conservados nos grupos analisados e que a maior parte da diferenciação entre eles foi de tamanho. Após corrigir para as diferenças de escala (tamanho) os micos-preto ainda são bastante distintos de todos os outros sagüis, mas curiosamente os sagüis-leãozinho são mais similares ao grupo de sagüis do leste brasileiro (*Callithrix*). Sagüis, micos-preto e sagüis-leãozinho, quando comparados com o resto dos macacos do novo mundo, apresentam pesos ao nascimento que desviam significativamente para baixo em relação ao valor esperado dado seu tempo de gestação. Estes resultados conjuntamente sugerem que a evolução dos sagüis como um clado de pequeno tamanho corporal, alcançando o extremo nos sagüis-leãozinho, ocorreu devido a um processo heterocrônico no qual as taxas de crescimento pré-natal foram reduzidas. Isto contrasta com outros mamíferos "anões" onde a miniaturização is primariamente pós-natal.

17.1 Introduction

New World Monkeys (NWM) show remarkable variation in body size (Marroig and Cheverud 2005). Marmosets, including pygmy marmosets, dwarf marmosets, Atlantic coastal forest and Amazonian marmosets (*Cebuella, Callibella, Mico,* and *Callithrix*), tamarins (*Leontopithecus* and *Saguinus*), and callimicos (*Callimico*) comprise a large radiation (the family Callitrichidae) in NWM showing an overall trend through evolutionary time to reduce their body sizes (Ford 1980; Marroig and

Cheverud 2005). For example, marmosets and pygmy marmosets are largely different in absolute size despite their close phylogenetic relationship, with the former should be weighing around 350 g while the latter only 110 g. Pygmy marmoset crania are, on average, only 70% of the marmoset's in size (Marroig and Cheverud 2005). Associated with this trend to reduce body sizes in callitrichids, many changes in habitat, locomotion, systems of mating, and diet took place. While interpreted by some in the past as a primitive group among NWM (Hershkovitz 1977), they are now widely conceived as a derived group where a secondary reduction in overall size took place (Ford 1980). In order to understand the evolution of these small anthropoids, we analyze size and shape variation in morphological data (skull measurements) and their potential association with life-history traits within the callimico and marmoset radiation. Herein, we use the terminology adopted in this volume; however, we remain skeptical of the validity of raising the argentata-group marmosets to a separate genus, as well as the validity of the newly named dwarf marmoset genus, *Callibella*. We believe further study is warranted. Despite the potential for disagreement in terms of how to classify these marmosets (into one, two or four genera), one thing is quite clear: they are indeed a very closely related monophyletic radiation (including *Callimico*, see Schneider et al. 2001; Cortés-Ortiz Chap. 1 this volume) with changes in body size and concomitant changes in their life history, ecology and morphology. This chapter will focus on the morphological evolution of this adaptive radiation and its interrelationships with life-history changes.

17.2 Materials and Methods

17.2.1 Data and Measurements

Our sample includes 18 species, and they are as follows: 5 species of the genus *Callithrix* (previously known as the Jacchus-group) (N=362), including the following species: *C. aurita* (N=14), *C. geoffroyi* (N=51), *C. jacchus* (N=65), *C. kuhlii* (N=148), *C. penicillata* (N=84); 11 species of the genus *Mico* (previously known as the Argentata-group) (N=183), including the following species: *M. argentata* (N=72), *M. chrysoleuca* (N=20), *M. emiliae* (N=13), *M. humeralifera* (N=29), *M. intermedia* (N=2), *M. leucippe* (N=11), *M. marcai* (N=3), *M. mauesi* (N=3), *M. melanura* (N=20), *M. nigriceps* (N=6), *M. saterei* (N=4); the single *Cebuella* species (N=67); and the single *Callimico* species (N=27). Three dimensional co-ordinates were recorded for 36 landmarks using a Polhemus 3Draw or a Microscribe 3DX digitizer in adult skulls (Fig. 17.1 and Table 17.1) deposited at several institutions: American Museum of Natural History (AMNH), Field Museum of Natural History (FMNH), Museu Nacional do Rio de Janeiro (MNRJ), Museu Paranaense Emílio Goeldi (MPEG), Museu de Zoologia da Universidade de São Paulo, and the National Museum of Natural History (USNM). A complete list of measured specimens sorted by species and museum can be obtained

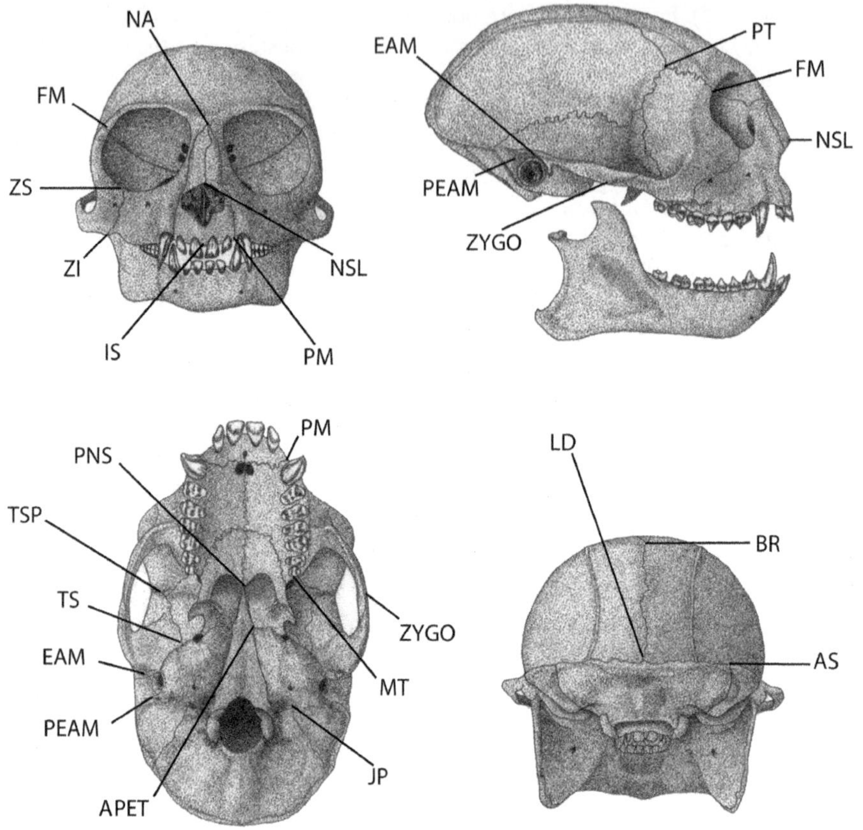

Fig. 17.1 Thirty-six craniofacial landmarks recorded from New World monkey skulls using three-dimensional digitizer

from the authors upon request. Only adult skulls without missing values were used in the subsequent analyses. Specimens were considered adult when they had totally erupted and functional dentition as well as closed or fused spheno-occipital and/or spheno-ethmoidal sutures. Three dimensional co-ordinates were transformed into 39 skull measurements (distances between landmarks, see Fig. 17.1 and Table 17.2) that are classified in functional/developmental groups following Cheverud (1995).

17.2.2 Size, Scaling and Allometry

Skull size was estimated for all specimens by using the first principal component (PC1) score of the natural log data as the overall size measure. This PC was extracted from the total variance/covariance matrix. The PC scores were then used as a measure of size and tested for differences among groups using an ANOVA followed by a Scheffé post hoc test.

Table 17.1 Landmarks recorded in Neotropical primates skulls using the three-dimensional digitizer

Landmark	Description	Position(s)
IS	Intradentale superior, A	Midline
PM	Premaxillary suture at the alveolus, A	Right, left
NSL	Nasale, A	Midline
NA	Nasion, A	Midline
BR	Bregma, AP	Midline
PT	Pterion, AP	Right, left
FM	Fronto-malare, A	Right, left
ZS	Zygomaxillare superior, A	Right, left
ZI	Zygomaxillare inferior, A	Right, left
MT	Maxillary tuberosity, A	Right, left
PNS	Posterior nasal spine, A	Midline
APET	Anterior petrous temporal, A	Right, left
BA	Basion, AP	Midline
OPI	Opisthion, AP	Midline
EAM	Anterior external auditory meatus, A	Right, left
PEAM	Posterior external auditory meatus, A	Right, left
ZYGO	Inferior zygo-temporal suture, A	Right, left
TSP	Temporo-spheno-parietal junction, A	Right, left
TS	Temporo-sphenoidal junction at the petrous, AP	Right, left
JP	Jugular process, AP	Right, left
LD	Lambda, P	Midline
AS	Asterion, P	Right, left

The designation A (anterior) or P (posterior) after the landmark name indicates in which position(s) the landmark was recorded. Landmarks are also identified in Fig. 17.1

Given the variation in the size of marmosets and callimicos and in the shape variation associated with those size differences, we apply a normalization technique to scale data and remove allometric effects (Marroig and Cheverud 2004). This method, which we will refer from now on as "Multivariate Allometric Size-Scaling (MASS)," is derived from theoretical equations of allometric growth removing all the information related to size, not only scaling all individuals to the same size, but also adjusting their shape to account for allometry (Lleonart et al. 2000). We adapted the Lleonart et al. (2000) method by using the first principal component (PC1) score as the overall size measure and regressing all 39 (ln transformed) traits onto PC1. The Lleonart et al. (2000) correction is:

$$Y_i^* = Y_i\left(\frac{X_0}{X_i}\right)^b$$

where Y_i and X_i are the values of a specific trait, and overall size (antilog$_e$ of the PC1 score) in individual 'i', respectively, Y^*_i is the theoretical value for the trait at the average size, X_0 is the average antilog$_e$ of the PC1 scores, and 'b' is the PC1 coefficient for each of the 39 traits. For example, specimen FMNH 92177 had an original value

Table 17.2 Thirty-nine linear skull measurements (distances between landmarks) and membership in the six functional/developmental groups and two major cranial regions

Functional/developmental group	Region	Trait
Oral	Face	ISPM
Nasal	Face	ISNSL
Oral, nasal	Face	ISPNS
Oral	Face	PMZS
Oral	Face	PMZI
Oral	Face	PMMT
Nasal	Face	NSLNA
Nasal	Face	NSLZS
Oral, nasal	Face	NSLZI
Cranial vault	Neurocranium	NABR
Orbit	Neurocranium	NAFM
Nasal	Face	NAPNS
Cranial vault	Neurocranium	BRPT
Cranial vault	Neurocranium	BRAPET
Orbit	Neurocranium	PTFM
Cranial vault	Neurocranium	PTAPET
Cranial vault	Neurocranium	PTBA
Cranial vault	Neurocranium	PTEAM
Zygomatic	Face	PTZYGO
Cranial vault, zygomatic	Neurocranium, Face	PTTSP
Orbit	Neurocranium	FMZS
Zygomatic	Face	FMMT
Oral	Face	ZSZI
Oral	Face	ZIMT
Zygomatic	Face	ZIZYGO
Zygomatic	Face	ZITSP
Oral	Face	MTPNS
Cranial base	Neurocranium	PNSAPET
Cranial base	Neurocranium	APETBA
Cranial base	Neurocranium	APETTS
Cranial base	Neurocranium	BAEAM
Zygomatic	Face	EAMZYGO
Zygomatic	Face	ZYGOTSP
Cranial vault	Neurocranium	LDAS
Cranial vault	Neurocranium	BRLD
Cranial vault	Neurocranium	OPILD
Cranial vault	Neurocranium	PTAS
Cranial base	Neurocranium	JPAS
Cranial base	Neurocranium	BAOPI

Table 17.1 defines each landmark and Fig. 17.1 shown their locations in a generalized Platyrrhine skull

(Y_i) of 4.808 for trait IS-PM and a PC1 score of 0.258 (1.295 after taking its anti-$\log_e$). The average size X_0 is 1.187 and the PC1 coefficient of IS-PM is 0.159. Therefore the corrected value (Y^*_i) for this specimen is 4.742. After this correction, the original data of all marmosets and callimicos are scaled to the same size, also adjusting their shape for allometric scaling effects. These scale-corrected (MASS) data were used to explore whether differences among marmosets and callimicos were size dependent in subsequent analyses.

Within groups, allometric patterns were estimated by linear regression of each log trait upon size (PC1). For the mono-specific *Callimico* and *Cebuella* this procedure is straightforward, but for *Mico* and *Callithrix*, to avoid mixing within-group (genus) and between-species differences, we first remove from the 39 log data set the differences among species within each of those two groups. This was accomplished using the general linear model in SYSTAT 11 with "species" as the independent factor and the 39 natural log traits as dependent variables and saving the residuals to use in the linear regression analyses.

17.2.3 Differentiation Analyses and Sexual Dimorphism

All analyses were done (except where noted below) using four a priori determined groups: *Callimico*, *Cebuella*, *Mico*, and *Callithrix*. We tested for differences between the groups, the sexes and an interaction between the sexes and groups using multivariate analysis of variance (MANOVA) upon both data sets, the original data, and the MASS data. To further explore differences among groups, we performed a discriminant function (DF) analysis, again upon both data sets. Mahalanobis D^2 morphological distances between group centroids were calculated from the DF analyses and then used as input in a cluster analysis (UPGMA with the average method) of morphological similarity among groups.

17.2.4 Life History

We also obtained life-history data from the literature (Hartwig 1996; Garber and Leigh 1997; Fleagle 1999; Lindenfors 2002; Porter and Garber 2004; Smith and Leigh 1998) for all NWM, particularly data on gestation length, body weight, and skull size (our own observations from museum specimens), age at first reproduction, and birth weight, all transformed to natural log scale to make their relationships linear. Fetal growth rate was estimated by dividing the natural log of birth weight by the natural log of gestation length. Postnatal growth rate was estimated by regressing the natural log of adult weight by the natural log of the age of first reproduction (in days) and using the regression slope as an estimate of the rate. Association among these variables was tested using the independent contrasts (IC) method to account for the nonindependence of phylogenetically structured data (Garland and Ives 2000). We used the module PDAP (Garland and Ives 2000) within the MESQUITE package (Maddison and Maddison 2003) to obtain the correlation among variables. The phylogenetic tree used is the same as in Marroig and Cheverud (2005) based on Schneider et al. (2001). Unfortunately, this phylogeny does not allow the separation between *Mico* and *Callithrix* and all analyses based upon it therefore pooled together *Mico/Callithrix* as one terminal in the tree. However, we think our conclusions are robust because there is only very minor variation between the two genera in life-history raw data (see Lindenfors 2002). For those unfamiliar with comparative analyses, the rationale here was to compare our focal group of taxa (callimico and marmosets) against all NWM in order to detect any significant departure, after accounting for their non-independency due to shared history, in those life-history traits in the focal group.

17.3 Results

17.3.1 Sexual Dimorphism

A MANOVA was performed on the 39 original measurements using sex, group, and sex by group interaction as independent variables in order to determine whether sexual dimorphism needs to be accounted for in the analyses. The analysis of 559 specimens shows significant multivariate (Wilks' $\Lambda=0.0012$, $df=117$; 1537, $p<10^{-6}$) differences among groups. All 39 traits had significant univariate ($P>1.0\times10^{-6}$) differences between the groups, even using a conservative Bonferroni level of significance ($p<0.05/39$ or 0.0013). Conversely, there is only a very weak multivariate sexual dimorphism detected (Wilks' $\Lambda=0.900$, $df=39$; 513, $p=0.04$) for only 3 traits (BR-PT, PT-ZYGO and ZI-TSP), showing univariate probabilities between the 1% and 5% level and none reaching significance under the Bonferroni threshold ($p<0.0013$). There is also significant, although weak, multivariate interaction between sex and group (Wilks' $\Lambda=0.756$, $df=117$; 1537, $p=0.03$) with 7 traits showing univariate probabilities between the 1% and 5% level and none reaching significance under the Bonferroni threshold ($p<0.0013$). Moreover, discriminant analyses within each group with sex as the independent factor show no significant multivariate differences between sexes within each group for the 39 original traits.

Another MANOVA was performed using the 39 MASS corrected data, again fitting the model of sex, group, and sex by group interaction as independent variables. The analysis of 559 specimens shows significant multivariate (Wilks' $\Lambda=0.022$, $df=117$; 1537, $p<10^{-6}$) differences among groups. Thirty six traits had significant univariate ($P>1.0\times10^{-6}$) differences between the groups, even using a conservative Bonferroni level of significance ($p<0.05/39$ or 0.0013), with only 1 trait (BA-EAM) not significant at the 5% level and two others (NSL-NA and BA-OPI) not reaching the Bonferroni level but showing P-values below 0.005. Conversely, there is no multivariate sexual dimorphism detected (Wilks' $\Lambda=0.918$, $df=39$; 513, $p=0.200$), with only 2 traits (BR-PT and BA-EAM) showing univariate probabilities between the 1% and 5% level and none reaching significance under the Bonferroni threshold ($p<0.0013$). There is also no significant multivariate interaction between sex and group (Wilks' $\Lambda=0.770$, $df=117$; 1537, $p=0.078$), with only 3 traits (IS-NSL, NA-BR and BR-PT) showing univariate probabilities between the 1%

and 5% level and none reaching significance under the Bonferroni threshold ($p<0.0013$). Furthermore, discriminant analyses within each group with sex as the independent factor show no significant multivariate differences between sexes for the 39 MASS corrected data.

An ANOVA performed on size (PC1 scores) to test for the effects of groups, sex, and sex by groups interaction shows significant differences between groups ($F=1621.2$, $p<10^{-5}$), no differences related to sex ($F=0.026$, $p=0.87$), and significant interaction between sex and group ($F=3.37$, $p=0.018$). Further ANOVAs within groups to test for sexual differences in size show no significant differences between sexes within *Mico* or *Callithrix*, significant differences for *Cebuella* ($p=0.017$), and marginally significant differences for *Callimico* ($p=0.09$). Therefore, the significant sex by gender interaction observed is basically because *Cebuella* females are slightly larger than males, while male callimicos are slightly larger than females. Overall, the results of the analyses indicate that sexual dimorphism is very small and therefore subsequent analyses were performed without regard to gender.

17.3.2 Differentiation Analyses

The linear discriminant analysis upon the original traits shows that all three functions are significant at least at $p<10^{-8}$, the first function accounting for 56.5% of the total variance, and corresponding figures for the second and third functions were 27.9 and 15.6%, respectively. The canonical correlations of the three functions were, from the first to the third 0.97, 0.94 and 0.90, respectively. Figures 17.2a–17.2c present the canonical scores plot of the DF and Table 17.3 shows the correlation between the DF scores and the original traits. There is a correlation of 0.96 ($p<10^{-5}$) between the DF1 and the PC1 scores, indicating that the first DF is basically a size vector. This interpretation of the DF1 as a size vector is also reinforced by the positive correlations of each trait with the DF1 (Table 17.3). The DF1 clearly separates *Cebuella* and *Callimico* from each other and from the two other marmoset genera, *Mico* and *Callithrix* (which mix together). The DF2 has significant contribution from both neural and facial traits and also a mixture of positive and negative correlations. Landmarks ZS, ZI and FM are involved in most traits showing significant correlations with DF2, and an examination of these correlations shows that specimens with larger values in DF2 also show the landmark ZS dislocated to a more posterior position, ZI to a more anterior and superior position, and FM downward along the rim of the orbit. The combination of the larger positive and negative trait correlations with DF2 suggests that this function could be interpreted as a factor where larger values are associated with a more vertical orientation of the orbit and a more outward (or rounded) zygomatic arch (including landmark MT). The DF2 clearly separates *Mico* and *Callithrix*, while *Cebuella* overlaps with both of them and *Callimico* is further separated from the remaining groups, with the largest values. DF3 further separates *Callimico* from the rest and can be interpreted as a factor contrasting face and neurocranium, which is dominated by positive values for

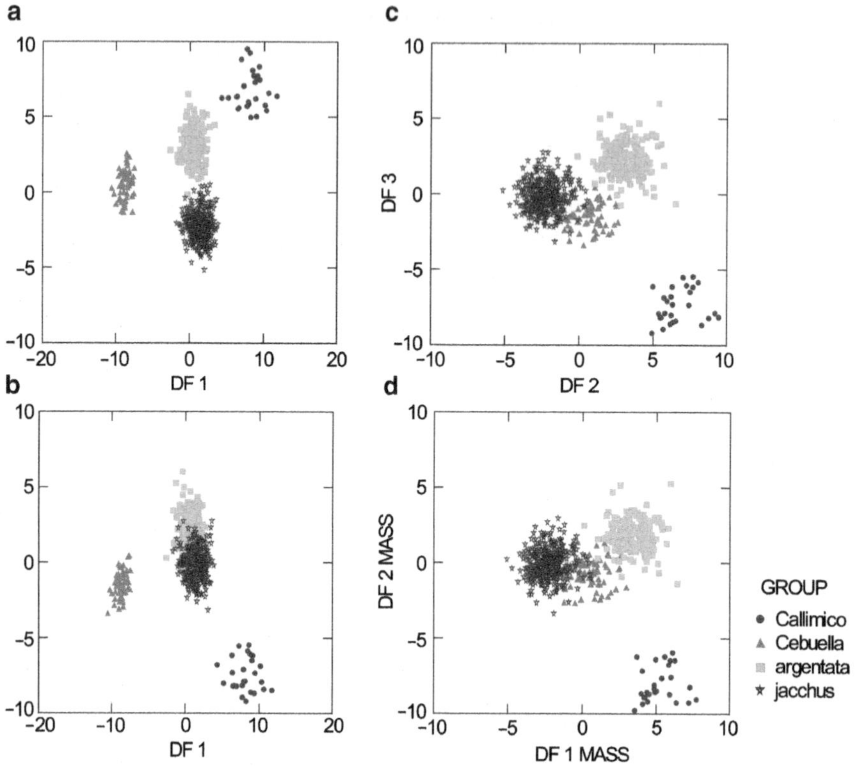

Fig. 17.2 Plot of marmoset and callimico scores against the discriminant functions (DF1, DF2 and DF3) obtained from original data and from the MASS corrected data (DF1_MASS and DF2_MASS). (circle - Callimico, triangle = Cebuella, square = Mico and Star = Callithrix)

facial traits and negative values for the neural region, meaning that specimens with larger scores (*Mico*) have somewhat enlarged nasal and oral traits, landmark PT dislocated upwardly, and again a smaller distance between the two landmarks in the orbit (FM and ZS). DF3 also indicates that landmark APET is dislocated to an anterior position for those specimens with larger scores in this function (given the contrast between negative PNS-APET and positive APET-BA). Virtually 100% of cases were classified correctly according to the DF, both in the original and jackknifed classification matrix (except for one *Mico* specimen allocated to *Callithrix* in the jackknifed matrix). Mahalanobis D^2 morphological distances are presented in the bottom diagonal in Table 17.3.

The linear discriminant analysis done with the MASS corrected data shows that the first two functions are significant at least at $p<10^{-8}$, with the first function accounting for 64.6% of the total variance, while the second for 34.3%. The third DF accounts for only 1.2% of the total variance, although still significant ($p<10^{-3}$). The canonical correlations of the three functions were, from the first to the third: 0.94, 0.89 and 0.33, respectively. Only the first 2 DF will be considered here.

Table 17.3 Pearson correlation of traits with the 3 original data based discriminant function (DF1-DF3) and with the MASS corrected data (DF1MASS and DF2MASS) along with their individual significance (Bonferroni corrected) are also presented

Trait	DF1	DF2	DF3	DF1MASS	DF2MASS
ISPM	0.64***	0	0.22***	0.02	0.19***
ISNSL	0.81***	–0.09*	0.28***	–0.13	0.32***
ISPNS	0.74***	–0.16***	0.38***	–0.22***	0.52***
PMZS	0.54***	0.63***	0.19***	0.79***	0.07
PMZI	0.87***	–0.27***	0.03	–0.64***	–0.1
PMMT	0.87***	–0.05	–0.11**	–0.2***	–0.43***
NSLNA	0.69***	0.14***	0	0.17***	–0.12
NSLZS	0.52***	0.65***	0.15***	0.8***	0.01
NSLZI	0.91***	–0.16***	0.13***	–0.53***	0.12
NABR	0.85***	–0.08*	0.08	–0.16**	–0.04
NAFM	0.49***	0.39***	0.1*	0.48***	–0.02
NAPNS	0.83***	–0.18***	0.25***	–0.36***	0.34***
BRPT	0.78***	–0.02	0	–0.05	–0.15*
BRAPET	0.92***	–0.19***	0.1**	–0.38***	0.01
PTFM	0.82***	–0.19***	0.13***	–0.33***	0.11
PTAPET	0.77***	0.14***	–0.07	0.19***	–0.31***
PTBA	0.86***	0.08	0.21***	0.22***	0.23***
PTEAM	0.81***	0.12**	0.27***	0.27***	0.29***
PTZYGO	0.78***	0.1**	0.19***	0.17***	0.14*
PTTSP	0.46***	0.39***	0.29***	0.46***	0.21***
FMZS	0.57***	–0.43***	–0.48***	–0.57***	–0.55***
FMMT	0.9***	–0.12**	–0.08*	–0.44***	–0.46***
ZSZI	0.8***	–0.39***	0.17***	–0.65***	0.19***
ZIMT	0.73***	–0.15***	–0.22***	–0.3***	–0.41***
ZIZYGO	0.72***	0.35***	0.11**	0.49***	0.01
ZITSP	0.77***	0.43***	0.19***	0.7***	0.11
MTPNS	0.73***	0.36***	0.17***	0.56***	0.08
PNSAPET	0.76***	0.18***	–0.26***	0.17***	–0.51***
APETBA	0.7***	–0.14***	0.48***	–0.13***	0.58***
APETTS	0.69***	0.18***	–0.14***	0.2***	–0.34***
BAEAM	0.88***	0.01	0.1**	0.01	–0.03
EAMZYGO	0.48***	0.25***	0.22***	0.31***	0.15*
ZYGOTSP	0.82***	–0.05	0.06	–0.12	–0.04
LDAS	0.6***	0.15***	–0.25***	0.13	–0.41***
BRLD	0.64***	0.14***	–0.16***	0.14*	–0.33***
OPILD	0.73***	0.16***	–0.12**	0.13*	–0.23***
PTAS	0.89***	0.1**	0.2***	0.26***	0.19***
JPAS	0.66***	–0.42***	0.19***	–0.6***	0.19***
BAOPI	0.45***	0.15***	–0.02	0.15**	–0.11

$^*p<0.05$, $^{**}p<0.01$, $^{***}p<0.001$

Figure 17.2d presents the canonical scores plot of the DF and Table 17.3 shows the correlation between the DF scores and the original traits. Interestingly, there is a correlation of 0.99 ($p<10^{-5}$) between the scores of the DF2 (extracted from the original data) and the DF1 (extracted from the MASS data), indicating that the DF1_MASS bears the same biological interpretation as DF2. The same pattern holds for DF3 and DF2_MASS showing a correlation of 0.98 ($p < 10^{-5}$) also indicating that the biological meaning of DF3 and DF2_MASS is the same. Furthermore, comparison of Figs. 17.2c and 17.2d clearly shows this similarity between DF2 × DF1_MASS and DF3 × DF2_MASS indicating essentially the same arrangement of the specimens in relation to the functions.

A reduced efficiency in the classification phase of this second DF analyses is evinced by the 95% (93% for the jackknifed) of the total cases being classified correctly according to the functions: 100% (100% jackknifed) of the callimico specimens were allocated correctly, but for the other groups, *Cebuella* 83% (79%), *Mico* 99% (97%) and *Callithrix* 96% (94%). Mahalanobis D^2 morphological distances are presented above the diagonal in Table 17.4. Interestingly, *Callithrix* and *Cebuella* overlap more in this second DF analysis than either does with *Mico*. The comparison of the cluster phenograms obtained for both D^2 distances (original and MASS corrected) are quite interesting (Fig. 17.3). The original data based D^2 shows that *Callithrix* and *Mico* group together, then with *Cebuella* and finally all three with *Callimico*. Conversely, in the MASS corrected data based D^2 *Callithrix* and *Cebuella* group together, then with *Mico* and finally all three with *Callimico*.

Table 17.4 Mahalanobis (D^2) morphological distances (original data) among groups are presented below diagonal and MASS corrected data based D^2 above diagonal

	Mico	*Callithrix*	*Cebuella*	*Callimico*
Mico	0	35.38	17.02	102.45
Callithrix	36.90	0	9.74	120.87
Cebuella	111.60	118.01	0	74.02
Callimico	167.99	181.97	361.48	0

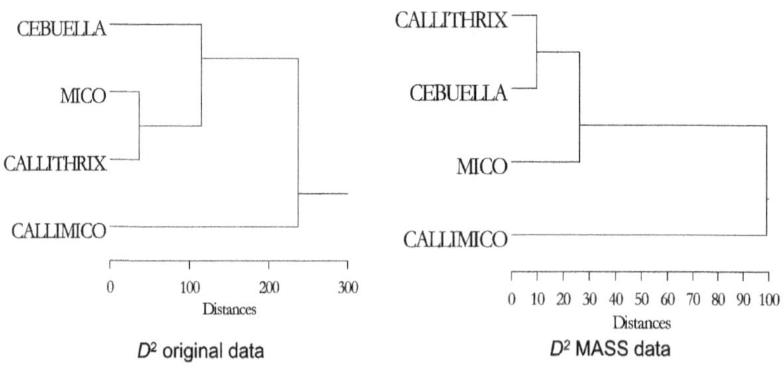

Fig. 17.3 Morphological similarity among the groups derived from UPGMA analysis of the original data and from the MASS corrected data Mahalanobis D^2 values

17.3.3 Allometry

Multivariate static allometric patterns were calculated for each group (after accounting for within group variation due to species differences in both *Mico* and *Callithrix*) and are presented in Table 17.5. Overall groups share the same basic allometric pattern as indicated by the vector correlations between groups (Table 17.6), with only one significant difference observed in the trait IS-PM between *Mico* (negatively allometric) and *Callithrix* (positively allometric) (Table 17.5). For the sake of brevity we present in Table 17.5 the regression coefficients, and not the allometric coefficients (AC's). To transform the regression coefficients into ACs it is necessary first to normalize the vectors to length 1 and then divide this normalized coefficients by $1/\sqrt{n}$, where n is the number of traits (39 in our case).

17.3.4 Life History

Figure 17.4 shows the regression between birth weights and gestation length, after correcting for nonindependence between points due to shared history (phylogeny). Marmosets (*Callithrix/Mico, Callimico* and *Cebuella*) deviate significantly from the regression line suggesting that birth weight in marmosets and callimicos is smaller than expected for their gestation lengths. Figure 17.5 shows the regression between the ages of first reproduction against body weight (the result is basically the same if we use skull size instead of body weight, but *Cebus* does not deviate significantly from the line in this latter case). Among NWM, only *Cebus* seems to deviate significantly from the regression line showing a delayed onset of reproduction for its size (using skull size instead, *Lagothrix* is the only significant deviant from the regression line – not shown). Similar analysis for gestation length against body size does not reveal any significant deviation in any callitrichid, indicating that marmosets and callimicos in particular present gestation lengths expected for their sizes (Fig. 17.6). Notice although that *Aotus* seems to deviate significantly from the regression line, indicating that night monkeys have a shorter gestation length than expected for a NWM of its size (using skull size instead of body weight *Ateles* and *Aotus* deviate significantly from the regression line). There is also nothing noteworthy about the regression of the age of weaning in relation to body size (results not shown).

17.4 Discussion

Sexual dimorphism in marmosets and callimicos is either nonexistent (for most traits) or is very small. After correcting for size-scale (MASS corrected data), there is no significant sexual dimorphism in our samples. This suggests that the small sexual dimorphism observed is due to size-scaling differences. One interesting pattern is the opposite trend shown by pygmy marmosets and callimicos with regard to skull size. Pygmy marmoset females are significantly larger than males while in callimicos the pattern is reversed (although not significant in our sample). Furthermore, males and females in all four groups present similar allometric patterns (results not shown). Considering these results, we decided to pursue all further analyses without regard to sex variation.

Species groups within the *Callithrix-Mico-Cebuella-Callimico* clade are clearly distinct in form (shape and size) (Figs. 17.2a–17.2c. *Cebuella*, *Callimico*, *Mico*-plus-*Callithrix* are clearly separated by size (DF1); this is particularly interesting as new weight data on wild callimicos suggest they are not much larger than marmosets in body weight (about 500 g; Porter and Garber 2004). The significant cranial/morphometric differences raise the question of whether or not marmoset and callimico morphological differences are size-related. Interestingly, while *Callimico* remains quite distinct from the other groups after accounting for size-scale differences, *Cebuella* mixes together with the other marmosets (Figs. 17.2c, 17.2d), especially with the *Callithrix* species. The two traditional marmoset groups (*Mico* and *Callithrix*) are also distinct, differences not being related to size variation but to the DF2. These interpretations are also reflected in the cluster diagrams (Fig. 17.3), where the MASS corrected diagram shows that the morphological distances between *Callithrix-Cebuella* are very small, with *Cebuella* actually more similar to *Callithrix* than to its phylogenetically closer group, *Mico*. Both phenograms represent their respective patterns in the D^2 distances matrices quite well, with the cophenetic correlation for the original data-based D^2 being 0.78 and the same figure for the MASS corrected data being a 0.93 correlation. Taken together these analyses suggest that pygmy marmosets are scaled down versions of other marmosets, but with morphology largely similar to species of the *Callithrix* group. While more field data is certainly needed, especially with respect to dietary habits, the morphological similarity of these two more gummivorous marmosets (*Callithrix* and *Cebuella*) suggests that this might be a morphological convergence of the pygmy marmoset to *Callithrix*-like morphology after phylogenetic divergence from *Mico*, despite the size reduction occurring in the pygmy marmoset's lineage. Conversely, gummivory might be a plesiomorphic feature of the marmoset radiation shared among *Callithrix* and *Cebuella* and later reduced in some or most lineages of *Mico* and some of *Callithrix* (like *Callithrix aurita*).

Postnatal growth period (age of first reproduction), age of weaning, and gestation time appear to be the same for all callitrichids, according to life-history data, after we account for body size and in comparison with the other Platyrrhini (Figs. 17.5 and 17.6). Additionally, skull growth rates as indicated by the allometric patterns are also the same for the marmosets and callimicos (Tables 17.5 and 17.6). Conservation of allometric patterns in marmosets, pygmy marmosets, and callimicos (both in the slope and the constant) suggest that they all follow the same growth patterns but evolved to attain different adult sizes. In a size-based scheme for heterochrony (see page 42 in McKinney and McNamara 1991), this would suggest that *Cebuella* evolved its small size by some sort of ontogenetic scaling or allometric progenesis.

Table 17.5 Results of the 39 traits regressions against skull size, showing the regression constant and slope along with the 95% confidence interval (CI)

	Mico		*Callithrix*		*Cebuella*		*Callimico*		*Mico*		*Callithrix*		*Cebuella*		*Callimico*	
Trait	Constant	CI	Constant	CI	Constant	CI	Constant	CI	Slope	CI	Slope	CI	Slope	CI	Slope	CI
ISPM	1.56	1.54–1.57	1.51	1.5–1.53	1.42	1.26–1.57	1.37	1.24–1.5	0.09	0.04–0.14	0.19	0.15–0.23	0.14	0.04–0.23	0.24	0.11–0.38
ISNSL	2.11	2.1–2.12	2.1	2.09–2.11	2.16	1.93–2.39	1.97	1.87–2.07	0.14	0.09–0.18	0.12	0.07–0.16	0.27	0.13–0.41	0.21	0.11–0.31
ISPNS	2.75	2.74–2.76	2.73	2.73–2.74	2.8	2.67–2.93	2.53	2.44–2.61	0.13	0.09–0.16	0.16	0.13–0.18	0.22	0.14–0.29	0.2	0.12–0.29
PMZS	2.3	2.28–2.31	2.12	2.11–2.13	2.43	2.25–2.61	2.23	2.14–2.31	0.12	0.08–0.16	0.13	0.1–0.16	0.26	0.16–0.37	0.21	0.13–0.3
PMZI	2.47	2.46–2.48	2.54	2.54–2.55	2.53	2.41–2.64	2.47	2.39–2.56	0.13	0.1–0.16	0.13	0.11–0.15	0.21	0.13–0.28	0.19	0.1–0.27
PMMT	2.45	2.44–2.46	2.48	2.47–2.48	2.44	2.33–2.55	2.55	2.48–2.62	0.11	0.08–0.14	*0.13*	0.11–0.15	0.15	0.08–0.21	0.14	0.08–0.21
NSLNA	1.86	1.84–1.88	1.8	1.79–1.82	1.7	1.44–1.97	2.05	1.85–2.24	0.10	0.03–0.16	*0.14*	0.08–0.19	0.08	−0.08–0.24	0.03	−0.17–0.22
NSLZS	2.51	2.5–2.52	2.35	2.35–2.36	2.51	2.39–2.62	2.48	2.41–2.55	0.09	0.06–0.12	*0.13*	0.1–0.15	0.16	0.09–0.23	0.17	0.1–0.24
NSLZI	2.84	2.83–2.84	2.87	2.86–2.87	2.83	2.74–2.92	2.8	2.74–2.86	0.12	0.1–0.14	0.13	0.12–0.15	0.19	0.13–0.24	0.18	0.12–0.24
NABR	3.17	3.16–3.18	3.19	3.18–3.2	3.11	3.03–3.2	3.24	3.16–3.33	0.07	0.04–0.1	0.09	0.07–0.12	0.1	0.05–0.16	0.05	−0.03–0.14
NAFM	2.58	2.57–2.59	2.52	2.51–2.53	2.57	2.49–2.65	2.56	2.51–2.61	0.11	0.08–0.14	0.10	0.07–0.12	0.09	0.04–0.14	0.12	0.07–0.17
NAPNS	2.66	2.65–2.67	2.67	2.67–2.68	2.76	2.6–2.92	2.58	2.52–2.63	0.13	0.1–0.16	0.14	0.12–0.16	0.24	0.15–0.34	0.16	0.1–0.21
BRPT	2.86	2.85–2.87	2.87	2.86–2.88	2.72	2.62–2.83	2.93	2.84–3.02	0.06	0.03–0.09	0.08	0.05–0.11	0.05	−0.01–0.12	0.07	−0.02–0.16
BRAPET	2.94	2.93–2.94	2.98	2.97–2.99	2.88	2.8–2.96	3	2.92–3.08	0.06	0.03–0.08	0.04	0.02–0.06	0.11	0.06–0.16	0.07	−0.01–0.14
PTFM	1.93	1.91–1.94	2	1.98–2.01	1.93	1.64–2.23	1.87	1.74–2.01	0.16	0.11–0.22	0.16	0.12–0.21	0.25	0.07–0.42	0.22	0.09–0.36
PTAPET	2.62	2.61–2.63	2.62	2.61–2.62	2.72	2.61–2.83	2.68	2.6–2.75	0.15	0.12–0.18	0.12	0.09–0.14	0.18	0.11–0.24	0.14	0.07–0.22
PTBA	3.12	3.12–3.13	3.11	3.1–3.11	3.2	3.12–3.27	3.06	3.01–3.11	0.14	0.12–0.17	0.11	0.09–0.12	0.18	0.13–0.22	0.16	0.12–0.21
PTEAM	2.7	2.68–2.71	2.66	2.65–2.67	2.8	2.68–2.91	2.6	2.51–2.68	0.16	0.13–0.2	0.12	0.1–0.14	0.22	0.15–0.29	0.2	0.11–0.28
PTZYGO	2.47	2.46–2.48	2.44	2.43–2.46	2.72	2.58–2.87	2.41	2.3–2.51	0.21	0.16–0.25	0.13	0.09–0.16	0.3	0.21–0.38	0.21	0.11–0.31
PTTSP	1.99	1.97–2.02	1.84	1.82–1.86	2.33	2–2.65	1.84	1.69–2	0.17	0.1–0.23	0.11	0.05–0.16	0.35	0.15–0.54	0.22	0.06–0.37
FMZS	1.85	1.83–1.87	2.06	2.05–2.07	1.71	1.5–1.92	2.25	2.13–2.36	0.02	−0.04–0.09	0.02	−0.02–0.06	−0.09	−0.22–0.03	−0.01	−0.13–0.1
FMMT	2.6	2.59–2.61	2.66	2.65–2.66	2.73	2.66–2.8	2.65	2.61–2.68	0.17	0.14–0.2	0.13	0.12–0.15	0.21	0.17–0.26	0.2	0.16–0.23
ZSZI	1.95	1.93–1.96	2.09	2.08–2.1	2.24	1.92–2.57	1.85	1.65–2.04	0.22	0.17–0.28	*0.14*	0.1–0.17	0.43	0.24–0.62	0.22	0.03–0.41
ZIMT	1.69	1.68–1.71	1.8	1.79–1.81	1.94	1.81–2.08	1.77	1.68–1.86	0.20	0.14–0.25	0.18	0.15–0.21	0.28	0.2–0.36	0.25	0.16–0.34
ZIZYGO	2.25	2.24–2.27	2.15	2.13–2.16	2.13	1.93–2.33	2.2	2.09–2.31	0.18	0.13–0.23	0.13	0.08–0.19	0.14	0.02–0.26	0.27	0.16–0.38
ZITSP	2.3	2.28–2.31	2.18	2.17–2.19	2.19	2.05–2.33	2.31	2.23–2.39	0.17	0.13–0.21	0.12	0.09–0.16	0.14	0.06–0.22	0.17	0.09–0.25
MTPNS	1.88	1.86–1.89	1.76	1.75–1.77	1.76	1.64–1.88	1.85	1.76–1.94	0.08	0.03–0.13	*0.14*	0.11–0.17	0.14	0.06–0.21	0.16	0.07–0.25
PNSAPET	1.96	1.94–1.98	1.98	1.96–1.99	1.85	1.61–2.09	2.22	2.12–2.33	0.14	0.07–0.2	0.09	0.05–0.13	0.07	−0.08–0.21	0.14	0.03–0.25

(continued)

Table 17.5 (continued)

	Mico		*Callithrix*		*Cebuella*		*Callimico*		*Mico*		*Callithrix*		*Cebuella*		*Callimico*	
Trait	Constant	CI	Constant	CI	Constant	CI	Constant	CI	Slope	CI	Slope	CI	Slope	CI	Slope	CI
APETBA	2.4	2.39–2.41	2.39	2.38–2.4	2.39	2.29–2.5	2.14	2.05–2.24	0.12	0.09–0.15	0.09	0.06–0.11	0.12	0.06–0.18	0.2	0.1–0.29
APETTS	1.88	1.87–1.9	1.88	1.87–1.89	1.75	1.64–1.87	1.93	1.85–2.01	0.10	0.06–0.14	0.13	0.1–0.16	0.05	–0.02–0.12	0.19	0.12–0.27
BAEAM	2.55	2.54–2.56	2.54	2.53–2.54	2.55	2.46–2.64	2.6	2.55–2.66	0.10	0.08–0.13	*0.12*	0.1–0.13	0.14	0.09–0.2	0.08	0.02–0.13
EAMZYGO	2.07	2.04–2.09	2	1.98–2.02	1.99	1.8–2.19	2.13	1.99–2.28	0.14	0.07–0.2	0.11	0.06–0.17	0.08	–0.04–0.2	0.02	–0.13–0.16
ZYGOTSP	1.85	1.83–1.86	1.9	1.89–1.91	2	1.72–2.27	1.87	1.77–1.97	0.23	0.18–0.28	*0.14*	0.1–0.18	0.28	0.12–0.44	0.21	0.11–0.3
LDAS	2.38	2.36–2.39	2.38	2.37–2.39	2.22	2.11–2.33	2.55	2.43–2.66	0.08	0.05–0.12	0.10	0.06–0.14	0.02	–0.04–0.09	0.08	–0.03–0.2
BRLD	3.02	3.01–3.03	2.98	2.96–2.99	2.93	2.82–3.05	3.08	2.99–3.17	0.05	0.01–0.09	*0.07*	0.02–0.11	0.05	–0.02–0.11	0.17	0.08–0.26
OPILD	2.41	2.38–2.44	2.44	2.42–2.46	1.8	1.3–2.3	2.55	2.29–2.82	0.08	–0.01–0.17	0.06	0.01–0.11	–0.15	–0.45–0.14	0.26	0–0.52
PTAS	3.15	3.14–3.16	3.11	3.11–3.12	3.1	3.02–3.18	3.16	3.1–3.21	0.09	0.06–0.11	0.10	0.08–0.12	0.12	0.07–0.17	0.1	0.05–0.16
JPAS	2.37	2.36–2.38	2.43	2.42–2.44	2.49	2.34–2.64	2.24	2.14–2.35	0.12	0.08–0.16	0.13	0.1–0.15	0.19	0.1–0.27	0.15	0.05–0.26
BAOPI	1.86	1.85–1.88	1.83	1.82–1.84	1.71	1.51–1.9	1.99	1.85–2.13	–0.02	–0.07–0.03	0.03	–0.01–0.07	–0.03	–0.14–0.09	–0.05	–0.19–0.09

Table 17.6 Correlations between marmosets and callimicos allometric vectors

	Mico	*Callithrix*	*Cebuella*	*Callimico*
Mico	1			
Callithrix	0.95	1		
Cebuella	0.92	0.88	1	
Callimico	0.94	0.94	0.86	1

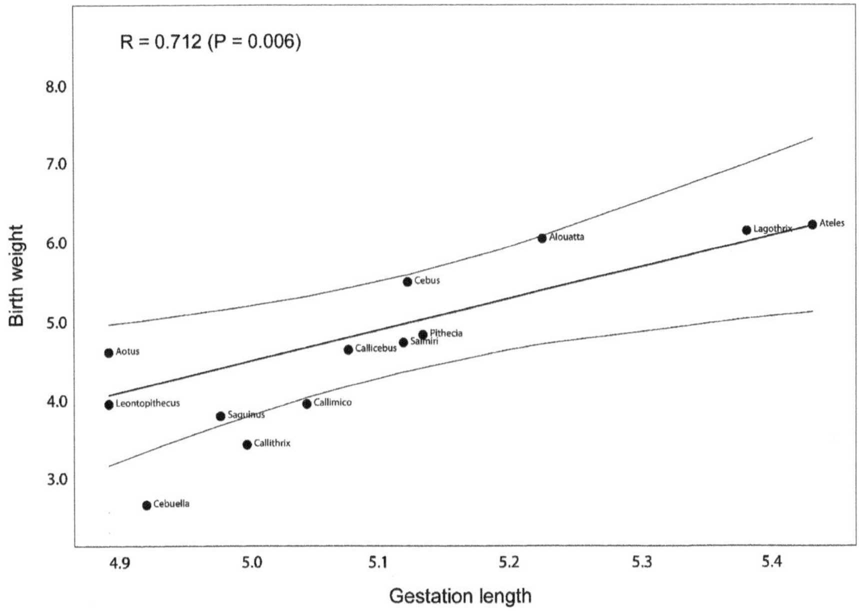

Fig. 17.4 Plot of the birth weight against gestation length. The regression line and 95% confidence limits were obtained from the method described in Garland and Ives (2000) and implemented in package PDAP in Mesquite. The correlation and associated probability is also given

However, as noted above, none of the Callitrichidae show any significant deviation for the age of first reproduction from what is expected for their sizes (Fig. 17.5).

One key piece to solve the puzzle of size evolution in the *Callimico-Callithrix-Mico-Cebuella* radiation is that provided in Fig. 17.4. All marmosets (*Callithrix/Mico, Callimico,* and *Cebuella*) are smaller at birth than expected from their gestation lengths (Fig. 17.4). Because their gestation lengths are not different from what is expected for a platyrrhine of their size, this small size of the marmosets at birth is probably due to a relatively slow intrauterine growth rate, which translates into smaller body sizes at birth. Furthermore, if we use the total litter body weight for the callitrichids that usually have twins (*Saguinus, Leontopithecus, Callithrix, Mico*, and *Cebuella*) and perform the same regression analysis controlling for

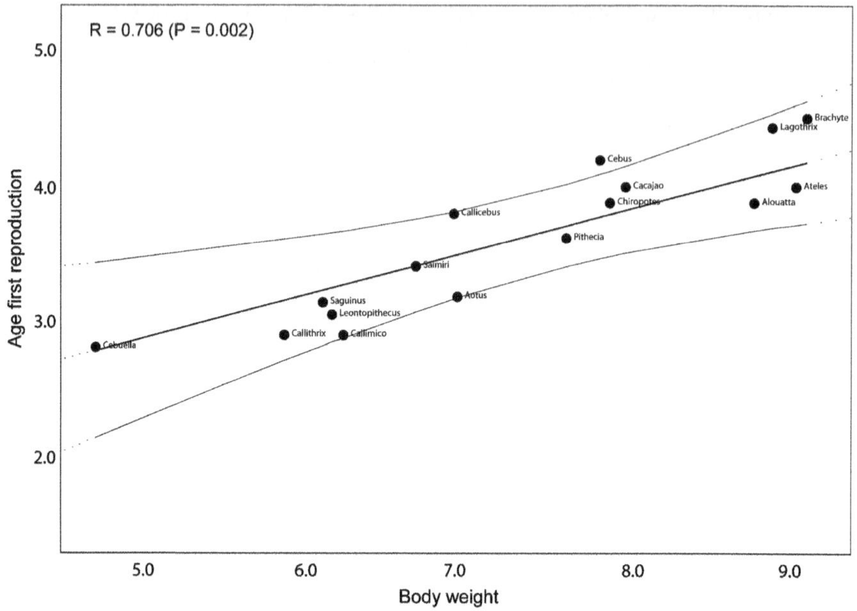

Fig. 17.5 Plot of the age of first reproduction against body weight. The regression line and 95% confidence limits were obtained from the method described in Garland and Ives (2000) and implemented in package PDAP in Mesquite. The correlation and associated probability is also given

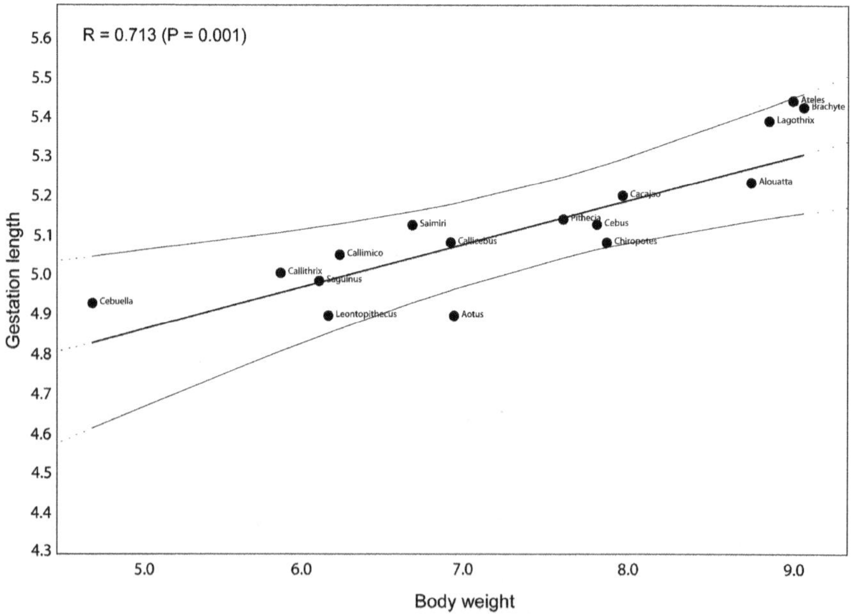

Fig. 17.6 Plot of the gestation length against body weight. The regression line and 95% confidence limits were obtained from the method described in Garland and Ives (2000) and implemented in package PDAP in Mesquite. The correlation and associated probability is also given

phylogenetic relationships as in Fig. 17.4, we observe that *Cebuella* and *Callimico* have smaller birth weights than expected from their gestation lengths in comparison with the rest of the NWM (Fig. 17.7). Finally, the whole clade comprising *Callimico-Callithrix-Mico-Cebuella* shows the smallest (Table 17.7) fetal growth rates among NWM, a trend exacerbated in *Cebuella.* This result is consistent even when two times the birth weight is used(as in Fig. 17.7) to estimate the fetal growth rates. Therefore, the evolution of small body sizes in the marmoset and callimico radiation was achieved by a reduction of intrauterine growth rates.

Another possibility is that *Cebuella*, and callitrichids in general, evolved its small size by decreases in both the rate and duration of the postnatal growth (Garber and Leigh 1997). With regard to the duration of the post-natal growth period our comparative analyses of all NWM taking into account the effects of shared history (phylogeny) does not support such an assertion, as already noticed in Fig. 17.5. Garber and Leigh (1997) also suggest that differences in adult body size among tamarins and marmosets are exclusively a product of shifts in postnatal growth rates. While our results at first do not conflict with changes in postnatal growth rates, our analyses clearly suggest that the clade formed by *Callimico/Mico/Callithrix/Cebuella* is unique in NWM by showing reduced prenatal growth rates in comparison to the rest of the Platyrrhini.

To help clarify the possible reduction in pre and postnatal growth rates in the marmosets/callimico clade we plot body weight (in LN-scale) against age in days (again in LN-scale) for all NWM with gestation length, age at first reproduction, birth weight and adult weight data available (Fig. 17.8). Therefore, the middle point corresponds to the birth weight/gestation length and the last point corresponds to the adult weight/age at first reproduction, while the origin is set to zero. This plot shows the LOWESS smooth for both periods pre and postnatal growth for each genus. Because these average growth trajectories of each genus are in a linear scale, the growth rate for both periods can be easily visualized by the slope of the lines. It is clear from the graph that the whole callitrichid clade is characterized by the lowest prenatal growth rates among NWM. Also, callitrichids do not show particularly slow postnatal growth rates. In fact, some of them, in particular *Callithrix* (including *Mico*) and *Callimico* are among the highest postnatal growth rates among NWM (see also Table 17.7). This suggests that the whole clade could be considered as morphologically pedomorphic or juvenile-like. Considering the phylogenetic relationships among NWM (Schneider et al. 2001; Cortés-Ortiz Chap. 1 this volume), these results suggest that this trend of reduced prenatal growth rate appeared early in the radiation of the Callitrichidae and became more pronounced in the later off-shoot clades (*Callithrix, Mico,* and *Cebuella*). Pygmy marmosets present by far the most extreme condition (derived) of this reduction in prenatal growth rates resulting in its remarkable pedomorphic morphology apparently attained by some sort of intra-uterine allometric post displacement. The position of the newly discovered and little known *Callibella* in this radiation, relative to the degree of development of reduction of fetal growth rate, remains to be determined.

The miniaturization process involved in the marmoset/callimico radiation is almost unique among mammals because it does not involve deceleration of the postnatal growth rates (Gould 1975; Shea and Gomez 1988; Plavcan and Gomez

1993a, b; Webster et al. 2004), but instead, a slowing down of the pre-natal growth rates. Gould (1975) proposed that mammalian phyletic dwarfs should display negative dental allometry, so that dwarf clades should have large postcanine dentition relative to their body size. Shea and Gomez (1988) found support for this proposition among human pygmy groups. The argument is that when miniaturization occurs by limiting postnatal growth, early determined features, such as tooth size and brain size, will evolve at a slower rate than body size resulting in large tooth size and brain size relative to body size. Plavcan and Gomez (1993a, b) failed to find relatively large post canine dentition in the callitrichids, noting that *Saguinus*, *Callithrix*, and *Cebuella* actually have relatively small teeth for their body size. They therefore rejected a rapid miniaturization in this clade. However, as we have suggested above, callitrichid miniaturization is due to prenatal rather than postnatal growth reductions. Therefore, we would not expect them to show relatively large post canine dentition because miniaturization is due to decrements in early growth. Thus, the dental evidence of Plavcan and Gomez (1993a, b) supports a model of slow prenatal growth as responsible for miniaturization in the marmosets.

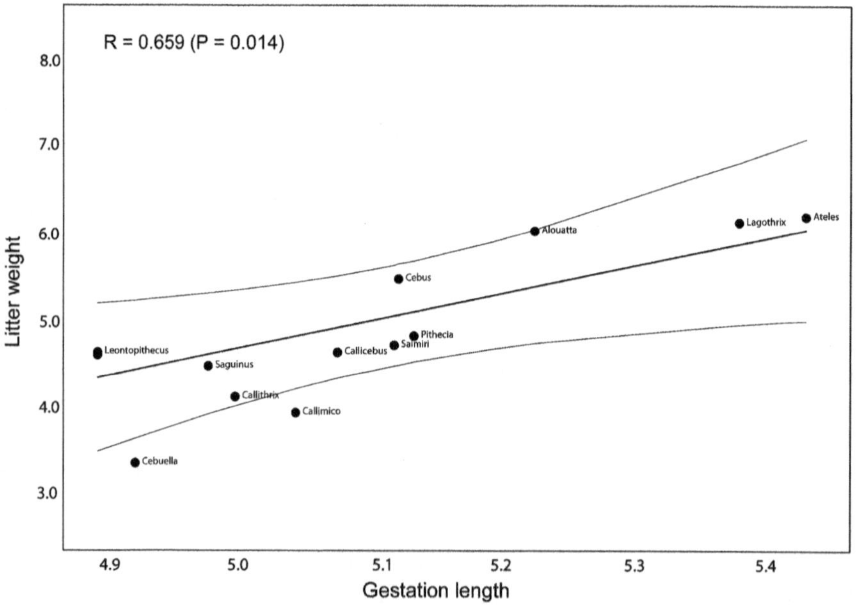

Fig. 17.7 Plot of litter weight (two times the birth weight for those genera which usually bear twins) against gestation length. The regression line and 95% confidence limits were obtained from the method described in Garland and Ives (2000) and implemented in package PDAP in Mesquite. The correlation and associated probability is also given

Table 17.7 Life-history data and pre- and post-natal growth rates

Genus	Age first reproduction (years)	Age first rep (days)	Adult weight	Gestation length	Age of weaning	Birth weight	Skull size	Pre-natal growth rate (regression)	Post-natal growth rate
Alouatta	4.00	1460.00	6404.2	186	369	407.7	1.55	1.150	1.337
Ateles	4.50	1642.50	8276.3	229	653	482.0	1.44	1.137	1.443
Brachyteles	7.50	2737.50	8840.0	225	639		1.60		
Lagothrix	7.00	2555.00	7150.0	218	340	450.0	1.35	1.135	1.124
Cacajao	4.50	1642.50	2893.8	180	547		1.01		
Chiropotes	4.00	1460.00	2632.5	160			0.76		
Pithecia	3.08	1125.42	2003.5	170	122	121.0	0.61	0.934	1.486
Callicebus	3.70	1350.50	997.3	160	192	100.0	–0.08	0.907	1.078
Cebus	5.50	2007.50	2475.1	168	477	234.6	1.11	1.065	0.950
Saimiri	2.50	912.50	786.9	167	51	109.0	–0.28	0.917	1.164
Aotus	2.00	730.00	1018.7	133	75	97.0	–0.06	0.935	1.381
Leontopithec	1.75	638.75	471.4	133	91	50.0	–0.47	0.800	1.430
Saguinus	1.92	699.58	444.4	145	79	43.0	–0.91	0.756	1.482
Callimico	1.50	547.50	505.0	155	65	50.0	–0.68	0.776	1.833
Callithrix	1.50	547.50	351.2	148	106	30.0	–1.07	0.681	1.881
Cebuella	1.38	501.88	108.5	137	91	14.0	–1.89	0.536	1.573

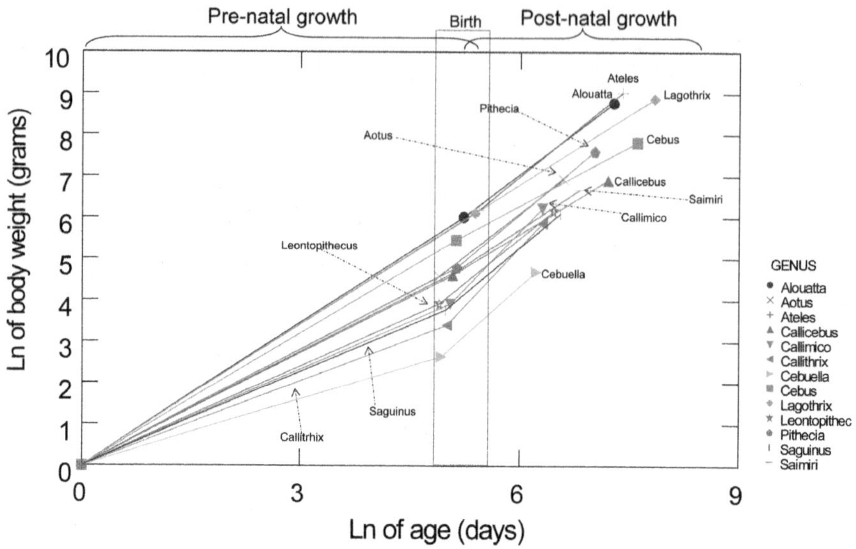

Fig. 17.8 Plot of body weight against age (all in Ln) for all NWM with available data. Pre-natal and post-natal growth rates are represented by the slopes of the LOWESS functions for each genus

References

Cheverud JM (1995) Morphological integration in the saddle-back tamarin (*Saguinus fuscicollis*) cranium. Am Nat 145:63–89

Cortés-Ortiz L (this volume) Molecular phylogenetics of the Callitrichidae with an emphasis on the marmosets and *Callimico*. In: Ford SM, Porter LM, Davis LC (eds) The smallest anthropoids: The marmoset/callimico radiation. Springer Press, New York, pp 3–24

Fleagle JG (1999) Primate adaptation and evolution. Academic Press, New York

Ford SM (1980) Callitrichids as phyletic dwarfs, and the place of the Callitrichidae in Platyrrhini. Primates 21:31–43

Garber PA, Leigh SR (1997) Ontogenetic variation in small-bodied New World primates: Implications for patterns of reproduction and infant care. Folia Primatol 68:1–22

Garland T Jr, Ives AR (2000) Using the past to predict the present: Confidence intervals for regression equations in phylogenetic comparative methods. Am Nat 155:346–364

Gould SJ (1975) On the scaling of tooth size in mammals. Am Zool 15:351–362

Hartwig WC (1996) Perinatal life history traits in New World Monkeys. Am J Primatol 40:99–130

Hershkovitz P (1977) Living New World monkeys, part 1 (Platyrrhini), with an introduction to primates. Chicago University Press, Chicago

Lindenfors P (2002) Sexually antagonistic selection on primate size. J Evol Biol 15:595–607

Lleonart J, Salat J, Torres GJ (2000) Removing allometric effects of body size in morphological analysis. J Theor Biol 205:85–93

Maddison WP, Maddison DR (2003) Mesquite: A modular system for evolutionary analysis. Version 1.5 http://mesquiteproject.org

Marroig G, Cheverud JM (2004) Cranial evolution in sakis (*Pithecia*, Platyrrhini) I: Interspecific differentiation and allometric patterns. Am J Phys Anthropol 125:266–278

Marroig G, Cheverud JM (2005) Size as a line of least evolutionary resistance: Diet and adaptive morphological radiation in New World Monkeys. Evolution 59:1128–1142

McKinney ML, McNamara KJ (1991) Heterochrony: The evolution of ontogeny. Plenum Press, New York

Plavcan JM, Gomez AM (1993a) Dental scaling in the callitrichinae. Int J Primatol 14:177–192

Plavcan JM, Gomez AM (1993b) Relative tooth size and dwarfing in callitrichines. J Hum Evol 25:241–245

Porter LM, Garber PA (2004) Goeldi's monkeys: A primate paradox? Evol Anthropol 13:104–115

Schneider H, Canavez FC, Sampaio I, Moreira MAM, Tagliaro CH, Seuanez HN (2001) Can molecular data place each neotropical monkey in its own branch? Chromosoma 109:515–523

Shea BT, Gomez AM (1988) Tooth scaling and evolutionary dwarfism: An investigation of allometry in human pygmies. Am J Phys Anthropol 77:117–132

Smith RJ, Leigh SR (1998) Sexual dimorphism in primate neonatal body mass. J Hum Evol 34:173–201

Webster AJ, Gittleman JL, Purvis A (2004) The life history legacy of evolutionary body size change in carnivores. J Evol Biol 17:396–407

Chapter 18
Cranial Morphology of the Dwarf Marmoset *Callibella* in the Context of Callitrichid Variability

John M. Aguiar and Thomas E. Lacher Jr.

Abstract The dwarf marmoset, first described as *Callithrix humilis* by van Roosmalen et al. (Goeldiana Zoologia 22:1-27, 1998), proved to be so unusual that van Roosmalen and van Roosmalen (Neotrop Primates 11(1):1-10, 2003) soon renamed it as the new genus *Callibella*. Occupying a tiny and indeterminate range in the central Amazon, *Callibella* has never been systematically studied in the wild and presents enigmas in nearly every aspect of its biology. Here we detail the chronology of its discovery, note some of the unresolved questions regarding its biology, and compare its morphology with that of the other callitrichids. We analyzed quantitative cranial and mandibular characters of *Callibella* to determine if it could be distinguished from other callitrichids in a genus-level analysis. We found that *Callibella* is clearly distinct from all other genera, in particular in the morphology of the lower jaw. In addition, our analysis demonstrated a strong separation of *Callimico* from all other callitrichids, indicating that this genus is morphologically distinct from its closest relatives, the marmosets.

Resumen El tití de corona negro, descrito por primera vez como *Callithrix humilis* por van Roosmalen et al. (1998), resultó ser tan inusual que van Roosmalen y van Roosmalen (2003) lo reclasificaron dentro de un nuevo género llamado *Callibella*. *Callibella* ocupa un área de distribución muy limitada e indeterminada en la Amazonía central, nunca ha sido estudiado sistemáticamente en vida silvestre, y presenta enigmas en cada aspecto de su biología. Aquí detallamos la cronología de su descubrimiento, mencionamos algunas preguntas sin resolver respecto a su biología, y comparamos su morfología con la de otros callitrícidos. Analizamos parámetros cuantitativos craneanos y mandibulares de *Callibella* para determinar si puede ser diferenciado de otros callitrícidos mediante un análisis a nivel de género. Encontramos como resultado que *Callibella* es claramente distinto de todos los demás géneros, especialmente en cuanto a la morfología mandibular. Adicionalmente,

J.M. Aguiar (✉)
Texas A&M University, Dept. of Wildlife and Fisheries Sciences, 210 Nagle Hall, 2258 TAMU, College Station, Texas 77843-2258, USA
e-mail: steelshard@new.tamu.edu

S.M. Ford et al. (eds.), *The Smallest Anthropoids*, Developments in Primatology: Progress and Prospects, DOI 10.1007/978-1-4419-0293-1_18,
© Springer Science+Business Media, LLC 2009

nuestro análisis demostró una fuerte separación de *Callimico* de todos los demás callitrícidos, lo que sugiere que este género también es morfológicamente distinto de sus parientes más cercanos entre los monos tití.

Resumo O sagüi-anão, originalmente descrito como *Callithrix humilis* por van Roosmalen et al. (1998), apresentou tão extraordinário que van Roosmalen e van Roosmalen (2003) o renomeiaram como o novo gênero *Callibella*. Ocupando uma área geográfica minúscula e mal-conhecida, *Callibella* nunca havia sido sistematicamente estudado no campo, e apresenta enigmas em quase todo aspeto da sua biologia. Aqui relatamos a história da sua descoberta, notamos algumas das perguntas persistentes sobre a biologia desta espécie, e fazemos uma comparação entre a morfologia de *Callibella* dos outros géneros dos calitriquídeos. Analisamos carácteres quantitativos do crânio e mandíbula de *Callibella* para determinar se pudesse ser diferenciado dos outros calitriquídeos numa análise ao nivel dos gêneros. Os resultados indicaram que o sagüi-anão é bastante distinto de todos os outros gêneros, destacando a morfologia da mandíbula. Além disso, a nossa análise demonstrou uma separação forte de *Callimico* de todos outros calitriquídeos, sugerindo que este gênero é também morfologicamente distinto do seus confamiliares mais próximos entre os sagüis.

18.1 Introduction

The dwarf marmoset (*Callibella humilis)* first appeared as one of seven new marmoset species discovered like a string of firecrackers in the final decade of the twentieth century. Of all these species, the dwarf marmoset was the most surprising: an atypical callitrichid which showed aspects of *Mico* (Amazonian marmosets) in a body scarcely larger than *Cebuella* (the pygmy marmoset), blending traits of each into an enigmatic whole. First heralded as a "missing link" between the two genera, or a deep-rooted ancestral form – some unknown ur-marmoset – the unexpected monkey was originally described as *Callithrix humilis* van Roosmalen et al. 1998. At the time, the Amazonian marmosets were still considered part of the genus *Callithrix*; having already discovered several other marmosets, the authors made the decision to present the new species as one more Amazonian form, during a period when new finds were almost routine.

According to the description, Marc van Roosmalen and his son Tomas first saw an infant dwarf marmoset on April 16, 1996 in the Amazonian town of Novo Aripuanã, on the eastern bank of the Rio Aripuanã where it flows into the Rio Madeira (Fig. 18.1, Loc. 6). This infant, the van Roosmalens were told, had been taken from its mother's back some distance upriver on the Aripuanã, from the community of Nova Olinda on the river's western bank (Fig. 18.1, Loc. 8). Bringing it to their home in Manaus, where for many years their family operated a facility for orphaned primates, the van Roosmalens showed it to Russell Mittermeier and Gustavo Fonseca of Conservation International, both of whom were experienced primatologists themselves. The consensus was that the infant monkey represented a new species – although of what, no one could be sure.

In July of that year the van Roosmalens returned to the Aripuanã to search for another specimen, without success; but on a third trip in November, Marc van Roosmalen found a group of dwarf marmosets at a settlement close to the community of Nova Olinda. Immediately thereafter, Mittermeier and Fonseca arrived at the site with David Quammen, a noted science writer who dramatized their journey in an article for *Sports Illustrated*, making for one of the most unusual citations in the callitrichid literature (Quammen 1997).

Appearing in print several months before the formal description, Quammen's article presented a more detailed account of the prelude to the first field sighting of dwarf marmosets. According to Quammen's timeline, van Roosmalen first saw a dwarf marmoset at the door of his own home in Manaus, where a *caboclo* – a resident of the interior Amazon – had brought it for adoption. At the time, van Roosmalen's facility for Amazonian primates was widely known, as well as his interest in discovering new species. The *caboclo* with the infant monkey may have been hoping for some compensation for his trouble, since (according to Quammen) the monkey had been captured somewhere on the Rio Madeira and then brought downriver on the ferry from Manicoré (Fig. 18.1, Loc. 2), a river town some 125 km to the southwest of Novo Aripuanã. Although reluctant at first, van Roosmalen accepted the orphan once he saw how unusual it was, and he showed it to Mittermeier soon afterwards.

Quammen does not mention the April 1996 trip to Novo Aripuanã reported in the formal description, nor does he give a date for the *caboclo*'s arrival in Manaus with the infant dwarf marmoset in hand. But he does detail the subsequent forays that year, culminating in Marc van Roosmalen's discovery of a semi-habituated group feeding on a morototó tree (*Didymopanax morototoni*) near the home of Antônio da Silva Pereira, close to the community of Nova Olinda. On many of these trips – including the one to Senhor Antônio's home – van Roosmalen traveled with an experienced boatman named Valquemar Souza de Araújo, also known as Gordo, whose intuition Quammen credits with clinching the discovery.

Quammen's article closes with that success; but according to the formal description, the van Roosmalens returned twice more to Nova Olinda and brought back a second specimen, an adult male that had been kept as a pet. Despite their care, it died soon afterwards in Manaus, and Marc van Roosmalen donated its skin and skull to the Museu Paraense Emílio Goeldi in Belém, Pará. Registered as MPEG 24769, the specimen serves as the holotype for the species initially described as *Callithrix humilis* by Van Roosmalen et al. 1998 (Fig. 18.2; see also Ford and Davis Chap. 21 this volume).

The van Roosmalens settled on that name after first experimenting with several others, believing for some time that their new monkey was simply another species of *Cebuella* gone astray. *Cebuella pygmaea*, the pygmy marmoset, is the world's smallest living anthropoid, and its substantial geographic range extends north from the Rio Madeira (Rylands et al. 1993; Rylands et al. Chap. 2 this volume). A new species of *Cebuella* living south of the Madeira would be exceptional enough; but

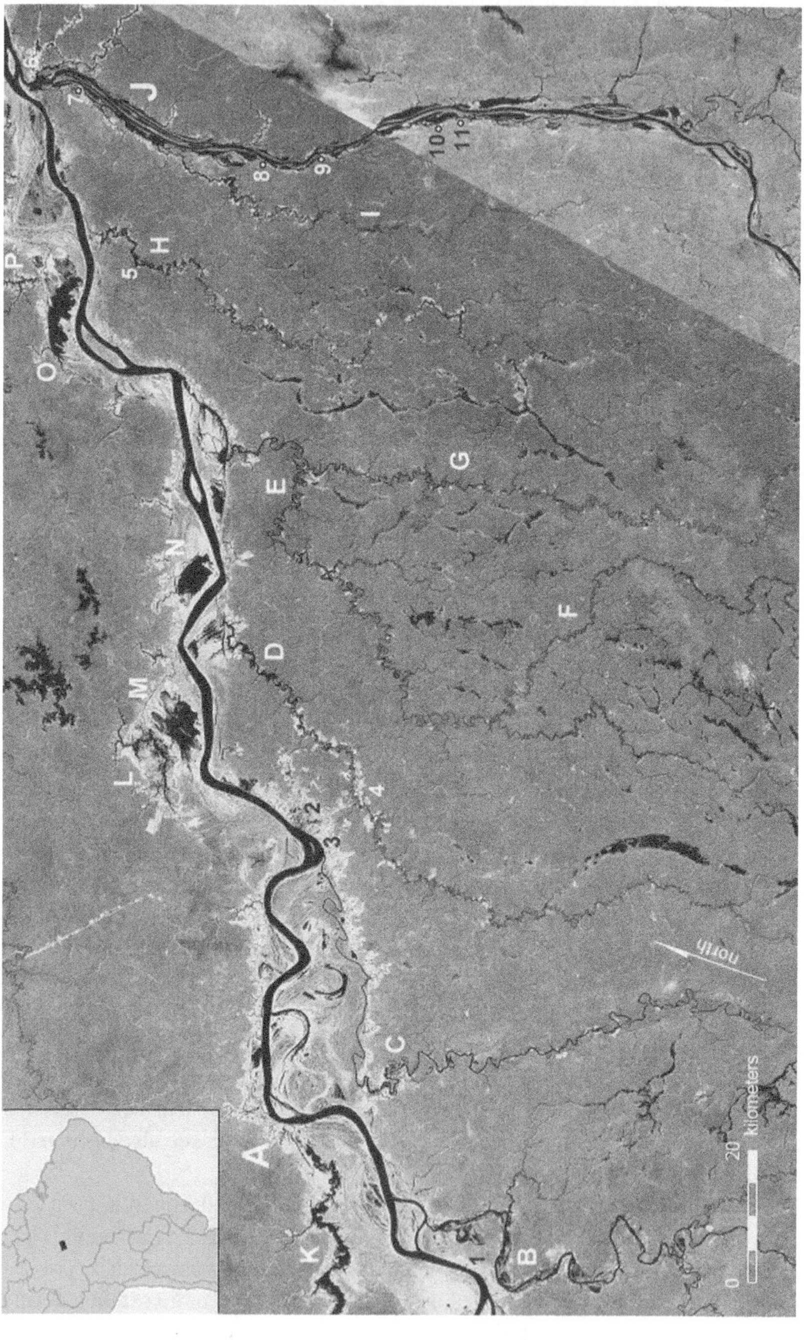
A
B
C
D
E
F
G
H
I
J
K
L
M
N
O
P
1
2
3
4
5
6
7
8
9
10
11
north
0
20
kilometers

by the time the van Roosmalens published their description with Mittermeier and Fonseca, they had witnessed the first infant grow to maturity, and observed the second adult as well – and they had seen enough to convince them that it was no ordinary *Cebuella*. In their 1998 description, van Roosmalen et al. listed five possibilities for its identity and origin: it was either a *Cebuella* that had somehow crossed the Madeira, or an intermediate between *Cebuella* and the Amazonian marmosets, or maybe some primitive form of *Callithrix* – all theories which they felt were improbable – or, more likely, it was either a strange, locally modified form of *Callithrix* (now *Mico*), or else an entirely new genus representing a distinct callitrichid radiation.

For the purposes of their initial description, the authors chose what they considered the conservative approach of naming the new monkey as another species of *Callithrix*, one which "just happens to be considerably smaller than any of its relatives" (van Roosmalen et al. 1998: 12). Other than this conservatism, they offered no quantitative criteria for selecting *Callithrix* as its genus, although they presented a list of behavioral attributes in which the new species overlapped more broadly with *Callithrix* than with *Cebuella*. The authors also emphasized "striking physical similarities" binding *C. humilis* to the genus *Callithrix* – chiefly aspects of its fur coloration which seemed to parallel those of *Callithrix jacchus*, the common marmoset from northeastern Brazil.

Of the list of behavioral attributes, presented in their Table 3 (van Roosmalen et al. 1998: 20), *C. humilis* shared nine with *Callithrix* and only one with *Cebuella*. (Their text on p. 12, however, states that "the new species shares only five out of 13 behavioral features with other Amazonian *Callithrix*….") The authors claimed that several of these attributes, including its lack of territoriality, marking behavior or pungent urine, were unique to *C. humilis*. These claims were presented without

Fig. 18.1 The Madeira-Aripuanã interfluvium, which covers the known range of the dwarf marmoset (*Callibella humilis*) and the Manicoré marmoset (*Mico manicorensis*). The Rio Madeira, flowing across the image from left to right, is a major tributary of the Amazon, and serves as the boundary for the genera *Mico* and *Callibella*, occurring to the river's southeast. By contrast, the third genus of Amazonian marmosets, *Cebuella*, only occurs north and west of the Madeira. The Rio Aripuanã flows from south to north at the far right of the image; most of the sightings of *Callibella humilis* have been made along its western bank. Smaller tributaries of the Madeira dissect the landscape, much of which remains thinly inhabited; lighter patterns along the margins of the Madeira indicate the settlements of *caboclos* and their fields, extending partway down the tributaries. Lakes and river-bays appear as sharp black patches; the broader grey features are *pantanal*, or swampy terrain. Those sightings of *Callibella humilis* which were reported with coordinates have been marked with cross-points; others are labeled in the approximate region. *Rivers and lakes*: *A* Rio Madeira; *B* Rio dos Marmelos; *C* Rio Manicoré; *D* Rio Atininga; *E* Rio Mataurá; *F* Rio Jatuarana; *G* Rio Uruá; *H* Rio Mariepauá; *I* Rio Arauá; *J* Rio Aripuanã; *K* Lago Capanã; *L* Lago Matupiri; *M* Lagoa do Acará; *N* Lagoa de Jenipapo; *O* Lagoa Xadá; *P* Lagoa Preta. *Towns and localities*: *1* Auxiliadora; *2* Manicoré; *3* Seringal São Luis (type locality for *Mico manicorensis*); *4* mid-reaches of the Atininga; *5* Santa Cruz; *6* Novo Aripuanã; *7* Guariúba; *8* Nova Olinda (type locality for *Callibella humilis*); *9* Monte Alegre; *10* Novo Oriente; *11* Terra Preta

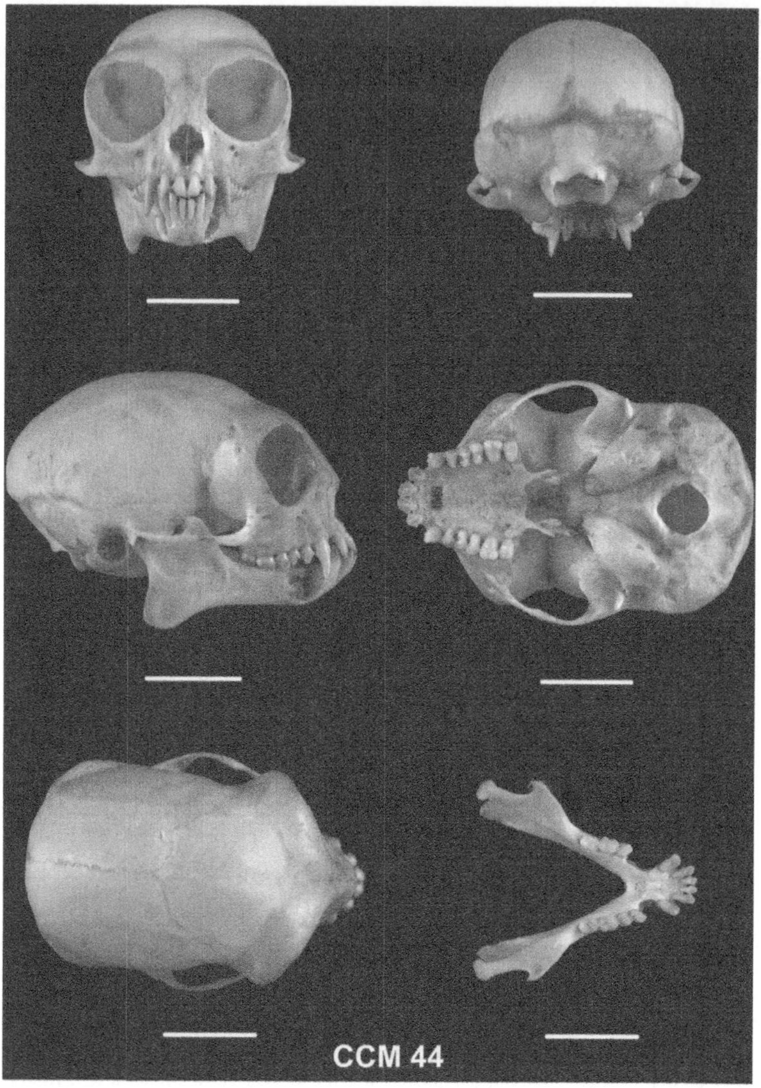

Fig. 18.2 Skull of the adult male holotype of the dwarf marmoset, *Callibella humilis* (MPEG 24769). Photographs by Stephen D. Nash; scale bar = 1.0 cm

supporting data, suggesting that at the time of publication the research had yet to be completed.

In the following years the van Roosmalens were able to acquire several more dwarf marmosets, at least two of whom gave birth in captivity. From close observation of these individuals, the van Roosmalens became convinced that they were in fact members of a new genus, which was first announced as *Callibella* at the 19th Congress of the International Primatological Society in Beijing (van Roosmalen

2002) and formally described the following year (van Roosmalen and van Roosmalen 2003).

The redescription of the dwarf marmoset as *Callibella humilis* relied primarily on a genetic analysis of mitochondrial DNA, supplemented by a long list of features which the authors considered "remarkable," and which together justified its presentation as a novel genus. A number of these characteristics, however, are not unique to the dwarf marmoset, and serve more to strengthen its similarity to *Cebuella* than to distinguish the two species – in particular its "diminutive" size, its "parking" of the young, and its strong reliance on exudate-gouging. In addition, they claimed that there is no pheromonal inhibition of female reproduction, and that more than one female may be reproductively active in a group. But they offered no observations to support this, gave no details on the number of groups and females observed in the wild, and did not address the fact that multifemale groups have been reported from other callitrichid species (see Yamamoto et al. Chap. 6 this volume) – all of which weakens the value of these traits as distinguishing features.

As in the 1998 description, other aspects of *Callibella*'s behavior, such as the lack of territoriality and the prevalence of singleton births, are presented without any observational context. Two of the most intriguing and potentially distinctive features, the unique vocal signature and the genital hypotrophy, are mentioned for the first time in the *Callibella* redescription without the most useful supporting evidence, namely comparative sonograms and anatomical illustrations. From the perspective of morphology and behavior, then, evaluating the species' potential status as a new genus is difficult with the information provided in its description and redescription alone.

The van Roosmalens' reliance on one subsection of mitochondrial DNA for their genetic analyses also raises questions. Their conclusions do not correspond with the results of other molecular analyses, in particular Tagliaro et al. (1997) and Schneider (2000), who argue that *Cebuella* is not a discrete genus, but rather part of a broadly interpreted genus *Callithrix*, which would also include the Atlantic Forest marmosets (*Callithrix*) and the Amazonian species (*Mico*)[1]. These latter authors were unable to include samples of *Callibella* in their analyses – but if *Cebuella* is subsumed into *Callithrix*, this might question the independence of *Callibella* as well. Here we do not address the conflicting molecular results, but focus only on the morphological aspects of the issue.

The redescription of *Callibella* was followed by a short paper in which we analyzed the cranial morphology of the dwarf marmoset, comparing *Callibella* with several other species of marmosets and tamarins (Aguiar and Lacher 2003). Although preliminary and based on a limited sample set, our analysis demonstrated that *Callibella* is distinct from other callitrichids, especially in the structure of its jaw. We were only able to include measurements from three specimens of *Callibella*, the holotype and two paratypes, which so far remain the only specimens

[1]Together with Marroig and Cheverud (Chap. 17, this volume), we are not convinced that the Amazonian marmosets merit a separate genus, but here we follow the convention established for this volume as a whole.

available. The van Roosmalens had at least five other dwarf marmosets in their private facility in Manaus, but the remainder of this group fell victim to an outbreak of yellow fever, which decimated the monkeys living there (M van Roosmalen, pers. comm.). Until these or other specimens become available, further morphological analyses will be restricted to the three individuals already examined.

18.2 Methods

We examined the extant specimens of *Callibella* in the context of a wider study involving callitrichid morphology and biogeography. The holotype of *Callibella humilis* is housed at the Museu Paraense Emílio Goeldi in Belém, Pará, where it was first catalogued with the name *Callithrix humilis*. The two paratypes are at the Instituto Nacional de Pesquisas Amazônicas in Manaus, Amazônas. The specimens representing the other genera analyzed here are kept at the American Museum of Natural History in New York, New York; the Field Museum of Natural History in Chicago, Illinois; the National Museum of Natural History in Washington, D.C.; the Los Angeles County Museum in Los Angeles, California; the Harvard Museum of Comparative Zoology in Cambridge, Massachusetts; the Museu Nacional de Rio de Janeiro and the Museu de Zoologia da Universidade de São Paulo, in Brazil; the Naturhistoriska Riksmuseet in Stockholm, Sweden; the Humboldt Museum für Naturkunde in Berlin, Germany; the Naturalis/Nationaal Natuurhistorisch Museum in Leiden, the Netherlands; and the Museum National de Histoire Naturelle in Paris, France. A full list of specimens and localities is available upon request from the senior author.

We measured all specimens to the nearest 0.01 mm with Mitutoyo Digimatic digital calipers, series/model 500-196. We chose a total of 32 standardized characters to measure from each specimen, although the actual number of data points often varied with damage, since we did not take partial measurements on damaged features. In order to avoid the complications of ontogenetic change, we only examined adult specimens; our criteria for determining adulthood were fully fused cranial sutures and completely descended upper canines, together with sharply defined superior temporal ridges. We analyzed the data using the Discriminant Analysis module of SPSS 13.0 (SPSS, Inc. 2003). For these analyses we chose not to attempt to substitute values for missing data; instead we removed those variables with less than 80% of the cases available, which we believe provides a cleaner dataset than using routines for missing value estimation. This resulted in a large dataset with specimens of all genera available for analysis.

For our analyses, we used measurements for the following variables: CL, OCP, SKW, OWC, BL, CONW, CW, PBG, NP, NL, CWJ, SGL-L, CJB-L, COR-L, SCN-L, SCOR-L and JWCY (Table 18.1, see also Aguiar and Lacher 2003). We used direct entry of all variables. Whenever two paired variables were symmetrical, we used the measurements from the left side. The classification procedure used the same cases; missing values were not substituted with mean samples, and the classification plots show the same cases as were used when deriving the discriminant functions.

Table 18.1 Cranial and mandibular variables used in the morphological analyses of callitrichid genera

Variable	Description
Cranial	
CL	Cranial length, as measured from prosthion (foremost tip between inner incisors) to the furthest reach of the skull's aft curvature
OCP	Distance from the left occipital condylion to the prosthion: essentially a measure of the underside of the skull
SKW	Skull width at the broadest span, usually far aft along the temporal flanges
OWC	Width across the eyes at the cyclosions, the widest span of the orbits
BL	Bregma to lambda: distance from the tripoint intersection of the parietals with occipital, at the aft of the skull to the bregma, or intersection of frontal and parietals
CONW	Width of the cranium at the condyles
CW	Canine width of the maxillary C1s
PBG	Prosthion to bregma – from the front tip of the skull to the top of the skull
NP	Nasion to prosthion – from the feature above the nares to the front tip of the skull
PL	Prosthion to lambda – from the foretip of the skull, between the front incisors, to the aft tripoint feature
Mandibular	
CWJ	Width across the molars and the canines of the lower jaw; parallel to CW
SGL-L	Measured from the symphysion (the mandibular equivalent of the prosthion) to the far edge of the gonion, the aft curve of the jaw
CJB-L	Condylion height as measured to the jaw base
COR-L	Height from the tip of the coronion to the jaw base
SCN-L	Symphysion to condylion: from the front tip of the jaw to the aft end of the right or left mandibular condyle
SCOR-L	Symphysion to coronion
JWCY	Jaw width measured across the condylia

Table 18.2 Number of cases per taxon for the discriminant and cluster analyses

Species	All variables	Skull only	Mandible only
Callimico goeldii	6	7	9
Cebuella pygmaea	8	9	11
Mico chrysoleucus	15	24	33
Callibella humilis	2	3	2
Leontopithecus spp.	2	2	11
Callithrix penicillata	11	16	20
Saguinus midas	14	18	20
Total	58	79	106

All the specimens included in this analysis were wild-caught, either shot by professional collectors or, in the case of *Callibella*, taken alive from their captivity as pets in riverside communities. (A list of species and sample sizes is given in Table 18.2.) The only exceptions are the specimen-sets of *Callimico* (callimicos) and *Leontopithecus* (lion tamarins), in which we have included a mixture of wild-sourced and zoo-bred individuals. The scarcity of wild-sourced specimens for these genera, especially *Callimico*, impelled us to include captive-born animals which we would have ordinarily rejected. For the same reason, our set of *Leontopithecus* is comprised of individuals from all four species; the cranial morphology of lion tamarins is distinctive enough that for purposes of a genus-level comparison, we expect that any interspecific variation will be overshadowed by the differences between genera.

18.3 Results

We compared the cranial and mandibular morphology of *Callibella humilis* with representatives of the other six recognized callitrichid genera: *Callimico*, *Cebuella*, *Leontopithecus*, *Saguinus*, and both the Atlantic (*Callithrix*) and Amazonian (*Mico*) clades of marmosets (Table 18.2). In these analyses our main interest was to evaluate morphological variation in the Callitrichidae at the generic level, with a particular emphasis on whether *Callibella* would stand out as equally distinct among the other genera, based upon the cranial and mandibular variables we selected.

We performed a discriminant analysis using a total of 17 cranial and mandibular characters to generate a classification matrix among the genera (Figs. 18.3 and 18.4; Tables 18.3 and 18.4). The first three discriminant functions accounted for 93.8% of the total variance among groups. The first function, accounting for 67.4%

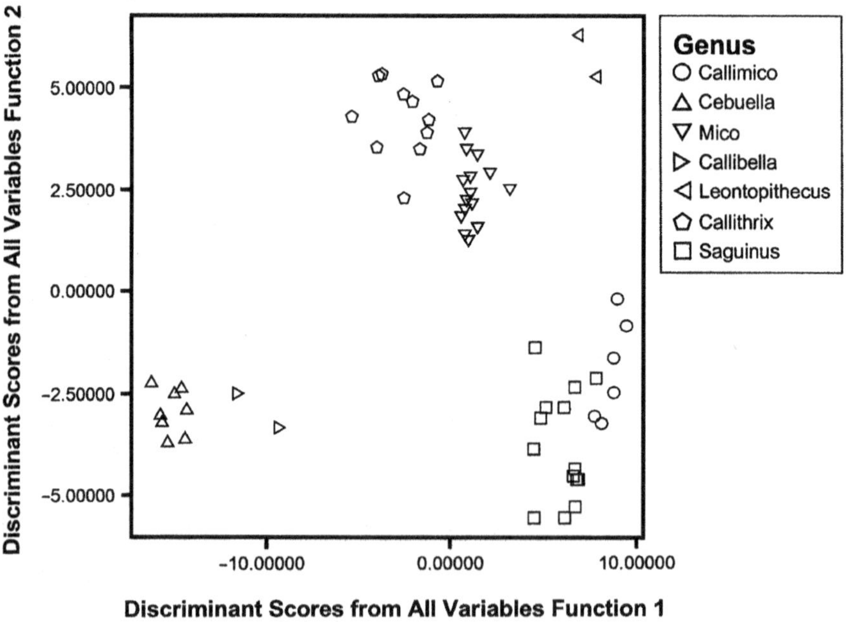

Fig. 18.3 The plot of cases for the seven genera of callitrichids, for discriminant functions 1 and 2 for all cranial and mandibular variables

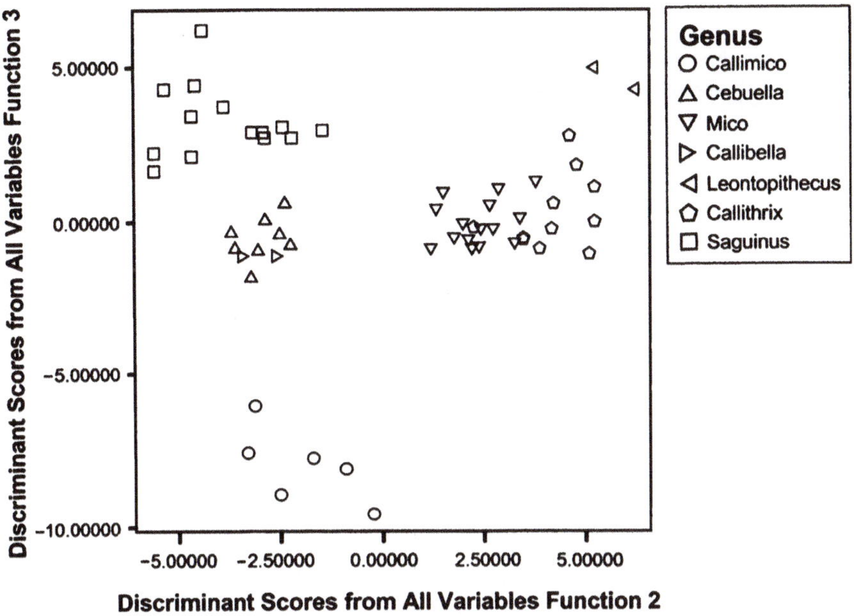

Fig. 18.4 The plot of cases for the seven genera of callitrichids, for discriminant functions 2 and 3 for all cranial and mandibular variables

Table 18.3 Results of the discriminant analyses for the comparisons among genera

Function	Eigenvalue	% of Variance	Cumulative (%)
Skull only			
1	33.697	71.2	71.2
2	6.272	13.2	84.5
3	5.432	11.5	96
4	1.639	3.5	99.5
5	0.167	0.4	99.8
6	0.091	0.2	100
Mandible only			
1	28.664	87.9	87.9
2	2.615	8	95.9
3	0.919	2.8	98.7
4	0.314	1	99.7
5	0.058	0.2	99.9
6	0.044	0.1	100
All variables			
1	62.325	67.4	67.4
2	13.004	14.1	81.5
3	11.317	12.2	93.8
4	4.952	5.4	99.1
5	0.571	0.6	99.7
6	0.243	0.3	100

Table 18.4 Standardized discriminant function coefficients for the generic comparisons. See Table 18.1 for variable codes

Variable	Function 1	Function 2	Function 3
Cranial variables			
CL	–0.199	0.932	–3.110
OCP	–0.399	0.267	1.440
SKW	–0.059	–0.309	–0.275
OWC	0.438	0.283	–0.942
BL	0.225	0.184	–0.319
CONW	0.192	0.139	0.018
CW	0.424	–1.106	0.179
PBG	0.488	0.688	0.001
NP	–0.644	0.215	0.308
PL	0.817	–0.897	2.718
Mandibular variables			
CMJ	0.344	0.432	0.809
SGLL	–0.109	–0.159	0.210
CJBL	0.1	–0.62	0.608
CORL	–0.041	0.703	–0.712
SCNL	0.117	–1.114	0.451
SCORL	0.517	0.742	–0.356
JWCY	0.521	0.038	–0.538
All variables			
CL	–1.428	–2.031	–2.334
OCP	0.179	0.862	1.641
SKW	–0.017	–0.105	–0.287
OWC	0.462	–0.437	–0.962
BL	0.433	–0.168	–0.034
CONW	–0.029	0.146	0.035
CW	–0.173	–0.992	0.985
PBG	0.766	0.548	0.032
NP	–0.736	0.483	–0.119
PL	1.292	1.471	2.001
CWJ	0.113	–0.277	0.086
SGLL	0.169	0.119	0.481
CJBL	0.012	0.190	0.499
CORL	0.042	–0.105	–0.670
SCNL	–0.072	0.469	–1.027
SCORL	0.301	–0.441	–0.598
JWCY	0.336	0.300	0.287

of the variance, separated genera largely on the basis of size, with larger taxa scoring higher on the positive side of Function 1 and the two smallest marmosets (*Cebuella* and *Callibella*) scoring on the negative side. This was not exclusively a size function, however, and coefficients indicated significant allometry as well (Table 18.4). Function 2, which accounted for an additional 14.1% of the variance, separated *Mico*, *Callithrix*, and *Leontopithecus* from the other genera. The combination of these two axes clearly separates the nongougers (*Leontopithecus*, *Saguinus* and *Callimico*) from the gougers (*Callithrix*, *Mico*, *Callibella* and *Cebuella*). Function 3 accounted for 12.2% of the total variance. Both Functions 2 and 3 separate the genera on the basis of form, and the plot of Functions 2 versus 3 (Fig. 18.4) shows strong separation among genera, especially of *Callimico* from all other genera. In the classification matrix, *Callibella* returned a 100% correct classification – as did all seven genera – showing no overlap with any other taxon in the full comparison. This strongly supports the classification of *Callibella* as a full genus, based on its clearly distinct morphology.

Our second discriminant analysis examined ten characters measured from the cranium alone (Figs. 18.5 and 18.6; Tables 18.3 and 18.4). The first three eigenvalues accounted for 96.0% of the variance, and from this perspective the relative overlap of the genera shifted most dramatically in *Callimico* and *Saguinus*. Here this pair overlaps completely on Function 1 but is now widely separated along Function 2, and *Callimico* shows no classification overlap with *Saguinus*. There is a size component on Function 2, but the key morphological traits involved in the separation appear to be larger canine width in *Saguinus* and a larger prosthion-to-bregma measure in *Callimico*. The plot of Function 2 versus Function 3 shows a clear and complete separation of *Saguinus* and *Callimico*, based upon cranial form. *Callibella* remains 100% distinct from the other genera, but now plots more closely

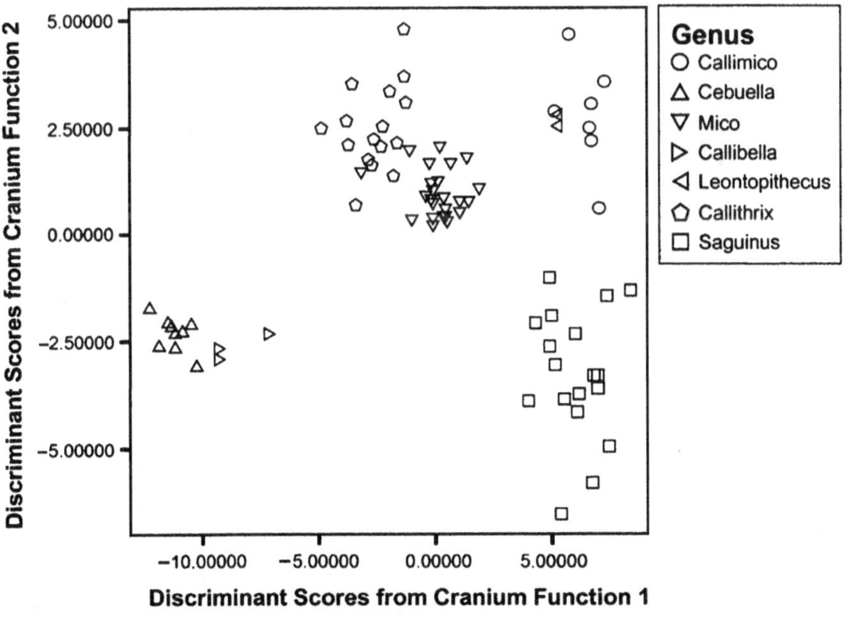

Fig. 18.5 The plot of cases for the seven genera of callitrichids, for discriminant functions 1 and 2 for all cranial variables

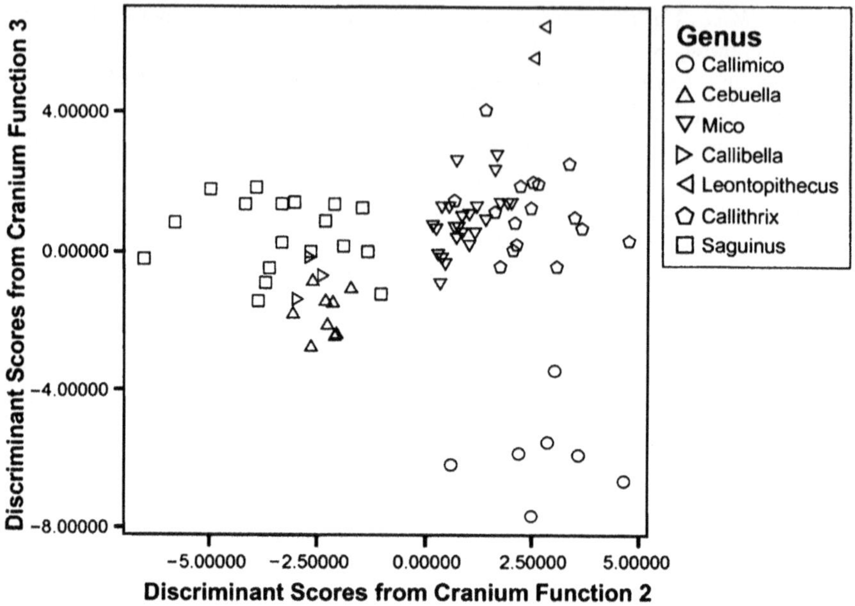

Fig. 18.6 The plot of cases for the seven genera of callitrichids, for discriminant functions 2 and 3 for all cranial variables

to *Cebuella* in both Figs. 18.5 and 18.6 – suggesting that the cranial differences between the two smallest marmosets are slight, and due more to size than any divergent functional pressures. In Fig. 18.6, where both Functions 2 and 3 represent differences in cranial form, *Saguinus* shows strong overlap with *Callibella* and *Cebuella*, even though there are significant size differences among the three genera.

Our third discriminant analysis used seven characters to compare features of the lower jaw (Fig. 18.7; Tables 18.3 and 18.4), with the first two functions accounting for 95.9% of the variance. Function 3 was not used in the analysis, accounting for only 2.8% of the variance among groups. Here the strong separation of *Callimico* and *Saguinus* is reversed: callimicos, together with the lion tamarins, overlap with *Saguinus* (11.1 and 9.1% misclassifications, respectively). *Mico* and *Callithrix* also show a small degree of mutual misclassification (between 3 and 10%), while *Cebuella* and *Callibella* are both classified as 100% distinct.

In all three discriminant analyses, the genera follow a strong gradient of size, with *Cebuella* and *Callibella* distinct but closely paired in every case. The Atlantic and Amazonian marmosets consistently appear as two well-defined but overlapping clouds, with *Mico chrysoleucus* (the golden-white tassel-ear marmoset) reliably larger than *Callithrix penicillata* (the black-tufted-ear marmoset). Both genera do show a greater degree of separation on the plots of Function 2 versus 3 for both the combined analysis (Fig. 18.4) and the cranial variables only (Fig. 18.6). *Callimico*,

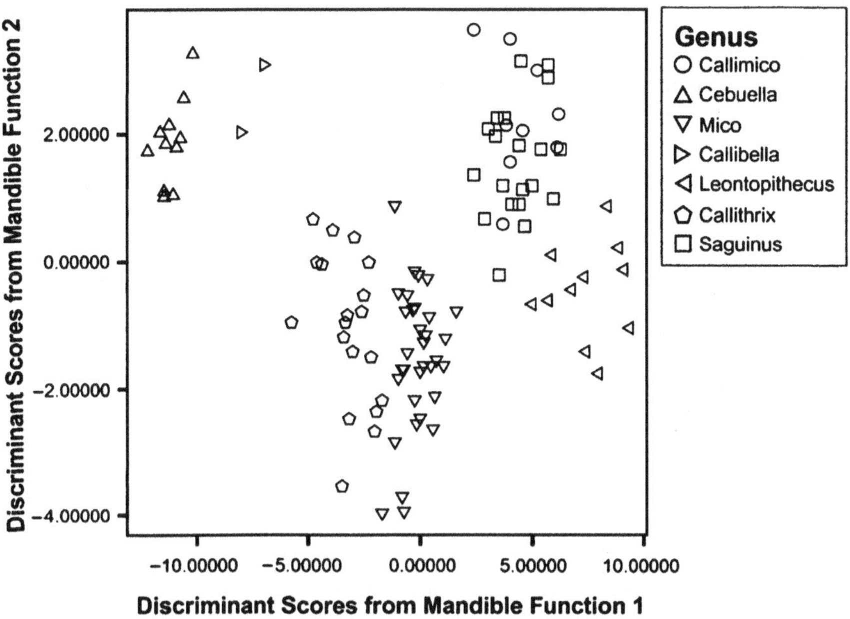

Fig. 18.7 The plot of cases for the seven genera of callitrichids, for discriminant functions 1 and 2 for all mandibular variables

Saguinus and *Leontopithecus* show a great deal of coarse overlap on Function 1, with a strong separation of *Leontopithecus* on Function 2; there is also a clear separation of *Callimico* and *Saguinus* on the plots of Functions 2 and 3 that separate the genera on the basis of shape (Figs. 18.4 and 18.6).

The strong separation of *Callimico* and *Saguinus* in cranial but not mandibular features in the overall comparison is one of the most interesting results from our discriminant analyses – suggesting that the primary differences in their skull morphology might stem from selective forces other than those involved in feeding behavior. Although completely overlapping on the axis of Function 1 for cranial variables (Fig. 18.5), showing a congruence of size, *Callimico* and *Saguinus* are completely separated in the plots of Functions 2 and 3 (Fig. 18.6), reflecting a strong disparity in cranial shape. Primarily plotting on the positive axis of Function 2, *Callimico*'s divergence in shape is driven by the strongest positive canonical coefficients, with a larger overall cranial length (CL, 0.932) and a greater distance from prosthion to bregma (PBG, 0.688). *Saguinus*, set apart by strongly negative values, is distinguished by the larger width across its upper canines (CW, –1.106) and the larger distance from prosthion to lambda (PL, –0.897). Taken together, these features indicate that the skull of *Callimico* is larger and higher in the crown, and more bulbous in the occipital region, while that of *Saguinus* is longer overall and wider in the front of its face. The skull of *Callimico* also appears to be broader across the cyclosia, perhaps indicating its eyes are more widely set apart.

A second gradient is clear in each of the discriminant plots, however, which is directly linked to feeding ecology: the sharp division between those species which gouge for exudates and those which do not. Easily seen in the overall comparison, the division between gouging and nongouging species is less evident in the analysis of cranial features alone. But the separation is stark in the structure of the jaws, as seen in the plot of Functions 1 and 2 (Fig. 18.7): a diagonal runs from the positive quadrant of Function 2 down to the positive quadrant of Function 1, evenly dividing *Cebuella*, *Callibella*, and both *Mico* and *Callithrix* from *Saguinus*, *Callimico*, and *Leontopithecus*. Along this diagonal, *Mico* is the least specialized for gouging, and *Callibella* and *Cebuella* the most. This is largely a size function, and might relate to size restrictions in the ability to use gums as a food resource, as recently suggested by Marroig and Cheverud (2005, Chap. 17 this volume).

18.4 Discussion

In Aguiar and Lacher (2003), we compared *Callibella* with representatives from four other callitrichid genera. In that limited dataset, all five groups returned 100% correct classification in our discriminant analyses, which convinced us that the dwarf marmoset deserved recognition as a unique genus. Here we have analyzed *Callibella* in the context of all the callitrichid genera, including *Callimico goeldii*, and we have added *Callithrix penicillata* to represent the Atlantic clade of marmosets.

In both our prior and current analyses, *Callibella* emerged as sharply distinct, but always closely associated with *Cebuella*, both of which were strongly separated from all other genera in our discriminant analyses. This is due primarily to their similarity in size, which is the major influence on Function 1 of all our discriminant analyses, and which appears as the most obvious gradient in the discriminant plots.

But *Callibella* and *Cebuella* share more than a coincidental perch as the world's two smallest anthropoids. They are similar enough, in fact, that when Marc van Roosmalen first saw the infant *Callibella* which the *caboclo* brought to his door, he thought it was simply another species of *Cebuella*. This was no momentary *deja vú*: the preliminary label on the holotype of *Callibella*, at the Museu Goeldi, was neatly pencilled "Cebuella humilii." During the initial search for groups in the wild, the underlying assumption was that this was an exceptional *Cebuella* (Quammen 1997), and even after the monkey was announced to the press, there were many who imagined it was a second pygmy marmoset. The creature's presumed identity as a new species of *Cebuella* only faded as the first specimen grew into adulthood, passing through a striking sequence of color patterns which helped mark it as unique.

Despite its differences in pelage, however, and its large half-bare ears, the dwarf marmoset is much like *Cebuella* in some respects. Like *Cebuella*, the dwarf marmoset is known from riparian areas; like *Cebuella*, it is not hunted intensively, but is still vulnerable to random target practice or live capture as a household pet. And like *Cebuella*, the dwarf marmoset apparently specializes in exudate-feeding to an extent unmatched by other Amazonian callitrichids – a habit made possible by the jaw structures of both species, and which speaks of either a close kinship or a remarkable convergence.

In our prior analysis of *Callibella*, we noted a second gradient in the discriminant plots, more subtle than the size gradient of Function 1, but not precisely following the shape-influenced pattern spread along Function 2 (Aguiar and Lacher 2003: 12). This gradient appeared most clearly in our comparison of mandibular features: a sharp boundary between the callitrichids which gouge trees and vines for exudates and those which do not. The latter genera, *Saguinus* and *Leontopithecus*, are joined in our present analyses by the enigmatic *Callimico*, which to our surprise was poorly discriminated from *Saguinus* in the shape and size of its jaw. Whatever their other differences, callimicos and at least some tamarins share a jaw structure which, aside from the matter of the third molar, is almost identical in its proportions. This, in turn, would suggest a lack of differential selective pressure on feeding behavior and the relevant anatomy, despite a strong separation in the design of their respective crania.

In our overall analysis of both cranial and mandibular characters, the third function accounted for 12.2% of the variance, which was substantial enough to justify plotting Functions 2 and 3 together. While Function 1 typically has a large size component, Functions 2 and 3 are principally driven by shape, and the results in this case were both surprising and challenging to interpret. *Cebuella* and *Callibella* overlapped completely, while *Callithrix* and *Mico* were fused against each other with some degree of overlap. *Saguinus* and *Callimico* also overlap strongly on Function 2 – but they are overwhelmingly separate on Function 3, and in fact *Callimico* is fundamentally removed from all other callitrichids on this axis. This division seems to be driven by three features in particular: the overall length of the skull, the width of the eyes, and the proportions of the posterior region of the jaw. The other callitrichids range from strongly positive to weakly negative on the axis of Function 3, but a well-defined lower limit keeps them firmly apart from *Callimico*.

When plotting the same functions derived from only cranial variables, the division between *Callimico* and *Saguinus* becomes absolute. Among all the callitrichids, the cranial shape of *Callimico* is at the furthest remove. Distinctions are less sharp-edged among the other genera on the plot of Function 2 against Function 3, and there is extensive overlap among the other callitrichids on Function 3. On Function 3, *Saguinus* overlaps with *Cebuella*, *Callibella*, *Callithrix* and *Mico*, but is entirely separate from *Callimico* – and *Callimico* is entirely distinct from all other callitrichids on this axis.

The strong positive vector for *Callimico* on Function 2, and its equally strong negative vector on Function 3, are both driven by the corresponding loadings for the variable CL, which represents overall cranial length. This is in strong opposition to PL, the distance between prosthion and lambda, which tracks a subtly different aspect of skull length and shows inverse loadings in *Saguinus*. The cumulative message of these results, then, is that *Saguinus* and *Callimico* are powerfully distinguished by differences in the shape of the lower skull – reflecting either a difference in the posterior curve of the skull, prognathism in the maxilla, or some unknown combination of the two.

Callimico and the tamarins, together with *Leontopithecus*, are strongly separated from the other callitrichids in our analyses, most dramatically in the shape of their lower jaws. Although not above collecting exudates from the wounds of trees caused by other organisms (or, in the case of saddleback tamarins, filching the efforts of true gougers), these three genera will not gouge trees themselves. In this they stand apart from the marmosets – pygmy, dwarf, Atlantic and Amazonian – who actively gouge for exudates and rely on them to greater or lesser degree (see review, Taylor et al. Chap. 19 this volume, Vinyard et al. Chap. 20 this volume). Of these, *Cebuella* is known for its dependence on exudates as the staple carbohydrates in its diet (Soini 1988), while the Amazonian marmosets – preferring areas of secondary forest dense with small fruits, large insects, and opportune vertebrates – are much less reliant on exudates for their primary nutrition, approaching them less as a staple and more of a last resort.

We find it no coincidence that *Cebuella* and the Amazonian marmosets occupy the opposite endpoints of a continuum of gouging behavior which is reflected in their mandibular morphology (Fig. 18.7). Much closer to the Amazonian marmosets behaviorally and morphologically, and overlapping them at their fringes, are the Atlantic marmosets – represented here by *Callithrix penicillata*, one of the most adaptable and widespread of all the marmosets, and one of the most exudate-dependent of the Atlantic clade (Rylands and de Faria 1987, 1993). These three marmoset groups are arranged along the diagonal continuum in our discriminant plots (Fig. 18.7) – *Cebuella* the most specialized gouger, *Callithrix penicillata* less so but still adept, and *Mico chrysoleucus* at the edge of the nongouging condition.

This continuum is also apparent in the plot of combined variables for Functions 1 and 2 (Fig. 18.3). When the results from the combined variables are plotted on Functions 2 and 3 – effectively viewing the dataplot from an orthogonal perspective – this continuum alters its order but not its composition (Fig. 18.4). The more intensively gouging marmosets – *Cebuella*, *Callibella* and *Callithrix* – now lie on either side of *Mico*, which might suggest that pure size was driving the gradient seen before; but although the positions of the nongouging callitrichids have shifted dramatically, *Mico* and *Callithrix* remain so tightly clustered that they should be considered as two lobes of the same datacloud. This suggests that the morphological support for the genus *Mico*, at least in its cranial and mandibular features, is tenuous at best (but see Ford and Davis Chap. 21 this volume, for strong support from postcranial data).

In our first examination of the type and paratypes of *Callibella humilis*, we noted that aside from its size – substantially smaller than *Callithrix* or *Mico*, and only slightly larger than *Cebuella* – there were no features of the dwarf marmoset's cranium that definitively set it apart. In both our current and prior discriminant plots of cranial characters, as expected, *Callibella* and *Cebuella* are barely divided on the size axis and essentially indistinguishable in shape. This would suggest, among other things, that equivalent pressures have molded equivalent skulls, or else an ancestral design has found no reason to change. This tight similarity makes it difficult to define either one in the context of the other. Hershkovitz (1977) relied on *Cebuella*'s small size to distinguish it from the other callitrichids; but with a second marmoset now in the same size range, some other physical characteristic is needed to separate the pair.

Although their cranial morphology is virtually identical, the structure of the mandible is clearly distinct between *Callibella* and *Cebuella*, as we detailed in our initial assessment. The mandible of *Cebuella* is a delicate thing, a wisp of recurved bone lighter than a paperclip yet strong enough to cut through the bark of a tropical tree. Unlike the jaws of the seed-predator sakis (*Pithecia* and kin), whose deep mandibles support the muscle mass needed to deliver their crushing force, the jaw of *Cebuella* leans out to the extreme, optimized to concentrate its strength at the tips of the incisors. In this design, the pygmy marmoset's jaw occupies the endpoint of a trend seen across the callitrichid genera: a reduction in the size of the ascending ramus and a lowering of the coronial and condylar processes. In *Cebuella*, this results in the condylion – the jaw's point of articulation with the cranium, and its natural pivot – lying directly in line with the tops of the molar and premolar teeth, resting within their occlusal plane. This design in *Cebuella* is far different from the arrangement of *Saguinus* and *Leontopithecus*, whose condylia rest at a high distance above the occlusal plane, with their coronia swept up like brandished scimitars.

The jaw of *Callithrix*, by contrast, shows a less aggressive coronion and a condylar pivot which lies nearer to the occlusal plane, reflecting the marmosets' greater emphasis on exudate-gouging, although not to *Cebuella*'s extreme. Unlike *Leontopithecus* and *Saguinus*, whose jaw base is relatively flat, the angular process of *Callithrix* often extends below the lower jawline as a rounded lobe. In *Cebuella* this lobe is much leaner, but proportionally projects far deeper, and overall the ramal assembly seems dorsally compressed and rotated back when compared to the *Saguinus* design.

In this context, *Callibella* appears intermediate between the moderate shape of *Callithrix* and *Cebuella*'s radical design, with a condylion which lies marginally above (but not precisely on) the occlusal plane; the coronion is not as bold and high as *Callithrix*, but fuller and more developed than the sharp light hook of *Cebuella*. This intermediate shape of the coronoid and condylar processes is combined with a uniquely protruding angular process, which projects even broader and deeper than in *Cebuella*.

As a unique morphological suite, these proportions of the aft mandible serve to isolate *Callibella* unmistakably from both pygmy and conventional marmosets. This design provokes a number of questions, however, as to the dwarf marmoset's evolutionary history and feeding ecology. From its position in our discriminant plots, lying between *Cebuella* and the Atlantic marmosets, the easy prediction is that *Callibella* is likewise specialized for intensive exudate-gouging, but perhaps to a lesser degree than *Cebuella*.

On the face of it, this fits well with what little has been reported of the dwarf marmosets' feeding behavior. Although the van Roosmalens kept several individuals at their primate facility in Manaus – both in cages and free in the house – they did not publish their feeding protocols, and thus the diet they used to raise these marmosets cannot be evaluated.

Moreover, the feeding behavior of dwarf marmosets in the wild has never been systematically documented. In their original description of *Callithrix humilis*, van Roosmalen et al. (1998) offered only hints of its diet and habits. Although they named half a dozen species of gum-producing trees growing around the human settlements, close to where the first semi-wild groups were found, the authors did not present observational data on feeding behavior. In their account of its discovery, echoed by David Quammen's earlier article, they noted that the dwarf marmosets gouged and fed from the morototó, *Didymopanax morototoni*. Relaying the observations of local people, van Roosmalen et al. (1998: 8) claimed that this species was the most important of the potential feeding trees in the area: "This tree is said to be [*Callibella*'s] principal exudate source all year round." But it is not clear whether these comments, presumably from Senhor Antônio's family, referred to the species in general or just the one group feeding on that one tree, and the authors did not present a feeding budget or other behavioral data. The authors did claim to have seen the marmosets feeding on fruit from trees around human houses, but without more detailed information, it is difficult to estimate the relative importance of fruit and gum in *Callibella*'s diet.

In their 2003 redescription of *Callibella* née *Callithrix*, the van Roosmalens made no mention of specific feeding behavior, except in a passing comment on the importance of exudate-gouging as a "keystone resource" for *Callibella* and its lesser importance to the sympatric Manicoré marmoset, *Mico manicorensis*. The morototó appears only once, in a species list of available fruit and exudate sources presented in the Appendix, and the authors no longer give it any special importance.

They do, however, make the new claim that *Callibella* is "almost totally dependent for survival on multispecies managed forests, fruit orchards and gardens" established and maintained by the *caboclos* who have settled on the fertile riverbanks (van Roosmalen and van Roosmalen 2003: 5). Marc van Roosmalen made the additional claim, during a presentation at the 2002 IPS congress in Beijing, that the dwarf marmoset would have gone extinct some four thousand years ago had it not developed a commensal relationship with humans in the Amazon. No evidence for this theory was offered, nor an explanation as to why *Callibella* as a species should be dependent on the proceeds of human cultivation, when the entire region is apparently suffused with *Inga*, *Spondias*, *Parkia*, *Enterolobium*, and *Didymopanax* itself.

This proposal also raises the question of why, if *Callibella* is a successful commensal of human settlements, it has not spread with them throughout the Amazon in its several thousand years of potential coexistence. Apart from the easy access to gardens and orchards, the prime draw for *Callibella* in this theory would be the ultrafertile soils of the *terra pretas* – the anthrosols of lost antiquity, which early Amazonians are thought to have created during thousands of years of small-scale occupation (Mann 2002; Lehmann et al. 2004). The *terra pretas* may occupy as much as 10% of the entire Amazon basin, representing the culmination of agricultural practices which were cut off abruptly with the European arrival (Wim Sombroek pers. comm.).

The issue of *Callibella*'s affinity for *terra pretas* is tied to the general assumption of its restricted range, which the van Roosmalens based on their having sighted the species primarily in a narrow corridor along the west bank of the Rio Aripuanã, extending some 65 km south of its confluence with the Rio Madeira (see Fig. 18.1). During their search for the dwarf marmoset, the van Roosmalens concentrated on the Rio Aripuanã, as well as the nearby territory on the northern margins of the Rio Madeira, where a trip to the Lago Matupiri, some 80 km southwest of Novo Aripuanã, revealed *Cebuella* but no *Callibella*.

In their original description, they report no explorations other than along the margins of the Rios Madeira and Aripuanã, apart from what seem to have been brief forays into the mouths of the Rios Mariepauá and Mataurá. In a proof note to the original description, van Roosmalen et al. (1998: 13) reported finding a "geographically isolated population" along the Rio Atininga, a small tributary to the immediate east of Manicoré. They claimed that this population was set apart from other dwarf marmosets by differences in fur color, in particular, "a more orange-ochraceous ventral coloration." They gave no information on how many individuals they had seen, nor exactly where on the Atininga they had been sighted, but they raised the possibility that this population was "yet another new taxon" of callitrichid.

In their 2003 redescription of *Callibella*, the location was narrowed to "ca. 10 km. east of the town of Manicoré," at 05°54'S, 61°15'W (van Roosmalen and van Roosmalen 2003: 3). (These coordinates plot to an area some 15 km south-southeast of Manicoré, although on their Fig. 1 this point is shown 20 km to the town's southwest.) According to the redescription, local people apparently collected a young female from this area and gave it to Marc van Roosmalen. On the female's death in captivity two years later, the remains became INPA 4090, which serves as the only representative of the Atininga population.

The first author (JMA) had the opportunity to examine INPA 4090 and to compare it with INPA 4091, another *Callibella* from the van Roosmalens' private facility, which presumably represents the baseline population of dwarf marmosets. Apart from a faint orange tinge to the ventral fur, the Atininga specimen is essentially identical to baseline *Callibella*, and the slight differences in color may be due to individual variation rather than the isolation of a breakaway population. No other information is available on this population, and a brief exploration of the region (Aguiar 2001) yielded no sightings.

By the time of their 2003 redescription, the van Roosmalens had made additional surveys at two sites along the west bank of the upper Aripuanã – approximately 100 and 150 km south of Novo Aripuanã – as well as at a site on the eastern bank of the Rio Manicoré, some 20 km southwest of the reported population at Atininga and approximately 130 km southwest of Novo Aripuanã. None of these surveys (cited as "van Roosmalen and Peres, in prep.") revealed any trace of *Callibella*, not even gouge marks, which convinced the van Roosmalens that the new genus did not occupy undisturbed forest or *igapó*. In the original description, they claimed the distinctive Atininga population occurred exclusively in *igapó*, but in 2003 they reported it from both *terra firme* and *igapó* habitat.

They did not, however, report surveys along the Rio Mariepauá beyond its mouth, nor anywhere along the Rios Uruá and Mataurá, which lie between the Aripuanã and the Atininga, and which – at least at some point in the past – dwarf marmosets must have occupied in order to diffuse from one area to the other. Why the van Roosmalens did not explore these rivers is not clear from their writings, although their efforts in searching the Aripuanã may have been daunting enough. More probably, though, they passed over the other three rivers because their upper reaches, in human terms, are thinly populated or completely uninhabited – making the surrounding forest uninviting to a species which, they believed, was dependent upon *terra pretas* and the human settlements which are often built on them. This conviction is firmly presented in the *Callibella* redescription: "We assume that *Callibella* nowadays occurs almost exclusively on bluffs along blackwater and clearwater streams and lakes, where generations of ancient Indian farmers once lived"- and who, intentionally or not, generated the *terra preta* anthrosols (van Roosmalen and van Roosmalen 2003: 4). The evidence for this association is not presented, other than the fact that *Callibella* was easily found close to human settlements (with their tempting concentration of fruit trees), and was not seen at three points in unoccupied forest.

Although undoubtedly true, this by itself hardly precludes dwarf marmosets from living quietly throughout the region and beyond, and – like *Cebuella* – skittishly avoiding human intruders, at least wherever there is no lure of planted fruit to lower their suspicions. Until more thorough and systematic surveys are conducted along all the small rivers and *igarapés* of the region, we cannot be certain that dwarf marmosets are so tightly confined as previously assumed.

In addition to its extraordinary morphology, *Callibella* is also unique in that it apparently shares its range with another marmoset, *Mico manicorensis*, discovered by the van Roosmalens in 1996 during their early search for the dwarf marmoset. Sympatry is known in rare cases between marmosets and other callitrichids, in particular between *Callithrix kuhlii* (Wied's black-tufted-ear marmoset) and *Leontopithecus chrysomelas* (golden-headed lion tamarin) in Bahia (Rylands et al. 1993; Rylands et al. Chap. 2 this volume). Ferrari et al. (1999, in press) have also reported sympatry between the Rondônia marmoset (*Mico* cf. *emiliae*) and *Saguinus fuscicollis weddelli* (Weddell's saddle-back tamarin), although their respective ranges may be different than first expected (Rylands et al. Chap. 2 this volume). Sympatry among tamarins, by contrast, is widespread and well-documented (Rylands et al. 1993), as well as between *Cebuella* and several species of tamarins (Soini 1988; Rylands et al. Chap. 2 this volume). But the apparent coexistence of *Callibella* with *M. manicorensis* is the first known instance of a marmoset sympatric with another marmoset - all the more interesting because, unlike the cases above, both species are adapted to actively gouge trees for exudate.

Like other recently discovered marmosets, such as *Mico acariensis* (the Rio Acarí Marmoset), *M. mauesi* (the Maués marmoset), and *M. saterei* (the Saterê marmoset), the Manicoré marmoset is known from only a bare handful of sites and specimens, and like others, its geographic range has been projected far beyond the current evidence. The holotype of *M. manicorensis*, INPA 2511, was taken alive from the "Seringal São Luis," a small patch of semi-managed forest on the outskirts of the river-town of Manicoré (van Roosmalen et al. 2000). A second individual, taken together with the holotype, was reported as a living paratype in the species description; no accession number was included and the whereabouts of this specimen, dead or alive, are unknown. A third specimen, INPA 2512, was reportedly from Santa Cruz, a settlement near the mouth of the Rio Mariepauá; this individual was kept by local residents for an unknown time and then acquired by van Roosmalen. A fourth specimen, INPA 3930, is also problematic: an adult female which the van Roosmalens claim had been kept in their private facility "for several years." Although their text gives no further details, the specimen tag at INPA lists the female's origin as the Seringal São Luis.

Thus this species is known from only two points along the Rio Madeira: the type locality, in the intensively modified vicinity of Manicoré; and the tiny community of Santa Cruz, located some 95 km to the northeast of Manicoré and approximately 10 km upstream on the Rio Mariepauá. The latter locality is not entirely certain; it is unclear whether the van Roosmalens visited the community themselves, and since the marmoset was being kept as a pet, it could have come from anywhere in the region.

In their description of the Manicoré marmoset, van Roosmalen et al. (2000: 6) mention "various groups observed in the wild" but give no information on *where* they were observed nor on their group size, composition or behavior. On the same page, the authors claim that the species is known from the mouth of the Rio Aripuanã "south to the confluence with the Rio Roosevelt" – a distance of over 270 km – but they provide no field observations, interviews, or other supporting evidence. In their redescription of *Callibella*, van Roosmalen and van Roosmalen (2003) mentioned sightings of *Mico manicorensis* during surveys they conducted on the west bank of the Rio Aripuanã, at two sites some 50 and 65 km south of its confluence with the Rio Madeira, but they gave no information on group size or other aspects of the species' behavior and ecology. Given this, the range maps for *Mico manicorensis* presented in its description (Figs. 1 and 2, van Roosmalen et al. 2000) appear to be based on inference rather than evidence. The most that can be said from what has been published to date, together with the three available specimens, is that the species occurs at two points on the southeast bank of the Rio Madeira, and has been observed at two sites on the west bank of the Rio Aripuanã.

However, it does follow that *Mico manicorensis* is sympatric with *Callibella*, since the one is known almost entirely from within the range of the other. But it is unclear whether the two species share specific habitats, and whether they interact directly – a key issue for evaluating the ecology and behavior of both.

In the original description of the dwarf marmoset, van Roosmalen et al. (1998) mentioned that *C. humilis* had been seen in both dense primary *terra firme* rainforest as well as the secondary disturbed forest around human settlements, locally known as *capoeira*. They also noted that the dwarf marmoset was sympatric with another species of *Callithrix*, but gave no further details. When that new species was named *Callithrix manicorensis* (now *Mico manicorensis*) by van Roosmalen et al. (2000), its habitat was described as dense primary *terra firme* rainforest and secondary disturbed forest. They also noted "very high densities" of *C. manicorensis* in *seringal* forest, and commented that these stands of rubber trees "appear to offer optimal habitat and a year-round food supply for small monkeys, such as marmosets, pygmy marmosets, dwarf marmosets, titi, and night monkeys" (van Roosmalen et al. 2000: 6). They repeated the assertion that *C. humilis* was sympatric with *C. manicorensis*, but cited no survey data or other supporting evidence.

In their redescription of *Callibella*, van Roosmalen and van Roosmalen (2003: 3) stated that they had seen dwarf marmosets "in disturbed primary and secondary *terra firme* rain forest" directly adjacent to human fields and gardens, and that their earlier report of *Callibella* in dense, distant primary rainforest was based on assumptions alone. After conducting the surveys at the two sites on the Rio Aripuanã, they reported "not a single sighting" of *Callibella* in pristine *terra firme* rainforest (van Roosmalen and van Roosmalen 2003: 3). They did, however, claim to see gouge marks made by *Mico manicorensis* at their survey sites in primary forest. They also noted that the Manicoré marmoset "share[s] with *Callibella humilis* a preference for disturbed forest near human settlements," but went on to say that they had seen *M. manicorensis* in secondary forest clearings well removed from human communities (van Roosmalen and van Roosmalen 2003: 4).

From the sum of their comments, it seems clear that *Mico manicorensis* is smoothly adaptable to a spectrum of forest types, from pristine *terra firme* rainforest to anthropogenic *seringal*. Given this ecological malleability, and its presumed affinity for *capoeira*, it seems strange that *M. manicorensis* has apparently never been sighted together with *Callibella*, especially if *Callibella*, as they claim, is found overwhelmingly in *capoeira* and *seringal* habitats. In 1999, Marc van Roosmalen indicated that *Callibella* could be seen together with *M. manicorensis* at the Seringal São Luis (pers. comm.), but there are no published observations to support this. The question of their coexistence bears directly on issues of niche overlap and potential competitive exclusion, which in turn may be central to an understanding of *Callibella*'s evolution. Until additional fieldwork can clarify their distribution, however, their mutual ecology will remain impossible to evaluate.

The dwarf marmoset presents a unique array of morphological features, in particular the distinctive structure of its lower jaw, which separates it from all other callitrichids. However, this structure does not exist in isolation, but rather in the context of a continuum of feeding behavior, with *Cebuella* most adapted for gouging exudates and the Amazonian marmosets least so. This potential difference in feeding ecology between *Callibella* and *Mico manicorensis* may explain their apparent sympatry in terms of niche partitioning, but confirming this will require evidence from field research.

Acknowledgements We would like to thank Susan Ford, Lesa Davis, and Leila Porter, organizers of the Marmoset and *Callimico* Symposium at the 2005 meeting of the American Association of Physical Anthropology, for inviting us to participate in the symposium and for the published volume which has resulted. We would also like to express our appreciation to M.G.M. van

Roosmalen for making the type skull of *Callibella humilis* available to examine. Thanks are also due to José de Sousa e Silva Júnior and Suely Aguiar of the Museu Paraense Emílio Goeldi in Belém; João Oliveira of the Museu Nacional in Rio de Janeiro; Mario de Vivo of the Museu de Zoologia da Universidade de São Paulo; and the staff of the Instituto Nacional de Pesquisas Amazônicas in Manaus, for access to their respective collections in Brazil.

The first author also appreciates the opportunity to examine specimens under the care of Linda Gordon and Richard Thorington, Jr. at the United States National Museum of Natural History; by Robert Randall at the American Museum of Natural History; by Maria Rutzmoser at the Harvard Museum of Comparative Zoology; and by Bruce Patterson at the Field Museum. Additional thanks are also due to Olavi Grönwall of the Naturhistoriska Riksmuseet in Oslo, Sweden; Manfred Ade of the Humboldt Museum für Naturkunde in Berlin, Germany; the staff of the Museum National de Histoire Naturelle in Paris, France; and most especially Chris Smeenk of the Rijksmuseum van Natuurlijk Historie in Leiden, the Netherlands. This research was supported by a grant from the Margot Marsh Biodiversity Foundation and by Conservation International, in cooperation with Fundação Biodiversitas in Belo Horizonte, Minas Gerais.

The authors are also grateful to Colin Groves, Duane Schlitter, and Michael Willig for constructive comments on the details of marmoset morphology and statistical analysis, and we very much appreciate the assistance of Edward Lohnes and Karyn Tabor in preparing several of the figures. We are particularly grateful to Daniel Juhn and Kellee Koenig of Conservation International, who helped the first author to locate and manipulate the satellite imagery that became Fig. 18.1. We are also grateful to Stephen D. Nash, scientific illustrator for Conservation International, for his kind permission to use his excellent photographs of the type skull of *Callibella*. The first author also thanks Mariella Superina for her Spanish translation of our abstract. Finally, many thanks must go to Ilmar Bastos-Santos of Fundação Biodiversitas for his invaluable support, both personal and professional, of the first author's research travel in Brazil.

References

Aguiar JM (2001) *Callithrix humilis*. Unpublished report to the Margot Marsh Biodiversity Foundation

Aguiar JM, Lacher TE Jr (2003) On the morphological distinctiveness of *Callithrix humilis* van Roosmalen et al., 1998. Neotrop Primates 11(1):11–18

Ferrari SF, Sena L, Schneider MPC (1999) Definition of a new species of marmoset (Primates: Callithrichinae) from southwestern Amazonia based on molecular, ecological and zoogeographic evidence. In: Mendes SL (ed) Livro de Resumos do IX Congresso Brasileiro de Primatologia, Museu de Biologia Prof. Mello Leitão, Santa Teresa, ES, 25–30 Julho 1999, pp 80–81

Ferrari SF, Sena L, Schneider MPC, Silva Jr JS (in press) Rondon's marmoset, *Mico rondoni* sp.n., from southwestern Brazilian Amazonia. Folia Primatologica, in press

Ford SM, Davis LC (this volume) Marmoset postcrania and the skeleton of the dwarf marmoset, *Callibella humilis*. In: Ford SM, Porter LM, Davis LC (eds) The smallest anthropoids: The marmoset/callimico radiation. Springer Press, New York, pp 411–448

Hershkovitz P (1977) Living New World Monkeys (Platyrrhini) with an Introduction to Primates, Vol. 1. The University of Chicago Press, Chicago

Lehmann J, Kern DC, Glaser B, Woods WI (eds) (2004) Amazonian dark earths: Origin, properties, management. Springer, New York

Mann CC (2002) The real dirt on rainforest fertility. Science 297:920–922

Marroig G, Cheverud JM (2005) Size as a line of least evolutionary resistance: Diet and adaptive morphological radiation in New World monkeys. Evolution 59:1128–1142

Marroig G, Cheverud JM (this volume) Size and shape in callimico and marmoset skulls: Allometry and heterochrony in the morphological evolution of small anthropoids. In: Ford SM,

Porter LM, Davis LC (eds) The smallest anthropoids: The marmoset/callimico radiation. Springer Press, New York, pp 331–353

Quammen D (1997) You looking for me? Sports Illustrated, 3 Feb 1997, pp 66–76

Rylands AB, de Faria DS (1987) Habitats and feeding ecology of the genus *Callithrix*. Int J Primatol 8:438

Rylands AB, de Faria DS (1993) Habitats, feeding ecology, and home range size in the genus *Callithrix*. In: Rylands AB (ed) Marmosets and tamarins: Systematics, ecology, and behavior. Oxford University Press, Oxford, pp 262–272

Rylands AB, Coimbra-Filho AF, Mittermeier RA (1993) Systematics, geographic distribution, and some notes on the conservation status of the Callitrichidae. In: Rylands AB (ed) Marmosets and tamarins: Systematics, ecology, and behavior. Oxford University Press, Oxford, pp 11–77

Rylands AB, Mittermeier RA, Coimbra-Filho AF (this volume) The systematics and distribution of the marmosets (*Callithrix, Callibella, Cebuella,* and *Mico*) and callimico (*Callimico*) (Callitrichidae, Primates). In: Ford SM, Porter LM, Davis LC (eds) The smallest anthropoids: The marmoset/callimico radiation. Springer Press, New York, pp 25–61

Schneider H (2000) The current status of the New World monkey phylogeny. Anais da Academia Brasileira de Ciências 72:165–172

Soini P (1988) The pygmy marmoset, genus *Cebuella*. In: Mittermeier RA, Rylands AB, Coimbra-Filho AF, da Fonseca GAB (eds) Ecology and behavior of Neotropical primates, vol 2. World Wildlife Fund, Washington, DC, pp 79–129

SPSS, Inc. (2003) SPSS 13.0. Chicago, Illinois

Tagliaro CH, Schneider MPC, Schneider H, Sampaio I, Stanhope M (1997) Marmoset phylogenetics, conservation perspectives, and evolution of the mtDNA control region. Mol Biol Evol 12:823–833

Taylor AB, Eng EM, Anapol FC, Vinyard CJ (this volume) The functional significance of jaw-muscle fiber architecture in tree-gouging marmosets. In: Ford SM, Porter LM, Davis LC (eds) The smallest anthropoids: The marmoset/callimico radiation. Springer Press, New York, pp 381-394

van Roosmalen MGM (2002) Conservation status of primates in the Brazilian Amazon. Presented at the XIXth Congress of the International Primatological Society, August 2002, Beijing, China

van Roosmalen MGM, van Roosmalen T (2003) The description of a new marmoset genus, *Callibella* (Callitrichinae, Primates), including its molecular phylogenetic status. Neotrop Primates 11(1):1–10

van Roosmalen MGM, van Roosmalen T, Mittermeier RA, da Fonseca GAB (1998) A new and distinctive species of marmoset (Callitrichidae, Primates) from the lower Rio Aripuanã, state of Amazonas, central Brazilian Amazonia. Goeldiana Zoologia 22:1–27

van Roosmalen MGM, van Roosmalen T, Mittermeier RA, Rylands AB (2000) Two new species of marmoset, genus *Callithrix* Erxleben, 1777 (Callitrichidae, Primates), from the Tapajós/Madeira interfluvium, south Central Amazonia, Brazil. Neotrop Primates 8(1):2–19

Vinyard CJ, Wall CE, Williams SH, Mork AL, Garner BA, de Oliveiro Melo LC, Valença-Montenegro MM, Valle YBM, Monteiro da Cruz MAO, Lucas PW, Schmitt D, Taylor AB, Hylander WL (this volume) The evolutionary morphology of tree gouging in marmosets. In: Ford SM, Porter LM, Davis LC (eds) The smallest anthropoids: The marmoset/callimico radiation. Springer Press, New York, pp 395–409

Yamamoto ME, Arruda M de F, Alencar AI, de Sousa MBC, Araújo A (this volume) Mating systems and female-female competition in the common marmosets, *Callithrix jacchus*. In: Ford SM, Porter LM, Davis LC (eds) The smallest anthropoids: The marmoset/callimico radiation. Springer Press, New York, pp 119–133

Chapter 19
The Functional Significance of Jaw-Muscle Fiber Architecture in Tree-Gouging Marmosets

Andrea B. Taylor, Carolyn M. Eng, Fred C. Anapol, and Christopher J. Vinyard

Abstract Marmoset species are specialized gummivorous callitrichids that gouge holes in trees to stimulate the flow of exudate. Recent experimental studies suggest that when common marmosets (*Callithrix jacchus*) gouge trees in the wild, they do so with jaw gapes approaching their maximum structural capacity for jaw opening. Common marmosets also have relatively elongated mandibular condyles (extending the radius of curvature) and glenoid articular surfaces, and low mandibular condyles relative to the height of the tooth row, features that are advantageous in improving the capacity to generate wide jaw gapes.

To determine if tree-gouging marmosets have jaw muscles that are architecturally suited to improving muscle stretch, and hence wide jaw gapes, we evaluated fiber architecture of the masseter and temporalis muscles in two tree-gouging marmosets (*Callithrix jacchus* and *Cebuella pygmaea*) and one nongouging tamarin (*Saguinus oedipus*). Common and pygmy marmosets have relatively longer masseter and temporalis fibers. As fiber length is proportional to muscle excursion, these findings indicate that tree-gouging marmosets have jaw-closing muscles that are well suited to facilitating muscle stretch, and thus large jaw gapes. Marmosets also have a lower tendon to muscle fiber ratio, which may facilitate greater neural control over this specialized feeding behavior. The masseter and temporalis physiological cross-sectional areas (PCSAs) are relatively smaller in tree-gouging marmosets compared to the tamarin. PCSA is directly proportional to the maximum tetanic tension that a muscle can generate. Therefore, tree-gouging marmosets have relatively lower force generating abilities compared to *S. oedipus*. The relatively smaller PCSAs support previous evaluation of the bony masticatory apparatus, which found no morphological evidence to suggest that marmosets have increased force generation or load resistance abilities compared to tamarins. Relatively longer fibers, and their capacity to enhance muscle stretch, suggest that masseter and temporalis fiber architecture function to facilitate the production of wide jaw gapes during tree gouging in the animal's natural environment.

A.B. Taylor (✉)
Departments of Community and Family Medicine and Evolutionary Anthropology, Duke University School of Medicine, Box 104002, Durham, NC, 27708, USA
e-mail: andrea.taylor@duke.edu

S.M. Ford et al. (eds.), *The Smallest Anthropoids*, Developments in Primatology: Progress and Prospects, DOI 10.1007/978-1-4419-0293-1_19,
© Springer Science+Business Media, LLC 2009

Resumen Las especies marmosets son gumnívoros calitricidos especializados que cavan oquedades en los árboles estimulados por corrientes exudadas. Estudios experimentales recientes sugieren que cuando los *Callithris jacchus* excavan árboles en la naturaleza, realizan ellos con las mandíbulas abiertas acercándose a su máxima capacidad estructural para la apertura de la mandíbula. Los *Callithrix jacchus* tienen también cóndilos mandibulares y superficies glenoides articulares relativamente alargadas, y cóndilos mandibulares inferiores relativamente altos en relación a la hilera de dientes, rasgos que generan ventajas al incrementar la capacidad de generar aperturas de mandíbula amplias.

Para determinar si los marmosets excavadores de árboles tienen músculos mandibulares que están arquitectónicamente colocados para incrementar la elasticidad muscular, y asociados a las aperturas amplias de mandíbula, evaluamos la arquitectura de fibra de los músculos masseter y temporales en dos marmosets (*C. jacchus* y *Cebuella pygmea*) y un tamarin no excavador (*Saguinus oedipus*). Los *C. jacchus* y *C. pygmaea* tienen fibras masseter y temporales relativamente más largos. Como el tamaño de la fibra es proporcional a la excursión muscular, los hallazgos indican que los marmosets excavadores de árboles tienen músculos para cerrar la mandíbula los cuales están bien situados para facilitar la elasticidad muscular, y por lo tanto aperturas de mandíbulas amplias. Asimismo, también muestran el tendón bajo para reforzar el radio de la fibra muscular, el cual puede facilitar un mayor control neutral sobre dicho comportamiento especializado de alimentación. Las áreas fisiológicas seccionales cruzadas PCSA de los masseter y los temporales son relativamente más pequeñas en los los *C. jacchus* y *C. Cebuella* comparados con el S. oedipus. El PCSA es directamente proporcional a la tensión máxima tetánica que un músculo puede generar. Por lo tanto, los *C. jacchus* y *C. Cebuella* muestran habilidades generadoras de fuerza relativamente baja comparadas con el *S. oedipus*. La PCSA relativamente más pequeña apoya evaluaciones previas relacionadas con el aparato de hueso masticatorio, el cual no encontró evidencia morfológica que sugiera que los marmosets han incrementado la generación de fuerza o habilidades de resistencia de carga comparada con los tamarins. Fibras relativamente más largas, y su capacidad para realzar la elasticidad muscular, sugiere que las fibras largas masseter y temporales funcionan para facilitar la producción de aperturas amplias de mandíbula durante la excavación de árboles en su ambiente natural.

Resumo Micos são espécies de calitriquídeos especializadas em comer goma, raspando buracos nas árvores para estimular o fluxo de exsudados. Estudos experimentais recentes sugerem que quando os micos-estrelas (*Callithrix jacchus*) raspam as árvores no campo, eles fazem com aberturas da mandíbula aproximando o máximo da capacidade estrutural de abertura da mandíbula. Micos-estrela também têm côndilos mandibulares (extendendo o raio da curvatura) e superfícies articulares glenóides relativamente longos, e côndilos mandibulares baixos em relação à altura dos dentes, caracteristicas que são vantajosas em aumentar a capacidade de mordidas amplas da mandíbula.

Para determinar se os micos "goivadores" de árvores têm músculos mandibulares que são arquiteturalmente adaptados para melhorar o estiramento muscular, e portanto amplas aberturas mandibulares, nós avaliamos a arquitetura de fibras do masseter e do temporal em dois micos (*Callithrix jacchus* e *Cebuella pygmaea*) e um mico não-goivador (*Saguinus oedipus*). Micos-estrelas e sagüis-leãozinho têm fibras massetéricas e temporais relativamente longas. Como o comprimento da fibra é proporcional à excursão do músculo, estas descobertas indicam que os micos goivadores-de-árvores têm músculos adutores da mandíbula que são bem ajustados para facilitar a estiramento muscular, e assim grandes aberturas da boca. Os micos também têm uma baixa proporção de tendão para fibra muscular, o que facilita mais ainda o controle neural(neurológico) sobre este comportamento alimentar especializado. As áreas de seção-transversal fisiológica (PCSAs) do masseter e temporal são relativamente pequenas nos micos-goivadores quando comparadas aos sagüis. A PCSA é diretamente proporcional a tensão tetânica que um músculo pode gerar. Portanto, micos-goivadores têm relativamente menor capacidade de gerar forças quando comparados à S. oedipus. As PCSAs relativamente menores suportam as avaliações prévias do aparato masticatório ósseo, que não mostraram nenhuma evidência morfológica que indicasse que os micos têm uma geração de força maior ou resistência a cargas quando comparados aos sagüis *(Saguinus)*. Fibras relativamente longas, e sua capacidade de potencializar a extensão múscular, sugerem que a arquitetura do masseter e do temporal funciona para facilitar a produção de grandes aberturas da boca durante a raspagem das árvores no ambiente natural.

19.1 Introduction

Among the gummivorous callitrichid monkeys, marmoset species are gum feeders that systematically gouge holes in trees to stimulate the flow of gums and saps. This specialized feeding behavior is accompanied by a number of morphological adaptations. Early investigators (e.g., Hill 1957; Napier and Napier 1967; Hershkovitz 1977) commented on the marmoset's 'short-tusked' condition, i.e., "with elongated lower incisors and incisiform lower canines" (Napier and Napier 1967: 371) and their conically-honed lower incisors characterized by a thinning of enamel on the lingual surface and concomitant hypertrophy of enamel on the labial surface (Hershkovitz 1977; Rosenberger 1978). More recently, it has been shown that marmosets differ from tamarins in their specialized gut morphology (Ferrari and Martins 1992) and in features of the lower jaw (Vinyard et al. 2003, Chap. 20 this volume). All of these differences have been functionally or adaptively linked to the unique feeding behavior of common marmosets.

Tree gouging is an important behavior that unites all species of marmosets and distinguishes them from tamarins (Garber 1992 and references cited therein). While several craniodental differences between tree-gouging and nongouging callitrichids have been documented, our knowledge of soft tissue variation is lacking. Marmosets routinely gouge trees, and performance data indicate that tree gouging involves generating relatively wide jaw gapes but not necessarily relatively large jaw forces (Vinyard et al. 2003, Chap. 20 this volume). With this in mind, knowledge of the architectural arrangement of the muscles that produce jaw forces and jaw movements would be particularly informative with regard to the biological role of the masticatory muscles during tree-gouging. In this chapter, we consider morphological differences between marmosets and tamarins in the fiber architecture of the jaw-closing muscles. We use these data to help elucidate the full range of functional and adaptive responses to tree gouging and to provide a more complete morphological profile of this unusual feeding behavior.

19.2 Forces and Movements During Tree Gouging

Although many callitrichids eat gums and saps, all marmosets, including common (*Callithrix jacchus*) and pygmy (*Cebuella pygmaea*) marmosets, are specialized gum feeders, actively gouging holes in trees to stimulate the flow of exudates (Coimbra-Filho and Mittermeier 1977). Behaviorally, tree-gouging marmosets anchor their upper jaws in the substrate while utilizing their lower jaws to gouge the trunk or branch surface. Gouging holes stimulates the release of gums, which the animals subsequently ingest. By contrast, callitrichids such as *Saguinus* (tamarins) and *Leontopithecus* (lion tamarins) are opportunistic gum feeders, exploiting gums and saps that have been released through mechanical damage by insects or other means (Rosenberger 1992).

In callitrichids, body size and the presence of claw-like nails facilitate vertical clinging postures, which have been functionally or adaptively linked to gummivory (Nash 1986; but see Rosenberger 1992). Tree-gouging, gummivorous callitrichids exhibit additional specializations. One such specialization is in the form of the incisor–canine relationship, which is characterized by canines and incisors of similar heights that provide an effective wedge for gouging trees (Hill 1957; Hershkovitz 1977). Another specialization is their pointed lower incisors with a thickened layer of enamel on the labial aspect but only a very thin layer of enamel on the lingual surface, which creates a wear gradient and helps maintain a chisel-like edge to facilitate gouging (Coimbra-Filho and Mittermeier 1977; Rosenberger 1978).

Until quite recently, it had been theoretically assumed that gouging holes in trees entails the generation of relatively large jaw forces (Szalay and Seligsohn 1977; Rosenberger 1992; Dumont 1997; Spencer 1999). Current and ongoing experimental studies, however, suggest that this may not be the case. *In vivo* experimental studies demonstrate that the forces generated by common marmosets during gouging are lower than the forces these animals can generate, for example, during transducer biting (Vinyard et al. 2001, Chap. 20 this volume). Likewise, there is no morphological evidence to indicate that tree-gouging primates, including common marmosets, have skull or jaw shapes that improve their capacity to generate or dissipate relatively large loads (Vinyard et al. 2003).

Although the functional requirements for tree gouging may not include producing large jaw forces, it appears that generating very wide jaw gapes is an important functional component of tree gouging (Vinyard et al. 2001, Chap. 20 this volume). Tree-gouging common marmosets also exhibit several morphological features that theoretically improve their capacity to generate relatively wide jaw gapes compared to nongougers. Common marmosets, for example, have relatively elongated mandibles, anteroposteriorly elongated condyles (which increase the condylar radius of curvature) and glenoid articular surfaces, and condyles positioned closer to the height of the mandibular toothrow (Vinyard et al. 2003). The theoretical benefits of this structural arrangement include a reduction in the amount of muscle stretch required to achieve a given jaw gape and the ability to maintain effective alignment of the lower anterior dentition relative to the substrate while gouging trees with wide jaw gapes (Herring and Herring 1974; Vinyard et al. 2003).

Here we address how fiber architecture of the jaw-closing muscles influences the capacity for jaw movements and force production in tree-gouging marmosets. We ask whether marmosets have jaw-closing muscles that facilitate the functional demands of tree gouging with wide jaw gapes and/or with large jaw forces. Before summarizing our results, we provide a brief overview of how skeletal muscle fibers are organized and the functional significance of how muscle fibers are arranged.

19.3 Implications of Fiber Architecture for Muscle Function

Skeletal muscles are designed for the production of force and movement as achieved through muscle contraction. Sarcomeres are the muscle's functional contractile unit and are composed of interdigitated myofilaments. The sliding filament model describes how the thin actin myofilaments slide past the thick myosin myofilaments, generating the tension necessary to produce muscle contraction. As the muscle shortens, cross-bridges form between overlapping myofilaments to generate muscle force (Gordon et al. 1966a, b; Lieber 2002).

The arrangement of muscle fibers has important implications for muscle function. The physiological cross-sectional area (PCSA) of a muscle is directly proportional to the maximum force that a muscle can generate, as PCSA reflects both the number of muscle fibers lying in parallel and the overlap of actin and myosin filaments. Increased overlap causes more crossbridges to form, and this generates tension within the muscle (Gordon et al. 1966a, b; Lieber 2002). When no further crossbridges can be formed, no additional force can be generated and, indeed, increased filament overlap beyond a certain point results in a decrease in muscle force. The PCSA of a muscle is calculated as the sum of the cross-sectional areas of all the muscle fibers within the muscle (Lieber 2002).

The angle at which muscle fibers lie in relation to the force-generating axis of a muscle has implications for the muscle's force and excursion potential. A parallel-fibered muscle contains fibers that lie parallel to the line of action of the muscle. By contrast, a pinnate-fibered muscle contains fibers that are angled relative to the muscle's line of action. Parallel-fibered muscles have many sarcomeres in longitudinal series, end-to-end (Gans and Bock 1965; Lieber 2002). The absolute excursion of a muscle (and, by extension, its contraction velocity) is equivalent to the unit excursions achieved by each sarcomere in series (Gans and De Vree 1987). Therefore, all things being equal, longer-fibered muscles are able to achieve greater excursions than shorter-fibered muscles.

Pinnate arrangements consist of shorter, more tightly packed fibers with more sarcomeres lying in parallel to one another (Gans 1982). Pinnation provides a means of packing more, but shorter, fibers in a given space, thereby increasing the potential force that can be generated by the muscle, though there will be some loss of force along the line of action of the muscle (Gans 1982; Lieber 2002). Pinnation has been explained as a means of increasing force production at the expense of muscle excursion, since in pinnate-fibered muscles, fewer sarcomeres will be placed in series, implying that the muscle can move through a more limited range of motion than would be possible with a parallel-fibered muscle of similar size (Gans and de Vree 1987, but see Herring 1992).

It has been argued that muscle fiber architecture represents a trade-off between increasing PCSA (and thus muscle force) or increasing fiber length (and thus muscle excursion/contraction velocity). We hypothesize that the masseter and temporalis muscles can be maximized for either muscle excursion or force production, but not both. Finally, it has also been shown that muscles involved in producing a specific action may be functionally partitioned, such that some muscles appear to be dedicated more to force production, while others are arranged more favorably for generating muscle contraction velocity and excursion (Anapol and Jungers 1986; Anapol and Barry 1996). We therefore explore functional partitioning of the masseter and temporalis muscles.

19.4 Fiber Architecture and Jaw Function in a Specialized Feeding Behavior

Comparative data on jaw forces and jaw movements indicate that common marmosets generate relatively wide jaw gapes, but not necessarily relatively large jaw forces, during tree gouging (Vinyard et al. 2001, Chap. 20 this volume). Based on these preliminary data, we tested the hypothesis that tree-gouging common and pygmy marmosets exhibit architectural features of the jaw-closing muscles that facilitate muscle excursion. Here we summarize some of our key findings related to this question (see Taylor and Vinyard 2004, 2008, and Taylor et al., 2009 for more details).

To date, we have evaluated fiber architecture of the superficial masseter and anterior temporalis muscles in two tree-gouging callitrichids, *Callithrix jacchus* (common marmosets, $n=18$) and *Cebuella pygmaea* (pygmy marmosets, $n=5$), and one nongouging callitrichid, *Saguinus oedipus* (cotton-top tamarins, $n=13$)

(Taylor and Vinyard 2004, 2008; Taylor et al., 2009). We investigated relative differences in the capacity to generate muscle stretch and muscle force by adjusting muscle architectural parameters by both jaw length and whole muscle length (Taylor and Vinyard 2004). Our results show that common and pygmy marmosets exhibit architectural features of these jaw-closing muscles that are functionally linked to facilitating muscle stretch and muscle excursion. Because relative differences in fiber architecture are the same regardless of the choice of independent variable, we only present results adjusted for jaw length.

For example, masseter and temporalis fiber lengths are relatively greater in both common and pygmy marmosets compared to cotton-top tamarins (Fig. 19.1a). Longer fibers indicate that common and pygmy marmosets have a greater active range through which their muscles can lengthen (and shorten). Thus, relatively longer fibers are advantageous to marmosets because their jaw-closing muscles require less stretch to achieve a given jaw gape compared to cotton-top tamarins. Longer fibers, therefore, may allow for the production of larger jaw gapes.

The PCSAs of the masseter and temporalis muscles are relatively smaller in the tree-gouging common and pygmy marmosets compared to cotton-top tamarins (Fig. 19.1b). Recall that PCSA is directly proportional to the maximum force that can be generated by these muscles. Therefore, findings of relatively smaller PCSAs indicate that marmosets have the potential to generate relatively smaller maximum jaw forces than tamarins. The smaller PCSAs combined with the longer muscle fibers observed in tree-gouging marmosets together provide empirical

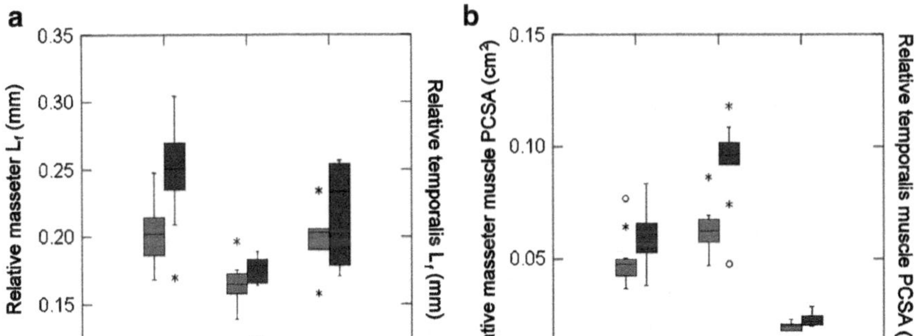

Fig. 19.1 Box plot comparing muscle fiber architecture in marmosets and tamarins. (**a**) Relative fiber lengths (L_f) of the masseter (left, light stipple) and temporalis (right, dark stipple) muscles. In all comparisons, fiber architecture parameters are adjusted by taking a ratio relative to jaw length. Tree gouging common (M-W *U*-tests; $p<0.01$) and pygmy (M-W *U*-tests; $p<0.05$) marmosets have relatively longer masseter and temporalis fibers than do nongouging tamarins. (**b**) Relative PCSAs of the masseter (left) and temporalis (right) muscles. Tamarins have relatively greater masseter and temporalis PCSAs than do marmosets (M-W *U*-tests; $p<0.05$). Collectively, these data indicate that tree-gouging marmosets have jaw-closing muscles that favor muscle stretch and muscle excursion over muscle force. The center vertical line marks the median of the sample. The box denotes the interquartile range and the asterisks indicate that one or more data points reside outside the fence for the individual data points

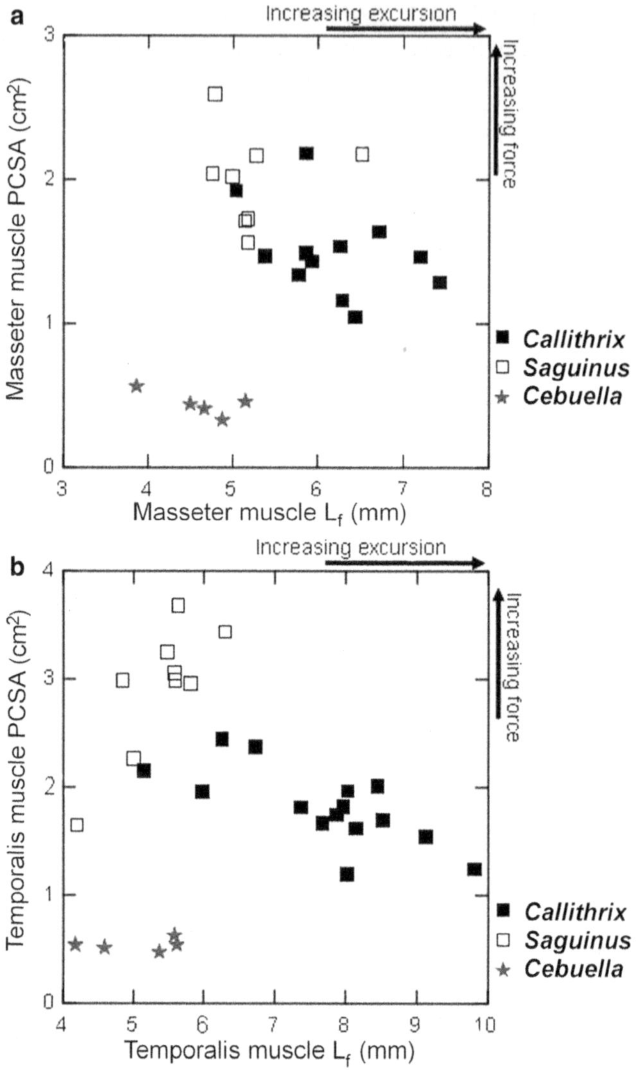

Fig. 19.2 Bivariate plot demonstrating the architectural trade-off between muscle force (PCSA) and muscle excursion (fiber length) (L_f) for (**a**) the masseter muscle ($r=-0.54$; $p=0.013$), and (**b**) the temporalis muscle ($r=-0.65$; $p=001$). Overlap between common marmosets (*Callithrix jacchus*) and cotton-top tamarins (*Saguinus oedipus*) is minimal. These data demonstrate that common marmoset masseter and temporalis, with relatively low PCSAs and long fibers, are suited more for muscle excursion (i.e., production of wide jaw gapes), while *S. oedipus* masseter and temporalis, with relatively larger PCSAs and shorter fibers, are better suited for generating larger muscle forces with smaller excursions. Note that *Cebuella* has jaw-muscle fiber lengths that are almost as long as those of *S. oedipus*, but at one quarter the body mass

support for the theoretical trade-off between muscle force and muscle excursion (Fig. 19.2). Thus, tree-gouging common marmosets appear to be maximizing muscle excursion in both the masseter and temporalis muscles, but at the expense of some muscle force, whereas the reverse is true for *S. oedipus* (Fig. 19.2).[1] We are careful to point out, however, that we do not have dietary or jaw force data to suggest that *S. oedipus* generates relatively greater maximum jaw forces compared to tree-gouging marmosets. We therefore interpret the relatively greater jaw-muscle PCSAs in *S. oedipus* as reflecting their relatively shorter, more pinnate fibers compared to those of tree-gouging marmosets.

Although the force-generating capacity of these two jaw-closing muscles is relatively lower in common and pygmy marmosets than in cotton-top tamarins, we can see that in all taxa, the temporalis muscle appears to have greater maximum force-generating capacity than does the masseter (Fig. 19.1b). However, if we consider the temporalis and masseter as part of a group comprising the muscles of mastication, we can also preliminarily address whether there are differences in the relative contributions of these two muscles to the overall force- and excursion-producing capabilities of this muscle group. If the PCSA of each muscle is expressed as a percentage of the combined PCSA for both muscles, it is apparent that the disparity between the masseter and temporalis muscles is greatest in *S. oedipus* compared to the two tree-gouging taxa (Fig. 19.3a). It therefore appears that in

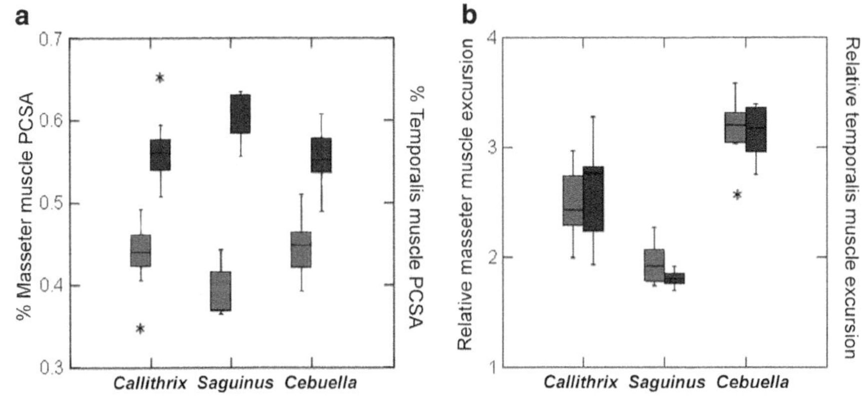

Fig. 19.3 (**a**) The PCSA of the masseter (left, light stipple) and temporalis (right, dark stipple) muscles expressed as a percentage of the combined PCSA for both muscles. The disparity between the masseter and temporalis muscles is greatest in *S. oedipus*, suggesting a division of labor between these two jaw-closing muscles in terms of their dedication to generating force. (**b**) Relative muscle excursion (excursion/cubed root of muscle mass) for the masseter (left, light stipple) and temporalis (right, dark stipple) muscles is greater in the two tree-gouging taxa compared to *S. oedipus*, while there is little disparity in the relative excursion abilities of these two muscles in any species

[1] Although *Cebuella pygmaea* does not fall along the trend for PCSA and fiber length as observed for *Callithrix jacchus* and *Saguinus oedipus*, *C. pygamea* has absolute fiber lengths that approach those of *S. oedipus*, at a quarter of their body mass.

cotton-top tamarins, the temporalis muscle potentially contributes relatively more to force production than does the masseter. The difference between the masseter and temporalis in terms of their relative contributions to force production is significant in comparisons between *S. oedipus* and *Cebuella pygmaea* ($\chi^2=3.93$, $df=1$, $p=0.047$), and approaches significance ($\chi^2=3.49$, $df=1$, $p=0.062$) in comparisons between *S. oedipus* and *Callithrix jacchus*. By contrast, relative muscle excursion for the masseter and temporalis muscles is greater in the two tree-gouging taxa compared to *S. oedipus*, whereas there is little disparity (Paired *t*-tests; $p>0.05$) in the relative excursion abilities of these two muscles in any species (Fig. 19.3b). Taken together, these data suggest the possibility of a greater division of labor between the masseter and temporalis muscles in terms of generating muscle force in *S. oedipus*. These findings should be interpreted cautiously, however, both because the masseter and temporalis muscles represent only two of the four muscles of mastication, and because the masticatory muscles are architecturally complex and contain regionally differentiated, task-specific portions (e.g., Herring et al. 1979; Weijs and Dantuma 1981; van Eijden et al. 1993; van Eijden and Turkawski 2001), not all of which have yet been evaluated in this analysis.

The relative proportion of tendon is smaller in marmosets compared to tamarins for both the masseter and temporalis muscles (Fig. 19.4). Previously, Taylor and Vinyard (2004, 2008) suggested that this architectural arrangement may also be advantageous to tree-gouging marmosets. Tendons are well-suited for storing and releasing elastic strain energy (Biewener 1998; Daley and Biewener 2003). However, although tendons passively transmit muscle force, the noncontractile properties of tendon are not suited to controlling change in muscle length (Biewener

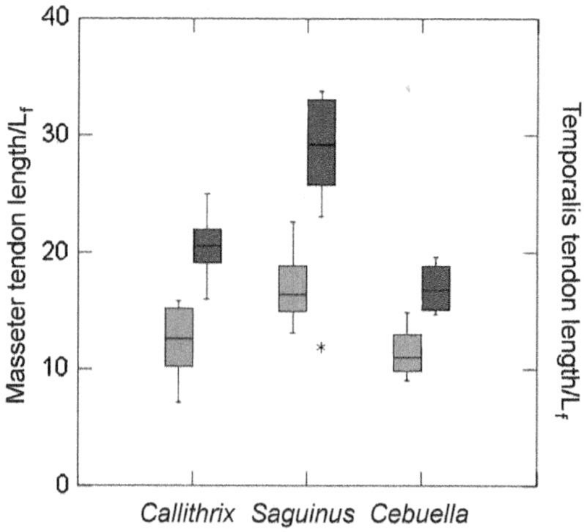

Fig. 19.4 The relative proportion of tendon is lower in common (M-W *U*-tests; $p<0.001$) and pygmy (M-W *U*-tests; $p<0.05$) marmosets for the masseter (left, light stipple) and temporalis (right, dark stipple) muscles. Lower ratios suggest the potential for greater neural control over jaw movement and jaw posture in marmosets

and Roberts 2000). Thus, increasing the ratio of tendon to muscle would potentially limit the capacity of the neuromuscular system to accurately position the lower jaw over a wider range of motion. Common marmosets actively gouge with jaw gapes that approach their maximum structural capacity for jaw opening. Therefore, achieving and maintaining accurate jaw positioning is likely to be of critical importance to tree gougers in order to minimize the risk of muscle strain, tearing, or temporomandibular joint dislocation. We speculate, therefore, that tree gouging is a highly modulated feeding behavior, and longer fibered muscles, comprising less tendon and more fiber, may facilitate greater neural control.

Collectively, relatively longer fibers coupled with lower PCSAs support field and experimental studies demonstrating that marmosets gouge trees with relatively wide jaw gapes. Thus far, our fiber architecture analyses provide no evidence to support the hypothesis that tree gouging requires generating relatively large muscle, and hence bite forces. The presence of relatively longer fibers, their capacity to enhance muscle stretch during jaw opening, and the performance data indicating that tree gouging involves the generation of wide jaw gapes, all suggest that longer masseter and temporalis fibers function to facilitate the production of wide jaw gapes during tree gouging in their natural environment (Bock and von Wahlert 1965).

19.5 Functional and Adaptive Implications of Fiber Architecture for Tree Gouging

Tree-gouging marmosets appear to be geared for generating relatively wide jaw gapes. We draw this conclusion from the jaw-muscle fiber architecture data presented here and elsewhere (Taylor and Vinyard 2004, 2008; Taylor et al., 2009), as well as observed differences in jaw morphology (Vinyard et al. 2003) between tree-gouging and nongouging callitrichids. During the generation of wide jaw gapes, which involves active lengthening of the jaw-closing muscles, it is likely that marmosets are stretching their jaw-closing muscle well beyond their resting length. Bite forces in primates diminish as jaw-muscle fibers are stretched beyond their resting length (Dechow and Carlson 1982, 1986, 1990), and gape-related decreases in bite force are exacerbated by relatively inefficient jaw leverage for producing bite forces at the anterior teeth. Alterations in both the bony and soft tissue structures of the masticatory apparatus suggest that tree-gouging marmosets are under pressure to minimize stretching of the jaw-closing muscles in order to generate adequate bite forces at wide jaw gapes.

It seems clear that relatively longer fibers are part of a larger morphological complex that favors muscle stretch and jaw opening. However, mammalian skeletal muscle is quite plastic; chronic muscle stretch can result in an increase in the number of sarcomeres in series and an adjustment of muscle resting length (Salmons and Vrbova 1969; Williams and Goldspink 1971, 1973; Williams et al. 1986). At present, we are unable to address whether the relative increases in masseter and temporalis fiber lengths represent an evolutionary adaptation to tree gouging (i.e., the result of natural selection on heritable variation), a phenotypically plastic response, or possibly both. We are currently approaching this question through investigation of the neurobiology of the masticatory system in tree-gouging and nongouging callitrichids (White et al. 2006). Whether differences in muscle fiber architecture reflect evolutionary adaptations or the plastic modulation of function and behavioral performance, our results provide a convincing functional link between fiber architecture of the masseter and temporalis muscles, the production of wide jaw gapes, and tree gouging in the natural environment.

Acknowledgments We thank Lesa Davis, Susan Ford, and Leila Porter for inviting us to contribute to this volume, and to participate in the symposium "Advances in Marmoset and Goeldi's Monkey (*Callimico*) Research: Anatomy, Behavioral Ecology, Phylogeny, and Conservation" held at the American Association of Physical Anthropologists Meetings on April 8, 2005. We are grateful to Elizabeth Curran (NEPRC), Amanda Trainor (WPRC), Suzette Tardiff (SFBR), and Donna Layne (SFBR) for their help in providing the muscle tissue. This research was supported by NSF (BCS-0412153), (BCS-0094666), and (BCS-0412153).

References

Anapol F, Barry K (1996) Fiber architecture of the extensors of the hindlimb in semiterrestrial and arboreal guenons. Am J Phys Anthropol 99:429–47

Anapol F, Jungers WL (1986) Architectural and histochemical diversity within the quadriceps femoris of the brown lemur (*Lemur fulvus*). Am J Phys Anthropol 69:355–375

Biewener AA (1998) Muscle-tendon stresses and elastic energy storage during locomotion in the horse. Comp Biochem Physiol B Biochem Mol Biol 120:73–87

Biewener AA, Roberts TJ (2000) Muscle and tendon contributions to force, work, and elastic energy savings: A comparative perspective. Exerc Sport Sci Rev 28:99–107

Bock WJ, von Wahlert G (1965) Adaptation and the form-function complex. Evolution 19:269–299.

Coimbra-Filho AF, Mittermeier RA (1977) Tree-gouging, exudate-eating and the "short-tusked" condition in *Callithrix* and *Cebuella*. In: Kleiman DG (ed) The biology and conservation of the Callitrichidae. Smithsonian Institution Press, Washington, DC, pp 105–115

Daley MA, Biewener AA (2003) Muscle force-length dynamics during level versus incline locomotion: A comparison of in vivo performance of two guinea fowl ankle extensors. J Exp Biol 206:2941–2958

Dechow PC, Carlson DS (1990) Occlusal force and craniofacial biomechanics during growth in rhesus monkeys. Am J Phys Anthropol 83:219–237

Dechow PC, Carlson DS (1982) Bite force and gape in rhesus macaques. Am J Phys Anthropol 57:179

Dechow PC, Carlson DS (1986) Occlusal force after mandibular advancement in adult rhesus monkeys. J Oral Maxillofac Surg 44:887–893

Dumont ER (1997) Cranial shape in fruit, nectar, and exudate feeders: Implications for interpreting the fossil record. Am J Phys Anthropol 102:187–202

Ferrari SF, Martins ES (1992) Gummivory and gut morphology in two sympatric callitrichids (*Callithrix emiliae* and *Saguinus fuscicollis weddelli*) from western Brazilian Amazonia. Am J Phys Anthropol 88:97–103

Gans C, Bock WJ (1965) The functional significance of muscle architecture: A theoretical analysis. Ergeb Anat Entwicklungsgesch 38:115–142

Gans C (1982) Fiber architecture and muscle function. Exer Sport Sci Rev 10:160–207

Gans C, de Vree F (1987) Functional bases of fiber length and angulation in muscle. J Morphol 192:63–85

Garber PA (1992) Vertical clinging, small body size, and the evolution of feeding adaptations in the Callitrichinae. Am J Phys Anthropol 88:469–482

Gordon AM, Huxley AF, Julian FJ (1966a) Tension development in highly stretched vertebrate muscle fibres. J Physiol (Lond) 184:143–169

Gordon AM, Huxley AF, Julian FJ (1966b) The variation in isometric tension with sarcomere length in vertebrate muscle fibres. J Physiol (Lond) 184:170–192

Herring SW, Herring SE (1974) The superficial masseter and gape in mammals. Am Nat 108:561–576

Herring S, Grimm AF, Grimm BR (1979) Functional heterogeneity in a multipinnate muscle. Am J Anat 154:563–576

Herring SW (1992) Muscles of mastication: Architecture and functional organization. In Davidovitch Z (ed) The Biological Mechanisms of Tooth Movement and Craniofacial Adaptation. The Ohio State University College of Dentistry, Columbus, Ohio, pp. 541–548

Hershkovitz P (1977) Living New World primates (Platyrrhini), with an introduction to primates, vol 1. University Chicago Press, Chicago

Hill WCO (1957) Primates, comparative anatomy and taxonomy, vol. 3: Pithecoidea: Platyrrhini. Edinburgh University Press, Edinburgh

Lieber RL (2002) Skeletal Muscle Structure, Function, and Plasticity. Baltimore: Lippincott Williams and Wilkins.

Napier JR, Napier PH (1967) A Handbook of Living Primates. Academic Press, London

Nash LT (1986) Dietary, behavioral, and morphological aspects of gumnivory in primates. Yrbk Phys Anthropol 29:113–137

Rosenberger AL (1978) Loss of incisor enamel in marmosets. J Mamm 59:207–208

Rosenberger AL (1992) Evolution of feeding niches in New World monkeys. Am J Phys Anthropol 88:525–562

Salmons S, Vrbova G (1969) The influence of activity on some contractile characteristics of mammalian fast and slow muscles. J Physiol (Lond) 201:535–549

Spencer MA (1999) Constraints on masticatory system evolution in anthropoid primates. Am J Phys Anthropol 108:483–506

Szalay FS, Seligsohn D (1977) Why did the strepsirhine tooth comb evolve? Folia Primatol 27:75–82

Taylor AB, Vinyard CJ (2004) Comparative analysis of masseter fiber architecture in tree-gouging (*Callithrix jacchus*) and nongouging (*Saguinus oedipus*) callitrichids. J Morphol 261:276–285

Taylor AB, Vinyard CJ (2008) The relationship between jaw-muscle fiber architecture and feeding behavior in primates: Tree-gouging and nongouging gummivorous callitrichids as a natural experiment. In: Vinyard CJ, Ravosa MJ, Wall CE (eds). Primate Craniofacial Function and Biology, p. 241–264. New York: Springer

Taylor AB, Eng CM, Anapol F, Vinyard CJ (2009) The functional correlates of jaw-muscle fiber architecture in tree-gouging and nongouging callitrichid monkeys. Am J Phys Anthropol.

Van Eijden TM, Blanksma NG, Brugman P (1993) Amplitude and timing of EMG activity in the human masseter muscle during selected motor tasks. J Dent Res 72:599–606

Van Eijden TM, Turkawski SJ (2001) Morphology and physiology of masticatory muscle motor units. Crit Rev Oral Biol Med 12:76–91

Vinyard CJ, Wall CE, Williams SH, Schmitt D, Hylander WL (2001) A preliminary report on the jaw mechanics during tree gouging in common marmosets (*Callithrix jacchus*). In: Brooks A (ed) Dental morphology 2001: Proceedings of the 12th international symposium on dental morphology. Sheffield Academic Press, Ltd, Sheffield, UK, pp 283–297

Vinyard CJ, Wall CE, Williams SH, Hylander WL (2003) A comparative functional analysis of the skull morphology of tree gouging primates. Am J Phys Anthropol 120:153–170

Vinyard CJ, Wall CE, Williams SH, Mork AL, Garner BA, de Oliveiro Melo LC, Valença-Montenegro, MM, Valle YBM, Monteiro da Cruz MAO, Lucas PW, Schmitt D, Taylor AB, Hylander WL (this volume) The evolutionary morphology of tree gouging in

marmosets. In: Ford SM, Porter LM, Davis LC (eds) The smallest anthropoids: The marmoset/callimico radiation. Springer Press, New York, pp 119–133
Weijs WA, Dantuma R (1981) Functional anatomy of the masticatory apparatus in the rabbit (*Oryctolagus cuniculus* L.). Neth J Zool 31:99–147
White LE, Jones KE, Vinyard CJ, Taylor AB (2006) Functional architecture of the brainstem trigeminal complex in two callitrichid species with divergent feeding behaviors (*Callithrix jacchus* and *Saguinus oedipus*). Am J Phys Anthropol Suppl 42:187
Williams PE, Goldspink G (1971) Longitudinal growth of striated muscle fibres. J Cell Sci 9:751–767
Williams PE, Goldspink G (1973) The effect of immobilization on the longitudinal growth of striated muscle fibers. J Anat 116:45–55
Williams P, Watt P, Bicik V, Goldspink G (1986) Effect of stretch combined with electrical stimulation on the type of sarcomeres produced at the ends of muscle fibers. Exp Neurol 93:500–509

Chapter 20
The Evolutionary Morphology of Tree Gouging in Marmosets

Christopher J. Vinyard, Christine E. Wall, Susan H. Williams, Amy L. Mork, Brooke A. Armfield, Leonardo César de Oliveira Melo, Mônica M. Valença-Montenegro, Yumma Bernardo Maranhão Valle, Maria Adélia Borstelmann de Oliveira, Peter W. Lucas, Daniel Schmitt, Andrea B. Taylor, and William L. Hylander

Abstract The marmosets, *Callithrix* spp. and *Cebuella pygmaea*, are unique among anthropoids in their habitual biting of trees with their anterior teeth to elicit exudate flow. This tree-gouging behavior is thought to offer certain ecological benefits to marmosets, such as routine access to an under-exploited resource, as well as have specific influences on their behavioral ecology.

In order to better understand morphological adaptations for tree gouging in the marmoset masticatory apparatus, we characterized the mechanics of this behavior in a laboratory setting and compared these data to field observations of common marmosets (*Callithrix jacchus*) in northeast Brazil. Common marmosets generate biting forces up to eight times their body mass when biting simulated tree substrates in the laboratory. When gouging, however, marmosets are not biting as forcefully as they can. Comparisons of the mechanical properties of laboratory substrates with trees gouged in the wild suggest that gouging forces observed in the laboratory are comparable to those that marmosets use during gouging in the wild. Moreover, marmosets use relatively wide jaw gapes during gouging both in the laboratory and in the wild. These wide jaw gapes during gouging approach the maximum structural capacity for jaw opening in common marmosets.

Morphological comparisons of masticatory apparatus form between gouging marmosets and nongouging tamarins corroborate these laboratory and field data. Marmosets do not exhibit morphologies that offer increased force generation or load resistance abilities. Furthermore, marmosets exhibit several morphologies of their masticatory apparatus that facilitate increased jaw opening. Previous suggestions that marmoset gouging involves relatively large bite forces likely misrepresent the nature of this behavior. Instead, marmosets need only to score the tree bark to elicit the tree's defense response of exudate flow.

C.J. Vinyard (✉)
Department of Anatomy and Neurobiology, NEOUCOM, 4209 St Rt 44, Box 95, Rootstown, OH, 44272, USA
e-mail: cvinyard@neoucom.edu

S.M. Ford et al. (eds.), *The Smallest Anthropoids*, Developments in Primatology: Progress and Prospects, DOI 10.1007/978-1-4419-0293-1_20,
© Springer Science+Business Media, LLC 2009

Resumen Los marmosets *Callithrix spp.* y *Cebuella pygmea*, son únicos entre los antropóides en su habitual mordisco en árboles con sus dientes anteriores para sacar corrientes exudadas. Este comportamiento de excavación de árboles se cree ofrece a ellos ciertos beneficios ecológicos, tal como el acceso rutinario a recursos poco utilizados, asi como tener influencias específicas en su comportamiento ecológico.

Con el propósito de entender adaptaciones morfológicas para excavar árboles en el aparato masticatorio de los marmoset, caracterizamos los mecanismos de dicho comportamiento en un laboratorio y comparamos nuestros datos de laboratorio con las observaciones de campo de *Callithrix jacchus* en el noreste de Brazil. Los *Callithrix jacchus* generan al morder hasta ocho veces su masa corporal al ser estimulados por substancias de árboles en el laboratorio. Sin embargo, al cavar, los *C. jacchus* no mordieron tan fuerte como ellos pueden. Comparaciones de las propiedades mecánicas de los substratos de laboratorio con árboles excavados en estado salvaje sugiere que las fuerzas de excavación observadas en el laboratorio son comparables con aquellos que los marmosets usarian durante la excavación en la selva. Alternativamente, los marmosets usan aperturas de mandíbulas relativamente amplias, medidas como la distancia entre las puntas incisivas superiores e inferiores, durante la excavación tanto en el laboratorio como en lo salvaje. Dichas aperturas de mandíbulas durante la excavación se acercan la capacidad máxima estructural de apertura de mandíbula en los marmosets comunes.

Comparaciones morfológicas de la forma de los aparatos masticatorios entre los marmosets excavadores y los tamarins no excavadores corroboran los datos de laboratorio y de campo. Los marmosets no muestran morfologías que ofrecen incremento en la generación de fuerza o habilidades de resistencia de carga. Alternativamente, los marmosets exhiben variadas morfologías en su aparato masticatorio que facilitan las habilidades de apertura de mandibula. Sugerencias anteriores de que la excavación marmoset muestra fuerzas de mordiscos relativamente grandes es probable que no representa la naturaleza de este comportamiento; los marmosets simplemente proponen marcar árboles para sacarle a los árboles su reacción de defensa.

Resumo Os sagüis, *Callithrix* spp. and *Cebuella pygmaea*, são únicos entre os antropóides no hábito de morder as árvores com seus dentes anteriores para induzir o fluxo de exsudados. Este comportamento de perfurar o tronco das árvores é concebido como algo que ofereceria benefícios ecológicos aos sagüis, tais como o acesso rotineiro a recursos sub-explorados, assim como traria influências específicas na sua ecologia comportamental.

Para entender melhor estas adaptações morfológicas para perfuração de árvores no aparato mastigação dos sagüis, nós caracterizamos a mecânica deste comportamento em laboratório e comparamos estes dados à s observações de campo de sagüis (*Callithrix jacchus*) no nordeste do Brasil. Sagüis geram forças de mordida de até oito vezes as suas massas corporais quando mordem substratos de árvores simulados no laboratório. Quando perfuram o tronco, entretanto, os sagüis não mordem tão forte quanto poderiam. Comparações das propriedades mecânicas dos substratos de laboratório com as árvores perfuradas no campo sugerem que as forças de mordida observadas em laboratório são comparáveis aquelas que os sagüis devem empregar na perfuração dos troncos no campo. Ademais, os sagüis fazem aberturas da boca relativamente amplas durante a perfuração tanto no laboratório como no campo. Estas aberturas amplas durante a perfuração dos troncos se aproximam da capacidade de abertura máxima estrutural da mandíbula nos sagüis.

Comparações morfológicas do aparato de mastigação entre micos e sagüis que perfuram e que não perfuram troncos corroboram com estes dados de laboratório e do campo. Sagüis não apresentam morfologias que permitam um aumento na geração de força ou na capacidade de resistência à carga. Além disso, sagüis exibem várias morfologias do aparato de mastigação que facilitam um aumento da abertura da boca. Sugestões prévias de que a perfuração dos troncos nos sagüis envolvem forças de mordidas relativamente grandes foram provavelmente fruto de uma má interpretação da natureza deste comportamento. Ao invés disso, os sagüis precisam somente raspar a casca das árvores para estimular su a resposta de defesa sob a forma de um fluxo de exsudado.

20.1 Introduction

Feeding on tree exudates, such as gums or saps, is common among primates[1]. Primatologists have observed at least 37 primate species eating tree exudates (Coimbra-Filho and Mittermeier 1977; Garber 1984; Nash 1986 and references therein). Additionally, this feeding behavior is widely dispersed among primates with most superfamilies having at least one exudate-eating member. A much smaller subset of exudate-eating primates actively elicits exudate flow by mechanically damaging trees with their anterior teeth. We define this type of biting behavior as tree gouging (Stevenson and Rylands 1988). Among anthropoids, the marmosets are the only habitual tree gougers (Nash 1986).

The development of this unusual feeding behavior in marmosets raises numerous questions related to how and in what context tree gouging evolved, whether there are essential morphological adaptations for gouging and if so, how these morphological features might be integrated during the adaptive evolution of this behavior. To begin to address these questions related to marmosets and gouging, we initially need to think about this behavior in the context of marmoset behavioral ecology. We also need to understand how this behavior works from a mechanical perspective in order to speculate what various morphological components do during gouging. In this chapter, we review data on laboratory experiments simulating tree gouging, field work studying both gouging and the mechanical properties of

[1]The terms gummivory and exudativory have both been used to describe feeding on tree exudates. Nash (1986) discusses differences in these terms. We use them synonymously to include references applying either term as our interest is primarily related to the associated feeding behavior.

trees as well as comparative morphometric analyses to discern possible morphological adaptations for tree gouging in marmosets. Finally, we combine these data with our current understanding of callitrichid phylogeny and evolution to speculate on the evolution of tree gouging in marmosets.

20.2 The Prevalence and Importance of Tree Gouging in Marmoset Behavioral Ecology

Before we examine the functional morphology of tree gouging, we need to ask how this behavior relates to marmoset ecology. Without some indication that gouging plays an important role in the behavioral ecology of marmosets, we cannot advance robust arguments for any derived morphologies being adaptations for this behavior.

We have already noted that eating exudates is common throughout primates. This is also true of callitrichids where most species have been observed to feed on gums opportunistically, seasonally or continually throughout the year (e.g., Izawa 1975; Kinzey et al. 1975; Coimbra-Filho and Mittermeier 1977; Ramirez et al. 1977; Garber 1980, 1984; Rylands 1981, 1984; Maier et al. 1982; Soini 1982, 1987; Lacher et al. 1981, 1984; Sussman and Kinzey 1984; Stevenson and Rylands 1988). The marmosets are the most frequent exudate feeders within the callitrichids. Additionally, field reports state that several marmoset species (including both the common marmoset, *Callithrix jacchus,* and the pygmy marmoset, *Cebuella pygmaea*) habitually gouge holes in trees with their anterior teeth to stimulate exudate flow (e.g., Kinzey et al. 1975; Coimbra-Filho and Mittermeier 1977; Ramirez et al. 1977; Rylands 1981, 1984; Maier et al. 1982; de Faria 1983; Soini 1982; Lacher et al. 1981, 1984; Fonseca and Lacher 1984; Sussman and Kinzey 1984; Stevenson and Rylands 1988; Yepez et al. 2005). Field data also indicate that marmosets spend a significant portion of their daily activity cycle gouging trees, sometimes in excess of 70% of their day, and that exudates comprise a major component of their diet (Ramirez et al. 1977; Coimbra-Filho et al. 1981; Lacher et al. 1981; Maier et al. 1982; Fonseca and Lacher 1984; Sussman and Kinzey 1984; Melo 2001; Yepez et al. 2005) (Fig. 20.1).

Exudates are hypothesized to provide marmosets several ecological benefits, including a dietary source of calcium and other minerals, a rich source of carbohydrates, a seasonally-stable food supply, and a means of "avoiding" interspecific competition by feeding on this under-exploited food source (e.g., Coimbra-Filho and Mittermeier 1977; Maier et al. 1982; Sussman and Kinzey 1984; Nash 1986; Stevenson and Rylands 1988; Garber 1992; Ferrari 1993; Power 1996). Additionally, certain aspects of their behavior including home range size, daily path length, and group sizes may be related to exudate feeding (Maier et al. 1982; Rylands 1984; Ferrari and Ferrari 1989; Garber 1992; Rylands and de Faria 1993; Harrison and Tardif 1994; Kinzey 1997). Collectively, the frequency of exudate feeding, the potential benefits, and the behavioral correlates suggest that exudate eating is important to callitrichid behavioral ecology.

Fig. 20.1 Gouge marks left in a Tambor tree (*Enterolobium contorticilicum*, Leguminosae) by common marmosets at Estação Ecológica do Tapacurá, Pernambuco, Brasil. Trees with similar numbers of gouging marks are common throughout the home ranges of various marmoset groups. Note the individual near the base of the tree for scale

20.3 Determining Masticatory Apparatus Function During Gouging

Several morphological features of callitrichids such as small body size, claw-like nails, robust lower incisors lacking lingual enamel, canine height reduction, reduced condylar height, longer jaw-muscle fibers, and/or expansion of the stomach or large intestines are suggested to be functionally and/or adaptively linked to exudate feeding and gouging in callitrichids (Kinzey et al. 1975; Hershkovitz 1977; Rosenberger 1978; Chivers and Hladik 1980; Coimbra-Filho et al. 1980;

Garber 1980; Martin 1990; Power 1996; Hamrick 1998; Vinyard et al. 2003; Taylor and Vinyard 2004, 2008; Taylor et al. Chap. 19 this volume). Furthermore, multiple researchers hypothesize that tree gouging involves relatively large jaw forces (Szalay and Seligsohn 1977; Rosenberger 1992; Dumont 1997; Spencer 1999) or jaw gapes (Vinyard et al. 2003; Taylor and Vinyard 2004; Taylor et al. Chap. 19 this volume). In most of these cases, there is little in vivo data demonstrating how these specific morphologies function during tree gouging and/or exudate feeding. Given that we cannot routinely determine function in primate skulls from simply observing skull form (e.g., Daegling 1993), we need empirical data demonstrating how the masticatory apparatus functions during gouging before we can make realistic interpretations of morphology in gouging marmosets.

With respect to the masticatory apparatus, both the forces required to indent, fracture, and peel barks during gouging as well as the jaw movements involved in this behavior must be considered from a mechanical perspective. We developed a simulated gouging apparatus (Vinyard and Schmitt 2004) that allows animals to freely gouge while we record the magnitude and timing of forces passing through the jaw to the gouging substrate. We use lateral-view video of jaw movements to capture the time and maximum amount of jaw opening, measured as a maximum linear gape between the upper and lower incisors, during gouging. We can use these data characterizing gouging mechanics in a laboratory setting to explore hypothesized morphological adaptations for tree gouging in marmosets.

To date, we have collected jaw force and movement data during several hundred gouges by four common marmosets (*Callithrix jacchus*) on our gouging apparatus (Vinyard et al. 2001, unpub. data). We only present summary data for the 25 gouges with the largest peak forces from the lower jaws because most adaptive hypotheses focus on the largest forces produced during gouging. The peak resultant force for the lower jaw averaged 28.0 N (SD = 3.1) for these largest gouges. Thus, these marmosets produced peak gouging forces that were up to eight times their body mass. Lower jaw forces lasted, on average, 121 ms (SD = 29) during a gouge. Peak superoinferior (SI) forces of 22.8 N (SD = 3.6) were significantly larger than peak anteroposterior (AP) forces of 11.9 N (SD = 3.5) (Paired *t*-test; $p < 0.001$). SI forces peaked significantly earlier than AP forces (Paired *t*-test; $p = 0.003$).

We can apply these results to build a preliminary, qualitative sketch of how marmosets produce jaw forces during gouging (Fig. 20.2). Marmosets anchor their upper jaws in the gouging substrate by pushing downward and into the substrate. Lower jaw forces begin as the animal pushes with its lower teeth both into and upwardly on the substrate. Figure 20.2 illustrates this as SI forces typically start when the AP force trace is briefly positive. The AP force trace changes direction and becomes negative, signifying the animal pulling away from the substrate with its incisors (Fig. 20.2). SI forces are often, but not always, larger than AP forces showing that the animal pushes harder upwardly as opposed to outwardly on the substrate (Fig. 20.2). Peak vertical (SI) forces usually occur as AP forces are declining. In summary, these marmosets appear to gouge by anchoring their upper jaws, then using their lower jaws to first penetrate the substrate and then "scoop" pieces of the

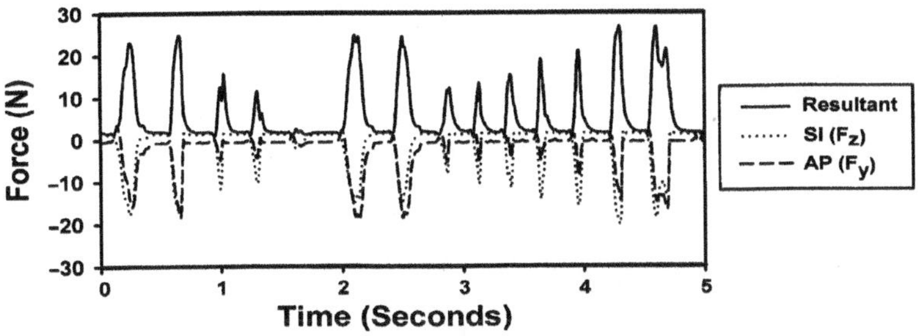

Fig. 20.2 Force traces for a typical gouging sequence of 13 gouges. These traces show the force (N) for the resultant (solid), AP (dashed) and SI (dotted) components. (The ML force was omitted to facilitate observation of the AP and SI forces). We arbitrarily represent the resultant force trace rising above zero and the AP (dashed line) and SI (dotted) forces extending down from zero. An increased negative value for the AP and SI forces represent the marmoset pulling away from the gouging block and pushing upwardly on the block, respectively

substrate by pushing up and pulling out. This final movement helps to propagate cracks during the isolation of a piece of tree substrate. Isolated pieces are often clamped between the teeth and pulled off the block in a subsequent event.

With respect to linear jaw gapes, we also focus on the largest 25 maximum linear gapes during gouging. The average of the largest 25 maximum jaw gapes during a gouge was 23.8 mm (SD=1.1). On average, jaw movements lasted 364 ms (SD=140) during a gouge. One novel finding from this work is that marmosets often use large gapes when gouging (Fig. 20.3). Marmosets gouged with gapes as large as 25.8 mm. These gapes are nearly as large as the maximum passive gape estimate of 29.5 mm, a measure of the structural capacity for jaw opening taken on several wild and laboratory-reared marmosets while they were anesthetized.

20.4 Evaluating the Relative Nature of Gouging Performance

We need comparative datasets to interpret the relative significance of jaw forces and movements during gouging. Towards this end, we compare peak gouging forces to peak vertical bite forces during anterior tooth biting in marmosets to test the hypothesis that gouging involves relatively high forces in the masticatory apparatus. Similarly, we compare maximum gapes during gouging to those during insect chewing to test the hypothesis that gouging involves relatively large gapes.

The comparison of gouging and transducer bite forces clearly shows that these marmosets were not gouging as hard as they could bite (Fig. 20.4a). The largest 25 vertical bite forces averaged 45.3 N (SD=6.4). These bite forces are significantly larger than the average peak force of 28.0 N (SD=3.1) from the 25 largest gouges (Mann-Whitney *U*-test; $p<0.001$) (Fig. 20.4a). This result offers no support for the hypothesis that gouging involves relatively high forces in the masticatory

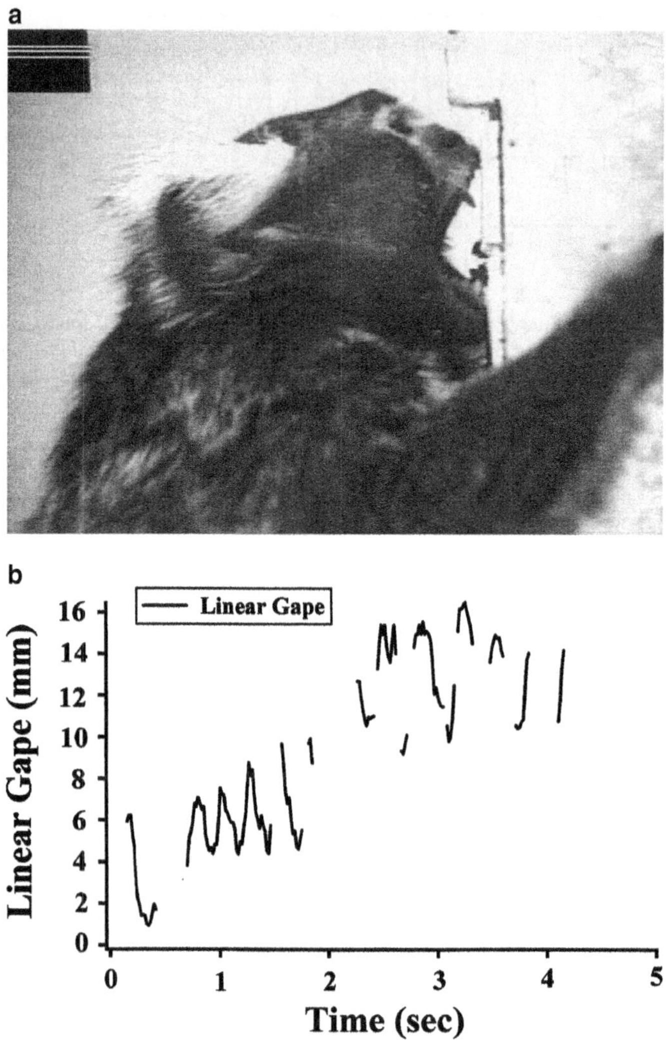

Fig. 20.3 Linear gapes during gouging. (**a**) Marmosets use large gapes when gouging. (**b**) Plot of linear gapes during a gouging sequence. Maximum gapes tend to increase across the sequence. Gape estimates are interrupted because incisor landmarks were not visible in every frame

apparatus (Vinyard et al. 2001). We remain cautious in concluding that marmosets do not generate relatively large forces during gouging because we are uncertain how closely transducer biting mimics behaviors that marmosets routinely perform in their natural habitat.

Comparison of maximum jaw gapes for gouging and cricket chewing shows that marmosets use much larger gapes when gouging (Vinyard et al. 2001) (Fig. 20.4b). The largest 25 maximum chewing gapes averaged 9.6 mm (SD=2.0). The average maximum gape during gouging, 23.8 mm (SD=1.1), was significantly larger (M-W U-test; $p<0.001$). The hypothesis that gouging involves relatively large gapes is

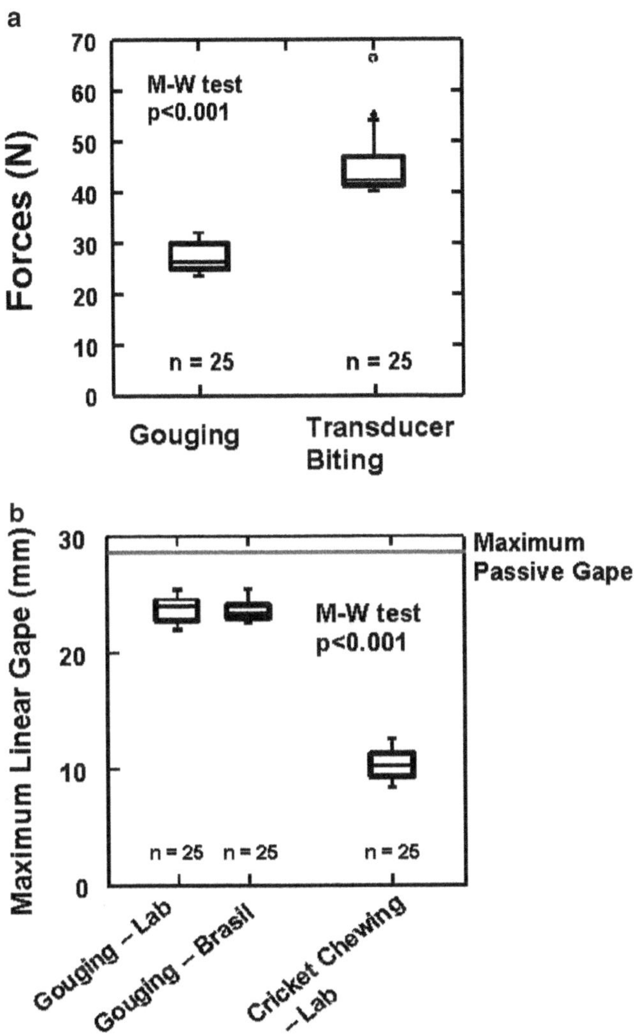

Fig. 20.4 (**a**) Maximum vertical bite forces were significantly larger than peak gouging forces. Transducer biting occurred at a gape of 12–15 mm. This gape is within the range of gapes during gouging. (**b**) Comparison of maximum jaw gapes during gouging in the laboratory, gouging in the field and cricket chewing. Mann-Whitney *U*-tests indicate that the largest maximum gapes during gouging in the laboratory and/or in the field are significantly larger than jaw gapes during cricket chewing. Furthermore, jaw gapes during gouging approach the average maximum passive gape estimate of 29.0 mm for marmosets (i.e., the line near the top of the figure). This result suggests that marmosets are approaching their maximum structural ability for jaw opening during gouging. This maximum passive gape is based on a sample of 32 individuals including an approximately equal number of laboratory-reared and wild individuals

supported by these data. Because the largest gapes during gouging approach the average maximum passive gape of *C. jacchus* (Fig. 20.4b), marmosets appear to be approaching their maximum structural capacity for jaw opening during gouging.

20.5 Linking Laboratory Data to Natural Environments: Giving the Laboratory a Biological Role

One of our key assumptions in determining morphological adaptations for tree gouging is that the laboratory setup accurately replicates the gouging environment marmosets normally experience in the wild. If the laboratory setup differs significantly from field conditions, then our interpretation of gouging performance may have little relevance to marmoset adaptations because of a lack of an appropriate environmental context. In other words, if our laboratory setup does not present a reasonable analog then it is unlikely that the mechanical events we observed during simulated gouging match the biological role of the masticatory apparatus in their environment.

We attempt to address this issue by comparing simulated gouging to tree-gouging behavior in free-ranging common marmosets at two sites in northeast Brazil: 1) Estação Ecológica do Tapacurá and 2) the campus of the Universidade Federal Rural de Pernambuco (Vinyard et al. 2004, unpub. data). Based on our behavioral observations, gouging in the laboratory appears qualitatively similar to marmoset behaviors in their natural environment. Thus, we conclude that we are eliciting a reasonable behavioral analog of tree gouging in the laboratory.

We cannot record jaw forces during gouging in the wild. However, because the properties of the trees that marmosets gouge influence the jaw forces elicited, we compared mechanical properties of trees gouged in the wild to those of laboratory substrates. Specifically, we compared four laboratory substrates, manufactured blocks of pine and red oak as well as fresh-cut pieces of sweetgum (*Liquidambar styraciflua*) and red maple (*Acer rubrum*), to properties of 17 trees gouged by common marmosets at these sites. We assessed the hardness and friction, as they relate to the initial indentation of the gouged substrate, the stiffness and toughness linked to the initial fracture, and the work to peel pieces of substrate from the tree or block. In general, the two fresh laboratory substrates (sweetgum and red maple) always fell within the range of properties from wild trees (Fig. 20.5a). Alternatively, the manufactured pine block often fell slightly outside the natural range while oak blocks were clearly different from naturally-gouged tree properties. Interestingly, laboratory marmosets showed little interest in gouging the oak substrate and typically used very low forces when they did.

We were able to estimate maximum gapes during gouging in the field by measuring the distance between the anchoring site of the upper teeth and the beginning of the lower jaw's scrape during single gouging sequences (Fig. 20.5b). The largest 25 maximum gapes measured in the field averaged 22.9 mm (SD = 1.1) and are significantly larger than those during insect chewing (Fig. 20.4b). Furthermore, these gapes are similar to maximum jaw gapes measured in the laboratory. This similarity validates our jaw movement data during gouging in the laboratory. In summary, the laboratory data (other than that on red oak) appear to provide a reasonable analog for marmoset gouging in the wild.

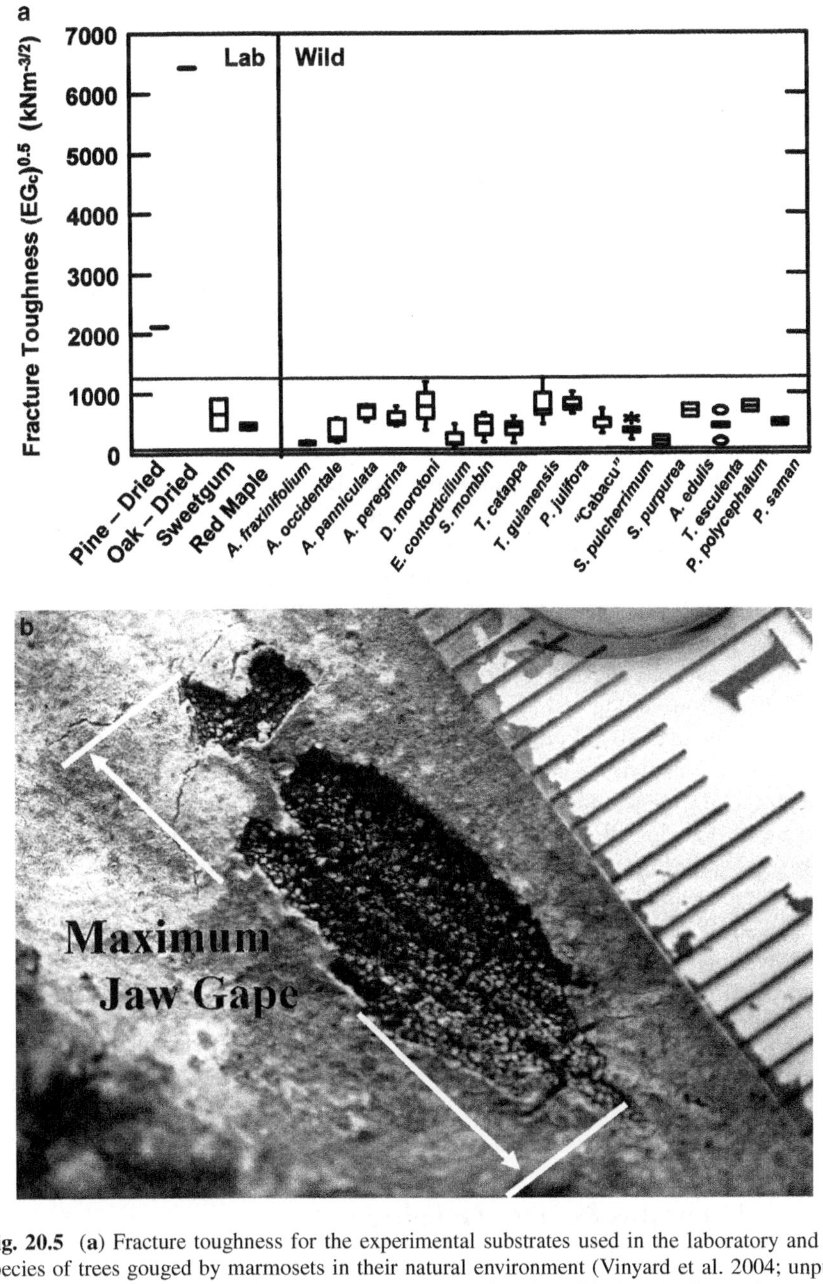

Fig. 20.5 (**a**) Fracture toughness for the experimental substrates used in the laboratory and 17 species of trees gouged by marmosets in their natural environment (Vinyard et al. 2004; unpub. data). Fracture toughness is a material constant measuring the critical stress for a given crack length at which fast fracture begins (Ashby and Jones 1980). We estimate fracture toughness as the square root of the product of stiffness (E) and toughness (G_c). The two manufactured substrates, pine and oak, require significantly more stress per unit crack length to generate fast fracture. Alternatively, the fracture toughness of sweetgum and red maple fall within the range of trees that are gouged in the field suggesting that these laboratory substrates have mechanical properties that mimic those of trees that marmosets gouge in the wild. (**b**) Measurement of maximum jaw gape during a gouge in the field

20.6 Morphology of the Marmoset Masticatory Apparatus and Tree Gouging

The combined laboratory and field data suggest that common marmosets use relatively large jaw gapes but do not necessarily generate relatively large bite forces during gouging. These results offer little support for hypotheses that marmosets generate large jaw forces during gouging and/or have morphological adaptations in the masticatory apparatus for creating or resisting large forces. Recent morphometric analyses comparing masticatory apparatus form in marmosets to nongouging tamarins support this conclusion in showing that marmosets likely do not have increased force generation or load resistance abilities in their bony skull or jaw muscles (Vinyard et al. 2003; Taylor and Vinyard 2004, Vinyard and Ryan 2006, Taylor et al. Chap. 19 this volume). We speculate that modifications of the lower anterior dentition such as increased labiolingual incisor thickness and the loss of lingual enamel help to increase the wedging ability and sharpness of these teeth, respectively, and thereby facilitate removing tree pieces with reduced bite forces.

Alternatively, morphological comparisons of the bony masticatory apparatus and jaw muscles indicate multiple features related to increasing jaw gapes as compared to nongouging tamarins. Specifically, marmoset jaws tend to have anteroposteriorly elongated glenoid articular surfaces and condyles (as a measure of the condylar radius of curvature) along with a condyle positioned closer to the toothrow. Along with a relatively longer mandible, these features all facilitate increased jaw gape (Vinyard et al. 2003). Similarly, the fiber architecture of marmoset jaw-closing muscles facilitate increased stretching and hence larger jaw gapes as compared to nongouging tamarins (Taylor and Vinyard 2004, 2008; Taylor et al. Chap. 19 this volume). We hypothesize that these morphological differences are linked to tree-gouging in marmosets (Vinyard et al. 2003). Furthermore, these modifications may provide functional and/or adaptive advantages related to an optimal alignment of the incisal cutting edge during gouging, increased jaw excursion during a gouge and/or increased bite force production at very large gapes when these jaw muscles are highly stretched.

20.7 Evolution of the Marmoset Masticatory Apparatus for Tree Gouging

It is fairly clear that features, such as claw-like nails and small body size, shared by gouging and nongouging callitrichids are unlikely to be adaptations specifically for tree gouging. It is possible that these morphological changes and their concomitant behavioral shifts acted as preadaptations for tree gouging in marmosets (Garber et al. 1996). Based on our in vivo and field data, we speculate that two components of the masticatory apparatus have adapted to this gouging behavior. The first component, which many researchers have recognized for some time, involves the anterior lower dentition. Morphological changes in the lower teeth that align the canine and incisal occlusal surfaces at the same level, increased the labiolingual thickness of the incisors and hypertrophied the labial enamel while reducing the lingual enamel (creating a sharp dentoenamel wear gradient), all appear functionally related to tree gouging (Rosenberger 1978). We speculate that these changes increased the mechanical efficiency of gouging and reduced the amount of force required to remove a given piece of substrate. The second component involves changes in the masticatory muscles and temporomandibular joint (TMJ) related to increasing jaw gapes. Changes in muscle and TMJ position theoretically reduce stretching of the jaw muscles when opening the jaw widely. These changes facilitate both increased gape and the ability to generate greater bite forces at these larger gapes (Vinyard et al. 2003). Additional changes in the TMJ facilitate joint rotation and translation that increase the ability to open the jaw widely (Vinyard et al. 2003). Finally, longer jaw-muscle fibers provide increased muscle stretching capacity for wide jaw opening (Taylor and Vinyard 2004, Taylor et al. Chap. 19 this volume). These observations linking tree gouging to morphological changes for increased jaw movements offer one of the first observed instances where jaw movements, rather than bite forces, appear to be the target of evolutionary change in a primate biting behavior.

Acknowledgements We thank Susan Ford, Lesa Davis and Leila Porter for inviting us to contribute to this volume and participate in the symposium "Advances in Marmoset and Goeldi's Monkey (*Callimico*) Research: Anatomy, Behavioral Ecology, Phylogeny, and Conservation" held at the American Association of Physical Anthropologists Meetings on April 8, 2005. We also thank B. Darvell, W. Horne, P. Lemelin and P. Vinyard for assistance and advice. This research was supported by grants from NSF (BCS-0094666), (BCS-0412153), (SBR-9701425), (SBR-940764), (BCS-99-04401), the L.S.B. Leakey Foundation, and the National Geographic Society.

References

Ashby MF, Jones DRH (1980) Engineering materials 1: An introduction to their properties and applications. Pergamon Press, New York

Chivers DJ, Hladik CM (1980) Morphology of the gastrointestinal tract in primates: comparisons with other mammals in relation to diet. J Morphol 166:337–386.

Coimbra-Filho AF, Mittermeier RA (1977) Tree-gouging, exudate-eating and the "short-tusked" condition in *Callithrix* and *Cebuella*. In: Kleiman DG (ed) The biology and conservation of the Callitrichidae. Smithsonian Institution Press, Washington, DC, pp 105–115

Coimbra-Filho AF, Da Cruz Rocha N, Pissinatti A (1980) Morfofisiologia do ceco e sua correlacao com o tipo odontologico em callitrichidae (Platyrrhini, Primates). Rev Brasil Biol 40:177–185

Coimbra-Filho AF, Mittermeier RA, Constable ID (1981) *Callithrix flaviceps* (Thomas, 1903) recorded from Minas Geraos. Brazil; (Callitrichidae, Primates). Rev Brasil Biol 40:177–185

Daegling DJ (1993) The relationship of in vivo bone strain to mandibular corpus morphology in (*Macaca fascicularis*). J Hum Evol 25:247–269

de Faria DS (1983) Uso de arvores gomiferas do cerrado por *Callithrix jacchus penicillata*. In: Mello MT (ed) A primatologica no Brasil. Sociedade Brasileira de Primatologica, Belo Horizonte, pp 83–96

Dumont ER (1997) Cranial shape in fruit, nectar, and exudate feeders: implications for interpreting the fossil record. Am J Phys Anthropol 102:187–202

Ferrari SF (1993) Ecological differentiation in the Callitrichidae. In: Rylands AB (ed) Marmosets and tamarins: systematics, behaviour, and ecology. Oxford Univ Press, Oxford, pp 314–328

Ferrari SF, Ferrari MA (1989) A re-evaluation of the social organisation of the Callitrichidae, with reference to the ecological differences between genera. Folia Primatol 52:132–147

Fonseca GAB, Lacher TE (1984) Exudate-feeding by *Callithrix jacchus penicillata* in the semideciduous woodland (Cerradao) in Central Brazil. Primates 25:441–450

Garber PA (1980) Locomotor behavior and feeding ecology of the Panamanian tamarin (*Saguinus oedipus geoffroyi*, Callitrichidae). Int J Primatol 1:185–201

Garber PA (1984) Proposed nutritional importance of plant exudates in the diet of the Panamanian tamarin, *Saguinus oedipus geoffroyi*. Int J Primatol 5:1–15

Garber PA (1992) Vertical clinging, small body size, and the evolution of feeding adaptations in the Callitrichinae. Am J Phys Anthropol 88:469–482

Garber PA, Rosenberger AL, Norconk MA (1996) Marmoset misconceptions. In: Norconk MA, Rosenberger AL, Garber PA (eds) Adaptive radiation of neotropical primates. Plenum Press, New York, pp 87–95

Hamrick MW (1998) Functional and adaptive significance of primate pads and claws: Evidence from New World anthropoids. Am J Phys Anthropol 106:113–127

Harrison ML, Tardif SD (1994) Social implications of gummivory in marmosets. Am J Phys Anthropol 95:399–408

Hershkovitz P (1977) Living New World primates (Platyrrhini), with an introduction to primates, vol 1. University Chicago Press, Chicago

Izawa K (1975) Foods and feeding behavior of monkeys in the upper Amazon basin. Primates 16:295–316

Kinzey WG (1997) Callithrix. In: Kinzey WG (ed) New World primates: Ecology, evolution, and behavior. Aldine de Gruyter, New York, pp 230–239

Kinzey WG, Rosenberger AL, Ramirez M (1975) Vertical clinging and leaping in a neotropical anthropoid. Nature 255:327–328

Lacher TE, Fonseca GAB, Alves C, Magalhaes-Castro B (1981) Exudate-eating, scent-marking, and territoriality in wild populations of marmosets. Anim Behav 29:306–307

Lacher TE, Fonseca GAB, Alves C, Magalhaes-Castro B (1984) Parasitism of trees by marmosets in a central Brazilian gallery forest. Biotropica 16:202–209

Maier W, Alonso C, Langguth A (1982) Field observations of *Callithrix jacchus jacchus*. Zeits fur Saugetier 47:334–346

Martin RD (1990) Primate origins and evolution: A phylogenetic reconstruction. Princeton University Press, Princeton

Melo LCO (2001) Seleção de recursos alimentares por *Callithrix jacchus* – Sagui-do-Nordeste: Um foco sobre a teoria de Otimização. Masters Thesis, Univ Federal de Pernambuco

Nash LT (1986) Dietary, behavioral, and morphological aspects of gumnivory in primates. Yrbk Phys Anthropol 29:113–137

Power ML (1996) The other side of callitrichine gummivory. In: Norconk MA, Rosenberger AL, Garber PA (eds) Adaptive radiation of neotropical primates. Plenum, New York, pp 97–110

Ramirez MF, Freese CH, Revilla J (1977) Feeding ecology of the pygmy marmoset, *Cebuella pygmaea*, in Northeastern Peru. In: Kleiman DG (ed) The biology and conservation of the Callitrichidae. Smithsonian Institution Press, Washington, DC, pp 91–104

Rosenberger AL (1978) Loss of incisor enamel in marmosets. J Mamm 59:207–208

Rosenberger AL (1992) Evolution of feeding niches in New World monkeys. Am J Phys Anthropol 88:525–562

Rylands AB (1981) Preliminary field observations on the marmoset, *Callithrix humeralifer intermedius* (Hershkovitz, 1977) at Dardanelos, Rio Aripuana, Mato Grosso. Primates 22:46–59

Rylands AB (1984) Exudate-eating and tree-gouging by marmosets (Callitrichidae, Primates). In: Chadwick AC, Sutton SL (eds) Tropical rain forest: The leeds symposium. Leeds Philosophical and Literary Society, Leeds, pp 155–168

Rylands AB, de Faria DS (1993) Habitats, feeding ecology, and home range size in the genus *Callithrix*. In: Rylands AB (ed) Marmosets and tamarins: Systematics, behaviour, and ecology. Oxford University Press, Oxford, pp 262–272

Soini P (1982) Ecology and population dynamics of the pygmy marmoset, *Cebuella pygmaea*. Folia Primatol 39:1–21

Soini P (1987) Ecology of the saddle-back tamarin *Saguinus fuscicollis illigeri* on the Rio Pacaya, northeastern Peru. Folia Primatol 49:11–32

Spencer MA (1999) Constraints on masticatory system evolution in anthropoid primates. Am J Phys Anthropol 108:483–506

Stevenson MF, Rylands AB (1988) The marmosets, Genus *Callithrix*. In: Mittermeier RA, Rylands AB, Coimbra-Filho AF, Fonseca GAB (eds) Ecology and behavior of neotropical primates. World Wildlife Fund, Washington, DC, pp 131–222

Sussman RW, Kinzey WG (1984) The ecological role of the Callitrichidae: A review. Am J Phys Anthropol 64:419–449

Szalay FS, Seligsohn D (1977) Why did the strepsirhine tooth comb evolve? Folia Primatol 27:75–82

Taylor AB, Vinyard CJ (2004) Comparative analysis of masseter fiber architecture in tree-gouging (*Callithrix jacchus*) and nongouging (*Saguinus oedipus*) callitrichids. J Morphol 261:276–285

Taylor AB, Vinyard CJ (2008) The relationship between jaw-muscle fiber architecture and feeding behavior in primates. Tree-gouging and nongouging gummivorous callitrichids as a natural experiment. In: Vinyard CJ, Ravosa MJ. Wall CE (eds) Primate Craniofacial Function and Biology, p. 241–264. New York: Springer

Taylor AB, Eng EM, Anapol FC, Vinyard CJ (this volume) The functional significance of jaw-muscle fiber architecture in tree-gouging marmosets. In: Ford SM, Porter LM, Davis LC (eds) The smallest anthropoids: The marmoset/callimico radiation. Springer Press, New York, pp 381–394

Vinyard CJ, Ryan TM (2006) Cross-sectional bone distribution in the mandibles of gouging and non-gouging platyrrhines. Int J Primatol 27:1461–1490

Vinyard CJ, Schmitt D (2004) A new technique for studying reaction forces during primate behaviors on vertical substrates. Am J Phys Anthropol 125:343–351

Vinyard CJ, Wall CE, Williams SH, Schmitt D, Hylander WL (2001) A preliminary report on the jaw mechanics during tree gouging in common marmosets (Callithrix jacchus). In: Brooks A (ed) Dental Morphology 2001: Proceedings of the 12th international symposium on dental morphology. Sheffield Academic Press, Ltd., Sheffield UK, pp 283–297

Vinyard CJ, Wall CE, Williams SH, Hylander WL (2003) A comparative functional analysis of the skull morphology of tree gouging primates. Am J Phys Anthropol 120:153–170

Vinyard CJ, Lucas PW, Valença-Montenegro MM, Melo LCO, Valle YM, Monteiro da Cruz MAO (2004) Where the wild things are: Linking lab and field work in studying tree gouging in common marmosets (*Callithrix jacchus*). Am J Phys Anthropol 38(Suppl):200–201

Yepez P, de la Torre S, Snowdon CT (2005) Interpopulation differences in exudate feeding of pygmy marmosets in Ecuadorian Amazonia. Am J Primatol 66:145–158

Chapter 21
Marmoset Postcrania and the Skeleton of the Dwarf Marmoset, *Callibella humilis*

Susan M. Ford and Lesa C. Davis

Abstract Four genera of marmosets are now recognized; these are *Callithrix, Mico, Cebuella,* and *Callibella*, the dwarf marmoset, first identified in 1998. These genera are distinctive in their pelage, DNA, and cranial anatomy, but there has been no comparative study of any aspect of their postcranium. This study examines post-cranial differences between the marmoset genera and their potential significance for understanding their phylogenetic relationships and their variation in positional behavior. It also provides the first description of the postcranium of *Callibella*.

Univariate and discriminate function analysis of the only known skeleton of *Callibella* was carried out in relation to members of all three other genera of marmosets, *Callimico goeldii*, and *Saguinus midas.* Marmosets form a unique clade morphologically. Each marmoset genus forms a discrete and highly distinctive group. While sharing general marmoset traits, *Callibella* shares no special affinity with the least specialized Atlantic marmosets of genus *Callithrix*. Amazonian *Mico* is characterized by narrow elbow articulations. *Callibella* lacks most of *Cebuella's* many highly derived hindlimb traits, sharing only a few, such as a narrow, short femoral head and narrow lateral knee articulation. *Callibella* exhibits several unique features of the forelimb and hindlimb; these serve to emphasize its distinctive form among marmosets.

These results support a three-way division of the Amazonian marmosets into separate *Mico, Callibella,* and *Cebuella* lineages, subsequent to their separation from the Atlantic *Callithrix* group. The marked differences in their skeletons suggest that extreme small size evolved independently in the pygmy and dwarf marmosets, associated with different postcranial adaptations, an interpretation that is in agreement with results of prior analysis of cranial anatomy. *Callibella* appears to be adapted for more quadrupedalism and to use more horizontal and low-angled supports than is typical of either *Mico* or, especially, *Cebuella.* Little is currently know of the positional behavior of most marmosets in the wild. As *Callibella*, in particular, lives in a very limited range characterized by a unique and at-risk forest, their study and preservation should be of high priority.

S.M. Ford (✉)
Department of Anthropology, Southern Illinois University, Carbondale, IL, 62901-4502, USA
e-mail: sford@siu.edu

S.M. Ford et al. (eds.), *The Smallest Anthropoids*, Developments in Primatology: Progress and Prospects, DOI 10.1007/978-1-4419-0293-1_21,
© Springer Science+Business Media, LLC 2009

Resumen Cuatro géneros de marmosets son reconocidos ahora los cuales incluyen el *Callithrix*, *Mico*, *Cebuella*, y el *Callibella*, el marmoset enano, identificado por primera vez en 1998. Estos géneros se muestran distintos en pelaje, ADN, y anatomía cranial; sin embargo, no han habido estudios comparativos de ningún aspecto de sus poscranium. El presente estudio examina las diferencias poscranium entre los géneros marmosets y su potencial significado para la filogénia y variación en comportamiento posicional. Asimismo, proporciona la primera descripción poscranium de *Callibella*.

El análisis funcional univariado y discriminado del único esqueleto conocido de *Callibella* fue llevado a cabo en relación a los miembros de todos los tres géneros de marmosets, *Callimico goeldii*, y *Saguinus midas*. Los marmosets forman una clade única morfológicamente. Cada género marmoset forma un grupo bastante discreto y distintivo, con *Callibella* cayendo en medio pero único entre todos. A pesar de compartir la mayoría de los rasgos marmosets, *Callibella* no comparte especial afinidad con los menos especializados marmosets del Atlántico del género *Callithrix*. El *Mico* Amazónico está caracterizado por articulaciones de codo estrechas. El *Callibella* carece de la mayoría de los rasgos de los miembros posteriores, compartiendo solamente algunos, tal como la cabeza femoral estrecha y la articulación de la rodilla lateral estrecha. Los *Callibella* muestran varios rasgos únicos en los miembros delanteros y traseros; ellos sirven para enfatizar su distinción entre los marmosets.

Los resultados apoyan las sugerencias de la división de tres formas del marmoset ancestro en los linajes separados *Mico*, *Callibella*, y *Cebuella*, relacionados a este grupo separados de el *Callithrix* del Atlántico. Las marcadas diferencias en sus esqueletos propone que el extremo tamaño pequeño evolucionó independientemente en los marmosets pigmeos y enanos, asociados con diferentes adaptaciones post-cranianas, todo ello en acuerdo con los resultados de previos análisis de anatomía cranial. El *Callibella* podria ser mayormente cuadrúpedo y utilizar mayores soportes horizontales y de ángulo bajo que los típicos Mico o, especialmente *Cebuella*. Poco se conoce en el presente sobre el comportamiento posicional de la mayoría de los marmosets en lo salvaje. *Callibella* en particular, vive en una extensión territorial verdaderamente limitada caracterizada por un bosque único y en riesgo; su estudio y preservación tiene que ser altamente prioritario.

Resumo Atualmente quatro gêneros de sagüis são reconhecidos; estes são *Callithrix, Mico, Cebuella*, e *Callibella*, o sagüi anão, primeiro identificado em 1998. Estes gêneros são distintos em sua pelagem, ADN, e anatomia craniana, mas não existem estudos comparativos de qualquer aspecto do pós-crânio. Estes estudo examina as diferenças no pós-crânio entre estes gêneros de sagüis e seu potencial significado para entendermos seu relacionamento filogenético e sua variação em comportamento postural. Também mostra a primeira descrição do pós-crânio de *Callibella*.

Análises univariadas e discriminantes do único esqueleto conhecido de *Callibella* foram conduzidas comparando este com exemplares de todos os três outro gêneros de sagüis, *Callimico goeldii*, e *Saguinus midas*. Sagüis compõe um clado único morfológicamente. Cada gênero de sagüis forma um grupo discreto e bastante distinto. Embora compartilhando as características gerais dos sagüis, *Callibella* não compartilha nenhuma afinidade especial com os sagüis Atlânticos menos especializados do gênero *Callithrix*. O gênero *Mico* amazônico é caracterizado por articulações do ombro estreitas. *Callibella* exibe várias características únicas dos membros anteriores e posteriores; isto serve para enfatizar sua forma distinta entre os sagüis.

Estes resultados suportam a divisão em três partes dos sagüis amazônicos nas linhagens saparadas *Mico, Callibella* e *Cebuella*, subsequente à sua separação do grupo Atlântico *Callithrix*. As diferenças marcadas nos seus esqueletos sugere que o tamanho pequeno extremo evoluiu independentemente nos sagüis-leãozinho e anão, associado com adaptações pós-cranianas diferentes, uma interpretação que está de acordo com os resultados da análise anterior da anatomia craniana. *Callibella* parece ser mais adaptado para o quadrupedalismo e ao uso mais horizontal e de suportes em ângulo-baixo do que é típico seja de *Mico* ou, especialmente, *Cebuella*. Pouco é conhecido do comportamento postural da maioria dos sagüis na natureza. Como *Callibella*, em particular, vive em uma distribuição bastante limitada caracterizada por um floresta única e em risco, seu estudo e preservação deve ter uma prioridade alta.

21.1 Introduction

Marmosets are a highly speciose group comprising the smallest living anthropoids. They are found in the Atlantic Coastal Forest of Brazil and large portions of Amazonia south of the Rios Amazonas/Japurá. One of their most distinctive attributes, in addition to their diminutive size and general callitrichid features, is their reliance on exudates as a food source and their associated, specialized anterior dentition (see Coimbra-Filho and Mittermeier 1976, 1978; Rosenberger 1984; Natori 1986; Rylands and de Faria 1987, 1993; Natori and Shigehara 1992; Garber 1992; Ferrari 1993). Very little is known of their skeletal differences, which is the subject of this study.

de Vivo (1991; see also Rylands and de Faria 1993), building on earlier pelage (Hershkovitz 1977) and dental work (Natori 1986), recognized two separate species groups of marmosets of the genus *Callithrix*, the jacchus-group including all species from coastal Brazil and the argentata-group from Amazonia. This separation was further supported by karyotypic (Nagamachi et al. 1992, 1999; Canavez et al. 1996) and molecular (Porter et al. 1997; Tagliaro et al. 1997, 2000; Canavez et al. 1999) data, which also suggested closer phylogenetic ties of the pygmy marmosets (*Cebuella pygmaea*) of western Amazonia to the argentata-group than to the jacchus-group. Groves (2001) followed the work of these geneticists in reducing *Cebuella* to subgenus level, as a third part of a tripartite marmoset radiation, with all three as subgenera of the genus *Callithrix*. Rylands et al. (2000, Chap. 2 this volume) have now placed all three groups of marmosets at the genus level, with *Cebuella* (the pygmy marmoset), *Mico* (the argentata-group), and *Callithrix* (the jacchus-group from coastal Brazil). Both *Callithrix* and, especially, *Mico* are viewed as highly speciose, and many new species of what is now *Mico* have been discovered and named within the last decade (Ferrari and Lopes 1992; Mittermeier et al. 1992; Alperin 1993; Silva and de Noronha, 1998; van Roosmalen et al. 1998, 2000; Ferrari et al. 1999; Sena et al. 2002). Most recently, one of these new species, a strikingly distinctive and diminutive form originally named *Callithrix humilis* (van Roosmalen et al. 1998), was raised to genus level as *Callibella humilis*, the dwarf marmoset (van Roosmalen and van Roosmalen 2003; see Aguiar and Lacher Chap. 18 this volume). Thus, at present, there are 6 recognized species of *Callithrix*, 14 of *Mico*, and the monotypic *Callibella* and *Cebuella* (Rylands et al. Chap. 2 this volume; see Table 21.1, Fig. 21.1).

The differences between the genera of marmosets are still being explored, since their recognition as distinct taxa is so recent. Although hybrid twins of *Callithrix jacchus* and *Cebuella pygmaea* have been reported in captivity (born 1998; Anzenberger et al. 2001; Neusser et al. 2005) with an intermediate chromosome number, their fertility is not yet tested nor are the possible implications for separate taxonomic status of fertile hybrids in captivity certain. The four named marmoset

Table 21.1 Marmoset species, outgroups, and sample sizes (genus totals in bold)

Genus	Species	N Male	N Female	N Unknown	N Total
Callithrix		**9**	**11**	**2**	**22**
	aurita[a]			1	1
	flaviceps				
	geoffroyi[a]	1	1	1	3
	jacchus[a]	6	6		12
	kuhlii				
	penicillata[a]	2	4		6
Mico		**5**	**4**	**3**	**12**
	acariensis				
	argentatus[a]		3	1	4
	chrysoleucus				
	emiliae				
	cf. *emiliae* (Rondonia)				
	humeralifer[a]			1	1
	intermedius				
	leucippe[a]	1			1
	manicorensis	1			1
	marcai				
	mauesi				
	melanurus[a]	3	1	1	5
	nigriceps				
	saterei				
Callibella	*humilis*	**1**			**1**
Cebuella	*pygmaea*[a]	**6**	**7**	**2**	**15**
(outgroups):					
Callimico	*goeldii*[a]	**6**	**6**	**2**	**14**
Saguinus	*midas*[a]	**3**	**5**	**4**	**12**

[a]Indicates species included in this study

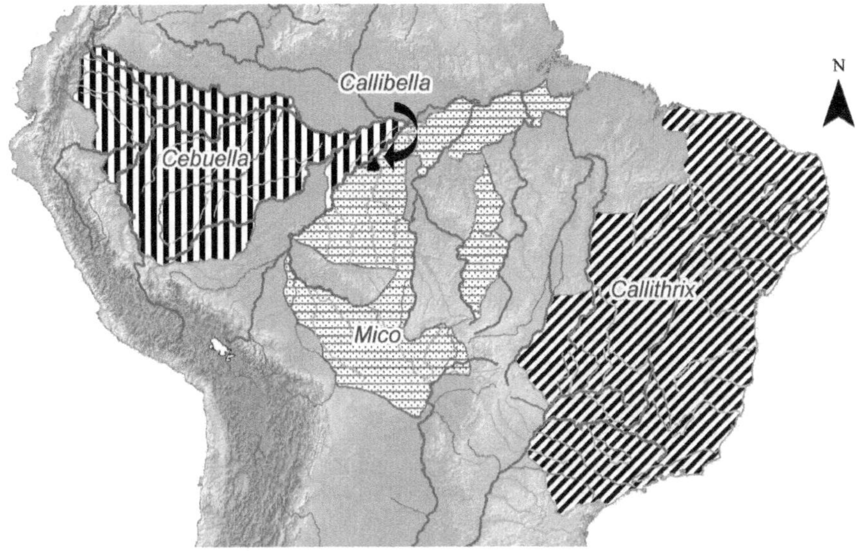

Fig. 21.1 Map of distributions of genera of marmosets. Shaded areas represent taxa included in this study. (adapted from van Roosmalen et al. 2000; van Roosmalen and van Roosmalen 2003)

genera maintain distinct genetic and chromosomal patterns in the wild (Rylands et al. 2000; Neusser et al. 2001; Cortés-Ortiz Chap. 1 this volume).

The dwarf marmoset, *Callibella humilis*, is found only in a very small area of Brazil south of the Amazon within the range of another recently described marmoset species, *Mico manicorensis* (the Manicoré marmoset, van Roosmalen et al. 2000). *Callibella* was described by van Roosmalen and van Roosmalen (2003) as distinctive in the following ways. First, it is morphologically intermediate in size between pygmy marmosets and other marmosets, with long hair tufts from the center of the ear pinna (unknown in marmosets). Second, it has a unique ontogeny, with three distinct stages of pelage color and length (see also van Roosmalen et al. 1998). Third, it has a distinctive mandibular structure (see Aguiar and Lacher 2003, Chap. 18 this volume). The van Roosmalen et al. original analysis of DNA, based on 902 bp mDNA control region, showed that *Mico* is the most genetically similar to *Cebuella* (11–12% difference), followed by *Callibella* (12–13%: *Callibella* and *Cebuella* have a 13–14% difference), and these Amazonian marmosets are least like *Callithrix* (12–15% divergence from others). Their study also showed a 3–7% divergence between species of *Mico*.

The van Roosmalens characterized *Callibella* as having distinctive behavior and ecology as well, including at least two examples of singleton births (although twin births were later seen – T. van Roosmalen, personal communication), an apparent lack of territoriality (seen in groups of up to 30 individuals), more than one reproductively active female in a group, and apparent lack of female reproductive inhibition by other females. Only biological mothers were seen carrying infants, with mothers' "parking" infants on a twig while they fed. Parking began when infants were only five days old (at least in captivity; in this case the infant stayed silent and immobile). As Aguiar and Lacher (Chap. 18 this volume) note, a number of these behaviors are shared with *Cebuella*. The van Roosmalens also reported a distinct vocal repertoire and extreme commensalism with humans and plants growing on human-altered soils (*terra pretas*). However, little comparative and supportive data have been provided on the degree of distinctiveness and affinities of *Callibella*, beyond the comparative cranial studies of Aguiar and Lacher (2003, Chap. 18 this volume).

At present, pelage, craniodental, vocalization, and chromosomal differences have been found among the genera of marmosets, supporting their distinctness seen in DNA analyses. These differences, and the studies noting them, are summarized in Table 21.2. However, given the very recent recognition of *Mico* and *Callibella* as separate genera of platyrrhine monkeys and the continuing discovery of new species belonging to the *Mico* group, our understanding of the differences between these marmoset genera, the history of their radiation, and the nature of their unique adaptations are still rudimentary. van Roosmalen et al. (2000, van Roosmalen and van Roosmalen 2003) have taken a first, and important, step in trying to elucidate some of these patterns. Recently, Sena et al. (2002) have suggested the link between *Cebuella* and *Mico* may not be as strong as previously argued, based on mitochondrial DNA (see review in Cortés-Ortiz Chap. 1 this volume).

We had the opportunity to clean and describe the first known skeleton of an individual dwarf marmoset, the holotype specimen, MPEG 24769, which was a wild-raised adult male, as well as the only known skeleton of *Mico manicorensis* (INPA 2512); we are grateful to Marc van Roosmalen for the opportunity to study these specimens. This study is an exploration of postcranial differences between the marmoset genera *Callithrix, Mico, Callibella,* and *Cebuella* and their potential significance for phylogeny and variation in positional behavior.

21.2 Materials and Methods

Skeletons of marmosets in general are rare, and there are none for most Amazonian *Mico* taxa. We took a total of 160 measurements on 48 skeletons of marmosets plus 26 skeletons of outgroup taxa. The measurements and their abbreviations are listed in Appendix 1. The marmoset specimens included 15 *Cebuella pygmaea*, 1 *Callibella humilis*, 12 *Mico* (sampling five species), and 22 *Callithrix* (four species); sample sizes are given in Table 21.1, and individual specimens are listed in Appendix 2. In addition, two outgroup taxa were analyzed. Twelve skeletons of the red-handed tamarin *Saguinus midas* were included because it is postcranially one of the most unspecialized of the tamarin radiation (Davis 2002). Finally, fourteen skeletons of *Callimico goeldii* (callimicos) were included. Despite strong morphological (Hill 1959; Rosenberger 1977; Ford 1986; Kay 1990; Rosenberger et al. 1990), ecological, and behavioral (Snowdon 1993; Garber 1994) evidence placing *Callimico* at the base of the entire callitrichid radiation, recent DNA studies (Schneider et al. 1993, 1996; Barroso 1995; Harada et al. 1995; Horovitz and

Table 21.2 Known differences between marmosets

Feature	*Callithrix*	*Mico*	*Cebuella*	*Callibella*	Source
Pelage general	Mottled or speckled, black with other color; Dorsal hairs banded (agouti)	All gray, white, or black with white flecks; Dorsal hairs not banded (not agouti)	Gray-brown; Dorsal hairs banded (agouti)	Dark olive brown above, orangish-to-grayish-yellow below; Dorsal hairs not banded (not agouti); Pelage changes with age	Coimbra-Filho and Mittermeier (1973), Hershkovitz (1977), Groves (2001), van Roosmalen and van Roosmalen (2003)
Pelage, hips	Hips/flanks undifferentiated	White hip patches, may extend to thigh	Hips/flanks undifferentiated	?	Hershkovitz (1977)
Pelage, face	White blaze; muzzle hairy	No blaze; muzzle naked	White moustache or dots by nostrils	Black triangular crown, white eyebrows, naked face	Hershkovitz (1977), van Roosmalen and van Roosmalen (2003)
Ears/Tufts	Ears small; Tufts marked, from around, in front, or inner ear	Ears large, notched; Bare or fringe from both aspects	No tufts; ears hidden by long cheek hairs	No tufts; naked ears with "streaks" of white hairs from central pinna	Hershkovitz (1977), Groves (2001), van Roosmalen and van Roosmalen (2003)
Tail	Strongly ringed gray and black	No true rings, uniform or ring-like from "lie of hair" (Groves, pg. 131)	Ringed, agouti	Not ringed or "obscurely orangish" ringed; nearly black, longer than *Cebuella*	Hershkovitz (1977), Groves (2001), van Roosmalen and van Roosmalen (2003)
Cranium	Small zygomatic in orbital margin	Large zygomatic in orbital margin	Intermediate zygomatic in orbital margin	?	Natori (1994)
Mandible	Sturdy;	Strong;	Gracile;	Less gracile;	Aguiar and Lacher (2003), this volume
	Condyle/coronoid nearly level;		Low condyle, sharply angled coronoid;	Low condyle, intermediate coronoid;	
	Strongly recurved inf. margin of horiz. ramus;	Straight, less arcuate horis. ramus;			
	Deep angular lobe	Angular lobe extends minimally below gnathion	Deep angular lobe, but very gracile	Very deep angular lobe	

(continued)

Table 21.2 (continued)

Feature	*Callithrix*	*Mico*	*Cebuella*	*Callibella*	Source
Dentition	Lower I_1 slightly smaller than I_2;	Lower I_1 much smaller than I_2;	Lower I_1 slightly smaller than I_2;	Lower I_1 smaller than I_2 (esp. breadth);	Natori (1986), Hershkovitz (1977), van Roosmalen and van Roosmalen (2003)
	Upper P^3 approx. square or wider than long;	Upper P^3 longer than wide;	Upper P^3 approx. square;	Upper P^3 ?;	
	Upper M^1 strong buccal flare;	Upper M^1 weak buccal flare, steep buccal slope;	Upper M^1 intermediate buccal flare/slope;	Upper M^1 ?	
	Lower C and I's closer in height	Lower C taller relative to I's	?	?	
Karyotype	$2n=46$	$2n=44$	$2n=44$	?	Hershkovitz (1977), Nagamachi et al. (1992, 1997), Canavez et al. (1996)
Vocalization	No trill reported	Trill with tongue in intragroup communication;	Trill with tongue in intragroup communication;	?	Snowdon, (1993), van Roosmalen and van Roosmalen (2003)
		Long calls with 2–3 notes/call of moderate note duration, lack of frequency;	Long calls with 2–3 notes/call of moderate note duration, lack of frequency;	Distinct double-tone contact or "distant" [=long] call	
		Caudal threat display	Caudal threat display		
Diversity	Moderately speciose; degree of hybridization debated	Highly speciose	Monotypic	Monotypic	Rylands et al. (2000), Groves (2001), van Roosmalen and van Roosmalen (2003)

Body size	250–430 g	??–340 g	110–130 g	150–185 g	Ford and Davis (1992), Groves (2001), van Roosmalen and van Roosmalen (2003)
DNA: Epsilon-globin gene	Distinct sequence	Distinct sequence, most like *Cebuella*	Distinct sequence, most like *Mico*	?	Porter et al. (1997)
mtDNA: D-loop control region	Distinct sequence	Distinct sequence, most like *Cebuella*	Distinct sequence, most like *Mico*	Distinct sequence, most like the *Cebuella/Mico* group (not closer to either)	Tagliaro et al. (1997), van Roosmalen et al. (2000), van Roosmalen and van Roosmalen (2003)
DNA: β_2-microglobulin	Distinct sequence	Distinct sequence, most like *Cebuella*	Distinct sequence, most like *Mico*	?	Canavez et al. (1999), van Roosmalen et al. (2000)
mtDNA: COII gene	Distinct sequence, most like *Mico*	Distinct sequence, most like *Callithrix*	Distinct sequence	?	Sena et al. (2002)
Genitalia	Male and female hypertrophied	Male and female hypertrophied	Male and female hypertrophied	Male and female NOT hypertrophied	van Roosmalen and van Roosmalen (2003)
Twinning	Most common	Most common	Most common	Singletons common? (but pers. comm. Tomas van Roosmalen indicates twins also)	van Roosmalen and van Roosmalen (2003)
Territoriality/group size	Territorial, small groups	Territorial, small groups	Territorial, small groups	Not territorial; large groups (up to 30+)	van Roosmalen and van Roosmalen (2003)
Exudativory	High levels	Lower levels	High levels	High levels	van Roosmalen and van Roosmalen (2003)
Infant parking	Not reported	Not reported	Yes	Yes	van Roosmalen and van Roosmalen (2003)
Paternal/other member infant carrying	Common	Common	Common	Not seen; only biological mothers observed to carry	van Roosmalen and van Roosmalen (2003)

Meyer 1995; Canavez et al. 1996, 1999; Porter et al. 1997; Horovitz et al. 1998; Pastorini et al. 1998; von Dornum and Ruvolo 1999; Singer et al. 2003; see review in Cortés-Ortiz, Chap. 1 this volume) support earlier protein work (Cronin and Sarich 1975, 1978) in placing *Callimico goeldii* as the closest relative of the marmosets.

All measurements were taken to the nearest 0.01 mm. Measurements were taken by both authors, with careful comparison of landmarks; a number of specimens were measured independently by both authors on different dates to determine consistency of data collection, and measurements were nearly always identical, indicating very little inter-observer error (both authors usually took more than one set of measurements on each specimen; means were used if there were varying measures but coefficients of variation were extremely low). Wild-caught adults were used whenever possible, to avoid both issues of ontogenetic changes and possible effects of captivity. However, as noted, skeletons of marmosets are generally rare; samples were increased for some taxa by the inclusion of data from captive adult animals. Statistical comparison of wild and captive marmoset samples indicates no significant difference in their measurements (Davis 2002). Sexes were combined for analysis; callitrichids exhibit little to no sexual dimorphism, including in the skeleton (Ford and Davis 1992, 1998; Davis 2002).

Data were then size corrected using ratios to a single measure shown to have a strong (high correlation coefficient) and nearly isometric relationship to body weight. This method was chosen over ratio to a geometric mean because specimens varied dramatically in which elements were present (particularly for hand and foot elements, frequently absent in small mammal skeletons). Recent work has suggested variation in the variables used to compute a geometric mean can dramatically affect the outcome of allometric correction and subsequent analyses (Coleman 2003; Vinyard 2003). Humeral head height (HHH) was found to be the best size estimator across callitrichids among the 160 measures (see Davis 2002 for more information). All analyses were done using JMP v.4.0.4 (SAS Institute Inc 2001).

Principle Components Analysis (PCA) using raw data and Discriminant Function/Canonical Variates Analysis (DFA) on size-corrected data were used to explore the data initially. These were used to search for patterns that might discriminate any of the six genera. If generic clusters were found, then the variables with large eigenvector values on those factors were noted and explored through univariate analyses. Many specimens were incomplete, which is typical of skeletons of small-bodied mammals; hands and feet, in particular, are often left in the skins during preparation. Only specimens with no missing data for included variables were used for each separate analysis. Therefore, a number of different analyses were performed with different combinations of variables in an attempt to maximize information content both through numbers of variables and numbers of specimens, and to try to include most specimens in at least one multivariate analysis.

Univariate analyses were performed on all size-corrected variables, using Analysis of Variance (ANOVA) and the post-hoc Tukey-Kramer Honestly Significantly Difference (HSD) test, which is designed to accommodate unequal sample sizes, to test for generic differences (Sokal and Rohlf 1981; Tukey cited in SAS Institute Inc 2001).

The Tukey-Kramer HSD test is more conservative than Student's *t*-test and decreases the likelihood of Type 1 experiment-wise errors. Emphasis was on ways marmosets differ from *Saguinus midas*, used to represent an unspecialized callitrichid "ancestral" morphology associated with arboreal quadrupedalism.

21.3 Results

21.3.1 Multivariate Analyses

In all Principle Components analyses (PCAs), using different subsets of specimens to minimize the effects of missing data, Factor 1 accounted for the largest amount of variation (at least 72%) and is assumed to represent size. The *Callithrix* and *Mico* specimens overlap, with *Mico* showing much greater variability in Factor 1 (overall size) and being generally larger. *Callibella* is lower for Factor 1, with *Cebuella* with the lowest values of all, in keeping with its diminutive size. *Saguinus midas* individuals are larger than the marmosets, but there is overlap with the largest marmoset in our sample, the single specimen of *Mico manicorensis*. *Callimico* had the highest values for Factor 1 in each analysis, although not significantly different from the *Saguinus midas* sample. Analyses looking at other factors from PCA showed similar patterns to those found in Discriminant Function analyses (DFA) but with less clear groupings, largely because many specimens are missing data, particularly for the hands and feet. Further discussion of multivariate analyses will therefore focus on the results of the DFA based on size-corrected data.

DFA was performed on size-corrected features of the scapula plus humerus, ulna plus radius, hand plus ulna, innominate, femur, tibia plus fibula, and foot (calcaneus plus astragalus). The scapula does not discriminate well between marmosets. The innominate separates *Callimico* and *Saguinus* from marmosets as a whole and the Atlantic *Callithrix* from the remaining marmosets; however, there is some overlap and there was not 100% correct post-analysis classification. Unfortunately, sample sizes for many hand and distal foot bones drop dramatically and left too few specimens for good discrimination based on those elements; they will only be considered univariately.

The remaining analyses all indicate discrete differentiation of all genera, with 100% accurate classification of specimens for all combinations of forelimb elements and the femur, and only a few misclassified individuals in other analyses (generally specimens of *Callithrix*). In all analyses, *Callibella* (the dwarf marmoset) is entirely distinct from all other taxa. The best 30 variables from these analyses (highest eigenvalues) were combined into a single analysis. In a plot of Canon1xCanon2 (Fig. 21.2a) *Saguinus midas* and *Callimico* overlap some, but no specimens were miscategorized, and both are very distant from the marmosets. All marmoset genera form highly distinct clusters, with Amazonian *Mico* most distant from Atlantic *Callithrix*. The dwarf marmoset, *Callibella*, falls between *Mico* and *Cebuella*, but it is completely distinct from both, with no overlap on either canonical variate.

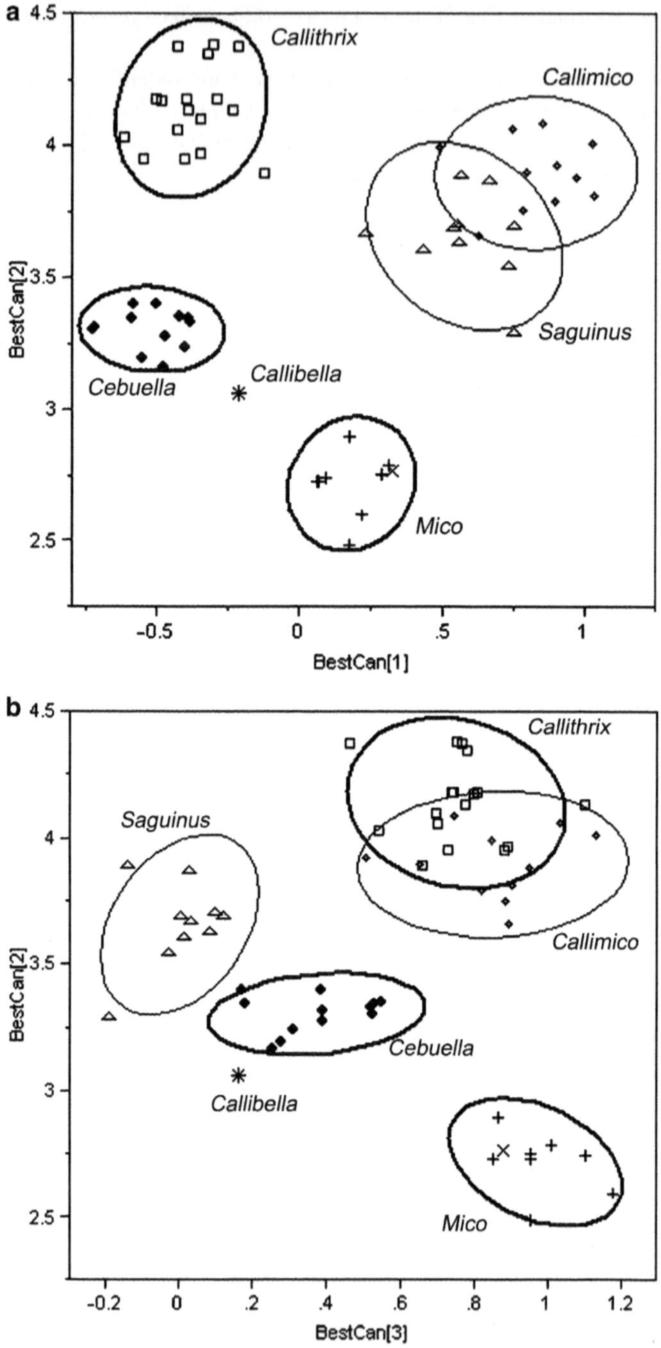

Fig. 21.2 Discriminate Function Analysis of 25 most discriminating variables (size corrected). (**a**) Canonical Variate 1×Canonical Variate 2. (**b**) Canonical Variate 2×Canonical Variate 3. *Cebuella* – solid diamonds; *Callibella* – star; *Mico* – crosses (+ and ×=*M. manicorensis*, the species sympatric with *Callibella*); *Callithrix* – squares; *Callimico* – open diamonds; *Saguinus midas* – triangles. Ellipses represent 90% density ellipse

The variables most strongly determining these distinctions include features of the humerus (anterior capitular, trochlear, and gutter widths, lateral trochlear height, and midshaft transverse diameter), radius (head breadth and width, anterior rim height, head facet breadth, and radiocarpal facet depth), ulna (neck width, olecranon process length, sigmoid notch depth, coronoid depth, proximal ulnar depth, midshaft sagittal diameter, radial facet height, and distal shaft transverse diameter), and femur (head breadth and width, lateral condyle width, and shaft width at the lesser trochanter). In Canon2xCanon3 (Fig. 21.2b), *Callithrix* partly overlaps with *Callimico*, but the marmoset genera themselves are each distinct from one another (and from *Saguinus*); again, while *Callibella* falls close to *Cebuella*, it is nonetheless clearly outside the 90% confidence ellipse for the pygmy marmosets.

Thus, multivariate analyses of postcranial features clearly indicate that each of the four marmoset genera occupies highly discrete morphological space, with pygmy marmosets roughly equidistant from the more separated *Callithrix* and *Mico*. The dwarf marmoset, *Callibella*, falls closest to *Cebuella* but distinct from it, and farthest from *Callithrix* in multivariate analysis. The other recently named marmoset from the Amazon, *Mico manicorensis*, whose range includes that of *Callibella*, firmly clusters within the *Mico* sample.

21.3.2 *Univariate Analyses*

Examination of the means and variability in each character, size-corrected, reveals that there are a number of defining marmoset traits among the genera *Callithrix, Mico, Cebuella,* and *Callibella,* relative to the nonmarmosets, particularly *Saguinus midas* (see Table 21.3). In addition, each marmoset genus is characterized by distinctive postcranial adaptations. While there are some conflicting similarities between genera, the strongest resemblance is between Amazonian *Mico* and the pygmy marmoset, *Cebuella*. We compare *Callibella* to these suites of defining features for marmosets.

21.3.2.1 Marmoset Traits

The marmosets form a derived clade, distinct from *Saguinus midas*, on the basis of five features, although for one feature the single specimen of *Callibella* falls at the edge of the range. These include three upper limb and two lower limb features. Marmoset arms display a medio-laterally narrower anterior humeral trochlea (measured by both HAT1 and HAT3), a narrow sagittal diameter of the ulnar midshaft (UMSS), and a mediolaterally narrow carpal facet on the radius (RCFW). Their legs share a large projection of the femoral lesser trochanter (FWLT, see Fig. 21.3; least marked in *Callibella*, which falls near the *Saguinus*-like edge of the marmoset range, with a slightly smaller lesser trochanter) and a short patella (FPL).

Table 21.3 Derived marmoset traits

Marmoset (compared to *Saguinus midas*)	*Callithrix*	*Mico+Cebuella* (Amazonian marmosets)	*Mico*	*Cebuella*	*Callibella*
HAT1 narrow +	RHW wide –	HPTW narrow +	HACW narrow –	IMI high –	HPTW very narrow
HAT3 narrow +	UNW wide –	HMEW narrow +	HMEW narrow –	UNL short –	RARH high
UMSS narrow +	CCFH tall –	UPUD shallow –	HAGW narrow –	MC3L short –	CPF3 very wide
RCFW narrow +		RHB narrow –	RNL short –	FHW narrow +	
FWLT large +		UNL short –	RNTD narrow –	FHB narrow +	
FPL short +		TTRW1 short +		FLCW narrow +	
				FPL short –	
				TPW narrow +	
				BDFH tall –	
				TDFH tall –	
				AMBH short +	
				APFL short +	
				APF1 short –	
				ATW1 narrow –	
				CML short +	
				CAL short +	
				CPF2 short –	
				XPOWR short +	
				XCUB short –	
				XT4L short –	

Abbreviations for features are defined in Appendix 1. +/– indicate that the feature is shared (+) or not shared (–) by *Callibella humilis*

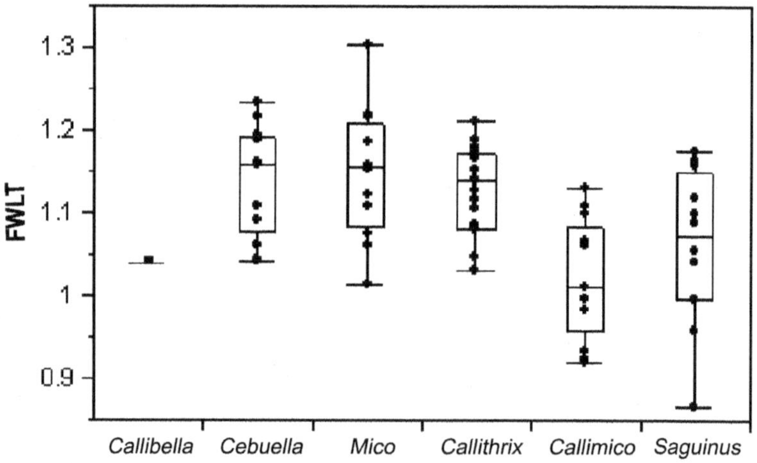

Fig. 21.3 Quartile box plot of lesser trochanter projection on the femur (FWLT), adjusted for size (see text). Increased projection characterizes marmosets

These features are consistent with increased vertical clinging and a decreased reliance on quadrupedalism in the marmosets. A medio-laterally narrow (or short) humeral trochlea is generally correlated with a propensity for vertical clinging across strepsirhines (Szalay and Dagosto 1980; Rose 1993) and callitrichids (Davis 2002), and it is markedly narrow in the marmosets. Previous studies have suggested that the opposite of the marmoset narrow ulnar shaft, a deep ulnar shaft, is associated with habitual quadrupedalism and/or forelimb first landings in primates (Preuschoft 1974; Fleagle 1976; Fleagle and Meldrum 1988; Davis 2002), and *Saguinus midas* is described as primarily quadrupedal (Mittermeier 1977; Fleagle and Mittermeier 1980; Youlatos 1995). Many callitrichids may in fact land forelimb first when vertically leaping (see Garber et al. Chap. 14 this volume). The narrow ulnar shaft and narrow distal carpal facet on the radius of marmosets may relate to dexterity in the wrist joint for versatility when claw clinging. In an earlier study of a large sample of callitrichid species (but lacking any *Mico* specimens), Davis (2002) identified an increase in the sagittal and transverse diameters of the distal radial shaft as uniquely characterizing the marmosets *Callithrix* and *Cebuella*. Here, we find that radial shaft width in *Mico* and *Callibella* actually spans the gap between *Callithrix*/*Cebuella* and *S. midas*, and the differences are not statistically significant with our larger marmoset samples, although the pattern persists. The projecting lesser trochanter, particularly as it is placed in a medial position on the femoral shaft, relates to increased strength and activity of the iliopsoas muscle in flexing the leg during clinging vertically (Ford 1990). The role of a shorter patella remains obscure.

Of these five defining marmoset features, *Callibella* shares four, and is at the margin of the range for the fifth (greater projection of the lesser trochanter). *Callibella* is thus morphologically a marmoset, sharing a structure built for increased vertical clinging.

21.3.2.2 *Callithrix* Traits

Callithrix species, the Atlantic coastal forest marmosets, share three postcranial features that are not seen in the other callitrichids. These include a wide radial head (RHW), a medio-laterally wide ulnar notch (UNW1), and a tall cuboid facet on the calcaneus (CCFH). The wide elbow features contrast with the narrowed features they share with other marmosets (see above), and may stabilize the elbow during rotation associated with more diverse positional behavior. With little to no positional data on any *Mico* species or *Callibella,* the interpretation of differences must remain speculative. However, it is clear that the Atlantic coastal marmosets are the least specialized postcranially of the radiation. Of the three unique *Callithrix* traits, *Callibella* shares none. Postcranially, there is little evidence of any close affinity between these genera.

21.3.2.3 *Mico* and *Cebuella* Shared Traits

The Amazonian marmosets of the genera *Mico* and *Cebuella* share six traits that appear derived from a common callitrichid/marmoset ancestor. These include a narrow posterior humeral trochlea (HPTW, see Fig. 21.4a), a narrow humeral medial

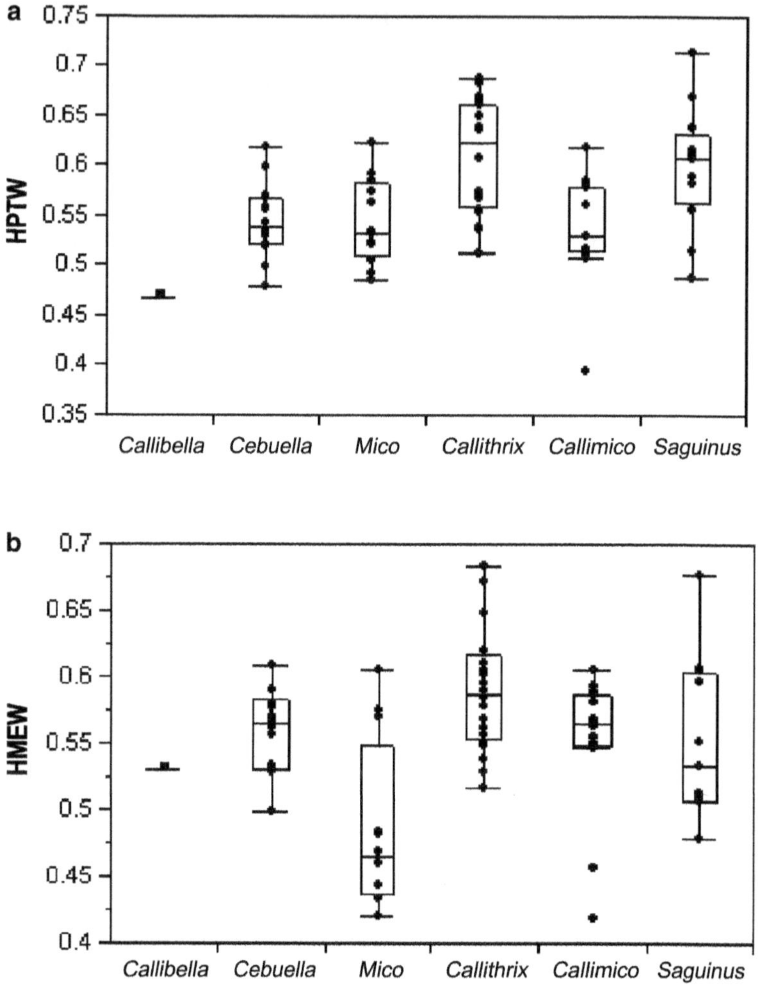

Fig. 21.4 Quartile box plots of two humeral traits characterizing Amazonian marmosets (each convergently with *Callimico*). (**a**) Posterior trochlear width (HPTW), adjusted for size (see text). The trochlea narrows further in *Callibella*. (**b**) Medial epicondylar width (HMEW), adjusted for size (see text). The medial epicondyle narrows further in *Mico*

epicondyle (HMEW, see Fig. 21.4b), a shallow proximal ulnar notch (UPUD), an antero-posteriorly narrow radial head (RHB), a short ulnar neck (UNL), and a medio-laterally narrow tibial trochlea (TTRW1). The distal tibial facet on the fibular shaft is also slightly elongated, but not quite to the level of statistical significance. From this ancestor, the *Mico* lineage went on to further specialize their forelimbs in the same direction, while *Cebuella* developed highly specialized hindlimbs (see below). The increasingly narrow and shallow elbow articulations may indicate a greater commitment to flexed clinging; the strength of adaptations to gouging/

exudativory may be variable, with some data suggesting that *Mico* has less dependence on gums than do *Callithrix* species or *Cebuella* (see overview in Table 21.2), but Veracini (Chap. 12 this volume) found her *Mico* more exudativorous than other *Mico*. Clearly, much more data on behaviors of Amazonian marmosets are needed. Likewise, the medio-laterally short tibial trochlea and the hint of closer appression of the distal tibial and fibular shafts shared by *Mico* and *Cebuella* suggest a stronger role for tight flexion in the ankle, consistent with greater clinging. However, these are not features which would enhance powerful leaping.

Callibella shares three of these features, the narrow humeral posterior trochlea, narrow humeral medial epicondyle and narrow tibial trochlear width (see Figs. 21.4 and 21.5), but the humeral traits appear to have developed also in *Callimico*, which otherwise is quite distinct postcranially despite the likely close phylogenetic ties to marmosets. While these traits may enhance tighter pronation/supination at the elbow and flexion at the ankle, *Callibella* lacks the fuller development of these complexes seen in the other two Amazonian marmoset clades. These few similarities may indicate a more distant relationship of *Callibella* to the other two Amazonian marmosets, or the characters may have even developed independently in *Callibella* as well as in *Callimico*.

21.3.2.4 *Mico* Traits

Mico species are distinguished in sharing five derivations: significantly narrowed medio-lateral dimensions of the distal humerus (including a narrow capitulum [HSCW], narrow trochlear groove or gutter [HAGW], and narrow medial epicondyle [HMEW], see Figs. 21.4b and 21.5), and a narrow and short radial neck [RNTD, RNL]. These traits, all indicating a highly narrowed and compressed elbow joint, intensify the shared marmoset narrowing of the anterior humeral trochlea.

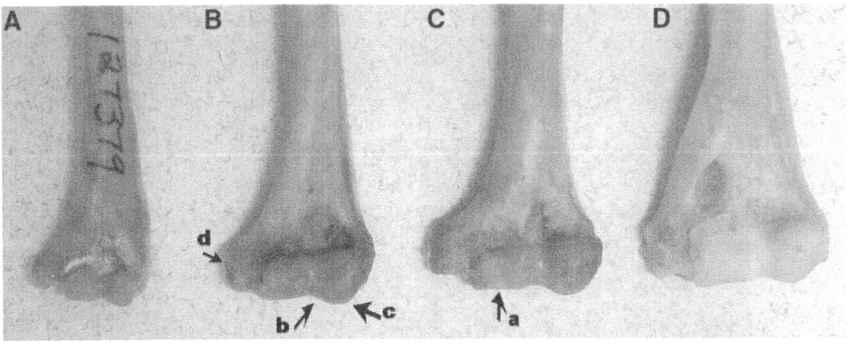

Fig. 21.5 Callitrichid distal left humeri, anterior view. (**A**) *Cebuella pygmaea*, (**B**) *Mico argentatus*, (**C**) *Callithrix jacchus*, (**D**) *Saguinus midas*. Features shown include: (**a**) narrow anterior trochlea (shared marmoset trait), (**b**) narrow trochlear groove or gutter (*Mico* only), (**c**) narrow capitulum (*Mico* only), and (**d**) narrow medial epicondyle (Amazonian marmosets, especially *Mico*). (For scale, maximum biepicondylar width of *Cebuella* (**A**) is 6.7 mm)

A significant narrowing of elbow features, similar to that seen in *Mico*, also characterizes the highly specialized manipulative forager *Leontopithecus* (Davis and Ford 2001; Davis 2002). The shortened radial neck of *Mico* is within the range reported for strepsirhines *Microcebus murinus* and *Galago moholi*, and is distinctly shorter than that known for other callitrichids (Davis 2002), including *Callithrix* and *Cebuella*. A short neck may indicate an adaptation for rapid elbow flexion in a supinated forearm, but in the absence of detailed behavioral data for *Mico*, the significance of this feature remains unclear. The opposite condition of a significantly elongated radial neck is seen in the habitual vertical clinger, *Callimico goeldii* (Davis 2002). Davis (2002) also found that *Callithrix* species (solely the Atlantic coastal forest species) exhibit some of the widest medial epicondyles in a large sample of callitrichid species, further demonstrating the contrasts between *Callithrix* and *Mico* in the postcranium. Large medial epicondyles (Aiello and Dean 1990) are generally associated with strong wrist and digital flexion and forearm pronation, which would be expected in a habitual clinger, and thus make sense in *Callithrix*. The reduced medial epicondyle of *Mico*, like the short radial neck, is thus surprising, and may suggest differences in the manner or frequency of clinging between *Mico* and both *Callithrix* and *Callimico*. Only detailed data on the positional behaviors of *Mico* will allow us to address this question.

Callibella, the dwarf marmoset, lacks all five of the unique *Mico* features, indicating it never developed *Mico*'s distinctive elbow morphology (see Fig. 21.6). Postcranially, there is no evidence of a close link between the dwarf marmoset and marmosets of the genus *Mico*, to the exclusion of the pygmy marmoset – either functionally or phylogenetically.

21.3.2.5 *Cebuella* Traits

Cebuella is clearly the most distinctive marmoset (and callitrichid), especially in the hindlimb (see also Davis 1998, 2002). Of the features examined in this study

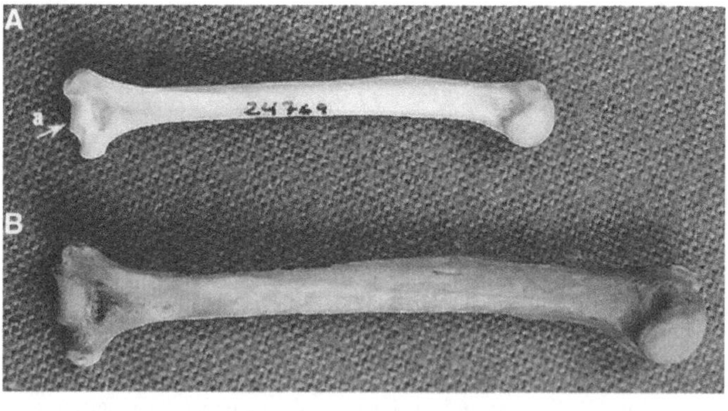

Fig. 21.6 Left humeri, posterior view. (**A**) *Callibella humilis,* (**B**) *Mico manicorensis*. Feature shown is: (**a**) narrow posterior trochlea (*Callibella* only). (For scale, maximum humeral length of *Callibella* (**A**) is 38.8 mm, and maximum biepicondylar width is 7.1 mm)

alone, *Cebuella* exhibits unique adaptations in twenty of them (see Table 21.3). Many features in the femur, tibia, astragalus, and calcaneus are significantly reduced in size in the pygmy marmoset, including femoral head dimensions (FHW, FHB), several condylar and tibial plateau dimensions (FLCW, TPW, FPL, see Fig. 21.7), width of the astragalar trochlea (ATW1), several ventral astragalar dimensions (AMBH, APFL, APF1), length of the calcaneus and of its anterior portion and posterior facet (CML, CAL, CPF2), length of the cuboid (XCUB), of the fourth metatarsal (XT4L), and of the combined power arm of the foot (XPOWR). In addition, the intermembral index is high (see Fig. 21.8), reflecting a shortened hindlimb/long forelimb. (Youlatos 1999, following Thorington and Thorington, 1989) suggests that this might enhance lateral placement of the forelimbs not seen in other vertical climbers (as described in Garber 1991). These various features might allow a more effective and tighter flexion in the reduced joint surfaces that are habitually held in tension during prolonged foraging bouts, but not necessarily for strength in leaping – rather, this may be a complex for postural clinging, with relatively longer arms in this diminutive anthropoid for grasping (arm/claw placement) and stabilizing.

In contrast to this general reduction in size of hindlimb features, the distal fibular facet on the tibia and its corresponding facet on the distal fibula are actually significantly elongated in many *Cebuella* specimens (BDFH, TDFH). This would correspond with tighter appression and stabilization in the distal tibio-fibular joint, often seen in hindlimb dominant leapers (Fleagle and Simons 1983, 1995). In the case of *Cebuella*, this tight appression is more likely to support the tight clinging postures, or to stabilize a leg otherwise poorly designed for short distance leaping.

Cebuella shows little specialization of the forelimb seen in *Mico*, with the exception of an even shorter ulnar neck (UNL) from the shortening shared with *Mico*, and

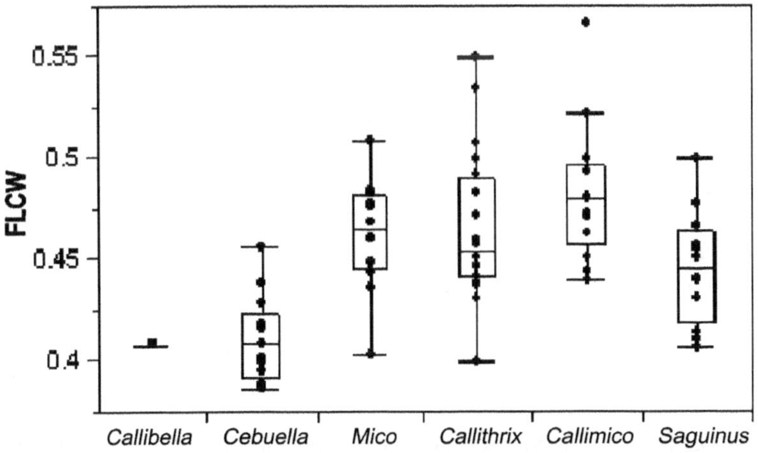

Fig. 21.7 Quartile box plot of lateral condyle width on the femur (FLCW), adjusted for size (see text). The condyle is narrow in the diminutive *Cebuella* and *Callibella*

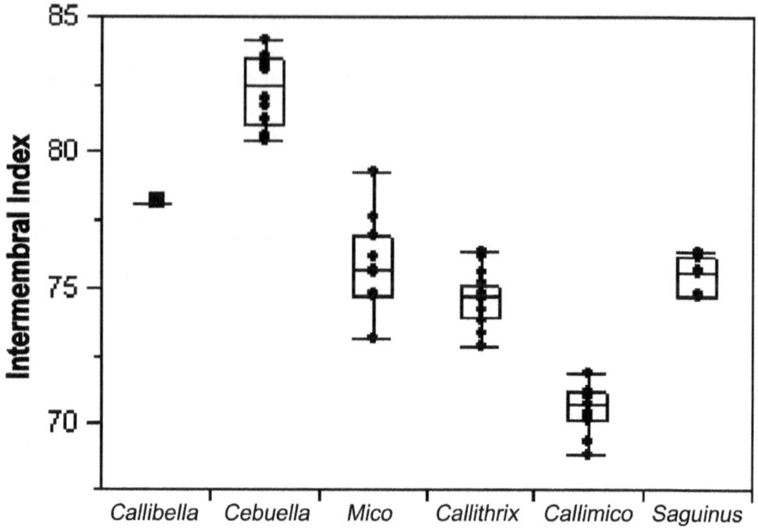

Fig. 21.8 Quartile box plot of intermembral index. A large value, indicating short legs to long arms, characterizes *Cebuella*

short metacarpals (as measured on metacarpal 3, MC3L). Having both shortened metacarpals and metatarsals may result in greater foot stability during claw clinging for the diminutive pygmy marmosets, who cannot grasp most of their preferred vertical feeding supports in a more typical primate fashion.

Not only is *Cebuella* unique, but it is often far more variable in each of these traits than the highly speciose *Mico* and *Callithrix*. This fact, given that museum specimens derive from a variety of localities, reinforces the view that the uniform and cryptic pelage of pygmy marmoset populations may disguise considerable population differentiation and possible speciation (see Rylands et al. Chap. 2 this volume).

The dwarf marmoset, *Callibella*, displays an intriguing pattern of both similarities to and differences from *Cebuella*. *Callibella* shares 9 of these 20 features, including a small femoral head (two dimensions – FHW, FHB), narrow femoral lateral condyle (FLCW, see Fig. 21.7) and tibial plateau (TPW), a short astragalar body (AMBH) with a short posterior facet for the calcaneus (APFL), and a short calcaneus with a short anterior extension (CML, CAL) (see Fig. 21.7 – FLCW or CAL). As discussed above, these may all indicate a hindlimb strongly adapted to vertical clinging but weakly designed for powerful leaping.

21.3.2.6 *Callibella* Traits

Callibella, the dwarf marmoset, lacks the other eleven distinctive features which characterize *Cebuella* as well as most of the shared *Cebuella-Mico* traits. In many of these features, *Callibella* not only lacks the specializations seen in the other

Amazonian marmosets but actually departs morphologically in a different direction. The dwarf marmoset is not simply a slightly larger pygmy marmoset. *Callibella* lacks the pygmy marmoset's relatively short legs and long arms (intermembral index – see Fig. 21.8), its short metacarpals and metatarsals, and its markedly elongate distal facet on the tibia and fibula for close appression, among many others.

Callibella is unique in its own right, with a very narrow posterior trochlear width on the humerus (even narrower than in *Mico*, HPTW, see Figs. 21.4a and 21.6), a tall anterior rim on the proximal radius for the ulna (also seen in callimicos, RARH), and a wide calcaneal posterior facet (CPF3). While there is only one specimen, it falls outside of the range of the other Amazonian marmosets, indeed of all marmosets, for these traits.

This very dispersed (anatomically) list of unique features is difficult to interpret functionally. Individually, they suggest an elbow built for close flexion and an ankle for broad contact at the calcaneo-astragalar joint, perhaps in tightly dorsiflexed clinging postures. However, the elbow and hindlimb adaptations which would reinforce such postures and which occur in cluster in other marmosets are absent, perhaps indicating that the arm-dominant (*Mico*) or leg-dominant (*Cebuella*) clinging behaviors of the other Amazonian marmosets are not as fully developed, or necessary, for the as yet unknown behavioral repertoire of *Callibella*.

21.3.2.7 *Callithrix* and Other Marmoset Traits

Last, there are notably few traits consistent with an alternative *Callithrix-Mico, Callithrix-Cebuella,* or *Callithrix-Callibella* link. In the first two cases, the range of variability in the "primitive" tamarin *Saguinus midas* broadly overlaps shared features found in the marmoset taxa, making any such link suspect. In the case of *Callithrix-Callibella*, there is only one shared feature, in opposition to more shared features seen among the Amazonian species and the great geographic distance separating the two.

21.4 Discussion

The multivariate DFA demonstrate that postcrania strongly differentiate the four genera of marmosets, both from one another and from other callitrichids. Univariate analyses reinforce the distinctiveness of *Mico* from a more generalized *Callithrix*, the likely phylogenetic ties of *Cebuella*, with *Mico,* and the highly derived nature of *Cebuella*. The recently described and little known *Callibella* appears equally distinctive, but its postcranium is consistent with that of a marmoset and is overall more similar to the Amazonian marmosets than to the more generalized Atlantic Coastal *Callithrix*. This is made more striking by the fact that a recently described *Mico* species which is sympatric with *Callibella*, *Mico manicorensis*, falls well within the range of variability of other *Mico* species. All of these findings confirm prior dental, cranial, vocal, chromosomal, and DNA analyses of the marmosets. The marked discrimination between all of the marmoset clades (especially considering how speciose two of the genera are) strongly supports the concept that these are each valid and distinct taxa on the generic level. The four named marmoset genera maintain distinct genetic and chromosomal patterns in the wild (Rylands et al. 2000; Neusser et al. 2001), and we argue that they should be considered valid genera.

All marmosets are reported to be highly exudativorous (up to 70% of feeding time), especially pygmy marmosets (see reviews in Rylands and de Faria 1987, 1993; Garber 1992; Garber et al. Chap. 14 this volume), but there are almost no data for the positional behavior of these primates. The best studied is the most specialized pygmy marmoset, *Cebuella*. In addition to qualitative and largely anecdotal accounts (Sanderson 1957; Tokuda 1969; Kinzey et al. 1975; Moynihan 1976; Ramirez et al. 1978; Soini 1988; Terborgh 1983; Buchanan-Smith et al. 2000), Youlatos (1999, Chap. 15 this volume) has presented a detailed, quantitative study of *Cebuella*'s positional behavior, and recently, data have been presented on positional behavior in captive *Cebuella* (Jackson and Ford 2006). In contrast, there is no information on positional behavior for any of the many *Mico* species, and only anecdotal and qualitative accounts for two *Callithrix* species, *C. penicillata* (Sanderson 1957; Lacher et al. 1981) and *C. jacchus* (Sanderson 1957; Maier et al. 1982; Leslie Digby personal communication in Davis 2002; see also Stevenson and Rylands 1988). These studies paint a portrait of small-bodied monkeys spending a significant amount of stationary time in either a vertical cling posture (feeding on exudates) or sit/lay/sprawl postures to rest. In movement, they appear to prefer quadrupedalism (both fast and slow), leaping (often horizontal) for gap-crossing, and some climbing, including claw climbing. *Cebuella* may spend even more time claw-climbing, with much higher frequencies of climbing, quadrupedal walk/bound, and nonvertical leaping rather than vertical leaping (Youlatos 1999, Chap. 15 this volume); they have been seen to run along the undersides of large branches, suspended upside-down (Jackson and Ford 2006; Garber personal communication), as well. However, vertical claw-clinging, followed by grasp-clinging, are the most common postures (Youlatos 1999, Chap. 15 this volume; see discussion in Davis 2002). They favor small supports for most activities, except scansorial movements (vertical claw climbing) and vertical leaping. With no solid comparative data, the degree to which pygmy marmoset behavior typifies that of all marmosets is uncertain, although Youlatos (1999: 549) states, "I believe that *Cebuella* would still use scansorial activities more than other marmosets," due to its small size.

Rather than firmly interpreting differences in anatomical structure based on known behaviors, we must use the anatomical differences to predict possible behavioral differences among marmoset taxa that need to be investigated in the field. We predict that *Callithrix* species will be the least specialized beyond a shared marmoset propensity for increased vertical clinging, with markedly increased use of flexed postures during feeding on exudates. *Mico* species should demonstrate an altered use of the arm, perhaps emphasizing increased pronation/supination while in a flexed, vertical position. It would be particularly interesting to look for foraging convergences with the specialized *Leontopithecus*. *Cebuella*, as suggested by Youlatos (1999), may demonstrate far higher frequencies of scansorial behavior and highly flexed hindlimb postures during both feeding and nonleaping locomotion than will be found in the other marmosets. We suspect that *Cebuella* will be found to exhibit lower frequencies of vertical or other leaping compared to other marmosets, given its relatively short hindlimbs.

Callibella may show a propensity to use clinging and flexed postures more frequently than in *Callithrix*, but will be less specialized in its arm use than *Mico*. The mixed similarities and differences with the highly specialized *Cebuella* are difficult to interpret; some may indicate shared derivations in the use of the hindlimb for tightly flexed clinging postures, while others may simply be the shared result of extreme size reduction within the callitrichid radiation. Aguiar and Lacher (2003, Chap. 18 this volume) suggest *Callibella* may prove to have less reliance on exudates than that seen in *Cebuella*. These behaviors may relate to the reported heavy use of a single tree species by dwarf marmosets, which has been associated with a unique forest (*terra pretas*) characterized by many vines and twining branches found in connection with human habitation (but see Aguiar and Lacher Chap. 18 this volume for possibly wider ecological settings for *Callibella*). In addition, dwarf marmosets often inhabit local gardens, backyards, and orchards (van Roosmalen and van Roosmalen 2003). We suggest dwarf marmosets are likely to be more quadrupedal and use more horizontal and low-angled supports than is typical of *Cebuella*, although they may be more adapted to hindlimb dominated vertical clinging than *Mico* (and all Amazonian marmosets more so than is *Callithrix*).

There may also be important species-level differences in positional behavior and anatomy. Within the eastern Atlantic coastal *Callithrix*, *C. penicillata* and *C. jacchus* are more exudativorous than the other species (Rylands and de Faria 1987, 1993). Based on dental differences, Natori (1986) predicted this difference, and with behavioral data, Natori and Shigehara (1992) demonstrated the strong connection between degree of exudativory and size of the lower incisors and canines. They predicted a high rate of exudativory in Amazonian *Mico*, but there are almost no solid dietary data and limited dental data on the many *Mico* species (but see Veracini Chap. 12 this volume). Positional behavior and anatomical differences may prove to accompany these posited dietary differences. Recent DNA work (Sena et al. 2002) suggests that *Callithrix penicillata* and *C. jacchus* may be among the most derived of the Atlantic group, leading us to further speculate that if these two share higher levels of exudativory and scansorial behavior with *Mico* species, the derivations would have developed in parallel in the coastal and the Amazonian marmosets. Substantial additional data, both anatomically and behaviorally, are needed to test these hypotheses.

Features shared by the Amazonian marmosets (*Mico, Cebuella*, and, to a lesser degree, *Callibella*) reinforce the view that Atlantic coastal *Callithrix* is the most generalized of the marmosets (Rylands et al. 2000; van Roosmalen et al. 1998; van Roosmalen and van Roosmalen 2003). The matrix of derivations among the Amazonian marmosets is complex: *Mico* and *Cebuella* share some traits not found in *Callibella*; *Callibella* and *Cebuella* share many traits; but *Callibella* departs markedly from *Cebuella* in other features. These results support suggestions elsewhere (van Roosmalen and van Roosmalen 2003; Aguiar and Lacher 2003) for a three-way division of the Amazonian marmoset ancestor into separate *Mico, Callibella,* and *Cebuella* lineages, subsequent to this group separating from the Atlantic *Callithrix* (see Fig. 21.9). The marked differences in their skeletons suggest that high dependence on exudates and perhaps extreme small size evolved independently in the pygmy and dwarf marmosets, associated with different postcranial adaptations – a specialized elbow in *Mico,* a highly specialized leg for clinging in *Cebuella,* and a combination in *Callibella* (but much less fully developed than in either of the other two). In all cases, however, these adaptations appear to favor a tightly compressed clinging posture but they do not favor strong, powerful leaping. Indeed, the studies of Garber et al. (Chap. 14 this volume) and of Youlatos (1999, Chap. 15 this volume) suggest that *Cebuella,* at least, is not capable of strong, powerful, long leaps but restricts itself to shorter leaps coupled with quadrupedal movements and a great deal of vertical clinging postures.

This study has allowed us to confirm the distinct identity of the four marmoset genera, and to hypothesize about their positional behavior, adaptations, and history. However, the near total absence of data on positional (or other) behavior for most species in the wild and the scarcity of anatomical specimens make these hypotheses impossible to test further. As this volume attests, we are making headway in understanding this group of primates, but there is much still to learn. As many marmosets, including the dwarf marmoset, live in very small ranges characterized by at-risk forests whose unique character is unknown, the study and preservation of marmosets should be of high priority.

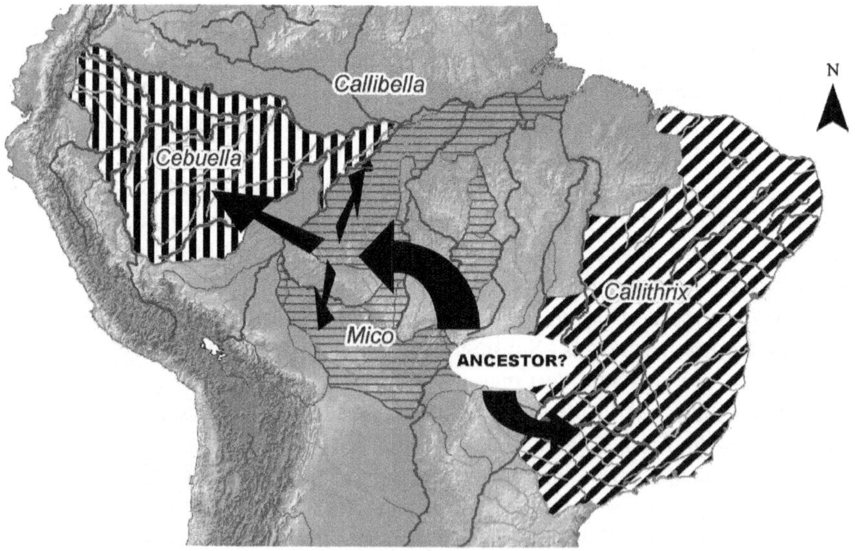

Fig. 21.9 Hypothetical marmoset biogeographical history

21.5 Summary and Conclusions

1. Postcranial data strongly support the recognition of four distinct marmoset genera, with separate adaptations, but a closer connection and shared history between the more western, Amazonian marmosets *(Mico, Cebuella,* and *Callibella).*
2. All marmosets share adaptations for greater vertical clinging and increased strength and support in more upright positions.
3. The Atlantic marmosets (jacchus-group, or *Callithrix*) are least specialized, although there may be important variation within this group in both degree of exudativory and scansorial behavior.
4. The more diverse Amazonian marmosets of the genus *Mico* have elbow specializations that may facilitate increased vertical orientations or other arm movements in feeding and foraging. We predict this accompanied increased exudate-feeding.
5. As pygmy marmosets (*Cebuella)* evolved and dwarfed in the far west, they have become the most specialized, particularly in their hindlimb morphology. It is likely that this reflects higher frequencies of scansorial and clinging behavior with a higher dependence on exudates.
6. The dwarf marmosets (*Callibella*) may represent a third marmoset radiation in the Amazon, with only one living member in a relict area. Its anatomy suggests less specialized clinging or foraging than in the other Amazonian marmosets, but marked similarities to the pygmy marmosets may represent shared or convergent features associated with extreme size reduction.
7. Quantified dietary and, especially, positional behavior data are badly needed for all marmosets.

Acknowledgments We gratefully acknowledge Jean Spence (AMNH), Linda Gordon (USNM), and Michi Schulenberg and Bill Stanley (FMNH) for assistance during data collection. In addition, we are particularly grateful to Marc van Roosmalen for the opportunity to study the skeletons of *Callibella humilis* and *Mico manicorensis*. Kevin Davie provided advice on the maps, Vijay Paniker assisted with digital photography, and several anonymous reviewers greatly improved the manuscript. This research was supported by the National Science Foundation (DBS 92-03884), the Smithsonian Institution, and Sigma Xi, The Scientific Research Society.

21.6 Appendix 1: Measurement Definitions and Abbreviations

21.6.1 Scapula

SGFH (glenoid facet height): maximum proximodistal height of the glenoid facet. Taken perpendicular to SGFW.

SGFW (glenoid facet width): maximum mediolateral width of the glenoid facet taken at the widest distal point of fossa. Taken perpendicular to SGFH.

SSSL (supraspinous length): maximum lateromedial length of supraspinous fossa taken parallel to the scapular spine. Extends from most latero-superior point of the scapular spine/neck juncture medially to the vertebral edge of the supraspinous fossa.

SSSH (supraspinous height): maximum height of supraspinous fossa, from most superior point to scapular spine, perpendicular to SSSL.

SISL (infraspinous length): length of infraspinous fossa parallel to lower base of inferior edge of spinous process.

SISH (infraspinous height): maximum height of infraspinous fossa, taken from the juncture of the scapular spine and the vertebral border to the most inferior point of the infraspinous fossa.

SAL (acromion length): maximum length of inferior edge of acromion.

SAW (acromion width): maximum super-inferior width of acromion.

SVH (scapular ventral height): maximum ventral height of subscapularis fossa (roughly parallel to vertebral scapular border).

SVL (scapular ventral length): maximum midline ventral length of subscapularis fossa, taken from the superior edge of the glenoid fossa (roughly parallel to scapular spine).

SVLB (ventral lateral border length): maximum ventral length of lateral border, from inferior edge of glenoid fossa to most inferior point on the scapula (measured on ventral aspect).

SALA (acromio-spine to lateral border angle): angle of scapular spine to lateral border.

SGLA (glenoid to lateral border angle): angle of glenoid fossa plane to lateral border.

21.6.2 Humerus

H3PL (humeral 3-point length): maximum length, superior (most proximal) point of head to 2-point distal capitulum/trochlea plane.

HDTL (deltoid tuberosity length): maximum proximodistal length of deltoid tuberosity.

HDSW (deltoid scar width): maximum mediolateral width of the roughened area marking the insertion of the deltoid muscle, generally measured near its proximodistal midpoint. Taken perpendicular to HDTL.

HHW (humeral head width): maximum transverse width of humeral head articular surface.

HHH (humeral head depth): maximum superoinferior depth of humeral head articular surface.

HMSD (midshaft transverse diameter): mediolateral width of humeral shaft at midpoint of humeral length.

HBEW (biepicondylar width): maximum mediolateral width of distal humerus, across epicondyles.

HAT1 (superior anterior trochlear width): maximum mediolateral width of trochlear articular surface (not including trochlear gutter), taken at anterosuperior portion of trochlea.

HAT3 (inferior anterior trochlear width): maximum mediolateral width of trochlear articular surface (not including trochlear gutter), taken at anteroinferior portion of trochlea.

HACW (anterior capitulum width): proximolateral edge of capitulum to lateral edge of trochlea, including gutter.

HMTH (medial trochlear height): proximodistal height of trochlea at medial edge, taken on anterior aspect.

HLTH (lateral trochlear height): proximodistal height of trochlea at lateral edge, taken on anterior aspect.

HACH (anterior capitulum height): maximum proximodistal height of capitulum, taken on anterior aspect.

HPTW (posterior trochlear width): maximum mediolateral width of trochlea, taken on posterior aspect.

HMEW (medial epicondyle width): trochlear edge to medial edge of medial epicondyle, taken on posterior aspect.

HLEW (lateral epicondyle width): trochlear edge to lateral edge of lateral epicondyle, taken on posterior aspect.

HDTA (deltoid tuberosity angle): angle of the deltoid tuberosity relative to the long axis of the humeral shaft.

21.6.3 Radius

RL (radial length): maximum length, including the styloid process.

RHB (radial head breadth): maximum anteroposterior breadth of head, taken on superior aspect.

RHW (radial head width): maximum mediolateral width of head, taken on superior aspect.

RHFB (radial head facet breadth): maximum anteroposterior breadth of head articular surface, taken on anterior aspect.

RHFW (radial head facet width): maximum mediolateral width of head articular surface, taken on anterior aspect.

RMRH (medial rim height): maximum height of rim on medial side of head (generally measured directly superior to radial tuberosity).

RLRH (lateral rim height): maximum height of rim on lateral side of head.

RARH (anterior rim height): maximum height of rim on anterior side of head.

RNTD (radial neck transverse diameter): mediolateral width of neck, taken on anterior aspect.

RNL (radial neck length): proximodistal length of radial neck, measured from distal edge of anterior rim of the head to proximal point of radial tuberosity.

RTL (radial tuberosity length): proximodistal length of radial tuberosity.

RCFD (radiocarpal facet depth): anteroposterior depth of distal articular facet measured at its mediolateral midpoint.

RCFW (radiocarpal facet width): mediolateral width of distal articular facet including the styloid process facet, measured at its anteroposterior midpoint; perpendicular to RCFD.

RDST (distal shaft transverse diameter): mediolateral width of the distal shaft, measured just proximal to the distal expansion of the shaft.

RDSS (distal shaft sagittal diameter): anteroposterior width of the distal shaft, measured just proximal to the distal expansion of the shaft; perpendicular to RDST.

21.6.4 Ulna

UL (ulnar length): maximum length, including the styloid process.

UNL (trochlear/sigmoid notch length): maximum proximodistal height of the trochlear notch measured along the long axis, from the midpoint of the proximal lip of the notch to the midpoint of the distal lip of the notch.

UNW1 (proximal notch width): maximum mediolateral width of the most proximal, inferiorly-facing facetal surface of the trochlear notch.

UNW2 (midpoint notch width): maximum mediolateral width of the midpoint, anteriorly-facing facetal surface of the trochlear notch.

UNW3 (distal notch width): maximum mediolateral width of the most distal, superoanteriorly-facing facetal surface of the trochlear notch.

UPUD (proximal ulna depth): anteroposterior depth of the proximal ulna, measured from the (anteriorly-facing) midpoint of the trochlear notch dorsally to the dorsal surface of the proximal ulna.

UCD (coronoid depth): anteroposterior depth of coronoid, taken to anterior surface of ulnar shaft.

URFH (radial facet height): maximum proximodistal height of proximal radial facet, from lateral edge of trochlear notch to distal point on radial facet.

URF1 (superior radial facet width): mediolateral width, taken at superior (proximal) point of facet.

URF2 (inferior radial facet width): mediolateral width, taken at inferior (distal) point of facet.

UOPL (olecranon process length): maximum proximodistal height of olecranon process, from proximal edge of trochlear notch.

UOLL (olecranon lever length): estimated length of olecranon process lever, measured from proximodistal midpoint of trochlear notch (estimated axis of rotation)to most superior (proximal) extension of olecranon.

UOPD (olecranon process depth): maximum anteroposterior depth of olecranon process.

UMST (midshaft transverse diameter): maximum mediolateral width, taken at midpoint of ulnar length.

UMSS (midshaft sagittal diameter): maximum anteroposterior depth, taken at midpoint of ulnar length.

UDST (distal shaft transverse diameter): mediolateral width of shaft, taken at distal end.

UDSS (distal shaft sagittal diameter): anteroposterior depth of shaft, taken at distal end.

21.6.5 Carpus

MC3L (third metacarpal length): maximum length of metacarpal III.

M3PP (proximal phalanx length of 3rd ray): maximum length of proximal phalanx of digit III.

M3MP (medial phalanx length of 3rd ray): maximum length of middle phalanx of digit III.

M3DP (distal phalanx length of 3rd ray): maximum length distal phalanx of digit III.

21.6.6 Innominate

IDPL (dorsal pelvis length): length of dorsal (proximal) edge of ilium to ischial tuberosity.

IICW (iliac crest width): maximum width of posterior iliac crest.

IPL (pubis length): length from acetabular center to superior (proximal) point of pubic symphysis.

ICRL (caudal ramus length): length from ischial tuberosity to angle of caudal ramus.

IIL (ischial length): length from acetabular center to ischial tuberosity.

ISYM (pubic symphysis length): proximodistal length of pubic symphysis.

IAH (acetabular height): maximum height of acetabulum, parallel to long axis of ilium.

IAW (acetabular width): maximum width of acetabulum, perpendicular to long axis of ilium.

ILPA (iliopubic angle): angle from plane of ilium to plane of pubis.

ISPA (ischiopubic angle): angle from plane of ischium to plane of pubis.

21.6.7 Femur

FL (femoral medial length): maximum femoral length, from proximal extension of head to distal extension of medial condyle.

FPST (proximal shaft transverse diameter): mediolateral shaft width, taken just below flare of lesser trochanter.

FMST (midshaft transverse diameter): mediolateral shaft width, taken at midpoint of shaft length (from FL).

FHW (femoral head width): maximum anteroposterior width of head.

FHB (femoral head breadth): maximum proximodistal breadth of head.

FHNL (femoral head/neck length): head-neck length, from craniomedial point of head to femoral neck crest (if present) or edge of fossa.

FNW (femoral neck width): proximodistal transverse width of neck at midpoint.

FNCL (femoral neck crest length): proximodistal length of femoral neck crest.

FLTH (lesser trochanter height): position of lesser trochanter on shaft, measured from superior point of lesser trochanter to fovea capitis.

FWLT (femoral width at lesser trochanter): transverse width of femur including lesser trochanter, in plane of lesser trochanter projection.

FIIL (iliopsoas insertion length): proximodistal length of facet on lesser trochanter.

FIIW (iliopsoas insertion width): maximum mediolateral length of facet on lesser trochanter.

F3TH (third trochanter height): position of third trochanter on shaft, measured from superior point of third trochanter to proximal point of greater trochanter.

FBCW (bicondylar width): maximum mediolateral width of distal epiphysis.

FBCD (bicondylar depth): maximum anteroposterior depth of distal epiphysis.

FSPG (superior patellar groove width): maximum width at widest part of superior portion of patellar groove.

FIPG (inferior patellar groove width): maximum width at widest part of inferior patellar groove, proximate to intercondylar notch.

FPGL (patellar groove length): length of patellar groove, measured from most proximal point of groove to beginning of intercondylar notch.

FMCH (medial condyle height): proximodistal height of medial condyle taken on posterior surface.

FLCH (lateral condyle height): proximodistal height of lateral condyle taken on posterior surface.

FMCW (medial condyle width): mediolateral width of medial condyle at midpoint, taken on posterior surface.

FLCW (lateral condyle width): mediolateral width of lateral condyle at midpoint, taken on posterior surface.

FINW (intercondylar notch width): mediolateral width of intercondylar notch taken on inferior aspect of femur.

21.6.8 Patella

FPL (patella length): maximum proximodistal length of patella.

FPW (patella width): maximum mediolateral width, taken at midpoint.

21.6.9 TIBIA

TL (medial tibial length): maximum medial length, measured from the superior surface of the medial tibial plateau to the most distal point of the tibia, including malleolus.

TMPD (medial plateau depth): maximum anteroposterior depth of medial plateau.

TLPD (lateral plateau depth): maximum anteroposterior depth of lateral plateau.

TMPW (medial plateau width): maximum mediolateral width of medial plateau.

TLPW (lateral plateau width): maximum mediolateral width of lateral plateau.

TPW (tibial plateau width): maximum mediolateral width of the total tibial plateau area.

PLAT (plateau angle): angle between plane of plateau and plane of anterior tibia.

TMST (midshaft transverse diameter): maximum mediolateral width of shaft at midpoint of length.

TMSS (midshaft sagittal diameter): maximum anteroposterior width of shaft at midpoint of length (same position as above).

TDST (distal shaft transverse diameter): distal mediolateral width of shaft at distal end, just before expansion.

TDSS (distal shaft sagittal diameter): distal anteroposterior width of shaft at distal end, just before expansion (same position as TDSS).

TDFD (distal fibular facet depth): maximum anteroposterior depth of distal fibular facet.

TDFH (distal fibular facet height): maximum proximodistal height of distal fibular facet.

TFCH (fibular contact area height): proximodistal height of distal contact area with fibula, from proximal point of fibular facet to proximal point of contact area.

TML (malleolus length): proximodistal length of medial malleolus, taken on internal (lateral) surface.

TMW (malleolus width): maximum anteroposterior width of medial malleolus.

TTRD (trochlear depth): maximum anteroposterior depth of tibial trochlea.

TRW1 (anterior trochlear width): maximum mediolateral width of anterior half of trochlea, not including malleolus.

TRW2 (posterior trochlear width): maximum mediolateral width of posterior half of trochlea, not including malleolus.

21.6.10 Fibula

BFL (maximum fibular length): maximum length.

BDFH (distal tibial facet height): maximum proximodistal height of distal tibial facet.

BDFW (distal tibial facet width): maximum anteroposterior width of distal tibial facet.

BAFH (astragalar facet height): maximum proximolateral to distomedial height of astragalar facet.
BAFW (astragalar facet width): maximum distolateral to proximomedial width of astragalar facet.

21.6.11 Astragalus

AML (astragalar length): maximum proximodistal length, between most anterior point of the head to two-point posterior plane (note – approximate anatomical length, parallel to long axis of trochlear rotation; not maximum length as often measured by others).
AHW (astragalar head width): ma.ximum width of astragalar head, taken along long axis of head (not necessarily mediolateral).
AHH (astragalar head height): height of astragalar head, taken perpendicular to AHW.
AMFL (medial facet length): distal edge of medial facet ("cup") to most proximal point ("tail").
AMCH (medial facet cup height): maximum superoinferior (dorsoplantar) height of "cup" of medial facet.
AMBH (astragalar medial body height): medial trochlear rim to inferomedial (plantarmedial) edge of posterior facet, as maximum superoinferior height.
ALFL (lateral facet length): distal edge of flare of lateral facet to most proximal point ("tail").
ALFH (lateral facet height): maximum height of lateral facet (without tail), from dorsal rim to plantar edge of flare.
ATRL (trochlear length): proximodistal (anteroposterior) length at midline of the astragalar trochlea.
ATW1 (anterior trochlear width): maximum anterior width of trochlea (excluding any extension of rims beyond central trochlear groove).
ATW3 (posterior trochlear width): most posterior width of trochea (excluding any extension of one rim beyond the other).
AVNL (ventral neck length): measured from distal extension of head to mediodistal edge of posterior calcaneal facet, taken on plantar (ventral) surface.
AAFL (anterior facet length): maximum length of anterior calcaneal facet from distal extension of head to proximal end of facet, taken on plantar (ventral) surface (includes separate medial calcaneal facet, if one is present).
APFL (posterior facet length): maximum length of posterior calcaneal facet.
APF1 (posterior facet anterior width 1): maximum anterior width of posterior calcaneal facet.
APF2 (posterior facet anterior width 2): width of posterior calcaneal facet at the midpoint.
APF3 (posterior facet posterior width 3): maximum posterior width of posterior calcaneal facet.

21.6.12 Calcaneus

CML (calcaneal medial length): maximum length of medial edge, along long axis of bone.

CCFW (cuboid facet width): mediolateral width of cuboid facet, taken along long axis of facet.

CCFH (cuboid facet height): plantardorsal (superoinferior) midline height of facet, taken perpendicular to CCFW.

CAFL (anterior facet length): proximodistal length of anterior astragalar facet.

CAL (anterior calcaneal length): taken from distal end of posterior astragalar facet to distal calcaneus, along long-axis midline.

CMW (calcaneal midpoint width): maximum width of calcaneus at midpoint of long axis, including sustentaculum.

CPFL (posterior facet length): proximodistal length of posterior astragalar facet.

CPF1 (posterior facet anterior width1): maximum anterior width of posterior astragalar facet.

CPF2 (posterior facet mid width2): midpoint width of posterior astragalar facet.

CPF3 (posterior facet posterior width3): maximum posterior width of posterior astragalar facet.

CTH (calcaneal tuberosity height): maximum plantardorsal height at tuberosity (posterior).

CTW (calcaneal tuberosity width): maximum mediolateral width of tuberosity (posterior).

CTL (calcaneal tuberosity length): extension of tuberosity, from posterior margin of posterior astragalar facet to posterior extension of tuberosity edge.

21.6.13 Distal Tarsus

CALTRO (XPOWR): length from center of astragalar trochlea to distal (anterior) end of calcaneus, in articulated ankle.

CUB4 (posterior cuboid to posterior base MT4): length from posterior edge of cuboid (calcaneocuboid facet) to posterior base of fourth metatarsal, in articulated foot.

XT4L (fourth metatarsal length): maximum length fourth metatarsal.

X4PP (proximal phalanx length of 4th ray): maximum length of proximal phalanx of digit IV.

X4MP (medial phalanx length of 4th ray): maximum length of medial phalanx of digit IV.

X4DP (distal phalanx length of 4th ray): maximum length of distal phalanx of digit IV.

21.7 Appendix 2: Specimens Examined

Callithrix: C. aurita – MCZ 439; *C. geoffroyi* – FMNH 134472, 134473, 140916; *C. jacchus* – FMNH 20226, USNM 397241, 398846, 398851, 399032, 399037, 503885, 503886, 503895, 503899, 518554, 518555; *C. penicillata* – AMNH 133692, 133694, 133698, 133702, 133703, 200756. *Mico: M. argentatus* – AMNH 184689, FMNH 104806, 140354, 140355; *M. humeralifer* – AMNH 188164; *M. leucippe* – AMNH 133712; *M. manicorensis* – INPA 2512; *M. melanurus* – FMNH 58989, 58990, 60737, 60766, 121553. *Callibella humilis:* MPEG 24769. *Cebuella pygmaea:* AMNH 244101, 244365, USNM 303037, 336325, 337322, 337323, 337325, 337326, 337328, 337329, 337330, 337947, 337948, 337949, 464990. *Callimico goeldii:* AMNH 183289, FMNH 57999, 58003, 98034, 134517, 134518, 134522, USNM 303323, 395455, 463933, 464991, 464993, 573934. *Saguinus midas:* AMNH 77693, 97316, 207726, 266480, 266481, FMNH 93239, 93515, 93516, MCZ 7110, UF 10175, USNM 267590, 362118.

AMNH American Museum of Natural History, New York; *FMNH* Field Museum of Natural History, Chicago; *INPA* Instituto Nacional de Pesquisas da Amazonia, Manaus; *MCZ* Museum of Comparative Zoology, Harvard University, Boston; *MPEG* Museu Paraense Emilio Goeldi, Manaus; *USNM* United States National Museum, Smithsonian Institution, Washington, DC.

References

Aguiar JM, Lacher TE Jr (2003) On the morphological distinctiveness of *Callithrix humilis* van Roosmalen, et al., 1998. Neotrop Primates 11:11–17

Aguiar JM, Lacher TE Jr (this volume) Cranial morphology of the dwarf marmoset *Callibella* in the context of callitrichid variability. In: Ford SM, Porter LM, Davis LC (eds) The smallest anthropoids: The marmoset/callimico radiation. Springer Press, New York, pp 355–380

Alperin R (1993) *Callithrix argentata* (Linnaeus, 1771): considerações taxonômicas e descrição de subespécie nova. Bol Mus Para Emílio Goeldi, Sér Zool 9:317–328

Anzenberger G, Muench M, Moisson P, Petit T (2001) Intergeneric marmoset hybrids (Cebuella pygmaea X Callithrix jacchus): An important case study for callitrichid taxonomy. [Abstract] Folia primatol 72(3): 153

Barroso CML (1995) Molecular phylogeny of the Callitrichinae. Neotrop Primates 3:186

Buchanan-Smith HM, Hardie SM, Caceres C, Prescott MJ (2000) Distribution and forest utilization of *Saguinus* and other primates of the Pando Department, Northern Bolivia. Int J Primatol 21:353–379

Canavez FC, Alves G, Fanning TG, Seuánez HN (1996) Comparative karyology and evolution of the Amazonian *Callithrix* (Platyrrhini, Primates). Chromosoma 104:348–357

Canavez FC, Moreira MAM, Ladasky JJ, Pissinatti A, Parnham P, Seuánez HN (1999) Molecular phylogeny of New World primates (Platyrrhini) based on β_2-microglobulin DNA sequences. Mol Phylogenet Evol 12:74–82

Coimbra-Filho AF, Mittermeier RA (1973) New data on the taxonomy of the Brazilian marmosets of the genus *Callithrix* Erxleben, 1777. Folia Primatol 20:241–264

Coimbra-Filho AF, Mittermeier RA (1976) Exudate-eating and tree-gouging in marmosets. Nature 262:630

Coimbra-Filho AF, Mittermeier RA (1978) Tree-gouging, exudate-eating and the "short-tusked" condition in *Callithrix* and *Cebuella*. In: Kleiman DG (ed) Biology and conservation of the Callitrichidae. Smithsonian Institution Press, Washington, DC, pp 105–115

Coleman MN (2003) An investigation of scaling relationships in sensory and masticatory systems of New World primates. Am J Phys Anthropol 36(Suppl):78

Cortés-Ortiz L (this volume) Molecular phylogenetics of the Callitrichidae with an emphasis on the marmosets and *Callimico*. In: Ford SM, Porter LM, Davis LC (eds) The smallest anthropoids: The marmoset/callimico radiation. Springer Press, New York, pp 3–24

Cronin JE, Sarich VM (1975) Molecular systematics of the New World monkeys. J Hum Evol 4:357–375

Cronin JE, Sarich VM (1978) Marmoset evolution: the molecular evidence. Prim Med 10:12–19

Davis LC (1998) Postcranial morphometrics of the pygmy marmoset, *Cebuella pygmaea*. Am J Phys Anthropol 26(Suppl):74

Davis LC (2002) Functional morphology of the forelimb and long bones in the Callitrichidae (Platyrrhini, Primates). Unpublished PhD Dissertation, Southern Illinois University

Davis LC, Ford SM (2001) Forelimb anatomy and foraging strategy in *Leontopithecus rosalia*. Am J Phys Anthropol 32(Suppl):57

de Vivo M (1991) Taxonomia de *Callithrix* Erxleben, 1777 (Callitrichidae, Primates). Fundação Biodiversitas, Belo Horizonte, Brasil

Ferrari SF (1993) Ecological differentiation in the Callitrichidae. In: Rylands AB (ed) Marmosets and tamarins: Systematics, behavior, and ecology. Oxford University Press, Oxford, pp 314–328

Ferrari SF, Lopes MA (1992) A new species of marmoset, genus *Callithrix* Erxleben 1777 (Callitrichidae, Primates) from western Brazilian Amazonia. Goeldiana Zoologia 12:1–3

Ferrari SF, Sena L, Schneider MPC (1999) Definition of a new species of marmoset (Primates, Callitrichinae) from southwestern Amazonia based on molecular, ecological, and zoogeographic evidence. Resumos do Congresso Brasileiro de Primatologia 9:80–81

Fleagle JG (1976) Locomotor behavior and skeletal anatomy of sympatric Malaysian leaf-monkeys (*Presbytis obscura* and *Presbytis melalophos*). Yearb Phys Anthropol 20:440–453

Fleagle JG, Meldrum DJ (1988) Locomotor behavior and skeletal morphology of two sympatric pitheciine monkeys, *Pithecia pithecia* and *Chiropotes satanas*. Am J Primatol 16:227–249

Fleagle JG, Mittermeier RA (1980) Locomotor behavior, body size, and comparative ecology of seven Surinam monkeys. Am J Phys Anthropol 52:301–314

Fleagle JG, Simons EL (1983) The tibio-fibular articulation in *Apidium phiomense*, an Oligocene anthropoid. Nature (London) 301:238–239

Fleagle JG, Simons EL (1995) Skeletal anatomy of *Apidium phiomense*, an early anthropoid from Egypt. Am J Phys Anthropol 97:235–289

Ford SM (1986) Systematics of the New World monkeys. In: Swindler DR, Erwin J (eds) Comparative primate biology, vol. 1: Systematics, evolution, and anatomy. Alan R Liss, New York, pp 73–135

Ford SM (1990) Locomotor adaptations of fossil platyrrhines. J Hum Evol 19:141–173

Ford SM, Davis LC (1992) Systematics and body size: implications for feeding adaptations in New World monkeys. Am J Phys Anthropol 88:415–468

Ford SM, Davis LC (1998) Sexual dimorphism in the postcranium of callitrichid primates. Am J Phys Anthropol 26(Suppl):104

Garber PA (1992) Vertical clinging, small body size, and the evolution of feeding adaptations in the Callitrichinae. Am J Phys Anthropol 88:469–482

Garber PA (1994) Phylogenetic approach to the study of tamarin and marmoset social systems. Am J Primatol 34:199–219

Garber PA (2001) A comparative study of positional behavior in three species of tamarin monkeys. Primates 32(1): 219–230

Garber PA, Sallenave A, Blomquist GE, Anzenberger G (this volume) A comparative study of the kinematics of trunk-to-trunk leaping in *Callimico goeldii, Callithrix jacchus* and *Cebuella pygmaea*. In: Ford SM, Porter LM, Davis LC (eds) The smallest anthropoids: The marmoset/callimico radiation. Springer Press, New York, pp 259–277

Groves C (2001) Primate Taxonomy. Smithsonian Institution Press, Washington, DC
Harada ML, Schneider H, Schneider MPC, Sampaio I, Czelusniak J, Goodman M (1995) DNA evidence on the phylogenetic systematics of New World monkeys : support for the sister-grouping of *Cebus* and *Saimiri* from two unlinked nuclear genes. Mol Phylogenet Evol 4:331–349
Hershkovitz P (1977) Living New World monkeys (Platyrrhini), vol 1. University of Chicago Press, Chicago
Hill WCO (1959) The anatomy of *Callimico goeldii* (Thomas). A primitive American primate. Trans Am Phil Soc 49(Part 5):1–116
Horovitz I, Meyer A (1995) Systematics of New World monkeys (Platyrrhini, Primates) based on 16S mitochondrial DNA sequences: a comparative analysis of different weighting methods in cladistic analysis. Mol Phylogenet Evol 4:448–456
Horovitz I, Zardoya R, Meyer A (1998) Platyrrhine systematics: a simultaneous analysis of molecular and morphological data. Am J Phys Anthropol 106:261–281
SAS Institute Inc (2001) JMP Version 4.0.4. SAS Institute Inc
Jackson CP, Ford SM (2006) Contexts of positional behavior in captive pygmy marmosets (*Cebuella pygmaea*). Am J Phys Anthropol 42(Suppl):109
Kay RF (1990) The phyletic relationships of extant and fossil Pitheciinae (Platyrrhini, Anthropoidea). J Hum Evol 19:175–208
Kinzey WG, Rosenberger AL, Ramirez M (1975) Vertical clinging and leaping in a Neotropical anthropoid. Nature 255:327–328
Lacher TE Jr, Fonseca GA, Bouchardet da Alves C, Magalhaes-Castro B Jr (1981) Exudate-eating, scent-marking, and territoriality in wild populations of marmosets. Anim Behav 29:306–307
Maier W, Alonso C, Langguth A (1982) Field observations on *Callithrix jacchus jacchus* L. Z Saugetierkunde 47:334–346
Mittermeier RA (1977) Distribution, synecology, and conservation of Surinam monkeys. Unpublished PhD Dissertation, Harvard University
Mittermeier RA, Schwarz M, Ayres JM (1992) A new species of marmoset, genus *Callithrix* Erxleben 1777 (Callitrichidae, Primates), from the Rio Maués region, state of Amazonas, Central Brazilian Amazonia. Goeldiana Zoologia 14:1–17
Moynihan M (1976) Notes on the ecology and behavior of the pygmy marmoset (*Cebuella pygmaea*) in Amazonian Colombia. In: Thorington RW Jr, Heltne PG (eds) Neotropical primates: field studies and conservation. National Academy of Sciences, Washington, DC, pp 79–82
Nagamachi CY, Pieczarka JC, Barros RMS (1992) Karyotypic comparison among *Cebuella pygmaea, Callithrix jacchus,* and *C. emiliae* (Callitrichidae, Primates) and its taxonomic implications. Genetica 85:249–257
Nagamachi CY, Pieczarka JC, Schwarz M, Barros RMS, Mattevi MS (1997) Comparative chromosomal study of five taxa of genus *Callithrix*, group *Jacchus* (Platyrrhini, Primates). Am J Primatol 41(1):53–60
Nagamachi CY, Pieczarka JC, Muniz JAPC, Barros RMS, Mattevi MS (1999) Proposed chromosomal phylogeny for the South American primates of the Callitrichidae family (Platyrrhini). Am J Primatol 49:133–152
Natori M (1986) Interspecific relationships of *Callithrix* based on the dental characters. Primates 27:321–336
Natori M (1994) Craniometrical variations among eastern Brazilian marmosets and their systematic relationships. Primates 35:167–176
Natori M, Shigehara N (1992) Interspecific differences in lower dentition among eastern Brazilian marmosets. J Mammal 73(3):668–671
Neusser M, Muench M, Anzenberger G, Mueller S (2005) Investigation of marmoset hybrids (Cebuella pygmaea x Callithrix jacchus) and related Callitrichinae (Platyrrhini) by cross-species chromosome painting and comparative genomic hybridization. Cytogenetic Genome Res 108(1-3): 191–196
Neusser M, Stanyon R, Bigoni F, Wienberg J, Mueller S (2001) Molecular cytotaxonomy of New World monkeys (Platyrrhini) - Comparative analysis of five species by multi-color chromosome

painting gives evidence for a classification of Callimico goeldii within the family of Callitrichidae. Cytogenetics Cell Genetics 94: 206–215

Pastorini J, Forstner MRJ, Martin RD, Melnick DJ (1998) A reexamination of the phylogenetic position of *Callimico* (Primates) incorporating new mitochondrial DNA sequence data. J Mol Evol 47:32–41

Porter CA, Czelusniak MJ, Schneider H, Schneider MPC, Sampaio I, Goodman M (1997) Sequences of the primate epsilon-globin gene: implications for systematics of the marmosets and other New World primates. Gene 205:59–71

Preuschoft H (1974) Body posture and mode of locomotion in fossil primates. Methods and example: *Aegyptopithecus zeuxis*. Proc Symp 5th Cong Int Primatol Soc 345–359

Ramirez MR, Freese CH, Revilla CJ (1978) Feeding ecology of the pygmy marmoset, *Cebuella pygmaea*, in northeast Peru. In: Kleiman DG (ed) Biology and conservation of the Callitrichidae. Smithsonian Institution Press, Washington, DC, pp 91–104

Rose MD (1993) Functional anatomy of the elbow and forearm in primates. In: Gebo DL (ed) Postcranial adaptations in nonhuman primates. Northern Illinois University Press, DeKalb, pp 70–95

Rosenberger AL (1977) *Xenothrix* and ceboid phylogeny. J Hum Evol 6:461–481

Rosenberger AL (1984) Aspects of the systematics and evolution of marmosets. In: Thiago de Mello M (ed) A Primatologia no Brasil. Universidade Federal, Brasilia, pp 159–180

Rosenberger AL, Setoguchi T, Shigehara N (1990) The fossil record of callitrichine primates. J Hum Evol 19:209–236

Rylands AB, de Faria DS (1987) Habitats and feeding ecology of the genus *Callithrix*. Int J Primatol 8:438

Rylands AB, de Faria DS (1993) Habitats, feeding ecology, and home range size in the genus *Callithrix*. In: Rylands AB (ed) Marmosets and tamarins: Systematics, ecology, and behavior. Oxford University Press, Oxford, pp 262–272

Rylands AB, Coimbra-Filho AF, Mittermeier RA (1993) Systematics, geographic distribution, and some notes on the conservation status of the Callitrichidae. In: Rylands AB (ed) Marmosets and tamarins: Systematics, ecology, and behavior. Oxford University Press, Oxford, pp 11–77

Rylands AB, Schneider H, Langguth A, Mittermeier RA, Groves CP, Rodriguez-Luna E (2000) An assessment of the diversity of New World primates. Neotrop Primates 8(2):61–93

Rylands AB, Mittermeier RA, Coimbra-Filho AF (this volume) The systematics and distribution of the marmosets (*Callithrix, Callibella, Cebuella,* and *Mico*) and callimico (*Callimico*) (Callitrichidae, Primates). In: Ford SM, Porter LM, Davis LC (eds) The smallest anthropoids: The marmoset/callimico radiation. Springer Press, New York, pp 25-61

Sanderson IT (1957) The monkey kingdom: An introduction to the primates. Hanover House, New York

Schneider H, Schneider MPC, Sampaio I, Harada ML, Stanhope M, Czelusniak J, Goodman M (1993) Molecular phylogeny of the New World monkeys (Platyrrhini, Primates). Mol Phylogenet Evol 2:225–242

Schneider H, Sampaio I, Harada ML, Barroso CML, Schneider MPC, Czelusniak J, Goodman M (1996) Molecular phylogeny of New World monkeys (Platyrrhini, Primates) based on two unlinked nuclear genes: IRBP intron 1 and epsilon-globin sequences. Am J Phys Anthropol 100:153–179

Sena L, Valinoto M, Sampaio I, Schneider H, Ferrari SF, Schneider MPC (2002) Mitochondrial COII gene sequences provide new insights into the phylogeny of marmoset species groups (Callitrichidae, Primates). Folia primatol 73:240–251

Silva JS Jr, de Noronha MA (1998) On a new species of bare-eared marmoset, genus *Callithrix* Erxleben 1777, from central Amazonia (Primates, Callitrichidae). Goeld Zool 21:1–28

Singer SS, Schmitz J, Schwieg K, Zischler H (2003) Molecular cladistic markers in New World monkey phylogeny (Platyrrhini, Primates). Mol Phylogenet Evol 26:490–501

Snowdon CT (1993) A vocal taxonomy of the callitrichids. In: Rylands AB (ed) Marmosets and tamarins: Systematics, behavior, and ecology. Oxford University Press, Oxford, pp 78–94

Soini P (1988) The pygmy marmoset, genus *Cebuella*. In: Mittermeier RA, Rylands AB, Coimbra-Filho AF, Fonseca GAB (eds) Ecology and behavior of neotropical primates, vol 2. World Wildlife Fund, Washington, DC, pp 79–129

Sokal RR, Rohlf FJ (1981) Biometry: The principles and practice of statistics in biological research, 2nd edn. WH Freeman and Co, New York

Stevenson MF, Rylands AB (1988) The marmosets, genus *Callithrix*. In: Mittermeier RA, Rylands AB, Coimbra-Filho AF, Fonseca GAB (eds) Ecology and behavior of neotropical primates, vol 2. World Wildlife Fund, Washington, DC, pp 131–222

Szalay FS, Dagosto M (1980) Locomotor adaptations as reflected on the humerus of Paleogene primates. Folia Primatol 34:1–45

Tagliaro CH, Schneider MPC, Schneider H, Sampaio I, Stanhope MJ (1997) Marmoset phylogenetics, conservation perspectives, and evolution of the mtDNA control region. Mol Biol Evol 14:674–684

Tagliaro CH, Schneider MPC, Schneider H, Sampaio I, Stanhope MJ (2000) Molecular studies of *Callithrix pygmaea* (Primates, Platyrrhini) based on transferrin intronic and ND1 regions: Implications for taxonomy and conservation. Genet Mol Biol 23:729–737

Terborgh J (1983) Five New World primates: A study in comparative ecology. Princeton University Press, Princeton

Thorington RW Jr, Thorington EM (1989) Postcranial proportions of *Microsciurus* and *Sciurillus*, the American pygmy tree squirrels. Adv Neotrop Mammal 1989:125–136

Tokuda K (1969) Group size and vertical distribution of New World monkeys in the Basin of the Rio Putumayo, the Upper Amazon. Proc VIIIth Intl Congr Anthropl and Ethnolog Sci 1:260–261

van Roosmalen MGM, van Roosmalen T (2003) The description of a new marmoset genus, *Callibella* (Callitrichinae, Primates), including its molecular phylogenetic status. Neotrop Primates 11:1–10

van Roosmalen MGM, van Roosmalen T, Mittermeier RA, de Fonseca GAB (1998) A new and distinctive species of marmoset (Callitrichidae, Primates) from the lower Rio Aripuanã, State of Amazonas, Central Brazilian Amazonia. Goeldiana Zoologia 22:1–27

van Roosmalen MGM, van Roosmalen T, Mittermeier RA, Rylands AB (2000) Two new species of marmoset, genus *Callithrix* Erxleben, 1777 (Callitrichidae, Primates), from the Tapajós/Madeira interfluvium, south central Amazonia, Brazil. Neotrop Primates 8:2–18

Veracini C (this volume) Habitat use and ranging behavior of the silvery marmoset (*Mico argentatus*) at Caxiuanã National Forest (eastern Brazilian Amazonia). In: Ford SM, Porter LM, Davis LC (eds) The smallest anthropoids: The marmoset/callimico radiation. Springer Press, New York, pp 221–240

Vinyard CJ (2003) Functional interpretations of jaw shapes: Beware of morphometricians bearing geometric means. Am J Phys Anthropol 36(Suppl):216

von Dornum M, Ruvolo M (1999) Phylogenetic relationships of the New World monkeys (Primates, Platyrrhini) based on nuclear G6PD DNA sequences. Mol Phylogenet Evol 11:459–476

Youlatos D (1995) Preliminary observations on the locomotion and postural behaviors of the red-handed tamarin (*Saguinus midas midas*) in French Guiana. Folia Primatol 64:94

Youlatos D (1999) Positional behavior of *Cebuella pygmaea* in Yasuni National Park. Ecuador Primates 40:543–550

Youlatos D (this volume) Locomotion, postures, and habitat use by pygmy marmosets (*Cebuella pygmaea*). In: Ford SM, Porter LM, Davis LC (eds) The smallest anthropoids: The marmoset/callimico radiation. Springer Press, New York, p279–297

Section V
Conservation

Leila M. Porter, Susan M. Ford, Lesa C. Davis

The marmosets and callimicos have several characteristics that potentially reduce their risk of extinction: they can reproduce rapidly as compared to other anthropoids, many can use disturbed and secondary forest, and due to their small size they are not regularly hunted by humans. Despite these favorable features, many species face considerable risk of extinction, a problem addressed by two authors in this last section of the book.

de la Torre and colleagues in Chap. 22 assess the conservation status of the pygmy marmoset (*Cebuella*) in Ecuador. They find that the densities of pygmy marmosets are negatively affected by human behaviors, an alarming discovery, given that the authors estimate that up to 85% of their habitat has been altered because of human disturbance. In addition, they find that although the pygmy marmosets initially appear to be able to cope with human induced habitat change, their populations decline over time if human activities persist over several years. In Chap. 23, Ferrari provides an overview of the conservation status of the rest of the marmosets and the callimicos, the particular problems each taxon faces, and the data needed to effectively plan for their protection. Ferrari's overview demonstrates that much additional data are needed to understand these monkeys' long term chances for survival. For example, Ferrari notes that it is extremely important that we determine the geographic distribution of the newly discovered species of marmosets in order to determine if they are endemic to small regions of the Amazon basin. Both chapters make it clear that in order to accurately assess a species' long-term viability, extensive data are required. This is particularly poignant given our limited knowledge of the ecology and behavior of most of the Amazonian marmosets, and the possibility that we have not even discovered all the species of marmosets and callimicos in this region yet.

Susan M. Ford
Department of Anthropology, Southern Illinois University, Carbondale, IL 62901, USA
e-mail: sford@siu.edu

Leila M. Porter
Department of Anthropology, Northern Illinois University, DeKalb, IL, 60115, USA
e-mail: lmporter@niu.edu

Chapter 22
Conservation Status of Pygmy Marmosets (*Cebuella pygmaea*) in Ecuador

Stella de la Torre, Pablo Yépez, and Charles T. Snowdon

Abstract Since 1996 we have studied eight populations of pygmy marmosets *Cebuella pygmaea* in northeastern Ecuador. Our data suggest that the habitat and diet specializations of pygmy marmosets make them particularly vulnerable to local extinction because of human activities. In periodical surveys we conducted of these populations, we found that pygmy marmosets were absent from disturbed habitats. We have also found a gradual decrease of population density and significant behavioral changes in pygmy marmosets living in areas affected by habitat destruction, human traffic, tourism, and human capture of marmosets. We evaluated the degree of habitat disturbance both at regional and local scales and have found that about 75–85% of the gallery forests inhabited by pygmy marmosets are affected by logging and agriculture. Based on our results, we propose to include pygmy marmosets in the list of Vulnerable species in Ecuador, based on the IUCN criterion A4 (c, d). This proposed change is one of the necessary steps we need to conserve Ecuadorian pygmy marmosets and their habitats. To achieve this aim we have also developed an on-going environmental education program directed to the children of the communities that live close to the pygmy marmoset populations as well as to children in urban areas.

Resumen Desde 1996 hasta el presente, hemos estudiado ocho poblaciones de leoncillos *Cebuella pygmaea* en el nororiente ecuatoriano. Nuestros datos sugieren que las especializaciones de hábitat y dieta de los leoncillos los hacen particularmente vulnerables a las actividades humanas. En censos realizados para estimar la densidad ecológica de los leoncillos en las poblaciones hemos registrado la ausencia de leoncillos en hábitats disturbados; también hemos registrado una disminución gradual de la densidad poblacional y cambios significativos en el comportamiento de leoncillos en áreas afectadas por alteración de hábitat, tráfico humano, turismo y captura de leoncillos. En una evaluación del grado de alteración

S. de la Torre (✉)
Universidad San Francisco de Quito, Quito, Ecuador
e-mail: stella@usfq.edu.ec

S.M. Ford et al. (eds.), *The Smallest Anthropoids*, Developments in Primatology: Progress and Prospects, DOI 10.1007/978-1-4419-0293-1_22,
© Springer Science+Business Media, LLC 2009

de los hábitats a una escala regional y local encontramos que alrededor del 75–85% de los bosques de galería habitados por los leoncillos están afectados por tala y agricultura. Con base en estos resultados, proponemos incluir a los leoncillos en la lista de especies Vulnerables en el Ecuador, según el criterio A4 (c, d) de la UICN. Esta propuesta es uno de los pasos que debemos dar hacia la conservación de estos primates ecuatorianos y de sus hábitats. Para lograr este objetivo hemos también desarrollado un programa de educación ambiental, en marcha, dirigido a los niños de las comunidades adyacentes a las poblaciones de leoncillos y a niños de las áreas urbanas.

Resumo Desde de 1996 nós temos estudado oito populações de sagüis-leãozinho *Cebuella pygmaea* no noroeste do Equador. Nossos dados sugerem que as especializações de habitat e dieta destes sagüis fazem deles particularmente vulneráveis à extinção local devido a atividades humanas. Em monitoramentos periódicos que conduzimos nestas populações, nós descobrimos que os sagüis-leãozinho estavam ausentes de habitats pertubados. Nós também encontramos um decréscimo da densidade populacional e mudanças comportamentais significativas nos sagüis vivendo em áreas afetadas por destruição de habitat, tráfego humano, turismo e captura humana dos sagüis. Nós avaliamos o grau de perturbação do habitat tanto na escala local como regional e descobrimos que cerca de 75–85% das florestas de galeria habitadas pelos sagüis-leãozinho foram afetadas por madeireiras ou agricultura. Baseados nos nosso resultados, nós propomos incluir os sagüis-leãozinho na lista de espécies vulneráveis no Equador, baseado no critério A4 (c, d) da IUCN. Esta mudança proposta é um dos passos necessários que necessitamos para conservar os sagüis-leãozinho equatorianos e seus habitats naturais. Para atingir este objetivo nós temos também desenvolvido um progama de educação ambiental em andamento direcionado as crianças das comunidades que vivem próximas as populações de sagüis-leãozinho tanto quanto para crianças em áreas urbanas.

22.1 Introduction

Any understanding of the behavioral ecology and conservation of marmosets must include information from species across the generalist-specialist continuum. Data from common marmosets (*Callithrix jacchus*) suggest that they are the most generalized and flexible species of their genus: they adapt to many habitats beyond their native range, including areas affected by human activities, such as urban parks in Rio de Janeiro, and they even breed well in research laboratories (Rylands et al. 1993; Stevenson and Rylands 1988). In contrast, the pygmy marmoset, *Cebuella pygmaea,* may be the most specialized of the marmosets, with high vulnerability to habitat change and human activities. Specialist species are at greatest risk from habitat alteration and destruction (Smith and Smith 2000); therefore, conservation efforts are critical for the pygmy marmosets.

The pygmy marmoset is the smallest Neotropical primate (120 g) and lives only in gallery forests in the Upper Amazon basin of Ecuador, Colombia, Perú, Bolivia and Brazil (Groves 2001; Soini 1988). In these habitats, pygmy marmosets feed mainly on exudates of certain plant species, with about 60–80% of the total feeding time devoted to exudate feeding and the remaining feeding time spent foraging for animal prey species. Fruit feeding is rare and occasional (Ramirez et al. 1977; Soini 1982, 1988; Yépez et al. 2005a).

Despite its relatively large geographic range, detailed information about the conservation status of the pygmy marmoset is largely unavailable. Some researchers have proposed that the pygmy marmoset may have benefited from the reduction in the populations of larger-bodied primate species (through human hunting), as larger-bodied primates may be competitors and even predators of the pygmy marmoset (Goldizen 1987; Soini 1988). This idea was supported by estimations of population sizes of pygmy marmosets in areas of Peru, Colombia, and Brazil made two decades ago (Coimbra-Filho 1984; Hernandez-Camacho and Defler 1985; Rylands et al. 1993; Soini 1982, 1988). However, no evaluations of the population status of the pygmy marmoset have been carried out in Ecuador and we are unaware of any recent evaluation of its population density and distribution in other countries.

In Ecuador, the pygmy marmoset inhabits gallery forests along rivers and lakes in the eastern lowlands, from about 220 to 450 m above sea level (de la Torre 2000; Yépez et al. 2005a). The pygmy marmoset is not included in any of the IUCN categories in the Red Data Book of Ecuadorian Mammals, and is only listed in Appendix II of the CITES (Tirira 2001). These categorizations, however, are not based on direct field studies of populations of pygmy marmosets but on estimations carried out in other countries many years ago, such as those cited above, and therefore may not reflect the current conservation status of this species.

Human settlements in the Ecuadorian Amazon (and elsewhere) have been established mainly in riparian habitats (Meggers 1989, pers. obs.), the habitats in which pygmy marmosets live. Thus, pygmy marmosets are likely to be particularly affected by human activities in this region. In this chapter, we present some of the data we have gathered during over 7 years of research with wild pygmy marmosets in northeastern Ecuador that show how the populations of this species are affected by human activities. These data suggest that the conservation status of pygmy marmosets in Ecuador, and in other countries, needs to be re-evaluated.

22.2 Methods

22.2.1 Study Areas

From 1996 to the present we have studied pygmy marmosets in six different populations in northeastern Ecuador (Fig. 22.1) (de la Torre et al. 2000; de la Torre and Snowdon 2002; Yépez et al. 2005a). The La Hormiga population (UTM, 18 Zone,

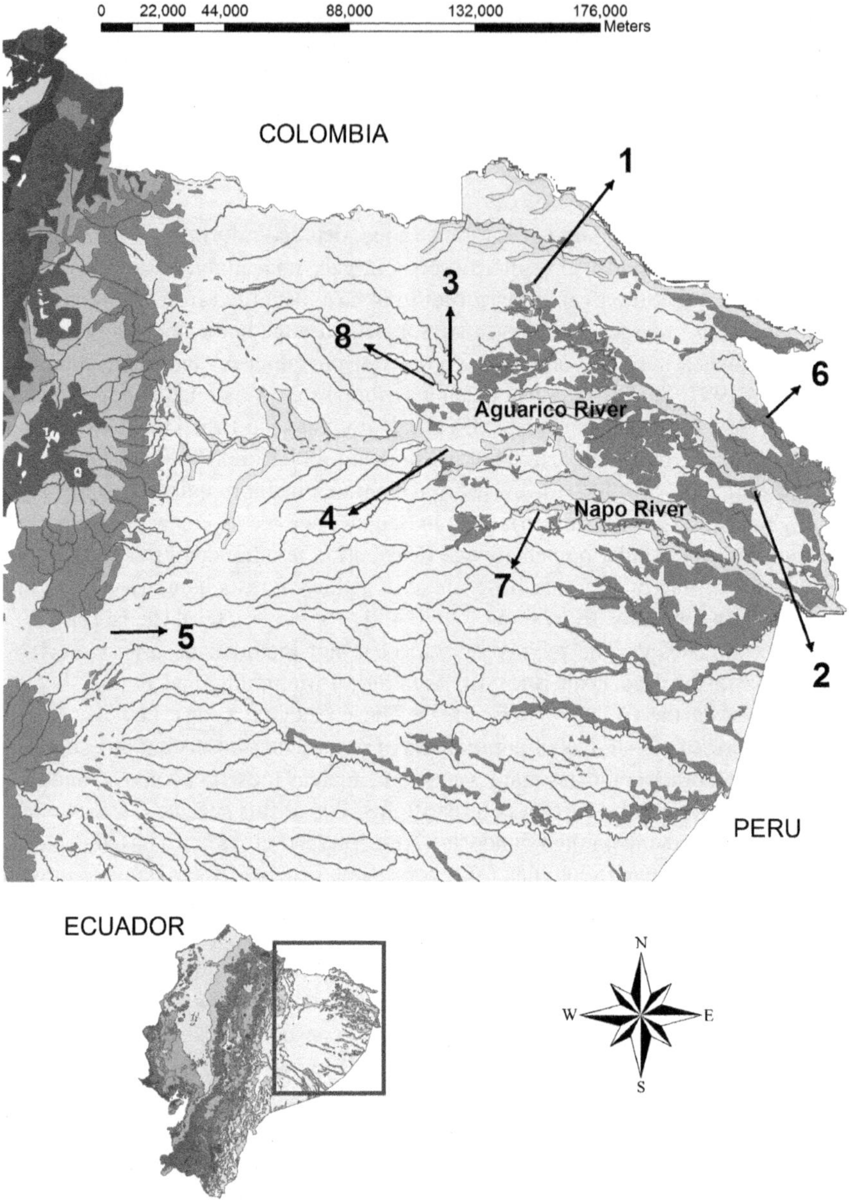

Fig. 22.1 Location of the studied populations of pygmy marmosets in Ecuadorian Amazonia (1 = La Hormiga, 2 = Zancudococha, 3 = San Pablo, 4 = Sacha, 5 = Amazoonico, 6 = Aguas Blancas, 7 = Tiputini, 8 = Shushifindi)

Datum PSAD 56 368315 E, 3685 N) is located on the margins of the Laguna Grande of the Cuyabeno hydrographic system. The Zancudococha population (UTM, 18 Zone, Datum PSAD 56 445459 E, 9933749 N) is located on the edges of the Zancudococha Lake. The San Pablo population (UTM, 18 Zone, Datum

PSAD 56 341767 E, 9969737 N) is located on the margins of the Aguarico River. The Sacha population (UTM, 18 Zone, Datum PSAD 56 337296 E, 9946861 N) is located on the margins of the Napo River. The Amazoonico population (UTM, 18 Zone, Datum PSAD 56 219290 E, 9883728 N) is located on the margins of the Arajuno River. Finally, the Aguas Blancas population (UTM, 18 Zone, Datum PSAD 56 449617 E, 9957865 N) is located on the margins of the Aguas Blancas River, a tributary of the Lagartococha River. In each of these areas, in addition to focusing our observations on 2–4 groups of marmosets, we carried out at least one annual survey of 2 days, in a transect of approximately 3 km along each river, to obtain data on the density of pygmy marmosets at each site. These annual surveys were also carried out in the Tiputini River (UTM, 18 Zone, Datum PSAD 56 371933 E, 9929635 N) and the Shushufindi River (UTM, 18 Zone, Datum PSAD 56 338694 E, 9968991 N). Thus, the areas from which we have data on the population status of pygmy marmosets in Ecuador represent an east–west transect of approximately 300 km and a north–south transect of approximately 150 km.

These study areas differ in their habitat, altitude, and human disturbance level (Table 22.1). Altitude varies from 230 to 360 m above sea level, with the Amazoonico area having the highest altitude. The habitats of the San Pablo, Sacha, Amazoonico, Aguas Blancas, and Tiputini are varzea forests (forests seasonally flooded by white-water rivers) (Palacios et al. 1999; Pires and Prance 1985) with different degrees of alteration. Varzea forests at San Pablo have been disturbed quite a lot by humans and are severely fragmented and surrounded by pasture and plantain agricultural plots. At Sacha, the varzea forests are currently regenerating as agricultural activities have stopped and have been replaced by tourism over the past 10 years. The varzeas at Amazoonico form a narrow fringe between the river and the terra firme forests located in the high hills of the area; these varzeas were previously used for agriculture and have regenerated in the past 5 years. The varzeas of Aguas Blancas and Tiputini are the least disturbed by human activities. The areas of Zancudococha, La Hormiga and Shushufindi are edge habitats between terra firme and igapo forests (forests seasonally flooded by black-water rivers) (Palacios et al. 1999; Pires and Prance 1985). The area of Shushufindi has been

Table 22.1 Forest types and relative intensity of human activities affecting pygmy marmosets in the study areas

Study area	Forest type	Human activities
Hormiga	Edge habitat of Terra Firme and Igapo	High tourism, high human traffic
Zancudo	Edge habitat of Terra Firme and Igapo	Low tourism, low human traffic
San Pablo	Varzea	Intense agriculture, intense logging, moderate human traffic
Sacha	Varzea	High tourism, very high human traffic
Amazoonico	Varzea	Moderate tourism, high human traffic
Aguas Blancas	Varzea	Low human traffic
Tiputini	Varzea	Low tourism, low human traffic
Shushufindi	Edge habitat of Terra Firme and Igapo	Sporadic agriculture, sporadic logging, low human traffic

affected by selective logging and sporadic agriculture in the past 10 years; the habitats of Zancudococha and La Hormiga have not been affected by agricultural activities, but the area of La Hormiga did have a high rate of tourism during the years when our research was carried out at that site (1996–1998) (de la Torre et al. 2000) (Table 22.1).

22.2.2 *Observations of Groups*

We studied a total of 19 groups of pygmy marmosets from six populations in a total of 2,680 h of direct observation. We carried out observations of three pygmy marmoset groups at La Hormiga (635 h of direct observation) from October 1996 through February 1997 and from March to May 1997. Three groups at Zancudococha (775 h of direct observation) were observed from June to August 1997 and from November 1997 to February 1998. We studied four groups of pygmy marmosets at San Pablo (July 2001, July 2002, June 2003, July 2004, July 2005; 431 h of direct observation), four groups at Sacha (August 2001, June 2002, August 2003; 507 h of direct observation), two groups at Amazoonico (July 2003, June 2004; 212 h of direct observation), and three groups at Aguas Blancas (August 2004, July 2005; 120 h of direct observation).

The marmosets in each group were classified by size and other morphological characters (e.g., presence of a white nasal stripe) into approximate age classes (Soini 1988). Sex determination was possible for adult and subadult animals. Group size and composition were easily determined by observations and counts of all group members early in the morning or late in the afternoon. We used scan samples (Martin and Bateson 1994) every hour to gather data on activity and use of habitat (every 20 min in La Hormiga and Zancudococha).

Home range size was estimated by connecting the extreme location points of group members during the study period; the periphery that enclosed all points was considered as the home range perimeter and the area inside the perimeter was calculated (de la Torre et al. 2000; Yépez et al. 2005a).

To estimate the degree of habituation to human presence of the groups, we obtained an observability coefficient by dividing the number of observed individuals/scan/day over the maximum possible number of individuals that could be observed/scan/day (determined by group size) (de la Torre et al. 2000).

22.2.3 *Population Censuses*

We carried out population censuses of eight populations, including six from which detailed observations were taken, and two others (Tiputini and Shushufindi). These censuses were conducted at La Hormiga (1996–1998); at Zancudococha (1997 and 1998); at San Pablo (2001–2005); in Sacha (2001–2003); in Amazoonico

(2003 and 2004); in Aguas Blancas (2004 and 2005); in Tiputini (2004); and in Shushufindi (2004 and 2005). Surveys were carried out in a paddle-canoe by two to three fieldworkers. We carried out one morning survey (06h00–10h00) and one afternoon survey (15h00–18h00) over 3 km on two days during the same climatic season; these times of the day coincide with activity peaks of the groups of marmosets when animals are more conspicuous (Yépez et al. 2005a). We began a 3 km river transect in a fixed area close to each of our base camps. Surveys were carried out in the Napo and Aguarico rivers, which have widths of more than 300 m, for 2 days on each side of the river. As groups in all areas lived at the edge of rivers or lakes, these surveys allowed us to estimate the ecological density of pygmy marmosets at each site. The transects included the home range areas of most of our studied groups.

In each survey, we searched for trees with exudate holes made by the marmosets and used the vocalizations of the marmosets to detect groups within a range of approximately 50 m in each side of the river. Since we did not have reliable counts of group members in all the groups observed in the censuses, we calculated the number of individuals/km of river margin by multiplying the number of groups found in each transect by the mean group size of that population; for the estimates of the Tiputini and Shushufindi populations we used the mean group size from all the studied populations.

22.3 Results and Discussion

22.3.1 Group Size and Composition

In the 19 groups studied, group size varied from 3 to 8 individuals, including infants (Table 22.2). Overall, the modal group size was 6 individuals (Mean = 5.36 ± 0.19). All groups included a reproductive couple (adult male and female), 0–3 subadults, 0–2 juveniles and 0–2 infants.

The largest mean group sizes were found in the Aguas Blancas population (6.67 individuals), whereas the smallest group sizes were found in the Amazoonico population (3.75 individuals). We found no significant differences in mean group size between populations living in igapo forests (La Hormiga and Zancudococha) and those living in varzea forests ($t = 0.45$, $df = 4$, $p = 0.68$). However, we did find differences within populations living in igapo and varzea forests: in both types of forests the largest group sizes occurred in the populations living in the areas that were least disturbed by human activities (Zancudococha for igapo and Aguas Blancas for varzea).

Changes in group size were due to births and losses of infants and juveniles, disappearances of subadults (due to death or emigration), and human capture. This last factor was reported not only in groups at La Hormiga (de la Torre et al. 2000) but also in groups at San Pablo (groups P1 in 2001, and P2 in 2004) and Sacha (groups S1 and S3 in 2003). Human capture usually involves the live capture of animals to raise them as pets; however, in some indigenous communities, pygmy

Table 22.2 Group size and home range size of the studied groups and populations of pygmy marmosets, averaged across years

Population	Group	Mean group size ± S.E.	Home range size (ha)	Population mean group size ± S.E.	Population mean home range size (ha) ± S.E.
La Hormiga	H1	5.5 ± 0.3	0.9	4.93 ± 0.3	0.84 ± 0.3
	H2	3.3 ± 0.3	1.09		
	H3	6 ± 0	0.53		
Zancudo	Z1	6.7 ± 0.3	0.73	6.39 ± 0.1	0.63 ± 0.3
	Z2	7.0 ± 1	0.4		
	Z3	5.5 ± 0.5	0.78		
San Pablo	P1	4.7 ± 0.3	0.45	5.40 ± 0.2	0.33 ± 0.3
	P2	6.7 ± 0.2	0.22		
	P4	6.3 ± 0.3	0.31		
	P5	4.0 ± 0.2	0.33		
Sacha	S1	4.7 ± 0.2	0.37	5.04 ± 0.1	0.58 ± 0.7
	S2	5.3 ± 0.4	1.2		
	S2B	4.5 ± 0.2	0.36		
	S3	5.7 ± 0.4	0.4		
Amazoonico	A1	4.5 ± 0.5	0.15	3.75 ± 0.3	0.28 ± 0.6
	A2	3 ± 0	0.4		
Aguas Blancas	L1	8 ± 0	0.4	6.70 ± 0.2	0.47 ± 0.4
	L2	6 ± 0	0.7		
	L3	6 ± 0	0.32		

marmosets are also captured for food or killed for target practice (de la Torre and Yépez 2003). The capture of animals significantly reduced the subsequent observability of the groups, their use of the lower strata of the forests, social behaviors such as play, and, possibly, their reproductive rate, as reported by de la Torre et al. (2000). Human capture may have also been related to the complete disappearance of some groups in the San Pablo population (see below).

22.3.2 *Home Range and Habitat Use*

Home range size varied from 0.15 to 1.2 ha (mean 0.9 ha ± 0.32). Overall, the largest home ranges were found in the La Hormiga population, whereas the smallest were found in the Amazoonico population (Table 22.2). We found a significant difference between the mean size of home ranges in populations living in igapo (mean home range size = 0.74 ha ± 0.1) and populations living in varzeas (mean home range size = 0.42 ha ± 0.07) ($t = -2.7$, $df = 4$, $p = 0.05$). We did not find a strong or significant correlation between group size and home range size. All of the group home ranges were relatively elongated fringes (30–100 m wide) of forest adjacent to a river, lake or swamp; the exceptions were the home ranges of groups P4 and P5 in San Pablo, which were located at approximately 200 m from the river edge, but still in areas of varzea.

The number of exudate sources in each home range varied from 2 to 8 (mode = 2), and were of 1–5 different species (mode = 2). We have reported a total of 18 different exudate species eaten by pygmy marmosets from Sacha, San Pablo, Amazoonico and Zancudococha populations (Yépez et al. 2005a). This number increases to 21 when exudate species used by pygmy marmosets at La Hormiga and Aguas Blancas are included. In our research, we have found that each of the populations differed in the total number of species used and in the preferred species. These differences did not appear to be related to the availability of these species in each population (Yépez et al. 2005a).

The plant species that are used as exudate feeding sources deserve protection in environmental plans in Ecuadorian Amazon and elsewhere. It is noteworthy, however, that our data on interpopulation differences in diet suggest that protecting exudate resources based on data from only one area will not be sufficient to preserve pygmy marmosets in all populations (Yépez et al. 2005a).

22.3.3 Population Densities

Averaged across annual censuses, the mean ecological densities of populations varied from 1.79 individuals/km of river edge, found along the Tiputini River, to 7.28 individuals/km, found along the Napo River in the Sacha population (Table 22.3). We found no significant differences between population densities in igapo and varzea forests. The greatest variability in population density occurred within the varzea populations and could be partly related to ecological factors such as the size of the rivers and the degree of anthropogenic disturbance. Varzeas of large rivers, like the Napo River, appear to have higher densities of pygmy marmosets than varzeas of smaller rivers, like the Tiputini. However, the varzeas of another large river, the Aguarico River, showed lower densities of pygmy marmosets than the varzeas of a small river such as the Aguas Blancas, the least disturbed area. Considering that the varzeas at San Pablo are continuously disturbed by human activities, these results suggest that anthropogenic factors may be related to the reduced population densities recorded in the area of the San Pablo population (see below).

Table 22.3 Mean ecological densities (individuals/km of river) in the study areas, averaged across years (groups were searched within a 50 m belt on both sides of the river)

Study area (river)	Years of survey	Individuals/km of river ± S.E.
Hormiga	1996, 1997, 1998	4.38 ± 0.2
Zancudo	1997, 1998	6.39 ± 0
San Pablo	2001, 2002, 2003, 2004, 2005	3.24 ± 0.2
Sacha	2001, 2002, 2003	7.28 ± 0.1
Amazoonico	2003, 2004	3.13 ± 0.3
Aguas Blancas	2004, 2005	5.6 ± 0.3
Tiputini	2004	1.8 ± 0
Shushufindi	2004, 2005	5.36 ± 0

22.3.4 Habitat Destruction

A strong relationship between the abundance of pygmy marmosets and the availability of suitable riparian habitats has already been proposed by other authors (Ramirez et al. 1977; Rylands et al. 1993). In a pilot study we carried out in 2004, we attempted to quantify the degree of human disturbance in the gallery forests inhabited by pygmy marmosets. We analyzed a segment of 200 km of the Napo River, which included some of our study areas, with GIS methodology to estimate the area of disturbed forests (by logging and agriculture) within a range of 1 km wide on both sides of the river edges (de la Torre et al. 2004). Disturbance was identified by differences in the color in the satellite images (LANDSAT 2001) and was confirmed by in situ observations. We found that the degree of disturbance along this segment of the Napo River was considerably high. By 2001 about 75% of the sampled area was altered by human activities (de la Torre et al. 2004).

Habitat destruction appears to be the greatest threat to pygmy marmoset populations. In our annual surveys in several rivers of northeastern Ecuador, including the Napo River, we have recorded the absence of pygmy marmosets in highly disturbed habitats (grasslands and areas of intense agriculture). Furthermore, in our studied populations, we have recorded the migration and disappearance of groups of pygmy marmosets from habitats that were affected by selective logging (e.g., group P2 at San Pablo). Other researchers have found pygmy marmosets in forest patches close to human settlements (e.g., Soini 1988), however as the loss of groups in disturbed areas does not necessarily occur immediately after the alteration, it is likely that these groups eventually disappeared. This finding is exemplified in the case study of the San Pablo population.

22.3.5 Human Impact on Pygmy Marmosets: The Case of the San Pablo Population

As we explained in our descriptions of the study areas, the San Pablo population inhabited disturbed varzea forests along the Aguarico River. Forest remnants in this area are fragmented and surrounded by pasture, plantain, and manioc agricultural plots. When we began our study of the groups of pygmy marmosets in this population in 2001, the combined area of forest patches, which included the home ranges of our four studied groups, around our base camp was approximately 3.0 ha. By 2005, this area was reduced by 85% to only 0.5 ha; all these forest remnants have been gradually converted into grasslands and agricultural fields.

This habitat alteration appears to have contributed to the gradual decrease in the population densities of pygmy marmosets in San Pablo across the years (Fig. 22.2). Even though none of the trees serving as exudate sources in the home ranges of the studied groups have been cut, three of the four groups that we studied in 2001 had disappeared from the area by 2005. The disappearance of pygmy marmoset groups

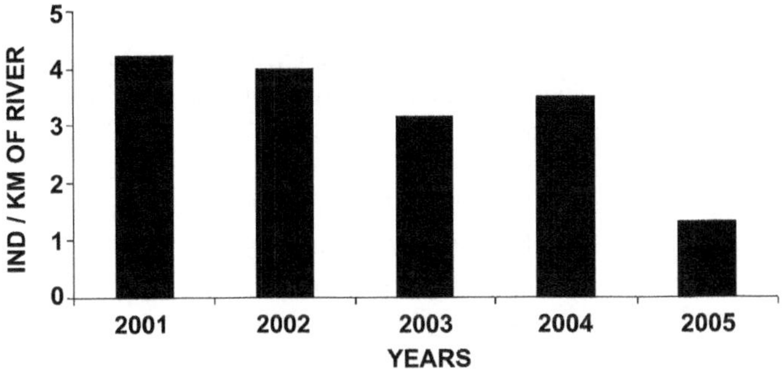

Fig. 22.2 Annual ecological density (individual/km of river) of pygmy marmosets in the San Pablo population

in disturbed areas could be partly explained by the depletion of their exudate sources and the absence of new sources in the vicinities of their original home ranges. Once resources were depleted, some groups may have migrated in search of new exudate sources, a difficult task considering the rate of habitat destruction in the area. This may be why we have not been able to find these lost groups in a radius of about 1 km from their original areas.

In addition to the effects of habitat alteration, the capture of young marmosets to raise as pets also has negatively impacted populations. Besides the obvious decreases in group size after the captures, capture may have also been related to the disappearances of group P1 in 2002 and of group P2 in 2005 that occurred immediately after these capture events. Group P1 was found again in 2003 but its observability coefficient (habituation of marmosets to human observers) decreased considerably, from 17% in 2001 to 5.6% in 2003 and 1.1% in 2004; by 2005 this group disappeared. The fate of the captured animals is also negative. Most of them die during the first 2 months after the capture (A. Payaguaje pers. com., pers. obs.).

22.4 Re-Evaluation of the Conservation Status of Pygmy Marmosets in Ecuador

In our studies we have reported the absence or disappearance of groups of pygmy marmosets in disturbed areas. We also have presented evidence that these primates are particularly sensitive to human traffic, ambient noise and life captures, altering their behavior and decreasing their group size and reproductive rate (de la Torre et al. 2000; de la Torre and Snowdon 2003, this chapter). We propose to include pygmy marmosets in the list of Vulnerable species in Ecuador, based on the IUCN criterion A4 (c, d) (Tirira 2001), considering that the rate of habitat destruction is one of the most important factors affecting pygmy marmoset populations in Ecuador, that the annual deforestation rate in Ecuador is around 150,000 ha

(Varea et al. 1997), and that the gallery forests along rivers in the Ecuadorian Amazon are being significantly altered by human activities (de la Torre et al. 2004). This criterion applies to species with an observed, estimated, inferred, projected or suspected population size reduction of more than or equal to 30% over any 10 year or three generation period, whichever is longer, where the time period includes both the past and the future, and where the reduction or its causes may not have ceased based on (c) a decline in area of occupancy, extent of occurrence or quality of habitat, and (d) actual or potential levels of exploitation.

Although we do not know the degree of habitat disturbance in pygmy marmoset ranges in other countries of the Western Amazon, the increased anthropogenic disturbances that we have described for Ecuador are equally likely to be occurring in other countries (Laurance 1998). We suggest that it is important to conduct similar surveys in Colombia, Brazil, Peru and Bolivia.

22.5 Some Concluding Remarks

The inclusion of pygmy marmosets in the list of Vulnerable species in Ecuador is only one of the necessary steps we need to take to conserve them and their habitats. In addition, we have taken steps to create public awareness about the importance of conserving the Ecuadorian primates, among them the pygmy marmosets. Several inter-site differences indicate that conservation of multiple populations is important. Pygmy marmoset populations differ from one another not only in the acoustic structure of some of their vocalizations, suggesting different dialects (de la Torre and Snowdon 2009), but also in their preference for exudate species in a way that cannot be explained only by the distribution and availability of the exudate resources (Yépez et al. 2005a). These interpopulation differences in feeding ecology and communication are comparable to population differences in tool use and feeding adaptations reported in chimpanzees and capuchin monkeys (Chapman and Fedigan 1990; Whiten et al. 1999). Given this interpopulation variability, the loss of even one population may imply the loss of a unique behavior.

We believe that some of the conservation problems in dealing with pygmy marmosets and other Ecuadorian primates could be mitigated with a program of environmental education. We began an education program in 2003 and have directed it to the children of the communities that live close to some of the pygmy marmoset populations we study (de la Torre and Yépez 2003). We have developed several interactive games to teach children about the importance of preserving Ecuadorian primates and their habitats, including a didactic poster, a booklet, a hand-puppet play and two videos. All this interactive material recreates aspects of the ecology and behavior of pygmy marmosets. The didactic poster, for example, represents a gallery forest and includes a feeding tree with real holes in the trunk, and the profiles of six animals of different ages and sex, representing a typical group of this species. The profiles of these animals are the basis of a puzzle of six wooden figures of the six different animals to be attached to the corresponding profiles by the children during a game in which we tell them about the habitat and social organization of pygmy marmosets. When we talk about the gum-feeding behavior of these primates, we let all the children experience how it might feel by licking a vitamin gel that we poured into the holes of the "gum tree" on the poster. Children later mimic insect-feeding by finding and eating candies hidden in the classroom (de la Torre and Yépez 2003). To date, we have reached about 250 children in all of the areas where we have conducted our research and about 100 children in urban schools. In each of our presentations, we have evaluated the improvement in the children's knowledge about the behavior, ecology, and conservation of primates and their habitats, and the acceptance of our program. In all these aspects we have had a positive evaluation of our efforts.

We have also tried to increase public awareness by writing popular articles about Ecuadorian Amazonia, primates, and pet trade in Ecuadorian magazines, and by presenting our videos to several audiences in nature festivals, seminars, and congresses. Finally, one of our last efforts to stop habitat degradation in the area of San Pablo has been to rescue the traditional knowledge that the indigenous Secoya peoples have about the use of plants through detailed ethnobotany research and the creation of an ethnobotany garden in a forest remnant in the Shushufindi river (close to San Pablo), where there are also groups of pygmy marmosets (Yépez et al. 2005b). We believe that helping the Secoya people, as well as other native ethnic groups, to reconnect with their culture and history and to rebuild the strong relationship they once had with the plants and the forest may be a very critical act for conservation in Ecuadorian Amazonia.

Acknowledgments We are grateful for the support from the following institutions: Ecuadorian Ministry of the Environment, Wisconsin Regional Primate Center (NIH Grant RR00167), Tiputini Biological Station, Transturi, Neotropic Turis, Selva Viva and Sacha Lodge. We thank Delfín Payaguaje, Alfredo Payaguaje, José Payaguaje, Daniel Payaguaje, Lucía de la Torre, Monserrat Bejarano, Fernanda Tomaselli, Hernán Castaneda, Santiago Molina, Beatriz Romero, Margarita Brandt, André Izurieta and Carolina Proaño for their help in the field work. We thank Susan Ford and Lucía de la Torre for their valuable help with the figures of this paper. Our research was supported by NIH Grant MH 29775 and National Geographic Society Grant 5806-96, with additional support from the Primate Action Funds/Conservation International, Milwaukee Zoological Society, the Tinker-Nave Fund, the University of Wisconsin Davis Fund and the Latin American Studies Program in American Universities, LASPAU.

References

Chapman CA, Fedigan LM (1990) Dietary differences between neighboring *Cebus capucinus* groups: local traditions, food availability or responses to food profitability? Folia Primatol 54:177–186

Coimbra-Filho AF (1984) The status of callitrichids in Brazil (Callitrichidae-Primates). Primatologia no Brasil 1:13–33

de la Torre S (2000) Primates de la Amazonía Ecuatoriana (Primates of Amazonian Ecuador). Proyecto PETRAMAZ/SIMBIOE, Quito

de la Torre S, Snowdon CT (2002) Environmental correlates of vocal communication of wild pygmy marmosets. Anim Behav 63:847–856 de la Torre S, Snowdon CT (2003) La Ecología del Comportamiento como herramienta para evaluar el impacto humano sobre poblaciones de primates. In: de la Torre S, Reck G (eds) Ecología y Ambiente en el Ecuador, Memorias del I Congreso de Ecología y Ambiente, Ecuador país megadiverso. Quito, CD, Universidad San Francisco de Quito

de la Torre S, Snowdon CT, Bejarano M (2000) Effects of human activities on pygmy marmosets in Ecuadorian Amazon. Biol Conserv 94:153–163

de la Torre S, Snowdon CT (2009) Dialects in pygmy marmosets? Population variation in call structure. Am J Primatol 71:333–342

de la Torre S, Yépez P (2003) Environmental education, a didactic tool for the conservation of pygmy marmosets *Cebuella pygmaea* in Ecuadorian Amazon. Neotrop Primates 11(2):73–75

de la Torre S, Yépez P, Snowdon CT, Pallares M (2004) Re-evaluación del estatus de conservación del leoncillo *Cebuella pygmaea* en Ecuador. Memorias del II Congreso de Ecología y Ambiente, Ecuador país megadiverso. Cuenca, CD Universidad del Azuay

Goldizen AW (1987) Tamarins and marmosets: communal care of offspring. In: Smuts B, Cheney D, Seyfarth R, Wrangham R, Struhsaker T (eds) Primate societies. University of Chicago Press, Chicago, pp 34–43

Groves C (2001) Primate taxonomy. Smithsonian Institution Press, Washington, DC

Hernandez-Camacho J, Defler TR (1985) Some aspects of the conservation of non-human primates in Colombia. Prim Conserv 6:42–50

Laurance WF (1998) A crisis in the making: responses of Amazonian forests to land use and climate change. Trends Ecol Evol 13:411–415

Martin P, Bateson P (1994) Measuring behaviour, an introductory guide. Cambridge University Press, Cambridge

Meggers BJ (1989) Amazonía: Hombre y cultura en un paraíso ilusorio. Siglo Veintiuno, Mexico

Palacios W, Cerón C, Valencia R, Sierra R (1999) Las formaciones naturales de la Amazonía del Ecuador. In: Sierra R (ed) Propuesta preliminar de un sistema de clasificación de vegetación para el Ecuador continental. Proyecto INEFAN/GEF-BIRF y Ecociencia, Quito, pp 109–119

Pires JM, Prance GT (1985) The vegetation types of the Brazilian Amazon. In: Prance GT, Lovejoy TE (eds) Amazonia, key environments. Pergamon Press, Oxford, pp 109–145

Ramirez M, Freese C, Revilla J (1977) Feeding ecology of the pygmy marmosets, *Cebuella pygmaea*, in northeastern Peru. In: Kleiman DG (ed) The biology and conservation of the Callitrichidae. Smithsonian Institution Press, Washington, DC, pp 91–104

Rylands AB, Coimbra-Filho AF, Mittermeier RA (1993) Systematics, geographic distribution, and some notes on the conservation status of the Callitrichidae. In: Rylands AB (ed) Marmosets and tamarins, systematics, behaviour and ecology. Oxford Science Publications, Oxford, pp 11–77

Smith RL, Smith TM (2000) Elements of ecology. Adison Wesley Longman, Inc., San Fransisco

Soini P (1982) Primate conservation in Peruvian Amazonia. Int Zoo Yearbk 22:37–47

Soini P (1988) The pygmy marmoset, Genus *Cebuella*. In: Mittermeier RA, Rylands AB, Coimbra-Filho AF, da Fonseca GAB (eds) Ecology and behavior of neotropical primates, vol 2. World Wildlife Fund, Washington, DC, pp 79–129

Stevenson MF, Rylands AB (1988) The marmosets, genus *Callithrix*. In: Mittermeier RA, Rylands AB, Coimbra-Filho AF, da Fonseca GAB (eds) Ecology and behavior of neotropical primates, vol 2. World Wildlife Fund, Washington, DC, pp 131–222

Tirira D (ed) (2001) Libro rojo de los mamíferos del Ecuador. SIMBIOE/Ecociencia/Ministerio del Ambiente/UICN. Serie Libros Rojos del Ecuador, tomo 1. Publicación Especial sobre los Mamíferos del Ecuador 4, Quito

Varea A, Barrera C, Maldonado AM, Endara L, Real B (1997) Ecologismo ecuatorial. Cedep, Abya-Yala, Quito

Whiten A, Goodall J, McGrew WC, Nishida T, Reynolds V, Sugiyama Y, Tutin CEG, Wrangham RW, Boesch C (1999) Cultures in chimpanzees. Nature 399:682–685

Yépez P, de la Torre S, Snowdon CT (2005a) Interpopulation differences in exudate feeding of pygmy marmosets in Ecuadorian Amazonia. Am J Primatol 66:145–158

Yépez P, de la Torre S, Cerón C, Palacios W (eds) (2005b) Al inicio del sendero: Estudios Etnobotánicos Secoya. Ed. Arboleda, Quito

Chapter 23
Conservation of the Marmosets and Callimicos

Stephen F. Ferrari

Abstract The marmosets and callimicos are a diverse group of primates distributed over much of tropical South America, south of the Amazon and east of the Andes. All but three species are endemic to Brazil. Given their small size and ecological flexibility, these primates are relatively resistant to the effects of habitat fragmentation, although the small geographic ranges of some species make them vulnerable to extensive habitat loss, a problem typical of the Atlantic Forest. For this reason, two species – *Callithrix aurita* and *Callithrix flaviceps* – are currently considered endangered, and the remaining Atlantic Forest endemics appear to be relatively vulnerable to extinction. The consolidation of the remaining forest and the management of marmoset metapopulations are the main priorities in this biome. In the Amazon, by contrast, while deforestation is still a distant threat to most species, recent discoveries have resulted in an increasing number of species with relatively small geographic ranges, which are potentially vulnerable to localized concentrations of human colonization. Recent trends indicate that the true diversity of the Amazonian marmosets (possibly including *Cebuella* and *Callimico*) is still unknown, and that further research is necessary before conservation strategies can be planned effectively. Overall, more data on the biological, ecological and genetic characteristics of all the species will be required in order to guarantee their conservation over the long term.

Resumen Los marmosets y los callimicos son un grupo diverso de primates distribuidos sobre la mayor parte de Sudamérica tropical al sur del Amazonas y al este de los Andes. Tres de las especies son endémicas de Brasil. Dado su pequeño tamaño y flexibilidad ecológica, dichos primates son relativamente resistentes a los efectos de fragmentación de habitat, a pesar de que los pequeños ámbitos geográficos de algunas de las especies los hacen vulnerables a una perdida extensiva de habitat, la cual es típica del bosque del Atlántico. Debido a ello, dos especies del bosque

S.F. Ferrari (✉)
Department of Biology, Universidade Federal de Sergipe, São Cristóvão, SE, Brazil
e-mail: ferrari@pq.cnpq.br

S.M. Ford et al. (eds.), *The Smallest Anthropoids*, Developments in Primatology: Progress and Prospects, DOI 10.1007/978-1-4419-0293-1_23,
© Springer Science+Business Media, LLC 2009

Atlántico – *Callithrix aurita* y *Callithrix flaviceps* – son actualmente consideradas en peligro, y los restantes endémicas parecen estar relativamente vulnerables. La consolidación del resto del habitat y el manejo de la metapoblaciones son la mayor prioridad en dicha bioma. En contraste, en el Amazonas, mientras la deforestación es todavía una amenaza distante para la mayoria de las especies, descubrimientos recientes han dado como resultado el incremento del número de especies con ámbitos relativamente reducidos, los cuales son potencialmente vulnerables a las concentraciones localizadas de colonización humana. Tendencias recientes indican que la verdadera diversidad de los marmosets amazónicos (posiblemente incluyendo *Cebuella* y *Callimico*) es todavía desconocida, e investigación adicional es necesaria antes de que estrategias de conservación puedan ser planificadas efectivamente. Principalmente, más información sobre las caracteristicas biológicas, ecológicas, y genéticas de todas las especies serán requeridas con el fin de garantizar su conservación a largo plazo.

Resumo Os saguis e *Callimico* formam um grupo diverso de primatas, distribuído ao longo de praticamente toda a América do Sul, a sul do Rio Amazonas e a leste dos Andes. Com exceção de três espécies, todas são endêmicas do Brasil. Devido ao seu tamanho reduzido e flexibilidade ecológica, estes primatas são relativamente resistentes aos efeitos de fragmentação de habitat, embora as distribuições geográficas pequenas de algumas espécies as tornam vulneráveis à perda de habitat, que é típica da Mata Atlântica. De acordo com isto, duas espécies da Mata Atlântica (*Callithrix aurita* e *Callithrix flaviceps*) já são consideradas ameaçadas de extinção, e as demais espécies endêmicas parecem estar relativamente vulneráveis. A consolidação do habitat remanescente, e o manejo de metapopulações são a prioridade maior neste bioma. Na Amazônia, por outro lado, o desmatamento ainda é uma ameaça distante para a maioria das espécies, mas as recentes descobertas têm resultado em um número cada vez maior de espécies com distribuições geográficas relativamente reduzidas, que são potencialmente vulneráveis a concentrações de colonização humana. As tendências recentes indicam que a diversidade verdadeira dos sagüis amazônicos (incluindo, possivelmente, *Cebuella* e *Callimico*) ainda é desconhecida, e que serão necessárias mais pesquisas para o planejamento efetivo de estratégias de conservação. De um modo geral, mais dados sobre as características biológicas, ecológicas e genéticas de todas as espécies serão imprescindíveis para garantir sua conservação a longo prazo.

23.1 Diversity and Conservation

If this book had been written 20 years ago, it would have had a completely different layout, and this overview of the conservation of the marmosets and callimicos would have been correspondingly dissimilar. At that time, the study of callitrichid biology was still incipient, and the taxonomic standard was Philip Hershkovitz's

(1977) classic review, in which there were just two genera and four species of marmoset, and *Callimico* was considered to be so distantly related that it was allocated to a distinct, monospecific family, the Callimiconidae. The only significant deviation from this scheme was even more conservative, with Rosenberger (1981) arguing for the inclusion of the pygmy marmoset in the genus *Callithrix*.

Following the taxonomic review of *Callithrix* by de Vivo (1990), in which twelve species were recognized, there was an "explosion" of diversification. Over the next 15 years, seven new marmoset species were recognized, *Callithrix* was divided into two genera (Rylands et al. 2000), a fourth marmoset genus was named (van Roosmalen and van Roosmalen 2003), and a consensus was reached on the close phylogenetic ties between marmosets and callimicos. While not all these modifications are universally accepted, and marmoset systematics continues to generate healthy debate (see Cortés-Ortiz this volume; Mendes this volume; Rylands et al. this volume), what is clear is that the group is far more diverse than it was considered to be only a few decades ago.

There are two main consequences of this process, and both have important implications for the conservation of the group's species. One corollary is that the complex and dynamic nature of current taxonomy generates a degree of uncertainty with regard to classifications and, in turn, the planning of conservation strategies. After all, even if the consensus view of Rylands et al. (2000) is taken as a definitive classification, there is still the possibility that new species will be discovered in the wild, even in Brazil's highly fragmented Atlantic Forest biome, where two species of small-bodied primate were discovered in the 1990s (Lorini and Persson 1990; Kobayashi and Langguth 1999). In the Amazon, where research efforts are so much thinner on the ground, the discovery of new species appears to be almost a certainty.

The second consequence is also fundamental to the conservation of the group as a whole, because the larger the number of species, the greater the overall difficulties for their conservation. According to Hershkovitz (1977), the marmoset-callimico group would consist of four species, each one with a geographic range of hundreds of thousands, or even millions of square kilometers. While some local populations (or subspecies) would be threatened, the species themselves would be in no great danger of extinction.

The greatest challenge may be the least expected, or predictable. Two monospecific genera – *Callimico* and *Cebuella* – have been conspicuous by their lack of revision over this same period. Some authors (e.g., van Roosmalen and van Roosmalen 1997; Buchanan-Smith et al. 2000; Tagliaro et al. 2000) have upheld the subspecies *Cebuella pygmaea niveiventris* Lönnberg, 1940, although it has not been widely accepted, especially as most recent reviews have concentrated on species-level diversity.

Both *Callimico* and *Cebuella* are distributed over a wide area of western Amazonia, a region known for its primate diversity. Within the same general area, for example, a third callitrichid genus – *Saguinus* – is represented by more than twenty species and subspecies, and no other platyrrhine genus is represented by only a single taxon. The obvious question here is why both *Callimico* and *Cebuella* should be so different, not only from the very diverse *Mico* and *Saguinus*, but also from the overall tendency, within the platyrrhines, of increasing taxonomic diversity with decreasing body size.

The answer may simply be that we do not know enough about the biological diversity of these two genera. The systematics of the platyrrhines in general, but of *Mico* and *Saguinus* in particular, is based primarily on the interpretation of variation in external morphology, especially pelage coloration, which is all but nonexistent in both *Callimico* and *Cebuella*.

An alternative approach would be to ask whether their lack of morphological variation actually reflects their species-level diversification. In the case of the pygmy marmosets, the answer to this question may be body size. As the smallest simian, *Cebuella* is the most vulnerable to predation, and its relatively drab (by marmoset standards) agouti coloration almost certainly reflects strong selective pressures favoring cryptic pelage.

As such pressure would be similar throughout its geographic range, it may have been a determining factor limiting the diversification of pelage coloration in this marmoset. Perhaps significantly, the only marked variation found in this genus is the lighter colored ventrum in the *niveiventris* form, which is obviously not visible during most normal postures. This conclusion on pelage coloration is also supported, tentatively, by the fact that *Callibella*, which is only slightly larger than *Cebuella*, is also relatively drab in comparison with most other marmosets from the same geographic area.

Even if predation pressure does limit the potential for variation in the pelage of *Cebuella*, other evidence suggests that this superficial homogeneity may hide underlying diversity. Recent genetic studies (Tagliaro et al. 1997, 2000; Meireles et al. 1989) have revealed considerable variability, compatible with species-level diversification in other genera. This obviously requires further investigation, in particular, the confirmation of systematic geographic variation in genetic characteristics, consistent with species-level diversity.

Assuming that *Cebuella* is not significantly different from *Mico* or *Saguinus* in its propensity for species-level diversification, its distinctive ecological characteristics would appear, if anything, to be relatively favorable to speciation. The available data indicate that pygmy marmosets are habitat specialists, occupying primarily floodplain habitats and forest edges (Terborgh 1983; Soini 1988; Buchanan-Smith et al. 2000; de la Torre et al. this volume). This degree of ecological specialization – which is unusual for a callitrichid – is almost certainly related to the unique challenge of competition with two sympatric genera (*Callimico* and *Saguinus*). By occupying marginal habitats, pygmy marmosets generally avoid contact with the larger callitrichids, normally found in *terra firme* forest. Pygmy marmoset populations thus tend to be relatively widely dispersed within a given area, and their limited capacity for dispersal, combined with the dynamic nature of floodplain habitats, would suggest that they are relatively more susceptible than sympatric tamarins to the effects of local selection pressures, as well as random processes such as genetic drift and bottlenecks. Whether this does in fact translate into species-level diversification remains to be seen.

The residual question is why – or how – *Callimico* is different. Callimicos are similar in size to tamarins, which would appear to proscribe the body size hypothesis, at least, as argued above in the case of *Cebuella*. Nevertheless, sympatry with tamarins has relegated *Callimico* to the lowest forest strata, where it is potentially more vulnerable to a range of terrestrial carnivores (Porter 2004). Its black pelage may thus reflect selection pressure favoring cryptic coloration against the shadows of the understory. As for *Cebuella*, recent studies (Vasarelhyi 2000, 2002) have indicated the existence of genetically distinct populations of *Callimico* from different areas within the genus' geographic range.

Whatever their "true" diversity, both *Callimico* and *Cebuella* challenge the current orthodoxy in callitrichid systematics: to consider all divergent morphological forms as distinct species. On the one hand, if both genera are, in fact, monospecific, the obvious question is, what is the functional significance of the morphological variation (pelage coloration) found in both sympatric *Saguinus* species groups? In other words, how can such similar animals have evolved in such different ways within the same ecological context?

On the other hand, if one or both of the other genera encompass cryptic species, the value of external morphology for the determination of species is brought into question. Either way, what is clear is that more data and more systematic analyses of the genetic and ecological variation of all three genera are required. While these are nothing more than philosophical questions to some, they are of fundamental importance for the planning of effective conservation strategies, making the definition of species-level diversity imperative.

23.2 Current Trends

All groups of nonhuman primates are under some threat of extinction, and the marmosets and callimicos are no exception, although their overall situation is relatively favorable in comparison with some other platyrrhines, such as the lion tamarins (*Leontopithecus*). There is, nevertheless, a major division between the Atlantic Forest species, half of which are "red listed" by the IUCN (2008), and the Amazonian forms, which are mostly at low risk or have undetermined status. Despite their current status, a specific set of problems facing the Amazonian species has potentially important implications for their long-term conservation.

In all cases, habitat loss is the main cause for concern, especially as the small size of these monkeys guarantees negligible hunting pressure. In contrast with most other primates, habitat fragmentation (rather than loss) may actually favor an increase in marmoset population density through factors such as an expansion in preferred habitat (disturbed, edge and secondary forest) and a reduction in competition from other primate species, which may become locally extinct (Ferrari 1993). In this case, the consolidation of the remaining habitat and metapopulation management are the main priorities for the conservation of the marmoset species.

Callimico goeldii is currently listed as "near threatened" by IUCN (2008), despite its vast geographic range. This classification is due primarily to its low densities and patchy distribution throughout most of this range. In the Pando of northern Bolivia, for example, callimicos were abundant at only one of the sites surveyed, whereas tamarins are relatively common throughout the region (Cameron et al. 1989; Porter 2006). This may make callimicos vulnerable to localized deforestation, although, like marmosets, they appear to prefer disturbed habitats (Rehg this volume), and may increase in density in bamboo forest (Porter 2006). As bamboo forest regenerates relatively quickly, and may often invade secondary forest, habitat fragmentation may even benefit callimico populations in the same way that it helps the marmosets.

The genus *Callithrix* presents both extremes of extinction risk. Two species, the common (*Callithrix jacchus*) and the pencil-tufted (*Callithrix penicillata*) marmosets, have ample geographic ranges covering hundred of thousands of square kilometers, which include both Atlantic Forest and savanna ecosystems, and almost invariably occur at high population densities not only in highly fragmented habitat, but also in plantations, and even urban environments. Populations of both species are also found well outside their natural ranges, and there are large captive breeding populations. Obviously, neither species is at any risk of extinction, and is unlikely to cause concern, even over the long term.

By contrast, the remaining four species, which are all endemic to the Atlantic Forest, have relatively restricted distributions, which coincide with areas that have lost more than 90% of their original forest cover. All but one of these species are "red listed" by the IUCN (2008). Of these, the buffy-headed marmoset (*Callithrix flaviceps*), which is listed as endangered, is the species closest to extinction in the wild. Its original range, which straddles the border between the Brazilian states of Espírito Santo and Minas Gerais, probably covered no more than 30,000 km^2.

The typical size of remnant forest fragments within this area is less than fifty hectares. Although the species is found in a number of much larger protected areas (most notably, the Augusto Ruschi Biological Reserve, in Espírito Santo, with 4,400 hectares of forest), and may survive in many small fragments (Ferrari and Diego 1993), it is clearly less flexible, in ecological terms, than either *C. jacchus* or *C. penicillata* (Ferrari 1993). Among other considerations, Ferrari (1988) recorded a home range of 35.5 ha in secondary forest, more than an order of magnitude than the smallest ranges recorded for *C. jacchus* or *C. penicillata* (Stevenson and Rylands 1988). While *C. flaviceps* may survive in forest patches smaller than this, it appears to be absent from many fragments within its original range, and records from the Augusto Ruschi Reserve (Ferrari and Mendes 1991) indicate that the species occurs at lower densities in primary forest in comparison with secondary habitat.

The endangered buffy-tufted-ear marmoset (*Callithrix aurita*) is ecologically similar to *C. flaviceps* (Ferrari et al. 1996; Martins and Setz 2000), and presents a comparable situation from a conservation standpoint. The principal difference between the two species is that *C. aurita* originally had a much larger geographic range, which has been better preserved, on the whole, in particular in São Paulo's Serra do Mar reserve complex. The evidence suggests that the numbers of *C. aurita* remaining in the wild are larger than those of *C. flaviceps*, and this is reflected in its recent reclassification as vulnerable by both IUCN (2008) and the Brazilian Environment Ministry (MMA 2003). own "red list."

The two other Atlantic Forest endemics – Geoffroy's marmoset (*Callithrix geoffroyi*) and Wied's marmoset (*Callithrix kuhlii*) – are less threatened overall, once again because of the size and preservation of their geographic ranges, as well as their increased ecological flexibility. In fact, the two species appear to be intermediate in many ways – both geographically and ecologically (Passamani and Rylands 2000) – between the southern "highland" species (*C. aurita* and *C. flaviceps*) and the ubiquitous northern forms, *C. jacchus* and *C. penicillata*. Only *C. kuhlii* is listed by IUCN (2008), as near threatened, whereas it was excluded by MMA (2003).

An additional problem facing the remnant populations of Atlantic Forest marmosets is hybridization. In some cases, habitat fragmentation appears to have altered or intensified existing hybrid zones, primarily through the isolation of populations (Ferrari and Mendes 1991). In others, exotic species have been introduced, often well outside their original ranges, by man. As crosses of all *Callithrix* species produce fertile hybrids (Coimbra-Filho et al. 1993), hybridization in isolated populations may threaten the morphological and ecological integrity of the original resident form significantly, with obvious implications for the conservation of the remnant populations.

Captive breeding populations have now been established for all four Atlantic Forest endemics, although, as might be expected, the most endangered species are the least represented. Captive populations constitute a potentially important tool for metapopulation management, although the only attempt to re-introduce marmosets (*C. geoffroyi*) into the wild proved surprisingly unsuccessful (Passmani et al. 1997).

All Amazonian species are protected – for the time being – by the relative isolation of their geographic ranges and the low deforestation rates in comparison with the Atlantic Forest. There is no room for complacency, however, as indicated by the example of the pied tamarin, *Saguinus bicolor*, the only Amazonian callitrichid listed as endangered by IUCN (2008). While this species is a typical callitrichid in the sense of its tolerance of habitat fragmentation, it has a relatively reduced geographic range – comparable to that of *C. flaviceps* – which coincides with the city of Manaus and the surrounding areas (Subirá 1998). In addition to a major loss of habitat, this species suffers intense anthropogenic pressure in an urban setting. A conservation program has been established (Gordo 2005), although the future of the species will depend primarily on adequate protection of the remaining habitat, which is extremely vulnerable to encroachment.

Despite the unique nature of this case, it serves as an example of what can happen where the geographic range of a species coincides with an area of intense anthropogenic pressure. No *Mico* species yet faces a situation of exactly this kind, although all the species with relatively small geographic ranges – including all of those discovered over the past 15 years – must be considered potentially vulnerable to this type of scenario. Possibly the most vulnerable is the black-headed marmoset (*Mico nigriceps*). Its relatively small range, which straddles the border between the states of Amazonas and Rondônia, is bisected by the Trans-Amazon highway, a major channel of human colonization in the region (Ferrari and Queiroz 1994). Other species, such as the golden-white bare-ear marmoset (*Mico leucippe*) and the intermediate marmoset (*Mico intermedius*), may be similarly threatened by their proximity to major highways such as the Trans-Amazon and the Santarém-Cuiabá.

There are two additional, related problems here – the lack of ecological and zoogeographic data, and the scarcity of protected areas. For example, a majority of species, both new and old, are known from less than a handful of localities, so the exact limits of their geographic ranges are unknown. In most cases, the only conclusion that can be drawn feasibly from the available evidence is that the species has a relatively small geographic distribution.

In spite of their ecological flexibility, marmosets may not always occupy all the forested habitat within a given area (see e.g., Ferrari 2004; Veracini this volume), so, without reliable ecological data – which are available for only two *Mico* species – it is risky to extrapolate from a small number of localities. The risks multiply in the case of a species with a small range, because any significant overestimation of the area occupied may result in equivocal and highly deleterious conclusions on variables such as population size and conservation status. The overall lack of reliable data is reflected in the classification of six species (*Mico acariensis, Mico chrysoleucus, Mico emiliae, Mico humeralifer, Mico marcai* and *Mico nigriceps*) as Data Deficient (IUCN 2008). It seems likely that other, more recently described species will be added to this list in due course.

All the new *Mico* species and *Callibella* occur in the Madeira-Tapajós interfluvium, a region with few protected areas. The lack of reserves is partly accounted for by the fact that the region is still relatively uninhabited. While it is likely that federal and regional governments will eventually implement more reserves, careful planning will be required if local marmoset populations are to be protected adequately. Reliable zoogeographic data are an obvious prerequisite here.

23.3 Perspectives for the Future

Unless a new arrangement alters their classification significantly, the conservation status of the callimicos and five marmoset species (*Callithrix jacchus, Callithrix penicillata, Cebuella pygmaea, Mico argentatus,* and *M. melanurus*) is unlikely to change in the near future. These species are protected primarily by their relatively ample geographic ranges, as well as their tolerance of anthropogenic habitat fragmentation.

Obviously, this situation could change significantly under a new taxonomic arrangement. This seems least likely to occur in the *Callithrix* species, which are relatively well defined, although recent advances in the systematics of the Amazonian marmosets – including the speculation on *Callimico* and *Cebuella* outlined above – suggest a potentially volatile situation, which may eventually result in the “creation” of new conservation problems, predominantly in the form of a larger number of species, each one with a relatively reduced geographic range. de la Torre et al. (this volume) have shown how the habitat preferences of *Cebuella* make pygmy marmosets vulnerable to human colonization on a local scale, a problem that would be exacerbated if the genus were segregated taxonomically.

A universal problem facing all the other species is a lack of reliable data on biological, ecological and genetic parameters, although the implications are clearly distinct when considering either the Atlantic Forest or the Amazonian species. In the case of the former group, geographic ranges are relatively well-defined, and the main priority is the management of the remnant populations. The keyword here is consolidation. While deforestation may not have ceased completely, it has been reduced to negligible levels, and it seems reasonable to conclude that most remaining marmoset habitats will be protected over at least the medium term.

The obvious models here are the lion tamarin conservation programs (e.g., Kleiman and Rylands 2002; Cullen et al. 2005), which integrate metapopulation management strategies with socioeconomic approaches, ranging from environmental education to the direct involvement of local landowners. Encouraging landowners to convert their reserves of natural habitat into private conservation units (*Reserva Particular de Patrimônio Natural* – RPPN) has proved to be an especially successful strategy.

The involvement of local landowners is a virtual necessity for the conservation of the Atlantic Forest, where the vast majority of habitat fragments are privately owned. One prominent recent event has been the creation of the RPPN Feliciano Miguel Abdalla, previously known as the Caratinga Biological Station, which protects almost one thousand hectares of forest. While the primary aim of the reserve is the conservation of its archetypal muriqui (*Brachyteles hypoxanthus*) population, it also protects a significant population of *Callithrix flaviceps* (Ferrari 1988).

Strategies such as the establishment of ecological corridors linking fragments will be an important component of metapopulation management, but where distances between fragments are prohibitive, translocations (see e.g., Médici et al., 2003) depending on population size and genetic variability may become necessary over time. The repopulation of vacant fragments may be especially important for the more endangered species, and the re-introduction of captive-born animals may also become an option, although this would require the development of more reliable procedures. The problem of hybridization also demands attention, although it affects only a limited number of populations, at least for the time being.

Reliable data on genetic variability will be an important complement of any management strategy. Up to now, there have been few studies of population-level variability in *Callithrix* species, although Nievergelt et al. (1998) provide an important baseline (see also Faulkes et al. this volume), and Gonçalves et al. (this volume) have shown that populations of *Mico argentatus* may be characterized by relatively reduced genetic variability, even under natural conditions; it is unclear, however, whether this is typical of all marmosets. These data will obviously be important for the assessment of problems, such as inbreeding depression (e.g., Dietz et al. 2000) in isolated remnant populations, and the planning of translocations and re-introductions.

In the case of the Amazonian marmosets (*Callibella* and *Mico*), the keyword, by contrast, is planning. The recent surge in taxonomic diversity has spawned a long list of species with relatively small geographic ranges, but little, if any, ecological data. Ideally, the first step in the planning of conservation strategies would be the definition of species-level diversity. While the identification of new species is an unpredictable process, a systematic, large-scale survey within the Madeira-Tapajós interfluvium might provide an adequate working data base.

Such a survey would also provide the much-needed data on geographic distribution, essential for the planning of protected areas, as well as first-hand information on the local trends in habitat fragmentation. Despite the overall lack of data, it is clear that some species, such as *Mico nigriceps* and *M. intermedius*, face more immediate threats than others, given their proximity to prominent areas of human colonization. However, the whole region is vulnerable to human encroachment of all kinds – most recently from soybean growers (Fearnside 2001) – and all species with relatively small ranges must be considered vulnerable to the establishment of local hubs of colonization.

An additional concern is the planning of protected areas. The reserve system in Brazilian Amazonia is characterized by the protection of relatively large blocks of forest, of the order of thousands of square kilometers. While reserves of this size may be adequate for the preservation of a representative sample of the fauna and flora of a given interfluvium, this strategy would almost certainly leave some, if not a majority, of marmoset species unprotected.

The area delimited by the Madeira-Tapajós interfluvium as far south as the Amazonas-Mato Grosso border, encompasses the geographic distribution of *Callibella*, and all or part of the ranges of all *Mico* species except *M. argentatus* (and possibly *Mico emiliae*). The only federal protected area within this polygon is the Amazônia National Park (ANP). While it covers a reasonably large area – almost 10,000 km² – this reserve protects the populations of two marmoset species (*Mico humeralifer* and *M. leucippe*), at most.

Given the nongame status of the marmosets, local Amerindian reserves undoubtedly offer a degree of protection to some of the other species However, two important questions arise here. One refers to the adequacy of the current reserve system. The polygon mentioned above covers an area of approximately 300,000 km², but the ANP protects only a relatively small and extremely localized proportion of the interfluvium, an area of high biodiversity, as exemplified so clearly by its marmosets.

The second question is how, exactly, to approach the problem of protecting so many marmoset species within the same area, especially considering their relatively low priority on a regional level. One possibility within the "single large" reserve policy of the federal government would be to protect a key stretch of the right bank of the Madeira, where many of the species are concentrated. While it may be unreasonable to expect the implementation of reserves specifically for the protection of given marmoset species, or even groups of species, it may be important for reserve planners to consider to what extent the diversity of these primates reflects that of the fauna and flora of the interfluvium in general.

Acknowledgments The author would like to thank Lesa Davis, Susan Ford, and Leila Porter for inviting this contribution. He also acknowledges Leila Porter for information on *Callimico*, and CNPq for supporting his research on marmosets through stipend no. 307506/20037.

References

Buchanan-Smith HM, Hardie SM, Caceres C, Prescott MJ (2000) Distribution and forest utilization of *Saguinus* and other primates of the Pando department, northern Bolivia. Int J Primatol 21:353–379

Cameron R, Wiltshire C, Foley C, Dougherty N, Aramayo X, Rea L (1989) Goeldi's monkey and other primates in northern Bolivia. Primate Conserv 10:62–70

Coimbra-Filho AF, Pissinatti A, Rylands AB (1993) Experimental multiple hybridism and natural hybrids among *Callithrix* species from eastern Brazil. In: Rylands AB (ed) Marmosets and tamarins: systematics, behaviour, and ecology. Oxford University Press, Oxford, pp 95–120

Cortés - Ortiz L (this volume) Molecular phylogenetics of the Callitrichidae with an emphasis on the marmosets and *Callimico*. In: Ford SM, Porter LM, Davis LC (eds) The smallest anthropoids: The marmoset/callimico radiation. Springer Press, New York, pp 3–24

Cullen L Jr, Alger K, Rambaldi D (2005) Land reform and biodiversity conservation in Brazil in the 1990s: conflict and the articulation of mutual interests. Conserv Biol 19:1–9

de la Torre S, Yépez P, Snowdon CT (this volume) Conservation statue of pygmy marmosets (*Cebuella pygmaea*) in Ecuador. In: Ford SM, Porter LM, Davis LC (eds) The smallest anthropoids: The marmoset/callimico radiation. Springer Press, New York, pp 451–464

de Vivo M (1990) Taxonomia de *Callithrix* Erxleben, 1777 (Callitrichidae, Primates). Fundação Biodiversitas, Belo Horizonte

Dietz JM, Baker AJ, Ballou JD (2000) Demographic evidence of inbreeding depression in wild golden lion tamarins. In: Young AG, Clarke GM (eds) Genetics, demography, and the viability of fragmented populations. Cambridge University Press, Cambridge, pp 203–211

Faulkes CG, Aruda MF, Monteiro da Cruz MAO (this volume) Genetic structure within and among populations of the common marmoset, *Callithrix jacchus*: implications for cooperative breeding. In: Ford SM, Porter LM, Davis LC (eds) The smallest anthropoids: The marmoset/callimico radiation. Springer Press, New York, pp 103–117

Fearnside PM (2001) Soybean cultivation as a threat to the environment in Brazil. Environ Conserv 28:23–38

Ferrari SF (1988) The behaviour and ecology of the buffy-headed Marmoset, *Callithrix flaviceps* (O. Thomas, 1903). PhD Dissertation, University of London

Ferrari SF (1993) Ecological differentiation in the Callitrichidae. In: Rylands AB (ed) Marmosets and tamarins: systematics, behaviour and ecology. Oxford University Press, Oxford, pp 314–328

Ferrari SF (2004) Biogeography of Amazonian primates. In: Mendes SDL, Chiarello AG (eds) A Primatologia no Brasil – 8. Sociedade Brasileira de Primatologia, Santa Teresa, pp 101–122

Ferrari SF, Diego VH (1993) Rethinking the status of *Callithrix flaviceps*. Neotrop Primates 1(3):2–3

Ferrari SF, Mendes SL (1991) Buffy-headed marmosets ten years on. Oryx 25:105–109

Ferrari SF, Queiroz HL (1994) Two new Brazilian primates discovered, endangered. Oryx 28:31–36

Ferrari SF, Corrêa HKM, Coutinho PEG (1996) Ecology of the southern marmosets: how different, how similar? In: Norconk MA, Rosenberger AL, Garber PA (eds) Adaptive radiations of neotropical primates. Academic Press, New York, pp 157–171

Gonçalves E, Ferrari SF, Coutinho PEG, Menezes E, Silva A, Schneider MPC (this volume) Limited dispersal and genetic structure of silvery marmosets (*Mico argentatus*) in the fragmented landscape of central Amazonia. In: Ford SM, Porter LM, Davis LC (eds) The smallest anthropoids: The marmoset/callimico radiation. Springer Press, New York, pp 205–220

Gordo M (2005) A conservação de *Saguinus bicolor* em fragmentos florestais na cidade de Manaus, Amazonas. Resumos do XI Congresso Brasileiro de Primatologia, Porto Alegre, p 50

Hershkovitz P (1977) Living New World monkeys, with an introduction to primates, volume 1. Chicago University Press, Chicago

IUCN (2008) www.redlist.org

Kleiman DG, Rylands AB (2002) Lion tamarins – biology and conservation. Smithsonian Institution Press, Washington, DC

Kobayashi S, Langguth A (1999) A new species of titi monkey, *Callicebus* Thomas, from northeastern Brazil (Primates, Cebidae). Rev Bras Zool 16:531–551

Lorini ML, Persson VG (1990) Nova espécie de *Leontopithecus* Lesson, 1840, do sul do Brasil (Primates, Callitrichidae). Bol Mus Nac 338:1–14

Martins MM, Setz EZF (2000) Diet of buffy tufted-eared marmosets (*Callithrix aurita*) in a forest fragment in southeastern Brazil. Int J Primatol 21:467–476

Médici P, Valadares-Pádua CB, Rylands AB, Martins CS (2003) Translocation as a metapopulation management tool for the black lion tamarin, *Leontopithecus chrysopygus*. Primate Conserv 19:23–31

Meireles CMM, Czelusniak J, Sampaio I, Schneider H, Ferrari SF, Coimbra-Filho AF, Pissinatti A, Muniz JACP, Ferreira HS, Schneider MPC (1998) Electrophoretic polymorphisms and their taxonomic implications in Callitrichini (Primates, Platyrrhini). Biochem Genet 36:229–244

Mendes SL, Vielliard JME, de Marco Jr P (this volume) The vocal identity of the *Callithrix* species (Primates, Callitrichidae). In: Ford SM, Porter LM, Davis LC (eds) The smallest anthropoids: The marmoset/callimico radiation. Springer Press, New York, pp 63–84

MMA (2003) www.mma.gov.br/port/sbf/index.cfm

Nievergelt CM, Mundy NI, Woodruff DS (1998) Microsatellite primers for genotyping common marmosets (*Callithrixjacchus*) and other callitrichids. Mol Ecol 5:1432–1434

Passamani M, Rylands AB (2000) Feeding behavior of Geoffroy's marmoset (*Callithrix geoffroyi*) in an Atlantic Forest fragment of south-eastern Brazil. Primates 41:27–38

Passmani M, Mendes SL, Chiarello AG, Laps R, Passmani JA (1997) Reintrodução do sagüi-da-cara-branca (*Callithrix geoffroyi*) em fragmentos de Mata Atlântica do Brasil. In: Ferrari SF, Schneider H (eds) A Primatologia no Brasil–5. Sociedade Brasileira de Primatologia, Belém, pp 119–128

Porter LM (2004) Differences in forest utilization and activity patterns among three sympatric callitrichines: *Callimico goeldii, Saguinus fuscicollis* and *S. labiatus*. Am J Phys Anthropol 124:139–153

Porter LM (2006) Distribution and density of *Callimico goeldii* in northwestern Bolivia. Am J Primatol 68:235–243

Rehg JA (this volume) Ranging patterns of *Callimico goeldii* (callimico) in a mixed species group. In: Ford SM, Porter LM, Davis LC (eds) The smallest anthropoids: The marmoset/callimico radiation. Springer Press, New York, pp 241–258

Rosenberger AL (1981) Systematics: the higher taxa. In: Coimbra-Filho AF, Mittermeier RA (eds) Ecology and behavior of neotropical primates, vol 1. Academia Brasileira de Ciências, Rio de Janeiro

Rylands AB, Schneider H, Langguth A, Mittermeier RA, Groves CP, Rodríguez-Luna E (2000) An assessment of the diversity of New World primates. Neotrop Primates 8:61–93

Rylands AB, Mittermeier RA, Coimbra-Filho AF (this volume) The systematics and distribution of the marmosets (*Callithrix, Callibella, Cebuella, and Mico*) and callimico (Callimico) (Callitrichidae, Primates). In: Ford SM, Porter LM, Davis LC (eds) The smallest anthropoids: The marmoset/callimico radiation. Springer Press, New York, pp 25–61

Soini P (1988) The pygmy marmoset, genus *Cebuella*. In: Mittermeier RA, Rylands AB, Coimbra-Filho AF, Fonseca GAB (eds) Ecology and behavior of neotropical primates, vol 2. WWF-US, Washington, DC, pp 19–129

Stevenson MF, Rylands AB (1988) The marmosets, genus *Callithrix*. In: Mittermeier RA, Rylands AB, Coimbra-Filho AF, Fonseca GAB (eds) Ecology and behavior of neotropical primates, vol 2. WWF-US, Washington, DC, pp 131–222

Subirá RJ (1998) Avaliação da Situação atual das Populações selvagens do Sauim-de-coleira *Saguinus b. bicolor* (Spix, 1823). Masters thesis, University of Brasília

Tagliaro CH, Schneider MPC, Schneider H, Sampaio IC, Stanhope M (1997) Marmoset phylogenetics, conservation perspectives, and evolution of the mtDNA control region. Mol Biol Evol 14:674–684

Tagliaro CH, Schneider MPC, Schneider H, Sampaio I, Stanhope M (2000) Molecular studies of *Callithrix pygmaea* (Primates, Platyrrhini) based on transferrin intronic and ND1 regions: implications for taxonomy and conservation. Gen Mol Biol 23:729–737

Terborgh J (1983) Five New World primates: a study in comparative ecology. Princeton University Press, Princeton

van Roosmalen MGM, van Roosmalen T (1997) An eastern extension of the geographical range of the pygmy marmoset, *Cebuella pygmaea*. Neotrop Primates 5:3–6

van Roosmalen MGM, van Roosmalen T (2003) The description of a new marmoset genus, *Callibella* (Callitrichinae, Primates), including its molecular phylogenetic status. Neotrop Primates 11:1–10

Vasarelhyi K (2000) Is *Callimico* monotypic? A reassessment in the light of new data. Dodô 36:20–29

Veracini C (this volume) Habitat use and ranging behavior of the silvery marmoset (*Mico argentatus*) at Caxiuanã National Forest (eastern Brazilian Amazonia). In: Ford SM, Porter LM, Davis LC (eds) The smallest anthropoids: The marmoset/callimico radiation. Springer Press, New York, pp 221–240

Vasarelhyi K (2002) The nature of relationships among founders in the captive population of Goeldi's monkey (*Callimico goeldii*). Evol Anthropol 11:155–158

Taxonomic Index

Subject Index

GPSR Compliance

The European Union's (EU) General Product Safety Regulation (GPSR) is a set of rules that requires consumer products to be safe and our obligations to ensure this.

If you have any concerns about our products, you can contact us on ProductSafety@springernature.com

In case Publisher is established outside the EU, the EU authorized representative is:

Springer Nature Customer Service Center GmbH
Europaplatz 3
69115 Heidelberg, Germany

Batch number: 10761599

Printed by Printforce, the Netherlands